AutoCAD
and Its Applications
ADVANCED

2016

by

Terence M. Shumaker
Faculty Emeritus
Former Chairperson
Drafting Technology
Autodesk Premier Training Center
Clackamas Community College
Oregon City, OR

David A. Madsen
President
Madsen Designs Inc.
Faculty Emeritus
Former Department Chairperson
Drafting Technology
Autodesk Premier Training Center
Clackamas Community College
Oregon City, OR
Director Emeritus
American Design Drafting Association

Jeffrey A. Laurich
Instructor, Mechanical Design
    Technology
Fox Valley Technical College
Appleton, WI

J.C. Malitzke
President
Digital JC CAD Services Inc.
Former Department Chair
Computer Integrated Technologies
Former Manager and Instructor
Authorized Autodesk Training Center
Moraine Valley Community College
Palos Hills, IL

Craig P. Black
Instructor, Mechanical Design
    Technology
Former Manager and Instructor
Autodesk Premier Training Center
Fox Valley Technical College
Appleton, WI

Publisher
The Goodheart-Willcox Company, Inc.
Tinley Park, IL
www.g-w.com

Library of Congress Cataloging-in-Publication Data

Shumaker, Terence M.
    AutoCAD and its applications. Advanced 2016 / by
    Terence M. Shumaker … [et al.]. -- 23rd Edition.
        pages cm
    Includes index.
    ISBN 978-1-63126-428-3
    1. Computer graphics. 2. AutoCAD. I. Madsen, David A.
    II. Title.
T385.S46123 2016
620'.0042028553--dc23

INTRODUCTION

AutoCAD and Its Applications—Advanced provides complete instruction in mastering three-dimensional design and modeling using AutoCAD. Topics are covered in an easy-to-understand sequence and progress in a way that allows you to become comfortable with the commands as your knowledge builds from one chapter to the next. In addition, *AutoCAD and Its Applications—Advanced* offers:

- Examples and discussions of industrial practices and standards.
- Professional tips explaining how to effectively and efficiently use AutoCAD.
- Exercises to reinforce the chapter topics. These exercises should be completed where indicated in the text as they build on previously learned material.
- Review questions at the end of each chapter for testing knowledge of commands and key AutoCAD concepts.
- A large selection of drawing problems supplements each chapter. Problems are presented as 3D illustrations, actual plotted drawings, and engineering sketches.

Fonts Used in This Text

Different typefaces are used throughout each chapter to define terms and identify AutoCAD commands. Important terms appear in *bold-italic face, serif* type. AutoCAD menus, commands, system variables, dialog box names, and toolbar button names are printed in **bold-face, sans serif** type. File names, folder names, and paths appear in the body of the text in Roman, sans serif type. Keyboard keys are shown inside of square brackets [ ] and appear in Roman, sans serif type. For example, [Enter] means to press the enter (return) key.

Other Text References

This text focuses on advanced AutoCAD applications. Basic AutoCAD applications are covered in *AutoCAD and Its Applications—Basics*, which can be ordered directly from Goodheart-Willcox Publisher. *AutoCAD and Its Applications* texts are also available for previous releases of AutoCAD.

Introducing the AutoCAD Commands

There are several ways to select AutoCAD drawing and editing commands. Selecting commands from the ribbon is slightly different from entering them at the keyboard. When a command is introduced, the command-entry methods are illustrated in the margin next to the text reference.

The example in the margin next to this paragraph illustrates the different methods of initiating the **CONE** command to draw a solid cone primitive while the 3D environment is active. The 3D environment consists of a drawing file based on the acad3D.dwt template and the 3D Modeling workspace current, as described in Chapter 1. This book assumes the 3D environment is current for all procedures and discussions.

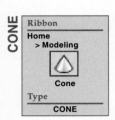

CONE

Ribbon
Home
> Modeling

Cone

Type
CONE

Flexibility in Design

Flexibility is the keyword when using *AutoCAD and Its Applications—Advanced*. This text is an excellent training aid for both individual and classroom instruction. It is also an invaluable resource for any professional using AutoCAD. *AutoCAD and Its Applications—Advanced* teaches you how to apply AutoCAD to common 3D design and modeling tasks.

When working through the text, you will see a variety of notices. These include Professional Tips, Notes, and Cautions that help you develop your AutoCAD skills.

PROFESSIONAL TIP

These ideas and suggestions are aimed at increasing your productivity and enhancing your use of AutoCAD commands and techniques.

NOTE

A note alerts you to important aspects of a command, function, or activity that is being discussed. These aspects should be kept in mind while you are working through the text.

CAUTION

A caution alerts you to potential problems if instructions or commands are incorrectly used or if an action can corrupt or alter files, folders, or storage media. If you are in doubt after reading a caution, always consult your instructor or supervisor.

AutoCAD and Its Applications—Advanced provides several ways for you to evaluate your performance. Included are:

- **Exercises.** The companion website contains exercises for each chapter. These exercises allow you to perform tasks that reinforce the material just presented. You can work through the exercises at your own pace. However, the exercises are intended to be completed when called out in the text.
- **Chapter reviews.** Each chapter includes review questions at the end of the chapter. Questions require you to give the proper definition, command, option, or response to perform a certain task. You may also be asked to explain a topic or list appropriate procedures.

- **Drawing problems.** There are a variety of drawing and design problems at the ends of chapters. These are presented as real-world CAD drawings, 3D illustrations, and engineering sketches. The problems are designed to make you think, solve problems, use design techniques, research and use proper drawing standards, and correct errors in the drawings or engineering sketches. Labels are used to represent the discipline to which a drawing problem applies.

  Mechanical These problems address mechanical drafting and design applications, such as manufactured part designs.

  Architectural These problems address architectural and structural drafting and design applications, such as floor plans, furniture, and presentation drawings.

  Piping These problems address piping drafting and design applications, such as tank drawings and pipe layout.

  General These problems address a variety of general drafting and design applications. These problems should be attempted by everyone learning advanced AutoCAD techniques for the first time.

NOTE

Some problems presented in this text are given as engineering sketches. These sketches are intended to represent the kind of material from which a drafter is expected to work in a real-world situation. As such, engineering sketches often contain errors or slight inaccuracies and are most often not drawn according to proper drafting conventions and applicable standards. Additionally, other drawings may contain errors or inaccuracies. Errors in these problems are *intentional* to encourage you to apply appropriate techniques and standards in order to solve the problem. As in real-world applications, sketches should be considered preliminary layouts. Always question inaccuracies in sketches and designs and consult the applicable standards or other resources.

Companion Website

The companion website is located at www.g-wlearning.com/CAD/. Select the entry for *AutoCAD and Its Applications—Advanced 2016* to access the material for this book. The companion website contains the exercises for each chapter and the appendix material for the textbook.

As you work through each chapter, exercises on the companion website are referenced. The exercises are intended to be completed as the references are encountered in the text. The solid modeling tutorial in Appendix A should be completed after Chapter 14. The surface modeling tutorial in Appendix B should be completed after Chapter 10. The remaining appendix material is intended as reference material.

Also included on the companion website are the exercise activities. The exercise activities are intended to supplement the exercises on the companion website. These activities are referenced within the appropriate exercises and can be completed as additional practice.

Additionally, full-color images of figures illustrating key topics in Chapters 16, 17, and 18 are provided in electronic format on the companion website. Figures with a corresponding color image on the companion website are identified with a reference above the figure caption. The full-color images are intended to enhance visualization and understanding of important topics relating to materials, lighting, and rendering and presentation.

About the Authors

Terence M. Shumaker is Faculty Emeritus, the former Chairperson of the Drafting Technology Department, and former Director of the Autodesk Premier Training Center at Clackamas Community College in Oregon City, Oregon. Terence taught at the community college level for over 28 years. He worked as a training consultant for Autodesk, Inc., and conducted CAD program development workshops around the country. He has professional experience in surveying, civil drafting, industrial piping, and technical illustration. He is the author of Goodheart-Willcox's *Process Pipe Drafting* and coauthor of the *AutoCAD and Its Applications* series and *AutoCAD Essentials*.

David A. Madsen is the president of Madsen Designs Inc. (www.madsendesigns.com). David is Faculty Emeritus and the former Chairperson of Drafting Technology and the Autodesk Premier Training Center at Clackamas Community College in Oregon City, Oregon. David was an instructor and a department chairperson at Clackamas Community College for nearly 30 years. In addition to teaching at the community college level, David was a Drafting Technology instructor at Centennial High School in Gresham, Oregon. David is a former member of the American Design Drafting Association (ADDA) Board of Directors. He was honored with Director Emeritus status by the ADDA in 2005. David has extensive experience in mechanical drafting, architectural design and drafting, and building construction. He holds a Master of Education degree in Vocational Administration and a Bachelor of Science degree in Industrial Education. David is coauthor of the *AutoCAD and Its Applications* series, *Architectural Drafting Using AutoCAD, Geometric Dimensioning and Tolerancing*, and other drafting and design textbooks.

Jeffrey A. Laurich has been an instructor in Mechanical Design Technology at Fox Valley Technical College in Appleton, Wisconsin, since 1991. He has also taught business and industry professionals in the Autodesk Premier Training Center at FVTC. Jeff teaches drafting, AutoCAD, Design of Tooling, GD&T, and 3ds Max. He created a certificate program at FVTC titled Computer Rendering and Animation that integrates the 3D capabilities of AutoCAD and 3ds Max. Jeff has professional experience in furniture design, surveying, and cartography. He holds an Associate degree in Applied Science in Natural Resources Technology and a Bachelor of Science degree in Career and Technical Education and Training. Together with other instructors at FVTC, Jeff created and is teaching a course in electric guitar building. The course is designed to introduce the concepts of manufacturing principles and STEM. After completing the course, students take away with them a beautiful, playable solid-body electric guitar they designed and built themselves.

J.C. Malitzke is president/owner of Digital JC CAD Services Inc. (digitaljccad.com). He is the former department chair of Computer Integrated Technologies at Moraine Valley Community College in Palos Hills, Illinois. J.C. was a professor of Mechanical Design and Drafting/CAD and taught for the Authorized Autodesk Training Center at Moraine Valley. J.C. has been actively using and teaching Autodesk products for nearly 30 years. He is a founding member and past chair of the Autodesk Training Center Executive Committee and the Autodesk Leadership Council. J.C. has been the coauthor and principal investigator on two National Science Foundation grants. He has won numerous awards, including: Drafting Educator of the Year by the Illinois Drafting Educators Association; the Instructor Quality Award by Autodesk; Autodesk University Instructor Award; Professor of the Year, Co-Innovator of the Year, and

Co-Master Teacher awards at Moraine Valley Community College; and the Illinois Outstanding Faculty Member of the Year awarded by the Illinois Community College Trustees Association. J.C. was also one of the recipients of the Top 10 Most Popular Autodesk University Classes Ever for his class Compelling 3D Features in AutoCAD. J.C. is an Autodesk Certified Instructor (ACI) and an Autodesk Certification Evaluator (ACE), and holds a Bachelor's degree in Education and a Master's degree in Industrial Technology from Illinois State University.

Craig P. Black is an instructor in the Mechanical Design Technology department and the former instructor and manager of the Autodesk Premier Training Center at Fox Valley Technical College in Appleton, Wisconsin. Craig has been teaching at FVTC since 1990. In 2009, he created the CAD Management certificate program at FVTC. In 2013, he created the Mechanical CAD Drafting Technical Diploma at FVTC. Craig has served two terms on the Autodesk Training Center Executive Committee (now known as the Autodesk Leadership Council) and chaired the committee in 2001. He has been working with Autodesk software products for more than 25 years. He has presented on various topics at a number of Autodesk University annual training sessions and has been contracted to teach training sessions on Autodesk products across the United States. In addition to teaching, Craig also does AutoCAD customization and AutoLISP and DCL programming for area businesses and industries. Prior to his current position, Craig worked in the civil, architectural, electrical, and mechanical drafting and design disciplines.

Acknowledgments

The authors and publisher would like to thank the following individuals and companies for their assistance and contributions.

Autodesk, Inc.
Fitzgerald, Hagan, & Hackathorn
Laser Design and GKS Services
Bill Strenth, Pittsburg State University School of Construction

Trademarks

Autodesk, the Autodesk logo, and AutoCAD are registered trademarks or trademarks of Autodesk, Inc., and/or its subsidiaries and/or affiliates in the USA and other countries.

Microsoft, Windows, Windows 8, and Windows 7 are registered trademarks of Microsoft Corporation in the United States and/or other countries.

ADDA Technical Publication

The content of this text is considered a fundamental component to the design drafting profession by the American Design Drafting Association. This publication covers topics and related material relevant to the delivery of the design drafting process. Although this publication is not conclusive, it should be considered a key reference tool in furthering the knowledge, abilities, and skills of a properly trained designer or drafter in the pursuit of a professional career.

BRIEF CONTENTS

Three-Dimensional Design and Modeling

Model Visualization and Presentation

EXPANDED CONTENTS

Three-Dimensional Design and Modeling

Chapter 14

Model Documentation, Analysis, and Point Clouds......343

Model Visualization and Presentation

Chapter 15

Visual Style Settings and Basic Rendering...............413

Chapter 16

Materials in AutoCAD 437

Companion Website Contents

www.g-wlearning.com/CAD/

Contents Tab

Exercises

Drawing Files
- Exercise Files
- Exercise Activity Files

Resources Tab

Appendices
- Appendix A Solid Modeling Tutorial
- Appendix B Surface Modeling Tutorial
- Appendix C Common File Extensions
- Appendix D AutoCAD Command Aliases
- Appendix E Advanced Application Commands

Reference Material
- Drafting Symbols
- Drawing Sheet Sizes, Settings, and Scale Parameters
- Standards and Related Documents
- Project and Drawing Problem Planning Sheet
- Standard Tables

Support Tab

Support and Resources

Introduction to Three-Dimensional Modeling

Learning Objectives

After completing this chapter, you will be able to:

✓ Describe how to locate points in 3D space.
✓ Describe the right-hand rule of 3D visualization.
✓ Explain the function of the ribbon.
✓ Identify the functions of the viewport controls and the view cube.
✓ Display 3D objects from preset isometric viewpoints.
✓ Display 3D objects from any desired viewpoint.
✓ Set a visual style current.

Three-dimensional (3D) design and modeling is a powerful tool for use in design, visualization, testing, analysis, manufacturing, assembly, and marketing. Three-dimensional models also form the basis of computer animations, architectural walk-throughs, parametric models used in building information modeling (BIM), and virtual worlds used in the entertainment industry and for gaming platforms. Drafters who can design objects, buildings, and "worlds" in 3D are in demand for a wide variety of positions, both inside and outside of the traditional drafting and design disciplines.

The first 14 chapters of this book present a variety of solid, surface, and mesh modeling techniques for drawing and designing in 3D. The skills you learn will provide you with the ability to construct any object in 3D and prepare you for entry into an exciting aspect of graphic communication.

To be effective in creating and using 3D objects, you must first have good 3D visualization skills. These skills include the ability to see an object in three dimensions and to visualize it rotating in space. Visualization skills can be obtained by using 3D techniques to construct objects and by trying to see two-dimensional sketches and drawings as 3D models. This chapter provides an introduction to several aspects of 3D drawing and visualization. Subsequent chapters expand on these aspects and provide a detailed examination of 3D drawing, editing, visualization, and display techniques.

Using Rectangular 3D Coordinates

In two-dimensional drawing, you see one plane defined by two dimensions. These dimensions are usually located on the X and Y axes and what you see is the XY plane. However, in 3D drawing, another coordinate axis—the Z axis—is added. This results in two additional planes—the XZ plane and the YZ plane. If you are looking at a standard AutoCAD screen after starting a new drawing using the default acad.dwt template, the positive Z axis comes directly out of the screen toward you. AutoCAD can only draw lines in 3D if it knows the X, Y, and Z coordinate values of each point on the object. For 2D drawing, only two of the three coordinates (X and Y) are needed.

Compare the 2D and 3D coordinate systems shown in **Figure 1-1**. Notice that the positive values of Z in the 3D coordinate system come up from the XY plane. In a new drawing based on the acad.dwt template, consider the surface of your computer screen as the XY plane. Anything behind the screen is negative Z and anything in front of the screen is positive Z.

The object in **Figure 1-2A** is a 2D drawing showing the top view of an object. The XY coordinate values of the origin and each point are shown. Think of the object as being drawn directly on the surface of your computer screen. However, this is actually a 3D object. When displayed in a pictorial view, the Z coordinates can be seen. Notice in **Figure 1-2B** that the first two values of each coordinate match the X and Y values of the 2D view. Three-dimensional coordinates are always expressed as (X,Y,Z). The 3D object was drawn using positive Z coordinates. Therefore, the object comes out of your computer screen when it is viewed from directly above. The object can also be drawn using negative Z coordinates. In this case, the object would extend behind, or into, the screen.

Study the nature of the rectangular 3D coordinate system. Be sure you understand Z values before you begin constructing 3D objects. It is especially important that you carefully visualize and plan your design when working with 3D constructions.

PROFESSIONAL TIP

All points in three-dimensional space can be drawn using one of three coordinate entry methods—rectangular, spherical, or cylindrical. This chapter uses the rectangular coordinate entry method. Complete discussions on the spherical and cylindrical coordinate entry methods are provided in Chapter 4.

Exercise 1-1

www.g-wlearning.com/CAD/

Complete the exercise on the companion website.

Figure 1-1.
A comparison of 2D and 3D coordinate systems.

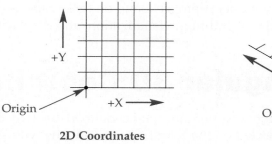

2D Coordinates   3D Coordinates

Figure 1-2.
A—The points making up a 2D object require only two coordinates. B—Each point of a 3D object must have an X, Y, and Z value. Notice that the first two coordinates (X and Y) are the same for each endpoint of a vertical line.

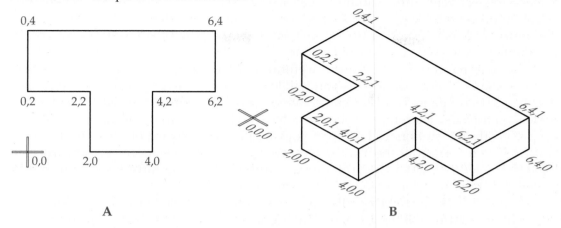

A   B

Right-Hand Rule of 3D Drawing

In order to effectively draw in 3D, you must be able to visualize objects in 3D space. The *right-hand rule* is a simple method for visualizing the 3D coordinate system. It is a representation of the positive coordinate values in the three axis directions. The AutoCAD world coordinate system (WCS) and a user coordinate system (UCS) are based on this concept of visualization.

To use the right-hand rule, position the thumb, index finger, and middle finger of your right hand as shown in **Figure 1-3**. Imagine that your thumb is the X axis, your index finger is the Y axis, and your middle finger is the Z axis. Hold your hand in front of you so that your middle finger is pointing directly at you, as shown in **Figure 1-3**. This is the plan view of the XY plane. The positive X axis is pointing to the right and the positive Y axis is pointing up. The positive Z axis comes toward you and the origin of this system is the palm of your hand.

The concept behind the right-hand rule can be visualized even better if you are sitting at a computer and the AutoCAD drawing window is displayed. Make sure the current drawing is based on the acad.dwt template. If the UCS icon is not displayed in the lower-left corner of the screen, turn it on by using the **UCSICON** command, or

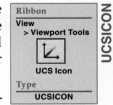

Ribbon
View
> Viewport Tools

UCS Icon

Type

UCSICON

UCSICON

Figure 1-3.
Positioning your hand to use the right-hand rule to understand the relationship of the X, Y, and Z axes.

select the **UCS Icon** button in the **Viewport Tools** panel of the **View** ribbon tab. Now, orient your right hand as shown in **Figure 1-3** and position it next to the UCS (or WCS) icon. Your index finger and thumb should point in the same directions as the Y and X axes, respectively. Your middle finger will be pointing out of the screen directly at you, representing the Z axis. See **Figure 1-4**. Notice the illustration on the right in the figure. This is the shaded UCS icon. It is displayed when the visual style is not 2D Wireframe. Visual styles are introduced later in this chapter.

The right-hand rule can be used to eliminate confusion when changing the orientation of the UCS. For example, as you will learn in Chapter 4, a simple way to change the UCS is to rotate it. The UCS can rotate on any of the three axes, just like a wheel rotates on an axle. Therefore, if you want to visualize how to rotate about the X axis, keep your thumb stationary and turn your hand either toward or away from you. If you wish to rotate about the Y axis, keep your index finger stationary and turn your hand to the left or right. When rotating about the Z axis, you must keep your middle finger stationary and rotate your entire arm.

If your 3D visualization skills are weak or you are having trouble visualizing different orientations of the UCS, use the right-hand rule. It is a useful technique for improving your 3D visualization skills. Rotating the UCS around one or more of the axes becomes easier once you begin drawing 3D objects. A complete discussion of UCSs is provided in Chapter 4.

Basic Overview of the Interface

AutoCAD provides different working environments tailored to either 2D drawing or 3D drawing or annotating a drawing. These environments are called *workspaces* and can be quickly restored. There are three default workspaces available in AutoCAD. The Drafting & Annotation workspace is designed for drawing in 2D and annotating a drawing. The workspace for basic 3D modeling and editing is called 3D Basics. The full range of 3D modeling features is found in the 3D Modeling workspace.

Figure 1-4.
Using the right-hand rule to visualize the X, Y, and Z axis directions.

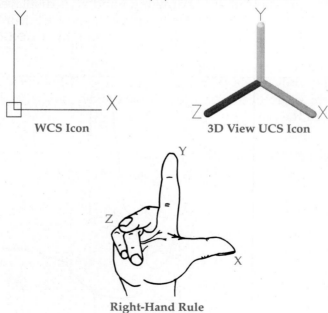

WCS Icon   3D View UCS Icon

Right-Hand Rule

Workspaces can be created, customized, and saved to allow a variety of graphical user interface configurations. The 3D Modeling workspace provides quick access to the full range of tools required to construct, edit, view, and visualize 3D models. It is the workspace used throughout this book. This section provides an overview of the 3D Modeling workspace and the layout of the ribbon and its panels.

NOTE

In order to use the default 3D "environment," you must start a new drawing file based on the acad3D.dwt template and set the 3D Modeling workspace current. All discussions in the remainder of this book assume that AutoCAD is in this default 3D environment.

Workspaces

A *workspace* is a drawing environment in which menus, toolbars, palettes, and ribbon panels are displayed for a specific task. A workspace stores not only which of these tools are visible, but also their on-screen locations. You can quickly change workspaces using the **Workspace Switching** button on the status bar or the **WSCURRENT** command. See **Figure 1-5**. The default 3D Modeling workspace is shown in **Figure 1-6**.

Ribbon Panels

Within the 3D Modeling workspace, the **Home** tab on the ribbon displays 12 panels. The tools in these panels provide all of the functions needed to design and view your 3D model. See **Figure 1-7A**. The panel title appears at the bottom of the panel. The up arrow button at the right-hand end of the tab names controls the display of the ribbon tabs, panels, and panel titles. Picking it cycles through the display of the full ribbon, tab titles and panel buttons, tab and panel titles only, or tab titles only. Picking the down arrow displays a drop-down menu containing the minimize options. See **Figure 1-7B**.

Many panels contain more tools than those displayed in the default view. These panels can be expanded by picking the flyout arrow to the right of the panel title. See **Figure 1-8A**. The flyout portion is displayed as long as the cursor remains over the panel. It retracts when the cursor moves off the panel. To retain the flyout display, pick the pushpin icon in the lower-left corner of the expanded panel. See **Figure 1-8B**. The panel continues to be displayed until you pick the pushpin icon again to release it.

Multiple panel flyouts can be displayed using the pushpin, but they may overlap in the process. This may obscure menu tools in adjacent panels. Simply pick on the panel you wish to use and its tools are displayed in full.

You can display only those panels that you need. Right-click anywhere on the ribbon to display the shortcut menu. Select **Show Panels** to display a cascading menu that contains a list of the panels available for that tab. See **Figure 1-9**. The panels that

Figure 1-5.
Switching workspaces using the **Workspace Switching** button on the status bar.

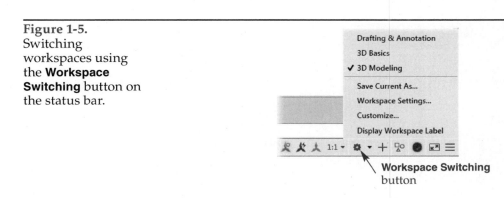

Figure 1-6.

The 3D Modeling workspace with a drawing file based on the acad3D.dwt template.

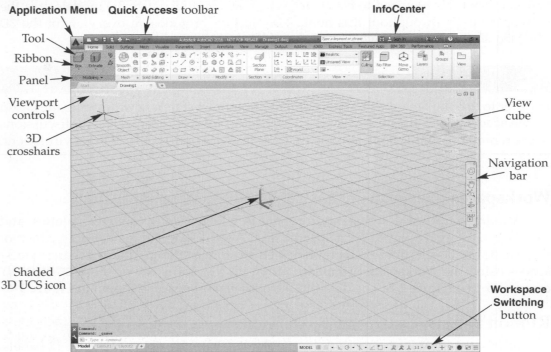

Figure 1-7.

A—The ribbon is composed of tabs and panels that provide access to 3D modeling and viewing commands without using toolbars and menus. B—The drop-down menu allows you to select a specific ribbon display.

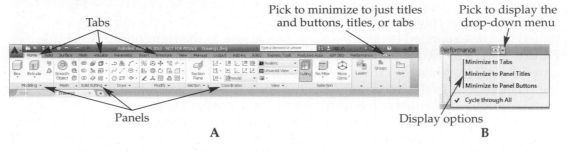

Figure 1-8.

A—Picking the flyout arrow expands the panel. The flyout is displayed as long as the cursor remains over the panel. It retracts when the cursor moves off the panel. B—Multiple panel flyouts can be displayed using the pushpin, but may overlap in the process. Pick the panel you wish to use to display its tools in full.

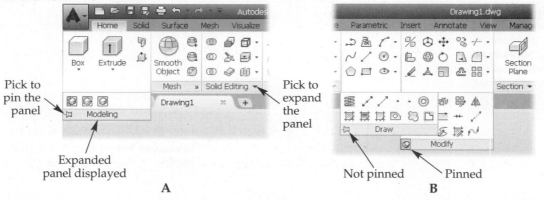

Figure 1-9.
Panels in a tab can be displayed or hidden using the shortcut menu.

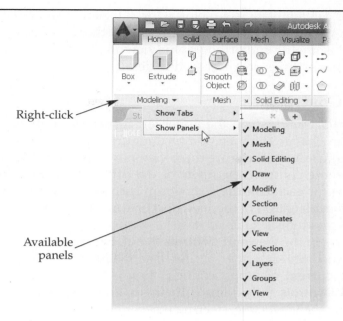

Right-click

Available panels

are currently displayed in the ribbon have a check mark next to their name. Select any of the checked panels that you wish to remove from the current display. Unchecked panels can be displayed by selecting their name. Each tab has its own set of available panels.

Each panel contains command tools. These are discussed in detail in the appropriate chapters of this book.

NOTE

The Drafting & Annotation, 3D Basics, and 3D Modeling workspaces display the ribbon by default. In each workspace, the ribbon displays a set of panels specifically related to that workspace. As soon as you select a different workspace or reload the current workspace, the panels associated with the workspace are displayed in the ribbon.

PROFESSIONAL TIP

All elements of the AutoCAD interface, including ribbon panels, toolbars, menus, and the drawing window appearance, can be customized using the customization functions in AutoCAD. In addition, custom commands can be created and assigned to interface elements. The **Options** dialog box, accessed with the **OPTIONS** command, and the **Customize User Interface** dialog box, accessed with the **CUI** command, are used to make customizations in AutoCAD. The AutoCAD help system provides detailed documentation and examples of AutoCAD customization.

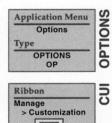

Displaying 3D Views

AutoCAD provides several methods of changing your viewpoint to produce different pictorial views. The default view in the 2D environment based on the acad.dwt template is a plan, or top, view of the XY plane. The default view in the 3D environment based on the acad3D.dwt template is a pictorial, or 3D, view. The *viewpoint* is the location in space from which the object is viewed. The methods for changing your viewpoint include preset isometric and orthographic viewpoints, the view cube, and camera lens settings. Camera settings are discussed in detail in Chapter 20.

Viewport Controls

The *viewport controls* in the upper-left corner of the drawing window provide a quick way to change the viewpoint, configure viewports, and manage the model display. See **Figure 1-10**. There are three viewport controls. When the cursor is moved over each of these controls, the name of the control is displayed in the tooltip. The current setting for each control is shown in brackets. Picking on a control displays a flyout. A brief explanation of the viewport controls is provided here to introduce you to the options and tools available.

The options in the **Viewport Controls** flyout, **Figure 1-10A**, allow you to work with viewports, which are described in detail in Chapter 5. The default setting (-) indicates that the current viewport configuration is "minimized" to a single viewport. The **Viewport Controls** flyout also contains options to display AutoCAD navigation tools, including the view cube, steering wheels, and the navigation bar. The view cube is introduced in this chapter, and other navigation tools are introduced in Chapter 3.

The **View Controls** flyout, **Figure 1-10B**, enables you to change the viewpoint to one of the preset orthographic or isometric viewpoints. This is discussed in more detail in the next section. The default setting when you start a new drawing with the acad3D. dwt template is Custom View. The other options in the **View Controls** flyout can be used to display a perspective or parallel projection or to access the **View Manager** dialog box, which is used to create named views. Creating named views is discussed in Chapter 3.

Figure 1-10.
Using the AutoCAD viewport controls. Picking on a control displays a flyout. A—The **Viewport Controls** flyout. B—The **View Controls** flyout. C—The **Visual Style Controls** flyout.

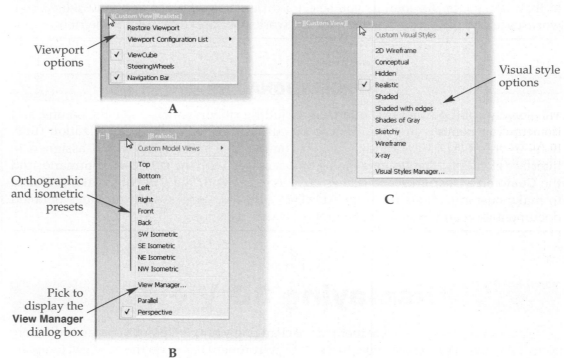

The **Visual Style Controls** flyout, **Figure 1-10C**, allows you to quickly display the drawing in one of 10 different visual styles. The default setting when you start a new drawing with the acad3D.dwt template is Realistic. Use of visual styles is introduced in this chapter and discussed completely in Chapter 15.

Isometric and Orthographic Viewpoint Presets

A 2D isometric drawing is based on angles of 120° between the three axes. AutoCAD provides preset viewpoints that allow you to view a 3D object from one of four isometric locations. See **Figure 1-11**. Each of these viewpoints produces an isometric view of the object. In addition, AutoCAD has presets for the six standard orthographic views of an object. The isometric and orthographic viewpoint presets are based on the world coordinate system (WCS).

The four preset isometric views are southwest, southeast, northeast, and northwest. The six orthographic presets are top, bottom, left, right, front, and back. To switch your viewpoint to one of these presets, pick the view name in the **View Controls** flyout in the viewport controls, as previously discussed. Refer to **Figure 1-10B**. You can also select one of the presets in the **View** panel in the **Home** tab of the ribbon. See **Figure 1-12**. The presets are also available in the **Views** panel in the **Visualize** tab.

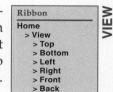

Ribbon
Home
> View
  > SW Isometric
  > SE Isometric
  > NE Isometric
  > NW Isometric
VIEW

Ribbon
Home
> View
  > Top
  > Bottom
  > Left
  > Right
  > Front
  > Back
VIEW

NOTE

You can also change the viewpoint to one of the orthographic or isometric presets by using the **VIEW** command to access the **View Manager** dialog box.

Once you select a view, the viewpoint in the current viewport is automatically changed to display an appropriate isometric or orthographic view. Since these presets are based on the WCS, selecting a preset produces the same view of the object regardless of the current user coordinate system (UCS).

A view that looks straight down on the current drawing plane is called a *plan view*. An important aspect of the orthographic presets is that selecting one not only changes the viewpoint, but, by default, it also changes the UCS to be plan to the orthographic view. All new objects are created on that UCS instead of the WCS (or previous UCS). Working with UCSs is explained in detail in Chapter 4. However, in order to

---

Figure 1-11.
There are four preset isometric viewpoints in AutoCAD. This illustration shows the direction from which the cube will be viewed for each of the presets. The grid represents the XY plane of the WCS.

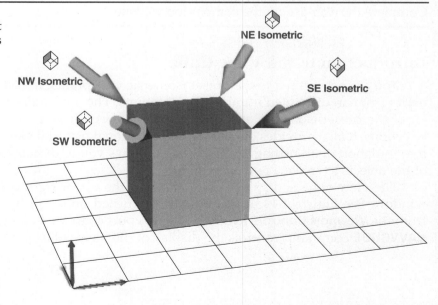

Figure 1-12.
Selecting preset views using the **View** panel in the **Home** tab of the ribbon.

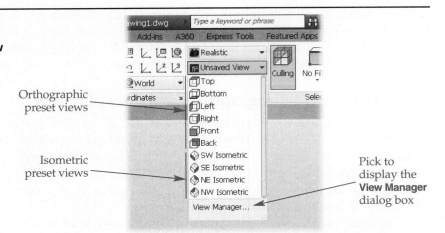

Orthographic preset views

Isometric preset views

Pick to display the **View Manager** dialog box

change the UCS to the WCS, type UCS to access the **UCS** command and then type W for the **World** option.

When an isometric or other 3D view is displayed, you can easily switch to a plan view of the current UCS by typing the **PLAN** command and selecting the **Current UCS** option. The **PLAN** command is discussed in more detail in Chapter 3.

NOTE

Selecting an orthographic view of a model using one of the methods described previously produces a plan view, but it may not achieve the results you desire. Three-dimensional models can be displayed in AutoCAD using either parallel or perspective projection. Displaying a plan view in either projection is possible. However, a true plan view, as used in 2D orthographic projections, can only be created when the model is displayed as a parallel projection. You can quickly change the display from perspective to parallel, or vice versa, by picking either **Parallel** or **Perspective** in the **View Controls** flyout in the viewport controls. The current projection is checked in the flyout.

Exercise 1-2

www.g-wlearning.com/CAD/

Complete the exercise on the companion website.

Introduction to the View Cube

You are not limited to the preset isometric viewpoints. In fact, you can view a 3D object from an unlimited number of viewpoints. The *view cube* allows you to display all of the preset isometric and orthographic views without making menu or ribbon selections. It also provides quick access to additional pictorial views and easy dynamic manipulation of all orthographic views. It allows you to dynamically rotate the view of the objects to create a new viewpoint.

The view cube is displayed by default in the upper-right corner of the drawing window. See **Figure 1-13**. It can be quickly turned on or off by selecting **ViewCube** from the **Viewport Controls** flyout in the viewport controls. Refer to **Figure 1-10A**. The **NAVVCUBE** command controls the display of the view cube.

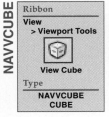

NAVVCUBE

Ribbon
View
> Viewport Tools

View Cube

Type
NAVVCUBE
CUBE

Figure 1-13.
Using the view cube to change the viewpoint.

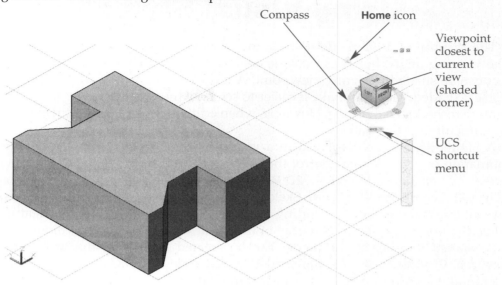

As the cursor is moved over the view cube, individual faces, edges, and corners are highlighted. If you pick one of the named faces on the view cube, that orthographic plan view is displayed. However, the UCS is not changed. If you pick one of the corners, a standard isometric view is displayed. Picking one of the four corners on the top face of the cube displays one of the preset isometric views. If you pick an edge, an isometric view is displayed that looks at the edge you selected. If you pick on the cube and drag the mouse, the view changes dynamically.

If you get lost while changing the viewpoint, just pick the **Home** icon (the small house shown in **Figure 1-13**) on the view cube to display the view defined as the home view. By default, the southwest isometric view is the home view. You can test this by selecting the **Home** icon and looking at the compass at the base of the view cube. Note that the left face of the cube aligns with the west point on the compass. The front surface aligns with the south point on the compass. Therefore, the top-front corner points to the southwest.

One edge, corner, or face of the view cube is always shaded or highlighted. Refer to **Figure 1-13**. This shading indicates the viewpoint on the cube that is closest to the current view.

The **NAVVCUBE** command has many options. This discussion is merely an introduction to the command. The command options and view cube features are covered in detail in Chapter 3.

PROFESSIONAL TIP

The **UNDO** command reverses the effects of the **NAVVCUBE** command.

Exercise 1-3

www.g-wlearning.com/CAD/

Complete the exercise on the companion website.

Introduction to 3D Model Display Using Visual Styles

The *display* of a 3D model is how the model is presented. This does not refer to the viewing angle, but rather colors, edge display, and shading or rendering. An object can be shaded from any viewpoint. A model can be edited while still keeping the object shaded. This can make it easier to see how the model is developing without having to reshade the drawing. However, when editing a shaded object, it may also be more difficult to select features.

A 3D model can be displayed in a variety of visual styles. A *visual style* is a combination of settings that control the display of edges and shading in a viewport. There are 10 basic visual styles—2D Wireframe, Conceptual, Hidden, Realistic, Shaded, Shaded with Edges, Shades of Gray, Sketchy, Wireframe, and X-ray. A *wireframe display* shows all lines on the object, including those representing back or internal features. A *hidden display* suppresses the display of lines that would normally be hidden.

A *shaded display* of the model can be created by setting the visual style to Conceptual, Realistic, or one of the shaded visual styles. The X-ray style presents the model in muted, translucent colors, and all hidden lines are displayed. This may be a good choice to use in the design of a model. The Realistic visual style is considered the most realistic *shaded* display. A more detailed shaded model, a *rendered display* of the model, can be created with the **RENDER** command. A rendering is the most realistic presentation.

Examples of the basic visual styles are shown in **Figure 1-14**. To change styles, select one from the **Visual Style Controls** flyout in the viewport controls. Refer to **Figure 1-10C**. You can also use the **VSCURRENT** command or the drop-down list in the **View** panel on the **Home** tab or the **Visual Styles** panel on the **Visualize** tab in the ribbon. See **Figure 1-15**.

In the default 3D environment based on the acad3D.dwt template, the default display mode, or visual style, is Realistic. In this visual style, all objects appear as solids and are displayed in their assigned layer colors. Other display options are available. These options are discussed in detail in Chapter 15, but are given here as an introduction.

- **2D Wireframe.** Displays all lines of the model using assigned linetypes and lineweights. The 2D UCS icon and 2D grid are displayed, if turned on. If the **HIDE** command is used to display a hidden-line view, use the **REGEN** command to redisplay the wireframe view.
- **Wireframe.** Displays all lines of the model. The 3D grid and the 3D UCS icon are displayed, if turned on.
- **Hidden.** Displays all visible lines of the model from the current viewpoint and hides all lines not visible. Objects are not shaded or colored.
- **Sketchy.** Edges appear hand-sketched.
- **Shades of Gray.** Gray shades are shown with highlighted edges.
- **Conceptual.** The object is smoothed and shaded with transitional colors to help highlight details.
- **Realistic.** The model is shaded and smoothed using assigned layer colors and materials.
- **Shaded.** A smooth-shaded model is displayed, but edges are not shown.
- **Shaded with Edges.** Edges are displayed on the smooth-shaded model.
- **X-Ray.** The model appears transparent.

A variety of options can be used to change individual components of each of these visual styles. A complete discussion of visual styles is given in Chapter 15. Detailed discussions on rendering, materials, lights, and animations appear in Chapters 15 through 20.

VSCURRENT

Ribbon
Home
> View
Visualize
> Visual Styles

Type
VSCURRENT
VS

Figure 1-14.
The default AutoCAD visual styles.

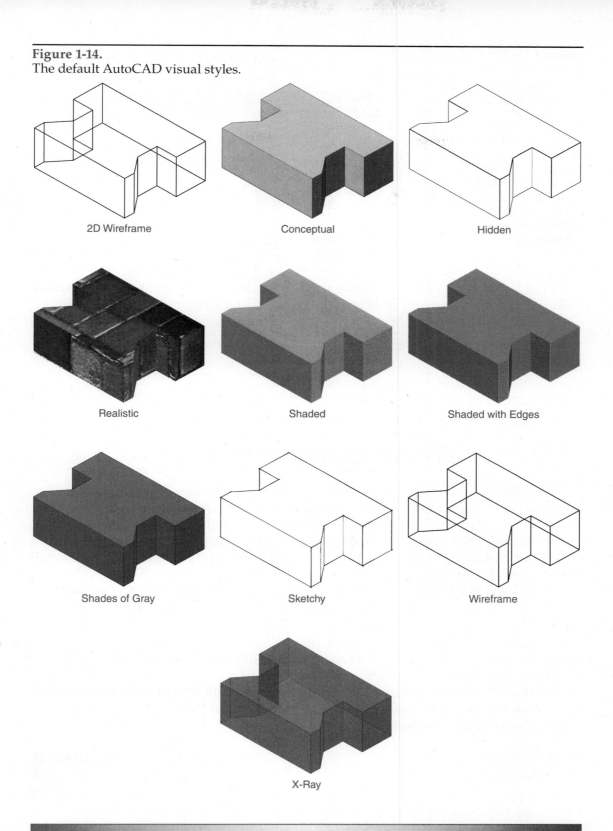

2D Wireframe

Conceptual

Hidden

Realistic

Shaded

Shaded with Edges

Shades of Gray

Sketchy

Wireframe

X-Ray

NOTE

When the visual style is 2D Wireframe, you can quickly view the model with hidden lines removed by typing HIDE. The **HIDE** command can be used at any time to remove hidden lines from a wireframe display. If **HIDE** is used when any other visual style is current, the Hidden visual style is set current.

Figure 1-15.
Visual styles can be managed using the **Visual Style Controls** flyout in the viewport controls or one of the **Visual Styles** drop-down lists in the ribbon. Shown is the **Visual Styles** drop-down list in the **Visual Styles** panel in the **Visualize** tab.

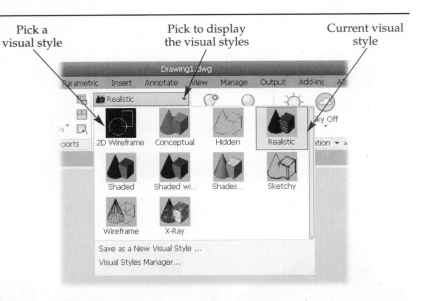

Pick a visual style

Pick to display the visual styles

Current visual style

Exercise 1-4

www.g-wlearning.com/CAD/

Complete the exercise on the companion website.

Rendering a Model

The **RENDER** command creates a realistic image of a model, **Figure 1-16**. However, rendering an image takes longer than shading an image. There are a variety of settings that you can use to fine-tune renderings. These include lights, materials, backgrounds, and rendering preferences. AutoCAD provides six rendering presets, the highest of which can require up to 12 hours to perform. Render settings are discussed in detail in Chapters 15 through 18.

When the **RENDER** command is initiated, the **Render** window is displayed and the image is rendered. See **Figure 1-17**. If the **Render** window closes, use the **RENDERWINDOW** command to redisplay the window. The rendering that is produced is based on a variety of render settings that are discussed in Chapters 15 through 18. If no lights have been created in the scene, AutoCAD uses default lighting to generate the rendering. The AutoCAD default lighting consists of one or two distant light sources that evenly illuminate all surfaces and generate shadows. If no materials have been applied to objects in the scene, objects are rendered with a matte material that is the same color as the object display color.

RENDER

Ribbon

Visualize
> Render

Render

Type

RENDER
RR

RENDERWINDOW

Ribbon

Visualize
> Render

Render Window

Type

RENDERWINDOW
RW

NOTE

If the image is rendered in the viewport, clean the screen using the **ZOOM** or **PAN** command. Setting the rendering destination as the viewport is discussed in Chapter 18.

Exercise 1-5

www.g-wlearning.com/CAD/

Complete the exercise on the companion website.

Figure 1-16.
Rendering produces
the most realistic
display of a scene.

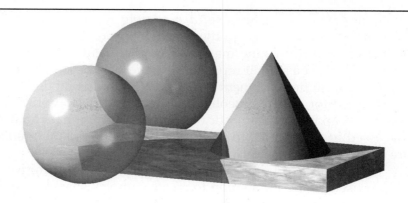

Figure 1-17.
The rendered model is displayed in the **Render** window.

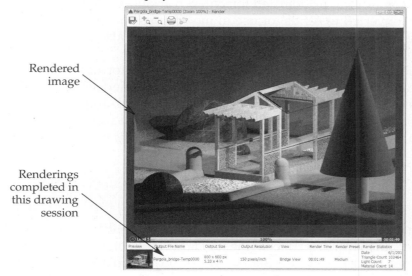

Rendered
image

Renderings
completed in
this drawing
session

3D Construction Techniques

Before constructing a 3D model, you should determine the purpose of your design. What will the model be used for—manufacturing, analysis, or presentation? This helps you determine which tools you should use to construct and display the model. Three-dimensional objects can be drawn as solids, meshes, or surfaces and displayed in wireframe, hidden-line removed, and shaded views.

A *wireframe object*, or model, is an object constructed of lines in 3D space. Wireframe models are hard to visualize because it is difficult to determine the angle of view and the nature of the surfaces represented by the lines. The **HIDE** command has no effect on a true wireframe model because there is nothing to hide. All lines are always visible because there are no surfaces or faces between the lines. True wireframe models have very limited applications.

Surface modeling represents solid objects by creating a skin in the shape of the object. However, there is nothing inside of the object. Think of a surface model as a balloon filled with air. A surface model looks more like the real object than a wireframe and can be used for rendering. Surface models are often constructed for applications such as civil engineering terrain modeling, automobile body design, sheet metal design and fabrication, and animation. Surface modeling techniques are discussed in Chapter 10.

Like surface modeling, *solid modeling* represents the shape of objects, but it also provides data related to the physical properties of the objects. Solid models can be analyzed to determine mass, volume, moments of inertia, and centroid. A solid model is not just a skin, it represents a solid object. Some third-party programs allow you to perform finite element analysis (FEA) on the model. Solid model files can also be exported for use in stereolithography, rapid prototyping, and 3D printing. These processes can produce a plastic or polymer prototype for analysis and testing. This is discussed in Chapter 14. In addition, solid models can be rendered. Most 3D objects are created as solid models.

In AutoCAD, solid models can be created from primitives. *Primitives* are basic shapes used as the foundation to create complex shapes. Some of these basic shapes include boxes, cylinders, spheres, and cones. Detailed shapes and primitives can be created using 3D mesh primitives, mesh modeling techniques, and surface modeling. A 3D *mesh object*, which is a type of surface model, can have a free-flowing shape because the size of the mesh can be adjusted to achieve various levels of smoothness. Mesh objects can be converted to solids for use in model construction. See Chapter 9 for a detailed discussion on 3D mesh modeling. Solid primitives also can be modified to create a finished product. See **Figure 1-18**.

Surface and solid models can be exported from AutoCAD for use in animation and rendering software, such as Autodesk 3ds Max®. Rendered models can be used in any number of presentation formats, including slide shows, black-and-white or color prints, and animations recorded to video files. Surface and solid models can also be used to create virtual worlds for entertainment and gaming applications.

3D Object Snaps

The construction and editing of a 3D model can be more efficient with the use of 3D object snaps. These work in the same manner as the standard 2D object snaps and can be set using the **3D Object Snap** tab of the **Drafting Settings** dialog box. If you use 3D object snaps, turn on only those options you need to construct the object. See **Figure 1-19**.

The **Vertex**, **Midpoint on edge**, **Center of face**, **Perpendicular**, and **Nearest to face** 3D object snap modes should be familiar to you from your work in 2D. The **Knot** 3D object

Figure 1-18.
A—The two cylinders and the box are solid primitives. B—With a couple of quick modifications, the large cylinder becomes a shaft with a machined keyway.

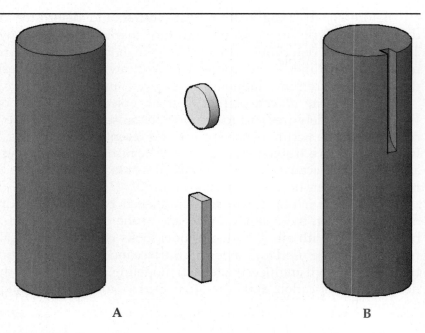

A   B

Figure 1-19.
The 3D object snaps can be set in the **Drafting Settings** dialog box.

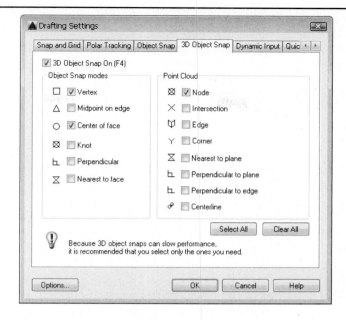

snap mode refers to a spline and the point at which one curve ends and the next curve begins. The point cloud object snap modes can be used when working with point cloud files. An introduction to the use of point cloud files is discussed in Chapter 14.

The 3D AutoSnap markers that appear when drawing can be displayed in a different color than 2D AutoSnap markers. Set the color for 3D AutoSnap markers by accessing the **Display** tab in the **Options** dialog box. Pick the **Colors...** button in the **Window Elements** area to access the **Drawing Window Colors** dialog box. Select the desired context in the **Context:** list and then select a color for the 3d Autosnap marker element in the **Interface element:** list.

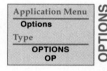

Sample 3D Construction

The example provided here illustrates how you can move from a layout sketch to an exact 2D drawing with parametric constraints, then quickly extrude the object into a 3D solid model. Techniques illustrated here are discussed throughout the book. The object shown is the plastic housing for a 24/15-pin cable connector. The final rendition of this object would include internal cutouts, rounded edges, and holes, but is not shown in this example.

Creating a 2D Sketch

A 2D layout can be drawn quickly using the Drafting & Annotation workspace. In doing so, you can create a quick sketch without close attention to exact dimensions or you can use direct distance entry and polar tracking for accuracy. If appropriate, constraints can then be used to apply accurate dimensions and geometric relationships, such as perpendicularity and parallelism, to the object. See **Figure 1-20A**. In this example, only half of the object is drawn, since it is symmetrical. Then the **MIRROR** command is used to complete the layout. See **Figure 1-20B**.

Converting to a Region

The object is still composed of just 2D lines. Therefore, it must be changed to an object that can be converted into a solid model. The **REGION** command is used for this purpose. After selecting this command, you are prompted to select objects. Be sure to

Figure 1-20.
A—Half of the 2D object is drawn using geometric constraints. B—The object is mirrored to create a full profile of the connector body. C—The object is now a "region" and is displayed in the southwest isometric viewpoint.

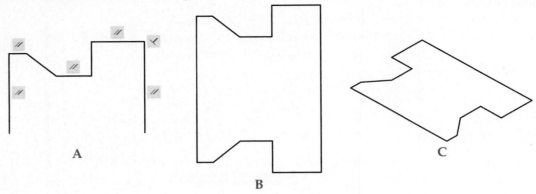

pick all of the lines of the object. After pressing [Enter], you are prompted that a loop is extracted, a region is created, and, if constraints were applied, that they are removed. The appearance of the object will not change.

Converting to a Solid

The object is now a "region" that can be quickly converted to a solid. First change to the 3D Modeling workspace, and then use the viewport controls or the view cube to display the object from the southwest isometric viewpoint. Working with models is more intuitive when a 3D display is used. See **Figure 1-20C.**

The region can be converted to a solid using either the **EXTRUDE** or **PRESSPULL** command. For example, pick the **Extrude** button in the **Modeling** panel on the **Home** tab of the ribbon and select the region. Using direct distance entry, enter a height of .55, as shown in **Figure 1-21**, and press [Enter].

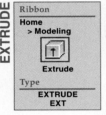

Ribbon

Home
> Modeling

Extrude

Type

EXTRUDE
EXT

Figure 1-21.
The object is extruded into a solid and displayed as a wireframe.

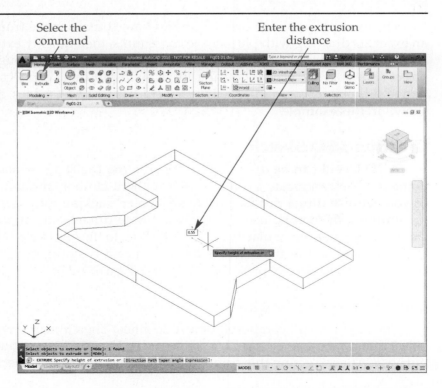

Using a Visual Style

The object you have created is now a 3D solid. If the drawing was based on the acad3D.dwt template, the model is displayed in the Realistic visual style. If the drawing was based on the acad.dwt template, the model is displayed in the 2D Wireframe visual style.

As mentioned previously, there are 10 different visual styles from which to choose. Select an option from the **Visual Style Controls** flyout in the viewport controls or the **View** panel on the **Home** tab of the ribbon. The Conceptual visual style provides a quick display of the model using shaded tones of the object color, as shown in **Figure 1-22**.

The techniques used in this example are just a brief introduction to the creation of 3D models. Detailed descriptions of modeling, display, and editing techniques are included in the following chapters.

Guidelines for Working with 3D Drawings

Working in 3D, like working with 2D drawings, requires careful planning to efficiently produce the desired results. The following guidelines can be used when working in 3D.

Planning

- Determine the type of final drawing you need and the manner in which it will be displayed. Then, choose the method of 3D construction that best suits your needs—wireframe, surface, mesh, or solid.
- If appropriate for the project, use 2D constraints to create a 2D layout or sketch.
- For an object requiring only one pictorial view, it actually may be quicker to draw an object in 3D rather than in AutoCAD's isometric mode. AutoCAD's 3D solid modeling tools enable you to quickly create an accurate model, and then display it in the required isometric format using preset views. The **VIEWBASE** command can then be used to create a 2D drawing of the model.
- It is best to use AutoCAD's 3D commands to construct objects and layouts that need to be viewed from different angles for design purposes.
- Construct only the features needed for the function of the drawing. This saves space and time, and makes visualization much easier.
- Use 2D or 3D object snap modes in a pictorial view in conjunction with UCS icon manipulation to save having to create new UCSs.
- Keep in mind that when the grid is displayed, the pattern appears at the current elevation and parallel to the XY plane of the current UCS.

Figure 1-22.
The Conceptual visual style provides a quick display of the model using shaded tones of the object colors.

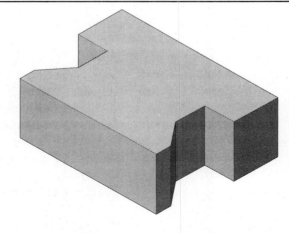

Editing

- Use the **Properties** palette to change the color, layer, or linetype of 3D objects.
- Use grips to edit a solid-modeled object (see Chapter 11).
- Do as much editing as possible from a 3D viewpoint. It is quicker and the results are immediately seen.

Displaying

- Use the **HIDE** command and visual styles to help visualize complex drawings.
- To quickly change views, use the preset isometric views, view cube, and **PLAN** command.
- Create and save 3D views for quicker pictorial views. This avoids having to repeatedly use the view cube or other methods to change the viewing angle.
- Freeze unwanted layers before displaying objects in 3D and especially before using **HIDE**. AutoCAD regenerates layers that are turned off, which may cause an inaccurate hidden display to be created. Frozen layers are not regenerated.
- You may have to slightly move objects that touch or intersect if the display removes a line you need to see or plot. However, be sure to move the objects back to maintain accuracy in the model.

Chapter Review

Answer the following questions using the information in this chapter.

1. What are the three coordinates needed to locate any point in 3D space?
2. In a 2D drawing, what is the value for the Z coordinate?
3. What purpose does the right-hand rule serve?
4. Which three fingers are used in the right-hand rule?
5. What is the definition of a *viewpoint*?
6. What is the function of the *ribbon* and its panels?
7. How do you turn the display of individual panels on or off in the ribbon?
8. How can you quickly change the display from perspective projection to parallel projection, or vice versa?
9. How many preset isometric viewpoints does AutoCAD have? List them.
10. How does changing the UCS impact using one of the preset isometric viewpoints?
11. List the six preset orthographic viewpoints.
12. When selecting a preset orthographic viewpoint, what happens to the UCS?
13. Which AutoCAD tool allows you to dynamically change your viewpoint using an on-screen cube icon?
14. Define *wireframe display*.
15. Define *hidden display*.
16. Define *surface model*.
17. Define *solid model*.
18. Define *primitive*.
19. Define *mesh object*.
20. What is a *visual style*?

2

Creating Primitives and Composites

Learning Objectives

After completing this chapter, you will be able to:

- ✓ Construct 3D solid primitives.
- ✓ Explain the dynamic feedback presented when constructing solid primitives.
- ✓ Create complex solids using the **UNION** command.
- ✓ Remove portions of a solid using the **SUBTRACT** command.
- ✓ Create a new solid from the common volume between two solids.
- ✓ Create regions.

Overview of Solid Modeling

In Chapter 1, you were introduced to the three basic types of 3D models—wireframe objects, solid models, and surface models. A solid model is probably the most useful and, hence, most common type of 3D model. A solid model accurately and realistically represents the shape and form of a final object. In addition, a solid model contains data related to the object's volume, mass, and centroid.

Solid modeling is very flexible. A model can start with solid primitives, such as a box, cone, or cylinder, and a variety of editing functions can then be performed. Think of creating a solid model as working with modeling clay. Starting with a basic block of clay, you can add more clay, remove clay, cut holes, round edges, etc., until you have arrived at the final shape and form of the object.

PROFESSIONAL TIP

Snaps can be used on solid objects. For example, using 2D object snaps, you can snap to the center of a solid sphere using the **Center** object snap. The **Endpoint** object snap can be used to select the corners of a box, apex of a cone, corners of a wedge, etc. Using 3D object snaps, you can snap to the vertex or midpoint of an edge or to the center of a face.

Constructing Solid Primitives

As you learned in Chapter 1, a primitive is a basic building block. The eight *solid primitives* in AutoCAD are a box, cone, cylinder, polysolid, pyramid, sphere, torus, and wedge. These primitives can also be used as building blocks for complex solid models. This section provides detailed information on drawing all of the solid primitives. All of the 3D modeling primitive commands can be accessed using one of several methods. You can use the **Primitive** panel in the **Solid** tab of the ribbon or the **Modeling** panel in the **Home** tab of the ribbon. You can also type the name of the solid primitive. Using the **Modeling** panel in the **Home** tab of the ribbon is shown in **Figure 2-1**.

The information required to construct a solid primitive depends on the type of primitive being drawn. For example, to draw a solid cylinder, you must provide a center point for the base, a radius or diameter of the base, and the height of the cylinder. A variety of command options are available when creating primitives, but each primitive is constructed using just a few basic dimensions. These are shown in **Figure 2-2**.

Certain familiar editing commands can be used on solid primitives. For example, you can fillet or chamfer the edges of a solid primitive. In addition, there are other editing commands that are specifically for use on solids. You can also perform Boolean operations on solids. These operations allow you to add one solid to another, subtract one solid from another, or create a new solid based on how two solids overlap.

PROFESSIONAL TIP

Solid objects of a more free-form nature can be created by first constructing a mesh primitive, editing it, and then converting it to a solid. This process is covered in detail in Chapter 9.

Using Dynamic Input and Dynamic Feedback

Dynamic input allows you to construct models in a "heads up" fashion with minimal eye movement around the screen. When a command is initiated, the command prompts are then displayed in the dynamic input area, which is at the lower-right corner of the crosshairs. As the pointer is moved, the dynamic input area follows it. The dynamic input area displays values of the cursor location, dimensions, command

---

Figure 2-1.
The **Modeling** panel in the **Home** tab of the ribbon.

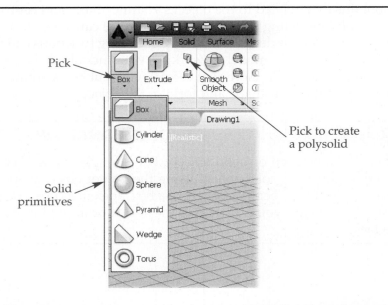

Figure 2-2.
An overview of AutoCAD's solid primitives and the dimensions required to draw them.

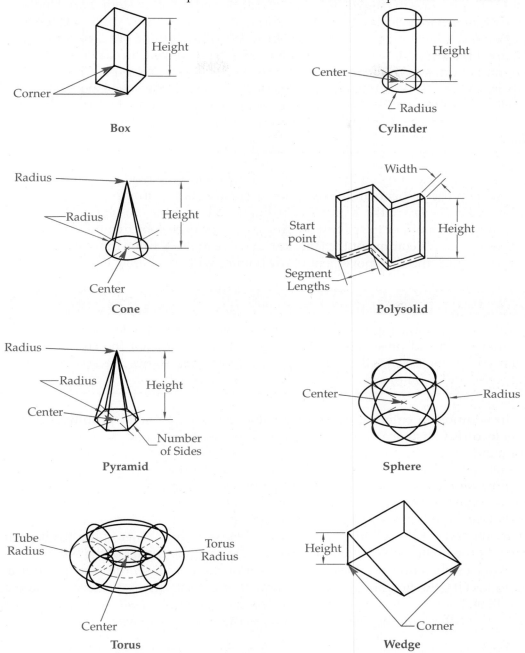

prompts, and command options (in a drop-down list). Coordinates and dimensions are displayed in boxes called *input fields*. When command options are available, a drop-down list arrow appears. Press the down arrow key on the keyboard to display the list. You can use the pointer to select the option or press the down arrow key until a dot appears by the desired option and press [Enter].

The command line provides an alternative to dynamic input. When a command is active, the available options are highlighted and can be selected directly at the command line. As the pointer is moved over an option, the pointer changes to a hand cursor and the option is selectable. This method requires no keyboard input.

AutoCAD provides a suggestion list of commands and system variables at the dynamic input area as you type. As additional characters are typed, the suggestion list changes. You can then use the pointer to select an item or press the down arrow key until the desired item is highlighted, and press [Enter]. In addition, AutoCAD recognizes certain synonyms of command names when you type a command. For example, typing SPIRAL displays the **HELIX** command in the suggestion list. Options controlling AutoComplete, AutoCorrect, and other AutoCAD input search functions can be accessed by right-clicking inside the command line window and selecting **Input Search Options…**.

As an example of using dynamic input, if you select the modeling command **BOX**, the first item that appears in the dynamic input area is the prompt to specify the first corner and a display of the X and Y coordinate values of the crosshairs. At this point, you can use the pointer to specify the first corner or type coordinate values. Type the X value and then a comma or the [Tab] key to move to the Y value input box. This locks the typed value and any movement of the pointer will not change it.

When using dynamic input to enter coordinate values from the keyboard, it is important that you avoid pressing [Enter] until you have completed the coordinate entry. When you press [Enter], all of the displayed coordinate values are accepted and the next command prompt appears.

In addition to entering coordinate values for sizes of solid primitives, you can provide direct distance dimensions. For example, the second prompt of the **BOX** command is for the second corner of the base. When you move the pointer, two dimensional input fields appear. Also, notice that a preview of the base is shown in the drawing area. This is the *dynamic feedback* that AutoCAD provides as you create a solid primitive. See **Figure 2-3A**. Once you enter or pick the first corner of the base, the dynamic input area changes to display X and Y coordinate boxes. In this case, the values entered are the X and Y coordinates of the opposite corner of the box base. If X, Y, *and* Z coordinates are entered, the point is the opposite corner of the box.

After establishing the location and size of the box base, the next prompt asks you to specify the height. Again, you can either enter a direct dimension value and press [Enter] or select the height with the pointer. See **Figure 2-3B**. AutoCAD provides dynamic feedback on the height of the box as the pointer is moved.

The techniques previously described can be used with any form of dynamic input. The current input field is always highlighted. You can always enter a value and use the [Tab] key to lock the input and move to the next field.

Ribbon
Home
> Modeling
Solid
> Primitive
Box
Type
BOX

Box

A *box* has six flat sides forming square corners. It can be constructed starting from an initial corner or the center. See **Figure 2-4**. A cube can be constructed, **Figure 2-4A**, as well as a box with unequal-length sides, **Figure 2-4B**.

Figure 2-3.
A—Specifying the base of a box with dynamic input on. Notice the preview of the base.
B—Setting the height of a box with dynamic input on. Notice the preview of the height.

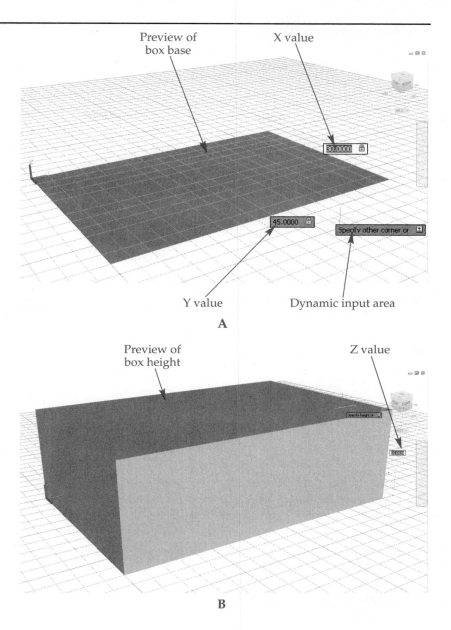

Preview of box base

X value

30.0000

45.0000

Specify other corner or

Y value

Dynamic input area

A

Preview of box height

Z value

Specify height or

B

Figure 2-4.
A—A box created using the **Cube** option. B—A box created by selecting the center point.

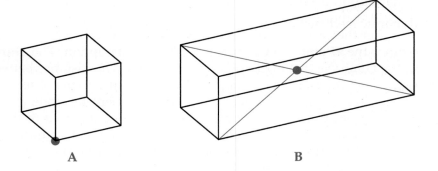

A

B

When the command is initiated, you are prompted to select the first corner or enter the **Center** option. The first corner is one corner on the base of the box. The center is the geometric center of the box, as shown in **Figure 2-4B**. If you select the **Center** option, you are next prompted to select the center point.

After selecting the first corner or center, you are prompted to select the other corner or enter the **Cube** or **Length** option. The "other" corner is the opposite corner of the box base if you enter an XY coordinate or the opposite corner of the box if you enter an XYZ coordinate. If the **Length** option is entered, you are first prompted for the length of one side. If dynamic input is on, you can also specify a rotation angle. After entering the length, you are prompted for the width of the box base. If the **Cube** option is selected, the length value is applied to all sides of the box.

Once the length and width of the base are established, you are prompted for the height, unless the **Cube** option was selected. Either enter the height or select the **2Point** option. This option allows you to pick two points on screen to set the height. The box is created. If the **Center** option was selected, the height will be applied in both the positive and negative Z directions.

PROFESSIONAL TIP

Using the UCS icon grips, you can select a surface that is not parallel to the current UCS on which to locate the object. Methods for working with user coordinate systems are discussed in detail in Chapter 4.

Exercise 2-1

www.g-wlearning.com/CAD/

Complete the exercise on the companion website.

Cone

CONE

Ribbon

Home
> **Modeling**
Solid
> **Primitive**

Cone

Type

CONE

A *cone* has a circular or elliptical base with edges that converge at a single point. The cone may be *truncated* so the top is flat and the cone does not have an apex. See **Figure 2-5**. When the command is initiated, you are prompted for the center point of the cone base, or to enter an option. If you pick the center, you must then set the radius of the base. To specify a diameter, enter the **Diameter** option after specifying the center.

The **3P**, **2P**, and **Ttr** options are used to define a circular base using three points on the circle, two points on the circle, or two points of tangency on the circle and a radius. The **Elliptical** option is used to create an elliptical base.

If the **Elliptical** option is entered, you are prompted to pick both endpoints of one axis and then one endpoint of the other axis of an ellipse that defines the base. If the **Center** option is entered after the **Ellipse** option, you are asked to select the center of the ellipse and then pick an endpoint on each of the axes.

After the base is defined, you are asked to specify a height. You can enter a height or enter the **2Point**, **Axis endpoint**, or **Top radius** option. The **2Point** option is used to set the height by picking two points on screen. The distance between the points is the height. The height is always applied perpendicular to the base.

The **Axis endpoint** option allows you to orient the cone at any angle, regardless of the current UCS. For example, to place a tapered cutout in the end of a block, first create a construction line. Refer to **Figure 2-6**. Then, locate the cone base and give a coordinate location of the apex, or axis endpoint. You can then subtract the cone from the box to create the tapered hole.

The **Top radius** option allows you to specify the radius of the top of the cone. If this option is not used, the radius is zero, which creates a pointed cone. Setting the radius to a value other than zero produces a *frustum cone*, or a cone where the top is truncated, and does not come to a point.

Figure 2-5.
A—A circular cone. B—A frustum cone. C—An elliptical cone.

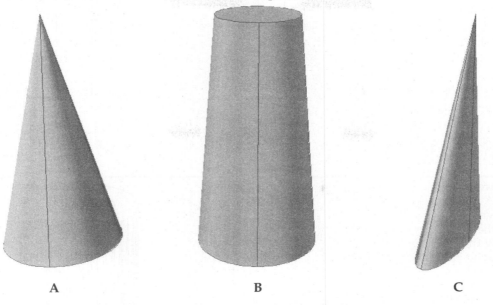

A   B   C

Figure 2-6.
A—Cones can be positioned relative to other objects using the **Axis endpoint** option. B—The cone is subtracted from the box.

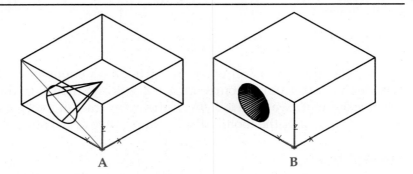

A   B

Cylinder

A *cylinder* has a circular or elliptical base and edges that extend perpendicular to the base. See **Figure 2-7**. When the command is initiated, you are prompted for the center point of the cylinder base, or to enter an option. If you pick the center, you must then set the radius of the base. To specify a diameter, enter the **Diameter** option after specifying the center.

The **3P**, **2P**, and **Ttr** options are used to define a circular base using three points on the circle, two points on the circle, or two points of tangency on the circle and a radius. The **Elliptical** option is used to create an elliptical base.

If the **Elliptical** option is entered, you are prompted to pick both endpoints of one axis and then one endpoint of the other axis of an ellipse defining the base. If the **Center** option is entered, you are asked to select the center of the ellipse and then pick an endpoint on each of the axes.

After the base is defined, you are asked to specify a height or to enter the **2Point** or **Axis endpoint** option. The **2Point** option is used to set the height by picking two points on screen. The distance between the points is the height. The **Axis endpoint** option allows you to orient the cylinder at any angle, regardless of the current UCS, just as with a cone.

The **Axis endpoint** option is useful for placing a cylinder inside of another object to create a hole. The cylinder can then be subtracted from the other object to create a hole. Refer to **Figure 2-8**. If the axis endpoint does not have the same X and Y coordinates as the center of the base, the cylinder is tilted from the XY plane.

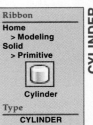

Ribbon
Home
> Modeling
Solid
> Primitive

Cylinder

Type
CYLINDER
CYL

CYLINDER

Figure 2-7.
A—A circular
cylinder. B—An
elliptical cylinder.

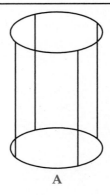

A   B

Figure 2-8.
A—A cylinder is
drawn inside of
another cylinder
using the **Axis
endpoint** option.
B—The large
cylinder has a hole
after **SUBTRACT** is
used to remove the
small cylinder.

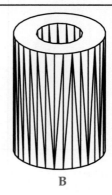

A   B

If polar tracking is on when using the **Axis endpoint** option, you can rotate the cylinder axis 90° from the current UCS Z axis and then turn the cylinder to any preset polar increment. See **Figure 2-9A**. If the polar tracking vector is parallel to the Z axis of the current UCS, the tooltip displays a positive or negative Z value. See **Figure 2-9B**.

Polysolid

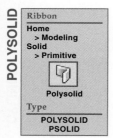

The *polysolid* primitive is simply a polyline constructed as a solid object by applying a width and height to the polyline. Many of the options used to create polylines are used with the **POLYSOLID** command. The principal difference is that a solid object is constructed using **POLYSOLID**.

When the command is initiated, you are prompted to select the start point or enter an option. By default, the width of the polysolid is equally applied to each side of the line you draw. This is center justification. Using the **Justify** option, you can set the justification to center, left, or right. The justification applies to all segments created in the command session. See **Figure 2-10**. If you select the wrong justification option, you must exit the command and begin again.

The default width is .25 units and the default height is 4 units. These values can be changed using the **Height** and **Width** options of the command. The height value is saved in the **PSOLHEIGHT** system variable. The width value is saved in the **PSOLWIDTH** system variable. Using these system variables, the default width and height can be set outside of the command.

The **Object** option allows you to convert an existing 2D object into a polysolid. AutoCAD entities such as lines, circles, arcs, polylines, polygons, and rectangles can be converted. The 2D object cannot be self intersecting. Some objects, such as 3D polylines, cannot be converted.

Once you have set the first point on the polysolid, pick the endpoint of the first segment. Continue adding segments as needed. The **Undo** option allows you to remove the last segment drawn and continue from the previous point without exiting the command. To complete the command, press [Enter].

(margin left, vertical) **POLYSOLID**

(margin box)
Ribbon
Home
  > Modeling
Solid
  > Primitive

Polysolid

Type
**POLYSOLID
PSOLID**

to 2047. **Figure 2-13B** displays spheres with 20 contour lines. It is best to use a lower number during construction and preliminary displays of the model and, if needed, higher settings for more realistic visualization. The contour lines setting is also available in the **Display** tab of the **Options** dialog box or by typing ISOLINES.

The Draw true silhouettes setting in the **2D Wireframe options** area of the **Visual Styles Manager** controls the display of silhouettes on 3D solid curved surfaces. The setting is either Yes or No. Notice the sphere silhouette in **Figures 2-13C** and **2-13D**. The Draw true silhouettes setting is stored in the **DISPSILH** system variable.

Torus

A basic *torus* is a cylinder bent into a circle, similar to a doughnut or inner tube. There are three types of tori. See **Figure 2-15**. A torus with a tube diameter that touches itself is called *self intersecting* and has no center hole. To create a self-intersecting torus, the tube radius must be greater than the torus radius. The third type of torus looks like a football. It is drawn by entering a negative torus radius and a positive tube radius of greater absolute value, such as –1 and 1.1.

Once the command is initiated, you are prompted for the center of the torus or to enter an option. If you pick the center, you must then set the radius of the torus. To specify a diameter, select the **Diameter** option after specifying the center. This defines a base circle that is the centerline of the tube. The **3P**, **2P**, and **Ttr** options are used to define the base circle of the torus using either three points, two points, or two points of tangency and a radius.

Once the base circle of the torus is defined, you are prompted for the tube radius, or to enter an option. The tube radius defines the cross-sectional circle of the tube. To specify a diameter of the cross-sectional circle, enter the **Diameter** option. You can also use the **2Point** option to pick two points on screen that define the diameter of the cross-sectional circle. If you are working conceptually, you can move the pointer and see the tube size change dynamically.

Ribbon	TORUS
Home > Modeling Solid > Primitive Torus	
Type	
TORUS TOR	

Wedge

A *wedge* has five sides, four of which are at right angles and the fifth at an angle other than 90°. See **Figure 2-16**. Once the command is initiated, you are prompted to select the first corner of the base or to enter an option. By default, a wedge is constructed by picking diagonal corners of the base and setting a height. To pick the center point,

Ribbon	WEDGE
Home > Modeling Solid > Primitive Wedge	
Type	
WEDGE WE	

Figure 2-15.
The three types of tori are shown as wireframes and with hidden lines removed.

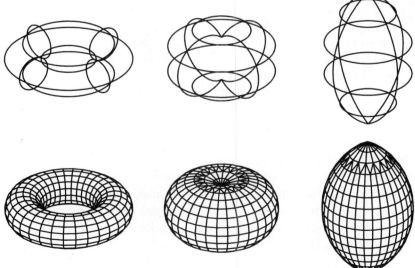

Figure 2-16.
A—A wedge drawn by picking corners and specifying a height. B—A wedge drawn using the **Center** option. Notice the location of the center.

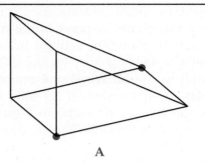

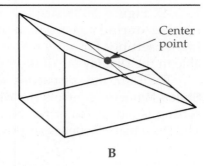

Center point

A

B

select the **Center** option. The center point of a wedge is the middle of the angled surface. You must then pick a point to set the width and length before entering a height.

After specifying the first corner or the center, you can enter the length, width, and height instead of picking a second corner. When prompted for the second corner, select the **Length** option and specify the length. You are then prompted for the width. After the width is entered, you are prompted for the height. Either enter the height or select the **2Point** option to pick two points on screen.

To create a wedge with equal length, width, and height, select the **Cube** option when prompted for the second corner. Then, enter a length. The same value is automatically used for the width and height.

Exercise 2-2

www.g-wlearning.com/CAD/

Complete the exercise on the companion website.

Constructing a Planar Surface

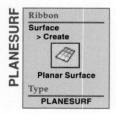

PLANESURF

Ribbon
Surface
> Create

Planar Surface

Type
PLANESURF

A *planar surface* primitive is an object consisting of a single plane and is created parallel to the current XY plane. The surface that is created has zero thickness and is composed of a mesh of lines. It is created with the **PLANESURF** command. The command prompts you to specify the first corner and then the second corner of a rectangle. Once drawn, the surface is displayed as a mesh with lines in the local directions, similar to the X and Y directions. See **Figure 2-17A**. These lines, called isolines, do not include the object's boundary. The **SURFU** and **SURFV** system variables determine how many isolines are created in the local U and V directions when the planar surface is drawn. The isoline values can be changed later using the **Properties** palette. The maximum number of isolines in either direction is 200.

The **Object** option of the **PLANESURF** command allows you to convert a 2D object into a planar surface. Any existing object or objects lying in a single plane and forming a closed area can be converted to a planar surface. The objects in **Figure 2-17B** are two arcs and two lines connected at their endpoints. The resulting planar surface is shown in **Figure 2-17C**.

Although a planar surface is not a solid, it can be converted into a solid in a single step. For example, the object in **Figure 2-17C** is converted into a solid using the **THICKEN** command. See **Figure 2-17D**. The object that started as two arcs and two lines is now a solid model and can be manipulated and edited like any other solid. This capability allows you to create intricate planar shapes and quickly convert them to a solid for use in advanced modeling applications. Model editing procedures are discussed in detail in Chapters 8, 11, and 12.

Figure 2-17.
A—A rectangular planar surface with four isolines in the U direction and eight isolines in the V direction. B—These two arcs and two lines form a closed area and lie on a single plane. C—The arcs and lines are converted into a planar surface. D—The planar surface is converted into a solid.

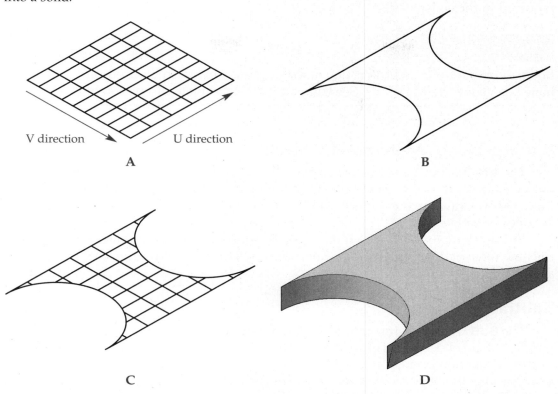

Creating Composite Solids

A *composite solid* is a solid model constructed of two or more solids, often primitives. Solids can be subtracted from each other, joined to form a new solid, or overlapped to create an intersection or interference. The commands used to create composite solids are found in the **Solid Editing** panel of the **Home** tab or the **Boolean** panel of the **Solid** tab in the ribbon. See **Figure 2-18.**

Introduction to Booleans

Three operations form the basis of constructing many complex solid models. Joining two or more solids is called a *union* operation. Subtracting one solid from another is called a *subtraction* operation. Forming a solid based on the volume of overlapping solids is called an *intersection* operation. Unions, subtractions, and intersections as a

Figure 2-18.
Selecting a Boolean command in the **Boolean** panel on the **Solid** tab of the ribbon. These commands are also located in the **Solid Editing** panel on the **Home** tab.

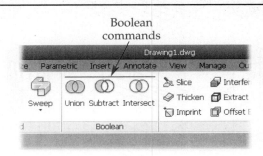

group are called **Boolean operations**. George Boole (1815–1864) was an English mathematician who developed a system of mathematical logic where all variables have the value of either one or zero. Boole's two-value logic, or *binary algebra*, is the basis for the mathematical calculations used by computers and, with respect to modeling, for those required in the construction of composite solids.

NOTE

Boolean operations used to create composite solids can also be used on meshes that have been converted to solids. See Chapter 9 for a complete discussion of meshes.

Joining Two or More Solid Objects

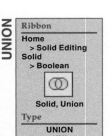

The **UNION** command is used to combine solid objects, **Figure 2-19**. The solids do not need to touch or intersect to form a union. Therefore, accurately locate the primitives when drawing them. After selecting the objects to join, press [Enter] and the action is completed.

In the examples shown in **Figure 2-19B**, notice that lines, or edges, are shown at the new intersection points of the joined objects. This is an indication that the features are one object, not separate objects.

Subtracting Solids

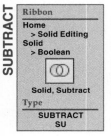

The **SUBTRACT** command allows you to remove the volume of one or more solids from another solid. Several examples are shown in **Figure 2-20**. The first object selected in the subtraction operation is the object *from* which volume is to be subtracted. The next object is the object to be subtracted from the first. The completed object will be a new solid. If the result is the opposite of what you intended, you may have selected the objects in the wrong order. Just undo the operation and try again.

Creating New Solids from the Intersection of Solids

When solid objects intersect, the overlap forms a common volume, a space that both objects share. This shared space is called an *intersection*. An intersection (common volume) can be made into a composite solid using the **INTERSECT** command. **Figure 2-21** shows several examples. A solid is formed from the common volume. The original objects are removed.

Figure 2-19.
A—The solid primitives shown here have areas of intersection and overlap. B—Composite solids after using the **UNION** command. Notice the lines displayed where the previous objects intersected.

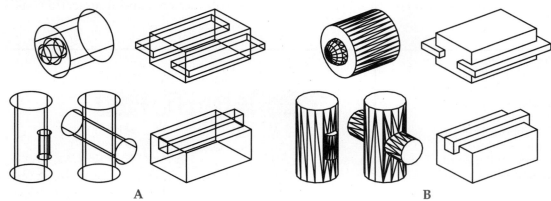

A   B

Figure 2-20.
A—The solid primitives shown here have areas of intersection and overlap. B—Composite solids after using the **SUBTRACT** command.

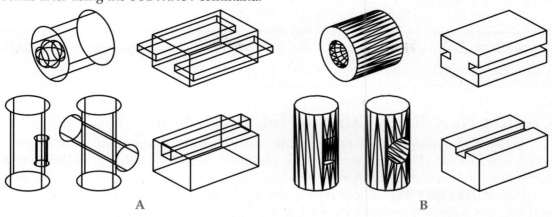

A   B

Figure 2-21.
A—The solid primitives shown here have areas of intersection and overlap. B—Composite solids after using the **INTERSECT** command.

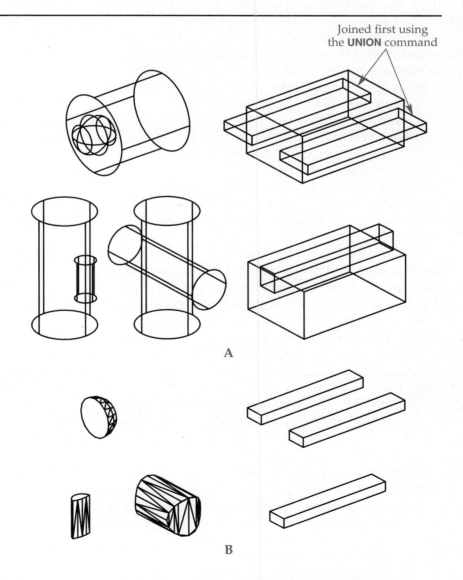

Joined first using the **UNION** command

A

B

The **INTERSECT** command is also useful in 2D drawing. For example, if you need to create a complex shape that must later be used for inquiry calculations or hatching, draw the main object first. Then, draw all intersecting or overlapping objects. Next, create regions of the shapes. Finally, use **INTERSECT** to create the final shape. The resulting shape is a region and has solid properties. Regions are discussed later in this chapter.

Exercise 2-3

www.g-wlearning.com/CAD/

Complete the exercise on the companion website.

Creating New Solids Using the Interfere Command

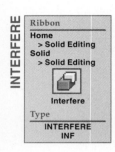

INTERFERE

Ribbon
Home
> Solid Editing
Solid
> Solid Editing

Interfere

Type
INTERFERE
INF

When you use the **SUBTRACT**, **UNION**, and **INTERSECT** commands, the original solids are deleted. They are replaced by the new composite solid. The **INTERFERE** command does not do this. A new solid is created from the interference (common volume) as if the **INTERSECT** command were used, but the original objects are retained and the new solid can be either deleted or retained.

Once the command is initiated, you are prompted to select the first set of solids or to enter an option. The **Settings** option opens the **Interference Settings** dialog box, which is used to change the visual style and color of the interference solid and the visual style of the viewport. The **Nested selection** option allows you to check the interference of separate solid objects within a nested block. A *nested block* is one that is composed of other blocks. When any needed options are set, select the first set of solids and press [Enter].

You are prompted to select the second set of solids or to enter an option. Entering the **Check first set** option tells AutoCAD to check the objects in the first set for interference. There is no second set when this option is used. Otherwise, select the second set of solids and press [Enter].

AutoCAD zooms in on the highlighted interference solid and displays the **Interference Checking** dialog box. See **Figure 2-22**. The visual style is set to a wireframe display by default and the interference solid is shaded in a color, which is red by default.

In the **Interfering objects** area of the **Interference Checking** dialog box, the number of objects selected in the first and second sets is displayed. The number of interfering pairs found in the selected objects is also displayed.

The buttons in the **Highlight** area of the dialog box are used to highlight the previous or next interference object. If the **Zoom to pair** check box is checked, AutoCAD zooms to the interference objects when the **Previous** or **Next** button is selected.

To the right of the **Highlight** area are three navigation buttons—**Zoom Realtime**, **Pan Realtime**, and **3D Orbit**. Selecting one of these navigation options temporarily hides the dialog box and activates the selected command. This allows you to navigate in the viewport. When the navigation mode is exited, the dialog box is redisplayed.

By default, the **Delete interference objects created on Close** check box is checked. With this setting, AutoCAD deletes the object(s) created by interference. In order to retain the new solid(s), uncheck this box.

An example of interference checking and the result is shown in **Figure 2-23**. Notice that the original solids are intact, but new lines indicate the new solid. The new solid is retained as a separate object because the **Delete interference objects created on Close** check box was unchecked. The new solid can be moved, copied, and manipulated just like any other object. **Figure 2-23C** shows the new object after it has been moved and a conceptual display generated.

When the **INTERFERE** command is used, AutoCAD compares the first set of solids to the second set. Any solids that are selected for both the first and second sets are automatically included as part of the first selection set and eliminated from the second. If

---

Figure 2-22.
The **Interference Checking** dialog box is used to check for interference between solids. To retain the interference solid, uncheck the **Delete interference objects created on Close** check box.

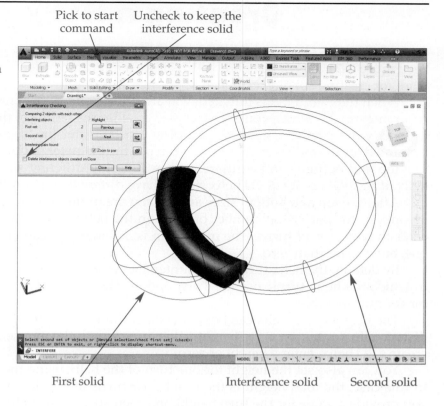

Pick to start command   Uncheck to keep the interference solid

First solid   Interference solid   Second solid

Figure 2-23.
A—Two solids form an area of intersection. B—After using **INTERFERE**, a new solid is defined (shown here in color) and the original solids remain. C—The new solid can be moved or copied.

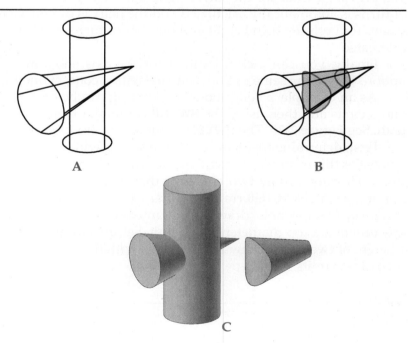

A   B   C

you do not select a second set of objects or the **Check first set** option is used, AutoCAD calculates the interference between the objects in the first selection set.

Exercise 2-4

www.g-wlearning.com/CAD/

Complete the exercise on the companion website.

Chapter 2 Creating Primitives and Composites   **39**

Creating a Helix

HELIX

Ribbon

Home
> Draw

Helix

Type

HELIX

A *helix* is a spline in the form of a spiral and can be created as a 2D or 3D object. See **Figure 2-24**. It is not a solid object. However, it can be used as the path or framework for creating solid objects such as springs and spiral staircases.

When the command is initiated, you are prompted for the center of the helix base. After picking the center, you are prompted to enter the radius of the base. If you want to specify the diameter, enter the **Diameter** option. After the base is defined, you are prompted for the radius of the top. You can use the **Diameter** option to enter a diameter. The top and bottom can be different sizes. Entering different sizes creates a tapered helix, if the helix is 3D. A 2D helix should have different sizes for the top and bottom.

After the top and bottom sizes are set, you are prompted to set the height or select an option. To specify the number of turns in the helix, select the **Turns** option. Then, enter the number of turns. The maximum is 500 and you can enter values less than one, but greater than zero.

By default, the helix turns in a counterclockwise manner. To change the direction in which the helix turns, use the **Twist** option. Then, enter CW for clockwise or CCW for counterclockwise.

The height of the helix can be set in one of three ways. First, you can enter a direct distance. To do this, type the height value or pick with the mouse to set the height. To create a 2D helix, enter a height of zero.

You can also set the height for one turn of the helix using the **Turn height** option. In this case, the total height is the number of turns multiplied by the turn height. If you provide a value for the turn height and then specify the helix height, the number of turns is automatically calculated and the helix is drawn. Conversely, if you provide values for both the turn height and number of turns, the helix height is automatically calculated.

Finally, you can pick a location for the axis endpoint using the **Axis endpoint** option. This is the same option available with a cone, cylinder, or pyramid.

As an example, a solid model of a spring can be created by constructing a helix and a circle and then using the **SWEEP** command to sweep the circle along the helix path. See **Figure 2-25**. The **SWEEP** command is discussed in detail in Chapter 7.

First, determine the diameter of the spring wire and then draw a circle using that value. For this example, you will create two springs, each with a wire diameter of .125 units. Therefore, draw two circles of that diameter, **Figure 2-26**. Their locations are not important. Next, determine the diameter of the spring and draw a corresponding helix. For this example, draw a helix anywhere on screen with a bottom diameter of one unit and a top diameter of one unit. Set the number of turns to eight and specify a height of two units. Draw another helix with the same settings, except make the top diameter .5 units.

Figure 2-24.
Three types of helices. From left to right, equal top and bottom diameters, unequal top and bottom diameters, and unequal top and bottom diameters with the height set to zero.

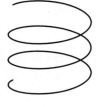 

Figure 2-25.
A helix can be used as a path to create a spring.

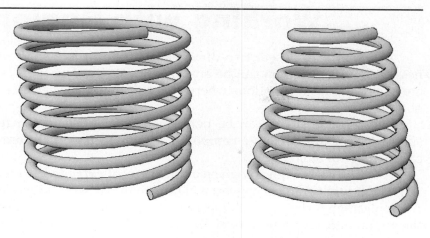

Figure 2-26.
To create a spring, first draw a circle the same diameter as the spring wire. Then, draw the helix and sweep the circle along the helix. Shown here are the two helices used to create the springs in Figure 2-25.

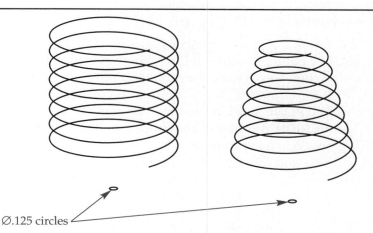

Ø.125 circles

Initiate the **SWEEP** command. You are first prompted to select the objects to sweep; pick one circle and press [Enter]. Next, you are prompted to select the sweep path. Select one of the helices. The first sweep, or spring, is completed. Repeat the procedure for the other circle and helix. The drawing is now composed of two swept solids. The two helices and the two circles are consumed by the **SWEEP** command.

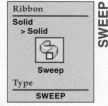

Ribbon
Solid
> Solid

Sweep
Type
SWEEP

SWEEP

NOTE

The **DELOBJ** system variable determines whether profile and path objects used to create 3D models are retained or deleted from the drawing. The default setting is 3. At this setting, all profile-defining and path-defining geometry is deleted. To retain the original objects used to create the model, set the **DELOBJ** system variable to 0.

Exercise 2-5

www.g-wlearning.com/CAD/

Complete the exercise on the companion website.

Working with Regions

A *region* is a closed, two-dimensional solid. It is a solid model without thickness (Z value). A region can be analyzed for its mass properties. Therefore, regions are useful for 2D applications where area and boundary calculations must be quickly obtained from a drawing.

Boolean operations can be performed on regions. When regions are unioned, subtracted, or intersected, a *composite region* is created. A composite region is also called a *region model*.

A region can be quickly given a thickness, or *extruded*, to create a 3D solid object. This means that you can convert a 2D shape into a 3D solid model in just a few steps. An application is drawing a 2D section view, converting it into a region, and extruding the region into a 3D solid model. Extruding is covered in Chapter 6.

Constructing a 2D Region Model

The following example creates, as a region, the plan view of a base for a support bracket. In Chapter 6, you will learn how to extrude the region into a solid. First, start a new drawing. Next, create the profile geometry in **Figure 2-27** using the **RECTANGLE** and **CIRCLE** commands. These commands create 2D objects that can be converted into regions. The **PLINE** and **LINE** commands can also be used to create closed 2D objects.

The **REGION** command allows you to convert closed, two-dimensional objects into regions. When the command is initiated, you are prompted to select objects. Select the rectangle and four circles and then press [Enter]. The rectangle and each circle are now separate regions and the original objects are deleted. You may need to switch to a wireframe visual style in order to see the circles. You can individually pick the regions. If you pick a circle, notice that a grip is displayed in the center, but not at the four quadrants. This is because the object is not a circle anymore. However, you can still snap to the quadrants.

In order to create the proper solid, the circular regions must be subtracted from the rectangular region. Using the **SUBTRACT** command, select the rectangle as the object to be subtracted *from*, and then select all of the circles as the objects to subtract. Now, if you select the rectangle or any of the circles, you can see that a single region has been created from the five separate regions. If you set the Conceptual or Realistic visual style current, you can see that the circles are now holes in the region. See **Figure 2-28**. Solids created in this manner can be given a thickness using either the **EXTRUDE** or **PRESSPULL** command.

Calculating the Area of a Region

A region is not a polyline. It is an enclosed area called a *loop*. Certain properties of the region, such as area, are stored as a value of the region. The **MEASUREGEOM** command can be used to determine the length of all sides and the area of the loop. This can be a useful advantage of a region.

Figure 2-27.
These 2D objects can be converted to regions in order to create a region model. The region model can then be made into a 3D solid.

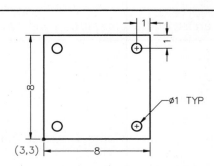

Figure 2-28.
Once the circular regions are subtracted from the rectangular region, they appear as holes. This is clear when a shaded visual style is set current.

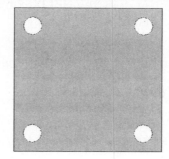

For example, suppose a parking lot is being repaved. You need to calculate the surface area of the parking lot to determine the amount of material needed. This total surface area excludes the space taken up by planting dividers, sidewalks, and lampposts because you will not be paving under these items. If the parking lot and all objects inside of it are drawn as a region model, the **MEASUREGEOM** command can give you this figure in one step using the **Object** option of the **Area** option. When the **Area** option is used, the calculated area is highlighted light green. Depending on the number of smaller regions inside the large region, the highlighting may appear incorrect, but the resulting calculation is correct. If a polyline is used to draw the parking lot, all internal features must be subtracted when the **MEASUREGEOM** command is used.

PROFESSIONAL TIP

Regions can prove valuable when working with many items:
- Roof areas excluding chimneys, vents, and fans.
- Bodies of water, such as lakes, excluding islands.
- Lawns and areas of grass excluding flower beds, trees, and shrubs.
- Landscaping areas excluding lawns, sidewalks, and parking lots.
- Concrete surfaces, such as sidewalks, excluding openings for landscaping, drains, and utility covers.

You can find many other applications for regions that can help in your daily tasks.

Exercise 2-6

www.g-wlearning.com/CAD/

Complete the exercise on the companion website.

Chapter Review

Answer the following questions using the information in this chapter.

1. What is a *solid primitive*?
2. How is a solid cube created?
3. How is an elliptical cylinder created?
4. Where is the center of a wedge located?
5. What is a *frustum pyramid*?
6. What is a *polysolid*?
7. Name at least four AutoCAD 2D entities that can be converted to a polysolid.

8. What type of entity does the **HELIX** command create and how can it be used to create a solid model?

9. What is a *composite solid*?

10. Which types of mathematical calculations are used in the construction of composite solids?

11. How are two or more solids combined to make a composite solid?

12. What is the function of the **INTERSECT** command?

13. How does the **INTERFERE** command differ from **INTERSECT** and **UNION**?

14. What is a *region*?

15. How can a 2D section view be converted to a 3D solid model?

Drawing Problems

Draw the objects in the following problems using the appropriate solid primitive commands and Boolean operations. When necessary, determine thickness dimensions for parts or use dimensions specified by your instructor. If appropriate, apply geometric constraints to the base 2D drawing. Do not add dimensions to the models. Save the drawings as **P2-***(problem number). Display and plot the problems as indicated by your instructor.*

1. General

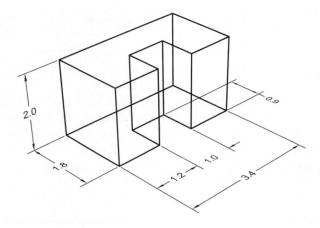

2. General

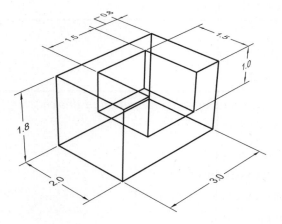

3. General

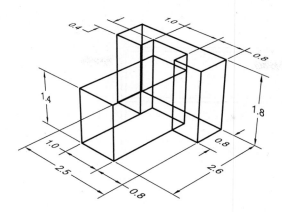

4. Mechanical

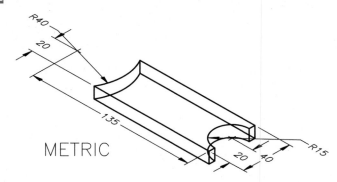

METRIC

5. Mechanical

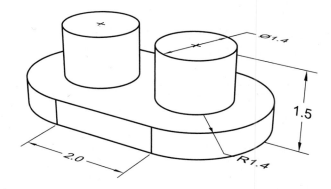

6. Mechanical

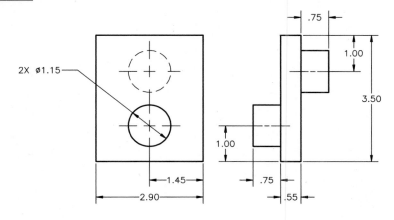

Chapter 2 Creating Primitives and Composites **45**

7. Architectural

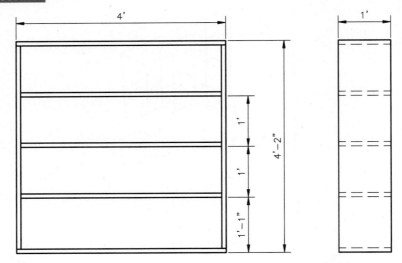

8. Architectural

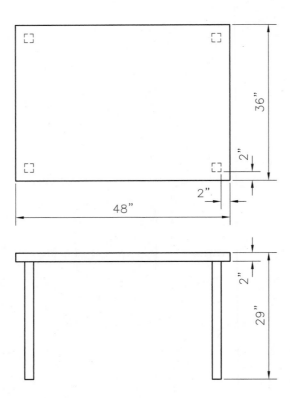

9. **Architectural**

36"

16" TYP

4 2" SQ LEGS

2"

18"

10. **Architectural**

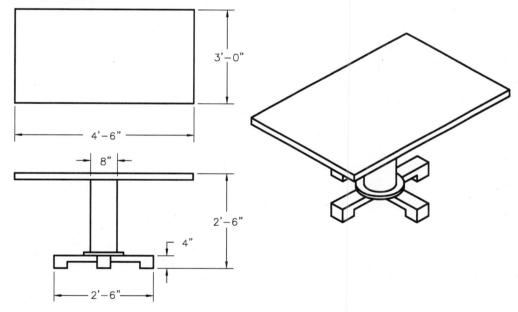

3'–0"

4'–6"

8"

2'–6"

4"

2'–6"

Chapter 2 Creating Primitives and Composites **47**

11. **Architectural** Draw the seat for a kitchen chair shown below. In later chapters, you will complete this chair. The seat is 1″ thick.

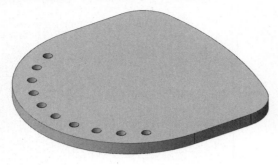

A

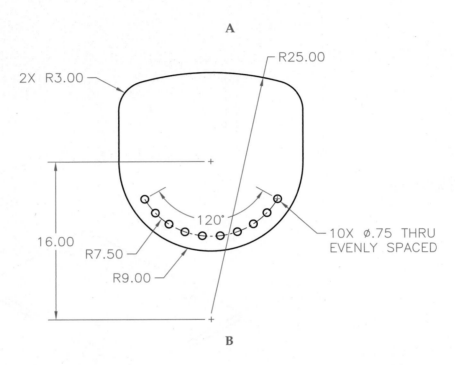

B

CHAPTER 3

Viewing and Displaying Three-Dimensional Models

Learning Objectives

After completing this chapter, you will be able to:

✓ Use the viewport controls to display views and control the display of view-navigation tools.
✓ Use the navigation bar to perform a variety of display manipulation functions.
✓ Create and save named views.
✓ Use the view cube to rotate the view of the model dynamically in 3D space.
✓ Use the view cube to display orthographic plan views of all sides on the model.
✓ Use steering wheels to display a 3D model from any angle.

AutoCAD provides several tools with which you can display and present 3D models in pictorial and orthographic views:

- The viewport controls, discussed in Chapter 1.
- Preset isometric viewpoints, discussed in Chapter 1.
- The view cube. This on-screen tool provides access to preset and dynamic display options.
- Steering wheels.
- The **3DORBIT**, **3DFORBIT**, and **3DCORBIT** commands. These commands provide dynamic display and continuous orbiting functions for demonstrations and presentations.

Once a viewpoint has been selected, you can enhance the display by applying visual styles. The **View** panel in the **Home** tab of the ribbon provides a variety of ways to display a model, including wireframe representation, hidden line removal, and simple shading. An introduction to visual styles is provided in Chapter 1 and complete coverage is provided in Chapter 15.

A more advanced representation can be produced by creating a rendering with the **RENDER** command. A rendering produces the most realistic image with highlights, shading, and materials, if applied. **Figure 3-1** shows a 3D model of a cast iron plumbing cleanout after using **HIDE**, setting the Conceptual visual style current, and using **RENDER**. Notice the difference in the three displays.

Figure 3-1.
A—Hidden display (hidden lines removed). B—The Conceptual visual style set current.
C—Rendered with lights and materials.

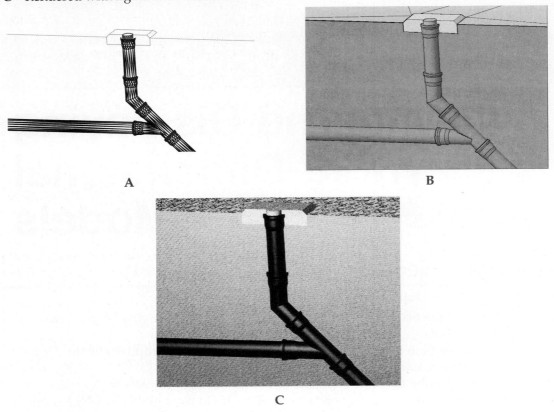

A

B

C

PROFESSIONAL TIP

The viewport controls, displayed in the upper-left corner of the AutoCAD drawing window, are discussed in Chapter 1. These controls provide quick access to preset isometric and orthographic views and settings for the view cube, navigation bar, and steering wheels. Use the viewport controls whenever possible to increase your drawing efficiency.

Keep in mind that undocked palettes may hide the viewport controls. If this happens, either dock the palette or move it out of the way.

Using the Navigation Bar to Display Models

By default, the navigation bar appears below the view cube on the right side of the screen. See **Figure 3-2**. It allows you to use the navigation tools described in this chapter, as well as the **ZOOM** and **PAN** commands and three orbit options. The **NAVBAR** command is used to turn on or off the display of the navigation bar. The navigation bar can be controlled using the viewport controls by selecting **Navigation Bar** in the **Viewport Controls** flyout. You can also select the **Navigation Bar** button in the **Viewport Tools** panel of the **View** ribbon tab. The appearance and location of the navigation bar can be customized, as discussed later in this chapter.

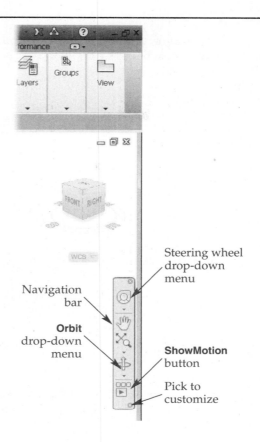

Figure 3-2.
The navigation bar appears below the view cube by default. This bar allows you to use the navigation tools quickly for drawing-display purposes.

Steering wheel drop-down menu

Navigation bar

Orbit drop-down menu

ShowMotion button

Pick to customize

Creating and Displaying Named Views

A *named view* is a single-frame display of a model from a given viewpoint. Named views are easy to create and have a wide variety of uses in AutoCAD. The **New View/Shot Properties** dialog box is used for creating named views. To access this dialog box, first pick the **ShowMotion** button in the navigation bar. Refer to **Figure 3-2**. When the **ShowMotion** toolbar appears at the bottom-center of the drawing window, pick the **New Shot...** button. See **Figure 3-3A**. The **New View/Shot Properties** dialog box appears, **Figure 3-3B**.

To create a named view, pick the **View Properties** tab. Enter a view name and select **Still** for the view type. Pick the **Current display** radio button to accept the model display as is. You can also pick the **Define window** radio button to return to the model and refine the display. Pick **OK** to complete the named view creation. Two images now appear above the **ShowMotion** toolbar. See **Figure 3-4A**. The large image is the view category and displays the default name of <None>. Above it is the thumbnail of the view just created with its name shown below. Regardless of the view that is displayed on the screen, you can always display a named view by picking the **Play** button in the view thumbnail. See **Figure 3-4B**.

The new view name will appear under the **Model Views** branch in the **View Manager** dialog box when using the **VIEW** command. Named views are saved with the drawing and can be displayed by selecting from the **Views** drop-down list in the **Views** panel of the **Visualize** ribbon tab.

A named view can be displayed at any time, and can be used for any show motion view. The functions of the view category and view thumbnails are much more obvious when working with a variety of different views. A complete discussion of show motion is provided in Chapter 19.

Ribbon
View
> Viewport Tools

ShowMotion
Type
NAVSMOTION
MOTION
Toolbar
Navigation Bar

ShowMotion

NAVSMOTION

Figure 3-3.
A—Pick the **New Shot...** button in the **ShowMotion** toolbar to display the **New View/Shot Properties** dialog box. B—The **New View/Shot Properties** dialog box provides options for creating a new named view.

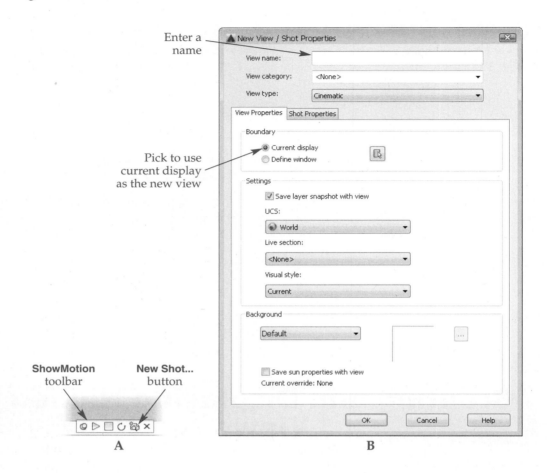

Figure 3-4.
A—Thumbnail images for the view category and named view are displayed above the **ShowMotion** toolbar. B— If the view has been changed, pick the **Play** button in the thumbnail of the desired view to redisplay the view.

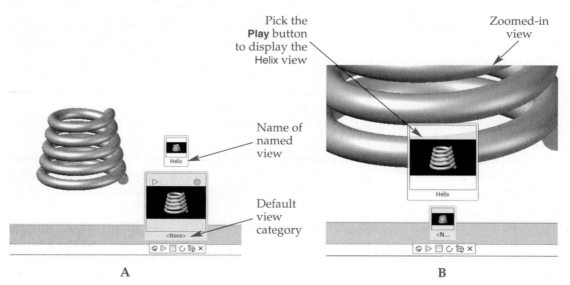

Dynamically Displaying Models with the View Cube

The *view cube* navigation tool allows you to quickly change the current view of the model to a preset pictorial or orthographic view or any number of dynamically user-defined 3D views. The tool is displayed, by default, in the upper-right corner of the drawing area. If the view cube is not displayed, select **View Cube** in the **Viewport Controls** flyout in the viewport controls. You can also pick the **View Cube** button in the **Viewport Tools** panel of the **View** ribbon tab.

The **View Cube Settings** dialog box offers many settings for expanding and enhancing the view cube. This dialog box is discussed in detail later in this chapter.

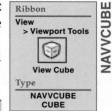

Ribbon

View
 > Viewport Tools

View Cube

Type
NAVVCUBE
CUBE

NAVVCUBE

PROFESSIONAL TIP

The location of the view cube can be quickly changed by picking the customize button on the navigation bar. Then, select **Docking positions** in the shortcut menu and select the location of your choice in the cascading menu.

Understanding the View Cube

The view cube tool is composed of a cube labeled with the names of all six orthographic faces. See **Figure 3-5**. A compass labeled with the four compass points (N, S, E, and W) rests at the base of the cube. The compass can be used to change the view. It also provides a visual cue to the orientation of the model in relation to the current user coordinate system (UCS). The *user coordinate system* describes the orientation of the X, Y, and Z axes. A complete discussion of user coordinate systems is given in Chapter 4.

Below the cube and compass is a button labeled WCS, which displays a shortcut menu. The label on this button displays the name of the current user coordinate system, which is, by default, the world coordinate system (WCS). This shortcut menu gives you the ability to switch from one user coordinate system to another. A user coordinate system is any coordinate system that is not the world coordinate system. The benefit of creating new user coordinate systems is the increase in efficiency and productivity when working on 3D models.

When the cursor is not over the view cube, the tool is displayed in its dimmed or *inactive state*. When the cursor is moved onto the view cube, the tool is displayed in its *active state* and a little house appears to the upper-left of the cube. This house is the **Home** icon. Picking this icon always restores the same view, called the *home view*. Later in this chapter, you will learn how to change the home view.

Figure 3-5.
The cube in the view cube tool is labeled with the names of all six orthographic faces. The view cube also contains a **Home** icon, compass, UCS shortcut menu, and shortcut menu button.

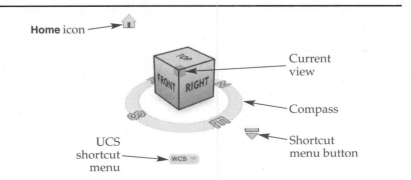

Home icon

Current view

Compass

UCS shortcut menu

WCS

Shortcut menu button

Notice that a corner, edge, or face on the cube is highlighted. In the default view of a drawing based on the acad3D.dwt template, the corner between the top, right, and front faces is shaded. The shaded part of the cube represents the standard view that is closest to the current view. Move the cursor over the cube and notice that edges, corners, and faces are highlighted as you move the pointer. If you pick a highlighted edge, corner, or face, a view is displayed that looks directly at the selected feature. By selecting standard views on the view cube, you can quickly move between pictorial and orthographic views.

PROFESSIONAL TIP

Picking the top face of the view cube displays a plan view of the current UCS XY plane. This is often quicker than using the **PLAN** command, which is discussed later in this chapter.

Dynamic Displays

The easiest way to change the current view using the view cube is to pick and drag the cube. Simply move the cursor over the cube, press and hold the left mouse button, and drag the mouse to change the view. The view cube and model view dynamically change with the mouse movement. When you have found the view you want, release the mouse button. This is similar to the **Orbit** tool in steering wheels, which are discussed later in the chapter.

NOTE

By default, the view cube does not move freely without snapping to the closest view. This action is reset by right-clicking on the view cube and picking **View Cube Settings...** in the cursor menu. In the **View Cube Settings** dialog box, uncheck **Snap to closest view** in the **When dragging on the View Cube** area.

When dragging the view cube, you are not restricted to the current XY plane, as you are when using the compass. The compass is discussed later in this chapter.

Pictorial Displays

The view cube provides immediate access to 20 different preset pictorial views. When one of the corners of the cube is selected, a standard isometric pictorial view is displayed. See **Figure 3-6A**. Eight corners can be selected. Picking one of the four corners on the top face of the cube restores one of the preset isometric views. You can select a preset isometric view using the viewport controls or the **Home** or **Visualize** tab of the ribbon. However, using the view cube may be a quicker method.

Selecting one of the edges of the cube sets the view perpendicular to that edge. See **Figure 3-6B**. There are 12 edges that can be selected. These standard views are not available on the ribbon.

Figure 3-6.
A—When one of the corners of the cube is selected, a standard isometric view is displayed.
B—Selecting one of the edges of the cube produces the same rotation in the XY plane as an
isometric view, but a zero elevation view in the Z plane.

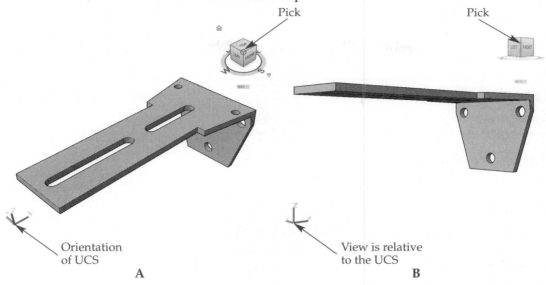

Pick

Pick

Orientation
of UCS

View is relative
to the UCS

A

B

PROFESSIONAL TIP

The quickest method of dynamically rotating a 3D model is achieved by using the
mouse wheel button. Simply press and hold the [Shift] key while pressing down and
holding the mouse wheel. Now move the mouse in any direction and the view rotates
accordingly. This is a transparent function that executes the **3DORBIT** command and
can be used at any time. Since it is transparent, it can even be used while you are in the
middle of a command. This is an excellent technique to use because it does not require
selecting another tool or executing a command.

Projection

The view displayed in the drawing window can be in one of two projections. The
projection refers to how lines are applied to the viewing plane. In a pictorial view, lines
in a *perspective projection* appear to converge as they recede into the background.
The points at which the lines converge are called *vanishing points*. In 2D drafting, it
is common to represent an object in pictorial as a one- or two-point perspective, especially in architectural drafting. In a *parallel projection*, lines remain parallel as they
recede. This is how an orthographic or axonometric (isometric, dimetric, or trimetric)
view is created.

To quickly change the projection using the viewport controls, pick **Parallel** or
Perspective from the **View Controls** flyout. To change the projection using the view
cube, right-click on the view cube to display the shortcut menu, or pick the shortcut
menu button at the lower-right of the view cube. Refer to **Figure 3-5**.

Three display options are given:
- **Parallel.** Displays the model as a parallel projection. This creates an ortho-
  graphic or axonometric view.
- **Perspective.** Displays the model in the more realistic, perspective projection.
  Lines recede into the background toward invisible vanishing points.

- **Perspective with Ortho Faces.** Displays the model in perspective projection when a pictorial view is displayed and parallel projection when an orthographic view is displayed. The parallel projection is only set current if a face on the view cube is selected to display the orthographic view. It is not set current if one of the preset orthographic views is selected from the viewport controls or the ribbon.

Figure 3-7 illustrates the difference between parallel and perspective projection.

Exercise 3-1

www.g-wlearning.com/CAD/

Complete the exercise on the companion website.

Orthographic Displays

The view cube faces are labeled with orthographic view names, such as Top, Front, Left, and so on. Picking on a view cube face produces an orthographic display of that face. Keep in mind, if the current projection is perspective, the view will not be a true orthographic view. See **Figure 3-8**. If you plan to work in perspective projection, but also want to view proper orthographic faces, turn on **Perspective with Ortho Faces** in the view cube shortcut menu. Another way to achieve a proper orthographic view is to turn on parallel projection.

When a face is selected on the view cube, the cube rotates to orthographically display the named face. The view rotates accordingly. In addition, notice that a series of triangles point to the four sides of the cube (when the cursor is over the tool). See **Figure 3-9A**. Picking one of these triangles displays the orthographic view corresponding to the face to which the triangle is pointing, **Figure 3-9B**. This is a quick and efficient method to precisely rotate the display between orthographic views.

When an orthographic view is displayed, two *roll arrows* appear to the upper-right of the view cube. Picking either of these arrows rotates the current view 90° in the selected direction and about an axis perpendicular to the view. See **Figure 3-9C**. Using the triangles and roll arrows on the view cube provides the greatest flexibility in manipulating the model between orthographic views.

Figure 3-7.
In a parallel projection, parallel lines remain parallel. In a perspective projection, parallel lines converge to a vanishing point. Notice the three receding lines on the box in the perspective projection. If these lines are extended, they will intersect.

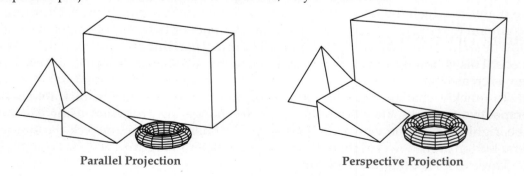

Parallel Projection   Perspective Projection

Figure 3-8.
A—To properly view orthographic faces using the view cube, turn on **Perspective with Ortho Faces** in the shortcut menu. B—When an orthographic view is set current with perspective projection on, the view is not a true orthographic view. Notice how you can see the receding surfaces.

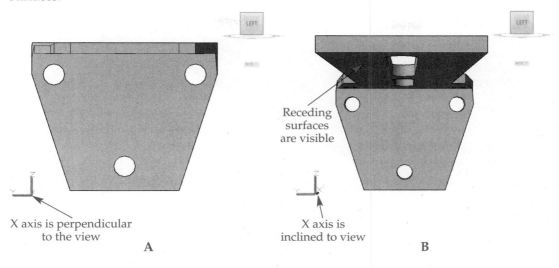

X axis is perpendicular to the view

A

Receding surfaces are visible

X axis is inclined to view

B

Figure 3-9.
A—The selected orthographic face is surrounded by triangles. Pick one of the triangles to display that orthographic face. B—The orthographic face corresponding to the picked triangle is displayed. Picking a roll arrow rotates the current view 90° in the selected direction. C—The rotated view.

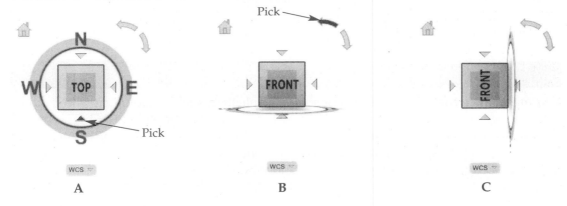

Pick

Pick

A

B

C

Setting Views with the Compass

The compass allows you to dynamically rotate the model in the XY plane. To do so, pick and hold on one of the four labels (N, S, E, or W). Then, drag the mouse to rotate the view. The view pivots about the Z axis of the current UCS. Try this a few times and notice that you can completely rotate the model by continuously moving the cursor off the screen. For example, pick the letter W and move the pointer either right or left (or up or down). Notice as you continue to move the mouse in one direction, the model continues to rotate in that direction.

You can use the compass to display the model in a view plan to the right, left, front, or back face of the cube. When the pointer is moved to one of the four compass directions, the letter is highlighted, **Figure 3-10A**. Simply single pick on the compass label that is next to the face you wish to view. See **Figure 3-10B**.

Figure 3-10.
A—When the pointer is moved over one of the four compass directions, the letter is highlighted. B—If you pick the letter, the orthographic view from that compass direction is displayed.

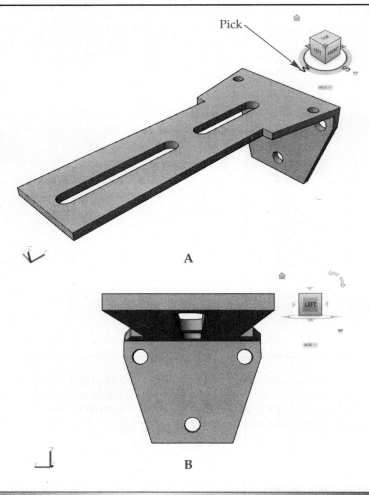

Pick

A

B

NOTE

You may notice that when different views are selected, such as isometric and front or side orthographic, that the drawing window background color changes. This is a result of the colors that are set for interface elements. Color settings for interface elements can be made in the **Drawing Window Colors** dialog box, which is accessed by picking the **Colors...** button in the **Display** tab of the **Options** dialog box.

Home View

The default home view is the southwest isometric view of the WCS. Picking the **Home** icon in the view cube always displays the view defined as the home view, regardless of the current UCS. You can easily set the home view to any display you wish. First, using any navigation method, display the model as required. Next, right-click on the view cube and pick **Set Current View as Home** from the shortcut menu. Now, when you pick the **Home** icon in the view cube, this view is set current. Remember, the **Home** icon does not appear until the cursor is over the view cube.

PROFESSIONAL TIP

Should you become disoriented after repeated use of the view cube, it is far more efficient to pick the **Home** icon than it is to use **UNDO** or try to select an appropriate location on the view cube.

UCS Settings

The UCS shortcut menu in the view cube lists the WCS and the names of all named UCSs in the current drawing. If there are no named UCSs, the listing is **WCS** and **New UCS**. See **Figure 3-11A**. To create a new UCS, pick **New UCS** from the shortcut menu. Next, use the appropriate **UCS** command options to create the new UCS. See Chapter 4 for complete coverage of the **UCS** command.

As new UCSs are created and saved, their names are added to the UCS shortcut menu. See **Figure 3-11B**. Now, if you wish to work on the model using a specific UCS, simply select it from the list. The UCS is then restored.

Exercise 3-2

www.g-wlearning.com/CAD/

Complete the exercise on the companion website.

View Cube Settings Dialog Box

The appearance and function of the view cube can be changed using the **View Cube Settings** dialog box, **Figure 3-12**. This dialog box is displayed by selecting **View Cube Settings…** from the view cube shortcut menu or by using the **Settings** option of the **NAVVCUBE** command. The next sections discuss the options found in the dialog box.

Type
**NAVVCUBE
CUBE**

NAVVCUBE

Figure 3-11.
The UCS shortcut menu below the view cube lists all named UCSs in the current drawing. A—The menu entries are **WCS** and **New UCS**. In this example, there are no saved UCSs in the drawing. B—The names of new UCSs are added to the UCS shortcut menu. The current UCS is indicated with a check mark.

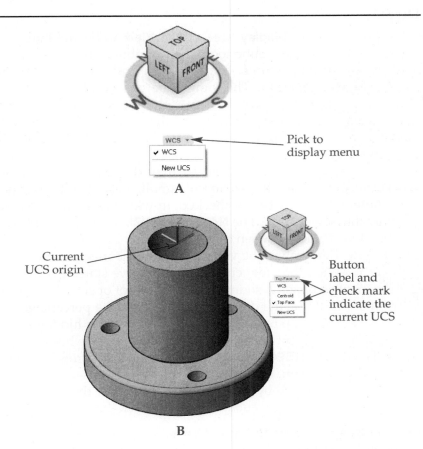

Figure 3-12.
The **View Cube**
Settings dialog box.

Set the size
of the
view cube

Set the
opacity
of the
view cube

Check to
display
the UCS
shortcut
menu
in the
view cube

Select a
location
for the
view cube

Preview
of view
cube

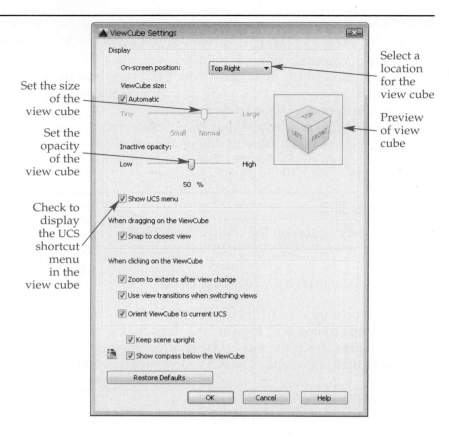

Display

Options in the **Display** area of the **View Cube Settings** dialog box control the appearance of the view cube tool. The thumbnail dynamically previews any changes made to the display options. There are four options in this area of the dialog box.

On-Screen Position. The view cube can be placed in one of four locations in the drawing area: top-right, bottom-right, top-left, or bottom-left corner. By default, it is located in the top-right corner of the screen. To change the location, select it in the **On-screen position:** drop-down list. The **NAVVCUBELOCATION** system variable controls this setting.

View Cube Size. The view cube can be displayed in one of four sizes. Use the **View Cube size:** slider to set the size to tiny, small, normal, or large. The slider is unavailable if the **Automatic** check box is checked, in which case AutoCAD sets the size based on the available screen area. The **NAVVCUBESIZE** system variable controls this setting.

Inactive Opacity. When the view cube is inactive, it is displayed in a semitransparent state. Remember, the view cube is in the inactive state whenever the pointer is not over it. When the view cube is in the active state, it is displayed at 100% opacity. Use the **Inactive opacity:** slider to set the level of opacity (transparency). The value can be from 0% to 100%; the default value is 50%. The percentage is displayed below the slider. A value of zero results in the view cube being hidden until the cursor is moved over it. If opacity is set to 100%, there is no difference between the inactive and active states. The **NAVVCUBEOPACITY** system variable controls this setting.

Show UCS Menu. By default, the UCS shortcut menu is displayed in the view cube. This menu is displayed by picking the button below the cube. If you wish to remove the UCS menu from the view cube display, uncheck the **Show UCS menu** check box.

When Dragging on the View Cube

By default, when you drag the view cube, the view "snaps" to the closest standard view that can be displayed by the view cube. This is because the **Snap to closest view** check box is checked by default. Uncheck this check box if you want the view to rotate freely without snapping to a preset standard view as you drag the view cube.

When Clicking on the View Cube

Options in the **When clicking on the View Cube** area control how the final view is displayed and how the labels on the view cube can be related to the UCS. There are three options in this area.

Zoom to Extents After View Change. If the **Zoom to extents after view change** check box is checked, the model is zoomed to the extents of the drawing whenever the view cube is used to change the view. Uncheck this if you want to use the view cube to change the display without fitting the model to the current viewport.

Use View Transitions When Switching Views. The default transition from one view to the next is a smooth rotation of the view. Although this transition may look nice, if you are working on a large model it may require more time and computer resources than you are willing to use. Therefore, if you are switching views a lot using the view cube, it may be more efficient to uncheck the **Use view transitions when switching views** check box. When this is unchecked, a view change just cuts to the new view without a smooth transition. This option does not affect the view when dragging the view cube or its compass.

Orient View Cube to Current UCS. As you have seen, the view cube is aligned to the current UCS by default. In other words, the top face of the cube is always perpendicular to the Z axis of the UCS. However, this can be turned off by unchecking the **Orient View Cube to current UCS** check box. The **NAVVCUBEORIENT** system variable controls this setting. When unchecked, the faces of the view cube are not reoriented when the UCS is changed. It may be easier to visualize view changes if the view cube faces are oriented to the current UCS.

When the **Orient View Cube to current UCS** check box is unchecked, WCS is displayed above the UCS shortcut menu in the view cube (unless the WCS is current). See **Figure 3-13**. This is a reminder that the labels on the view cube relate to the WCS and not to the current UCS.

Additional Options

Three additional options are available near the bottom of the **View Cube Settings** dialog box. These options are located below the **When clicking on the View Cube** area.

Keep Scene Upright. If the **Keep scene upright** check box is checked, the view of the model cannot be turned upside down when selecting a face, edge, or corner of the view cube. When unchecked, the view may be rotated so it is upside down. This can be confusing, so it is best to leave this box checked.

Show Compass below the View Cube. The compass is displayed by default in the view cube. However, if you do not find the compass useful, you can turn it off. To hide the compass, uncheck the **Show compass below the View Cube** check box.

Figure 3-13.
When the **Orient View Cube to current UCS** option is off in the **View Cube Settings** dialog box, the UCS shortcut menu has WCS displayed above it.

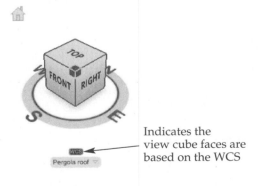

Indicates the view cube faces are based on the WCS

Restore Defaults. After making several changes in the **View Cube Settings** dialog box, you may get confused about the effect of different option settings on the model display. In this case, it is best to pick the **Restore Defaults** button to return all settings to their original values. Then change one setting at a time and test it to be sure the view cube functions as you intended.

Exercise 3-3

www.g-wlearning.com/CAD/

Complete the exercise on the companion website.

Creating a Continuous 3D Orbit

3DCORBIT

Ribbon

View
> Navigate

Continuous Orbit

Type

3DCORBIT

The **3DCORBIT** command provides the ability to create a continuous orbit of a model. By moving your pointing device, you can set the model in motion in any direction and at any speed, depending on the power of your computer. An impressive display can be achieved using this command. The command is located in the **Orbit** drop-down list in the **Navigate** panel of the **View** tab on the ribbon. See **Figure 3-14A**. If the **Navigate** panel is not displayed, right-click on the ribbon to access the shortcut menu and select **Navigate** from the **Show Panels** cascading menu. Additionally, the **3DCORBIT** command can be selected on the navigation bar.

Once the command is initiated, the continuous orbit cursor is displayed. See **Figure 3-14B**. Press and hold the pick button and move the pointer in the direction that you want the model to rotate and at the desired speed of rotation. Release the button when the pointer is moving at the appropriate speed. The model will continue to rotate until you pick the left mouse button, press [Enter] or [Esc], or right-click and pick **Exit** or another option in the shortcut menu. At any time while the model is orbiting, you can left-click and adjust the rotation angle and speed by repeating the process for starting a continuous orbit.

Figure 3-14.
A—Selecting the **3DCORBIT** command. B—This is the continuous orbit cursor in the **3DCORBIT** command (or **Continuous** option of the **3DORBIT** command). Pick and hold the left mouse button. Then, move the cursor in the direction in which you want the view to rotate and release the mouse button.

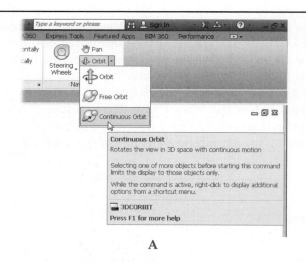

A   B

Plan Command Options

The **PLAN** command, introduced in Chapter 1, allows you to create a plan view of any user coordinate system (UCS) or the world coordinate system (WCS). The **PLAN** command automatically performs a **ZOOM Extents**. This fills the drawing window with the plan view. The command options are:

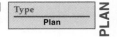

- **Current UCS.** This creates a view of the object that is plan to the current UCS.
- **UCS.** This displays a view plan to a named UCS. The preset UCSs are not considered named UCSs.
- **World.** This creates a view of the object that is plan to the WCS. If the WCS is the current UCS, this option and the **Current UCS** option produce the same results.

This command may have limited usefulness when working with 3D models. The dynamic capabilities of the view cube are much more intuitive and may be quicker to use. However, you may find instances where the **PLAN** command is easier to use, such as when the view cube is not currently displayed.

Displaying Models with Steering Wheels

Steering wheels, or *wheels,* are dynamic menus that provide quick access to view-navigation tools. A steering wheel follows the cursor as it is moved around the drawing. Each wheel is divided into wedges and each wedge contains a tool. See **Figure 3-15**. The **NAVSWHEEL** command is used to display a steering wheel, but a quicker method is to pick **Steering Wheels** from the **Viewport Controls** flyout in the viewport controls. The steering wheel drop-down menu on the navigation bar can also be used. All of the steering wheel formats discussed in this chapter can be selected in the drop-down menu on the navigation bar.

AutoCAD provides two basic types of wheels: View Object and Tour Building. Most of the options contained in these two wheels are included in the Full Navigation wheel. Each of these three types can be used in a full-wheel display or a minimized format. This section discusses the Full Navigation wheel and the two basic wheels in their full and mini formats. There is also a 2D Wheel that is displayed in paper (layout) space or when the **NAVSWHEELMODE** system variable is set to 3. The 2D Wheel provides basic view-navigation tools for 2D drafting.

Figure 3-15.
A—Full Navigation wheel. B—View Object wheel. C—Tour Building wheel.

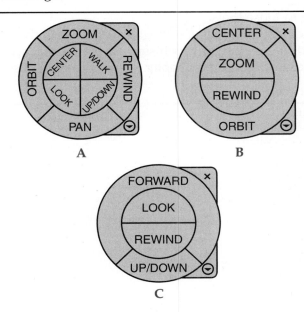

Using a Steering Wheel

Once a steering wheel is displayed, move the cursor around the screen. Notice as you move the cursor, the wheel follows it. When you stop the cursor, the wheel stops. If you move the cursor anywhere inside of the wheel, the wheel remains stationary. Note also that as you move the cursor inside of the wheel, a wedge (tool) is highlighted. If you pause the cursor over a tool, a tooltip is displayed that describes the tool.

To use a specific tool, simply pick and hold on the highlighted wedge, then move the cursor as needed to change the view of the model. Once you release the pick button, the tool ends and the wheel is redisplayed. Some options display a *center point* about which the display will move. The **Center** tool is used to set the center point. These features are all discussed in the next sections.

To change wheels, right-click to display the shortcut menu. Then, select **Full Navigation Wheel** from the menu to display that wheel. Or, select **Basic Wheels** to display a cascading menu. Select either **View Object Wheel** or **Tour Building Wheel** from the cascading menu to display that wheel. Refer to **Figure 3-16**.

PROFESSIONAL TIP

The view cube is a valuable visualization tool, especially when used in conjunction with a steering wheel. As you use the wheel options to manipulate the model, watch the view cube move. This provides you with a dynamic "map" of where your viewpoint is at all times in relation to the model, the UCS, and the view cube compass.

Exercise 3-4
www.g-wlearning.com/CAD/

Complete the exercise on the companion website.

Full Navigation Wheel

The Full Navigation wheel contains many of the tools available in the View Object and Tour Building wheels. Refer to **Figure 3-15A**. These tools are discussed in the following sections. The Full Navigation wheel also contains the **Pan** and **Walk** tools, which are not available in either of the other two wheels. These tools are discussed next. The tools shared with the View Object and Tour Building tools are discussed in the sections corresponding to those tools.

In addition, a number of settings are available to change the appearance of the steering wheel and the manner in which some of the tools function. Refer to the Steering Wheel Settings section later in this chapter for a complete discussion of these settings.

Figure 3-16.
The shortcut menu allows you to switch between steering wheels.

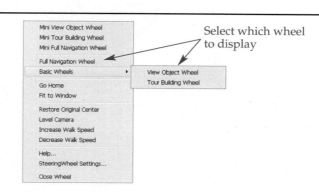

Pan

The **Pan** tool allows you to move the model in the direction that you drag the cursor. This tool functions the same as the AutoCAD **RTPAN** command. When you pick and hold on the tool, the cursor changes to four arrows with the label Pan Tool below the cursor.

If you use the **Pan** tool with the perspective projection current, it may appear that the model is slowly rotating about a point. This is not the case. What you are seeing is merely the effect of the vanishing points. As you pan, the relationship between the viewpoint and the vanishing points changes. You can quickly test this by closing the wheel, right-clicking on the view cube, and picking **Parallel** from the shortcut menu. Now display the wheel again and use the **Pan** tool. Notice the difference. The model pans without appearing to rotate.

Walk

The **Walk** tool is used to simulate walking toward, through, or away from the model. When you pick and hold on the tool, the center circle icon is displayed at the bottom-center of the drawing area. See **Figure 3-17**. The cursor changes to an arrow pointing away from the center of the circle as you move it off the circle. The arrow indicates the direction in which the view will move as you move the mouse. This gives the illusion of walking in that direction in relation to the model.

The farther away from the center tool you move the pointer, the faster the walk. Move the pointer away from the center tool until you have achieved the walk speed desired, then hold the pointer. Release the pick button when the desired view is displayed.

NOTE

If you begin moving the cursor at a point far away from the center circle icon, the walk speed will be substantial and the model may quickly disappear from view. If this occurs, close the steering wheel and pick the **Home** icon in the view cube to restore the home view. Display the steering wheel again, move it to the bottom-center of the drawing area, and then select the **Walk** tool. The cursor will be closer to the center circle icon, and the walk speed will be slower.

If you hold down the [Shift] key while clicking the **Walk** tool, the **Up/Down** slider is displayed. This allows you to change the screen Y axis orientation of the view. This equates to elevating the camera view relative to the object. Releasing the [Shift] key returns you to the standard walk mode. The up and down arrow keys can also be used to change the "height" of the view. The speed of walking can be increased with the plus key (+).

The **Up/Down** slider is also used with the **Up/Down** tool on the Tour Building wheel. It is explained in the Tour Building Wheel section.

Figure 3-17.
The **Walk** tool displays the center circle icon. As you move the cursor around the icon, an arrow is displayed to indicate the direction in which the view is being moved.

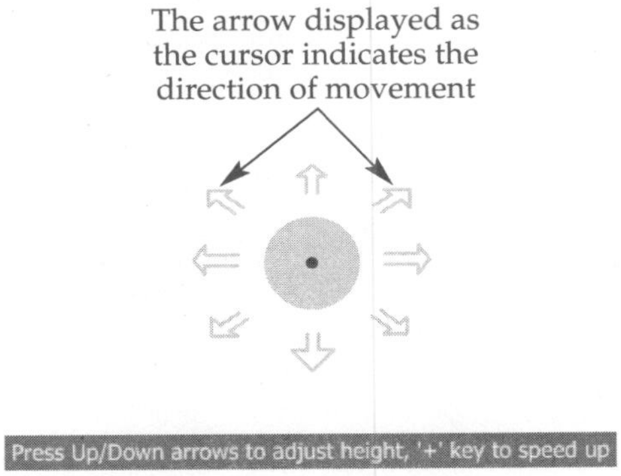

The arrow displayed as the cursor indicates the direction of movement

Press Up/Down arrows to adjust height, '+' key to speed up

View Object Wheel

If you think of your model as a building, the tools in the View Object wheel are used to view the outside of the building. This wheel contains four navigation tools: **Center**, **Zoom**, **Rewind**, and **Orbit**. Refer to **Figure 3-15B**.

Center

The **Center** tool is used to set the center point for the current view. Many tools, such as **Zoom** and **Orbit**, are applied in relation to the center point. Pick and hold the **Center** tool, then move the cursor to a point on the model and release. The display immediately changes to center the model on that point. See **Figure 3-18**. The selected point must be on an object. Notice that the center point icon resembles a globe with three orbital axes. These axes relate to the three axes of the model shown on the UCS icon.

Zoom

The **Zoom** tool is used to dynamically zoom the view in and out, just as with the AutoCAD **RTZOOM** command. The tool uses the center point set with the **Center** tool. When you pick and hold the **Zoom** tool in the wheel, the center point is displayed at its current location. The cursor changes to a magnifying glass with the label **Zoom Tool** displayed below it. See **Figure 3-19**. There are three different ways to use the **Zoom** tool, as discussed in this section.

When the **Zoom** tool is accessed in the Full Navigation wheel, the center point is relocated to the position of the steering wheel. If you wish to zoom on the existing center point when using the Full Navigation wheel, first press the [Ctrl] key and then access the **Zoom** tool. This prevents the tool from relocating the center point. You can also move the center point using the **Center** tool, then switch to the View Object wheel and access the **Zoom** tool from that wheel. In either case, the zoom is relative to the location of the center point.

Figure 3-18.
The **Center** tool allows you to select a new center point for the current view. This becomes the point about which many steering wheel tools operate.

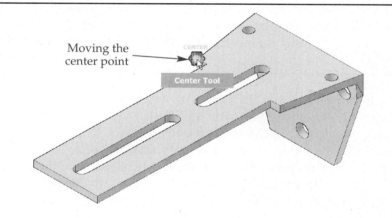

Moving the center point

Center Tool

Figure 3-19.
The **Zoom** tool operates in relation to the center point. If the tool is selected from the Full Navigation wheel, the center point is automatically relocated to the cursor location.

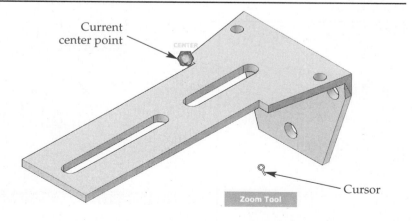

Current center point

Cursor

Zoom Tool

Pick and Drag. Pick and drag the cursor to dynamically zoom. This is similar to performing a realtime zoom. As the cursor is moved up or to the right, the viewpoint moves closer (zooms in). Move the pointer to the left or down and the viewpoint moves farther away (zooms out). The zooming is based on the current center point. When you have achieved the appropriate zoom location, release the pointer button.

Single Click. If you select the **Zoom** tool with a single click (using the View Object wheel), the view of the model zooms in by an incremental percentage. Each time you single click on the tool, the view is zoomed in by 25%. The zoom is in relation to the center point.

In order for this function to work when the **Zoom** tool is accessed from the Full Navigation wheel, you must check the **Enable single click incremental zoom** check box in the **Steering Wheels Settings** dialog box. This dialog box is discussed in detail later in the chapter.

Shift and Click. If you press and hold the [Shift] key and then pick the **Zoom** tool (using the View Object wheel), the view is zoomed out by 25%. As with the single-click method, the **Enable single click incremental zoom** check box must be checked in order to use this with the Full Navigation wheel.

Rewind

The **Rewind** tool allows you to step back through previous views. A single pick on this tool displays the previous view. If you pick and hold the tool, a "slide show" of previous views is displayed as thumbnail images. See **Figure 3-20**. The most recent view is displayed on the right-hand side. The oldest view is displayed on the left-hand side. The slide representing the current view is highlighted with an orange frame and a set of brackets. While holding the pick button, move the cursor to the left. Notice that the set of brackets moves with the cursor. As a slide is highlighted, the corresponding view is restored in the viewport. Release the pick button when you find the view you want and it is set current. When the brackets are hovered over two adjacent views, the orange highlighting is dimmed around each view. Selecting a view at this point enables you to display a view that is "between" two slides.

Figure 3-20.
A single pick on the **Rewind** tool displays the previous view. A "slide show" of previous views is displayed as thumbnail images if you press and hold the pick button.

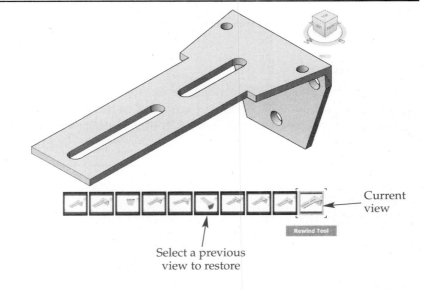

Current view

Rewind Tool

Select a previous view to restore

By default, thumbnail images of previous views outside of a steering wheel are not immediately displayed in the navigation history. The **Steering Wheels Settings** dialog box allows you to control when thumbnail images are generated in the navigation history. This dialog box is discussed later in the chapter.

The navigation history is maintained in the drawing file and is different for each open drawing. However, it is not saved when a drawing is closed.

Orbit

The **Orbit** tool allows you to rotate your point of view completely around the model in any direction. The view pivots about the center point set with the **Center** tool. When using the Full Navigation wheel, the center point can be quickly set by pressing and holding the [Ctrl] key, then picking and holding the **Orbit** tool. Next, drag the center point to the desired pivot point on the model and release. Now you can use the **Orbit** tool.

To use the **Orbit** tool, pick and hold on the tool in the steering wheel. The current center point is displayed with the label Pivot. Also, the cursor changes to a point surrounded by two circular arrows. See **Figure 3-21**. Move the cursor around the screen and the view of the model pivots about the center point. If this is not the result you wanted, just reset the pivot point using the **Center** tool.

PROFESSIONAL TIP

Since the **Orbit** tool is most often used to move your viewpoint quickly to another side of the model, it is more intuitive to locate the pivot point somewhere on the model. First, use the **Center** tool to establish the pivot point. Then, use the **Orbit** tool to rotate the view around the pivot point to the desired viewpoint. If the center point is not set, it defaults to the center of the screen.

CAUTION

When the **Zoom** tool is selected from the Full Navigation wheel, the center point is changed to the steering wheel location. Therefore, that center point is used as the pivot point for the **Orbit** tool. When the **Zoom** tool is selected from the View Object wheel, the center point is not changed. Therefore, it is best to switch to the View Object wheel to use the **Zoom** tool or to press the [Ctrl] key before accessing the tool in the Full Navigation wheel.

Tour Building Wheel

Where the tools in the View Object wheel are used to view the outside of the "building," the tools in the Tour Building wheel are used to move around inside of the "building." This wheel contains four navigation tools. The **Forward**, **Look**, and **Up/Down** tools

---

Figure 3-21.
The **Orbit** tool allows you to rotate your point of view completely around the model in any direction. This is the cursor displayed for the tool.

are discussed here. The **Rewind** tool is addressed in the View Object Wheel section, discussed earlier.

Forward

The principal tool in the Tour Building wheel is the **Forward** tool. It is similar to the **Walk** tool in the Full Navigation wheel. However, it only allows forward movement from the current viewpoint. This tool requires a center point to be set on the model from within the tool. The existing center point cannot be used.

First, move the cursor and steering wheel to the point on the model that will be the target (center point). Next, pick and hold the **Forward** tool. The pick point becomes the center point and a drag distance indicator is displayed. See **Figure 3-22**. This indicator shows the starting viewpoint, the center point, and the surface of the model that you selected. Hold the mouse button down while moving the pointer up. The orange location slider moves to show the current viewpoint relative to the center point. As you move closer to the model, the green center point icon increases in size, which also provides a visual cue to the zoom level.

Look

The **Look** tool is used to rotate the view about the center of the view. When the tool is activated, the cursor appears as a half circle with arrows. See **Figure 3-23**. As you move the cursor down, the model moves up in the view as if you are actually tilting your head down. Similarly, as you "look" away from the model to the right or left, the model appears to move away from your line of sight. If the vertical movement does not seem intuitive to you, it is easy to change. Open the **Steering Wheels Settings** dialog box and check the **Invert vertical axis for Look tool** check box. Now, when using the **Look** tool, moving the cursor up or down also moves the object in the same direction.

The distance between you and the model remains the same and the orientation of the model does not change. Therefore, you would not want to use this tool if you wanted to see another side of the model.

Up/Down

As the name indicates, the **Up/Down** tool moves the view up or down along the Y axis of the screen, regardless of the orientation of the current UCS. Pick and hold on the tool and the vertical distance indicator appears. See **Figure 3-24**. Two marks on this indicator show the upper and lower limits within which the view can be moved. The

---

Figure 3-22.
The drag distance indicator is displayed when using the **Forward** tool. This indicator shows the start point of the view and the selected surface of the model. The slider indicates the current view position relative to the starting point.

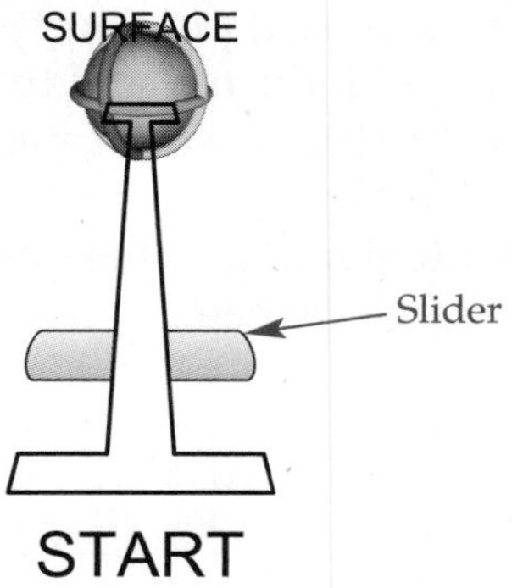

SURFACE

Slider

START

Figure 3-23.
The **Look** tool cursor appears as a half circle with arrows.

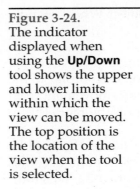

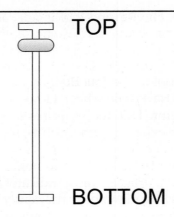

Figure 3-24.
The indicator displayed when using the **Up/Down** tool shows the upper and lower limits within which the view can be moved. The top position is the location of the view when the tool is selected.

TOP

BOTTOM

orange slider shows the position of the view as you move the cursor. When the tool is first activated, the view is at the top position. The **Up/Down** tool has limited value. The **Pan** tool is far more versatile.

PROFESSIONAL TIP

Use steering wheel tools such as **Forward**, **Look**, **Orbit**, and **Walk** to manipulate a view and fully explore the model. The **Rewind** tool can then be used to replay all of the previous views saved in the navigation history. This process reveals model views that can be saved as named views for later use in model construction or for shots created with show motion. Using show motion is discussed in detail in Chapter 19. Remember, views created using steering wheels are not saved with the drawing file. Therefore, if there is a possibility they will be needed later, it may be a time-saver to create and save named views as previously discussed in this chapter.

Exercise 3-5

www.g-wlearning.com/CAD/

Complete the exercise on the companion website.

Mini Wheels

The three wheels discussed previously were presented in their *full wheel* formats. As you gain familiarity with the use and function of each wheel and its tools, you may wish to begin using the abbreviated formats. The abbreviated formats are called *mini wheels*. See **Figure 3-25**. The mini wheels can be selected from the navigation bar or the steering wheel shortcut menu. When you select a mini wheel, it follows the pointer in the same manner as a full wheel.

Figure 3-25.
In addition to full-size steering wheels, mini wheels can be used. A—Mini Full Navigation wheel. B—Mini View Object wheel. C—Mini Tour Building wheel.

Current tool

Name of tool

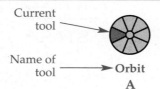
Orbit
A

Pan
B

Walk
C

Slowly move the mouse in a small circle and notice that each wedge of the wheel is highlighted. The name of the currently highlighted tool appears below the mini wheel. When the tool you need is highlighted, simply click and hold to activate the tool. All tools in the mini wheels function the same as those in the full-size wheels. The only difference is the appearance of the wheels.

Steering Wheel Shortcut Menu

A quick method for switching between the different wheel formats is to use the steering wheel shortcut menu. Pick the menu arrow at the lower-right corner of a full wheel to display the menu. See **Figure 3-26**. To select a wheel or change wheel formats, simply select the appropriate entry in the menu. Note that a check mark is *not* placed by the current wheel.

In addition to wheel formats, the shortcut menu provides options for viewing the model. These additional options are described as follows:

- **Go Home.** Returns the display to the home view. This is the same as picking the **Home** icon in the view cube.
- **Fit to Window.** Resizes the current view to fit all objects in the drawing inside of the window. This is essentially a zoom extents operation.
- **Restore Original Center.** Restores the original center point of the drawing using the current drawing extents. This does not change the current zoom factor. Therefore, if you are zoomed close into an object and pick this option, the object may disappear from view by moving off the screen.
- **Level Camera.** The camera (your viewpoint) is rotated to be level with the XY ground plane.
- **Increase Walk Speed.** The speed used by the **Walk** tool is increased by 100%.
- **Decrease Walk Speed.** The speed used by the **Walk** tool is decreased by 50%.
- **Help....** Displays the help system documentation for steering wheels.
- **Steering Wheel Settings....** Displays the **Steering Wheels Settings** dialog box, which is discussed in the next section.
- **Close Wheel.** Closes the steering wheel. This is the same as pressing [Esc] to close the wheel.

Exercise 3-6

www.g-wlearning.com/CAD/

Complete the exercise on the companion website.

Figure 3-26.
Select the down arrow to display the steering wheel shortcut menu. Both full-size wheels and mini wheels can be displayed using this menu.

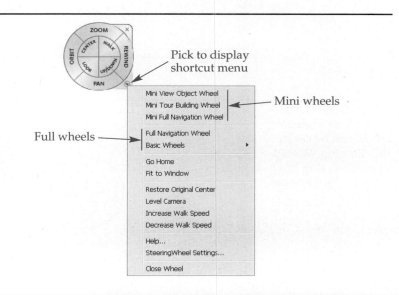

Steering Wheel Settings

The **Steering Wheels Settings** dialog box provides options for wheel appearance. See **Figure 3-27**. It also contains settings for the display and operation of several tools. It is displayed by selecting **Steering Wheel Settings...** in the steering wheel shortcut menu.

Changing Wheel Appearance

The two areas at the top of the **Steering Wheels Settings** dialog box allow you to change the size and opacity of all wheels. The settings in the **Big Wheels** area are for the full-size wheels. The settings in the **Mini Wheels** area are for the mini wheels. The **Wheel size:** slider in each area is used to display the wheels in small, normal, or large size. See **Figure 3-28**. The mini wheel has a fourth, extra large size. These sliders set the **NAVSWHEELSIZEBIG** and **NAVSWHEELSIZEMINI** system variables.

The **Wheel opacity:** slider in each area controls the transparency of the wheels. These sliders can be set to a value from 25% to 90% opacity. The sliders control the **NAVSWHEELOPACITYBIG** and **NAVSWHEELOPACITYMINI** system variables. The appearance of the full wheel in three different opacity settings is shown in **Figure 3-29**.

Figure 3-27.
The **Steering Wheels Settings** dialog box provides options for wheel appearance and the display and operation of several tools.

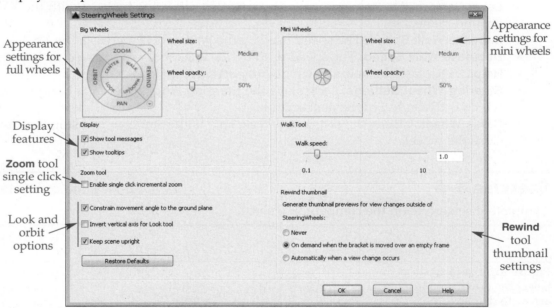

Appearance settings for full wheels

Display features

Zoom tool single click setting

Look and orbit options

Appearance settings for mini wheels

Rewind tool thumbnail settings

Figure 3-28.
The size of a steering wheel can be set in the **Steering Wheels Settings** dialog box or by using a system variable.

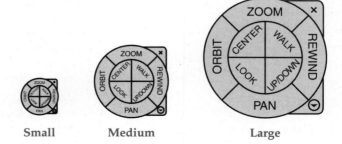

Small   Medium   Large

Figure 3-29.
The opacity of a
steering wheel can
be changed.
A—Opacity of 25%.
B—Opacity of 50%.
C—Opacity of 90%.

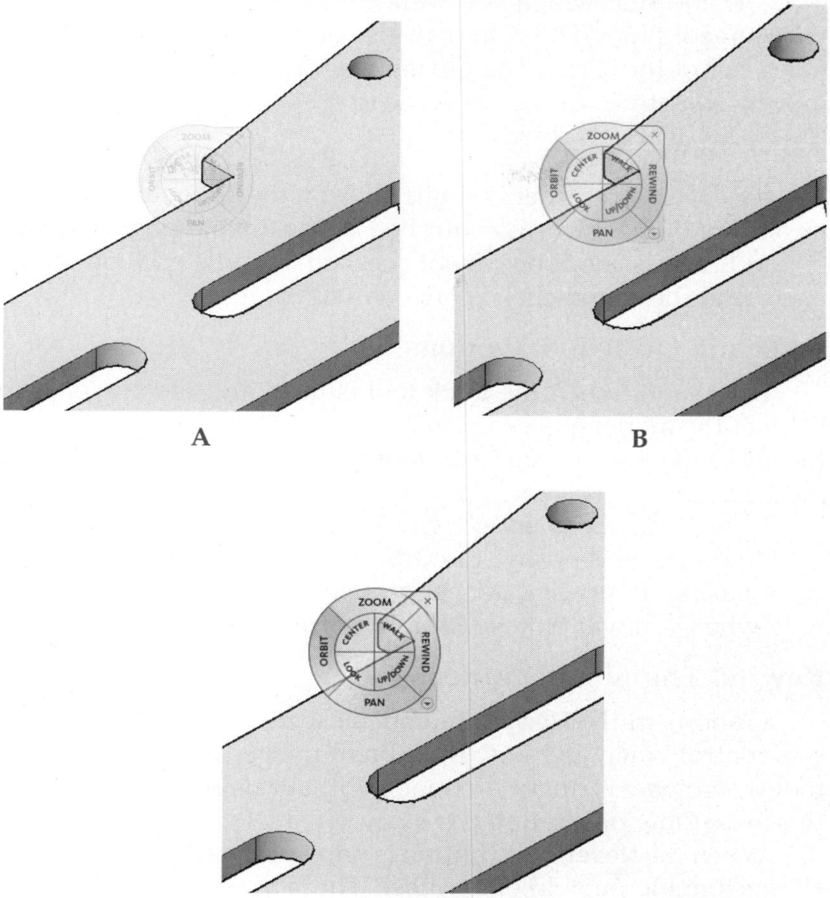

A

B

C

Display Features

The **Display** area of the **Steering Wheels Settings** dialog box controls two features of the wheel display. These settings determine whether messages and tooltips are displayed.

When a tool is selected, its name is displayed below the cursor. In addition, some tools have features or restrictions that can be indicated in a tool message. To see these messages, the **Show tool messages** check box must be checked. Otherwise, the messages are not displayed, but the restrictions remain in effect.

A tooltip is a short message that appears below the wheel when the cursor is hovered over the wheel. When the **Show tooltips** check box is checked, you can hold the cursor stationary over a tool for approximately three seconds and the tooltip will appear. Then, as you move the cursor over tools in the wheel, the appropriate tooltip is immediately displayed.

Walk Tool

By default, when the **Walk** tool is used, you move parallel to the ground plane. This is because the **Constrain movement angle to the ground plane** check box is checked. Test this by selecting the **Walk** tool and then move the cursor toward the top of the screen. You appear to be "walking" over the top of the model. Now, open the **Steering Wheels Settings** dialog box and uncheck the **Constrain movement angle to the ground plane** check box. This allows you to "fly" in the direction the cursor is moved when using the **Walk** tool. Exit the dialog box and again select the **Walk** tool. Move the cursor toward the top of the screen. This time it appears that you are flying directly toward or into the model.

The speed at which you walk through or around the model is controlled by the **Walk speed:** slider. The value can also be changed by typing in the text box at the right-hand end of the slider. The greater the value, the faster you will move as the cursor is moved away from the center circle icon.

Zoom Tool

As discussed earlier, a single click on the **Zoom** tool zooms in on the current view by a factor of 25%. This is controlled by the **Enable single click incremental zoom** check box. When this check box is not checked, a single click on the tool has no effect. Some users find the single-click zoom confusing.

Look and Orbit Tool Options

By default, when the **Look** tool is used and the cursor is moved downward, the view of the model moves up, just as if you were moving your eyes down. If you check the **Invert vertical axis for Look tool** option, this movement is reversed. In this case, the model moves in the same direction as the cursor.

If the **Keep scene upright** check box is not checked, it is possible to turn the model upside down while using the **Orbit** tool. This may not be desirable because it can be disorienting. To prevent this, be sure to leave the option checked. Uncheck this option only when you want to use **Orbit** in a "free-floating" mode.

Rewind Thumbnail Options

Options in the **Rewind thumbnail** section of the **Steering Wheels Settings** dialog box control when and how thumbnail images are generated for use with the **Rewind** tool when view changes are made without using a wheel. There are three options in this area. Only one option can be on.

When the **Never** radio button is on, thumbnail images are never generated for view changes made outside of a wheel. Thumbnail images are created for view changes made with a wheel, which is the case for all three options.

When the **On demand when the bracket is moved over an empty frame** radio button is on, thumbnail images are not automatically generated for view changes made outside of a wheel. The frames for these views display a double arrow icon when the **Rewind** tool in a wheel is used. However, as the brackets are moved over these frames, thumbnail images are generated. This is the default setting. See **Figure 3-30**.

When the **Automatically when a view change occurs** option is selected, a thumbnail image is generated any time a view change is made outside of a wheel. When the **Rewind** tool in a wheel is used, the frames for these views automatically display thumbnail images.

Figure 3-30.
A—By default, when the **Rewind** tool is selected, frames representing view changes made outside of a steering wheel display double-arrow icons. B—As the brackets are moved over the blank frames, thumbnail images are generated of the views created outside of a wheel.

View change made outside of a wheel

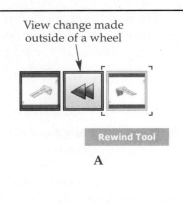

Rewind Tool

A

Thumbnail is generated

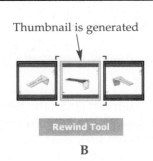

Rewind Tool

B

Restore Defaults

Picking the **Restore Defaults** button in the **Steering Wheels Settings** dialog box returns all of the settings in the dialog box to their default values. Select this when at any time you are not sure how the settings are affecting the appearance and function of the steering wheel tools. Then, make changes one at a time as needed.

Exercise 3-7

www.g-wlearning.com/CAD/

Complete the exercise on the companion website.

Controlling the Display of the Navigation Bar

The navigation bar is a flexible tool that can be customized or positioned to suit the needs of your drawing project. The three principal navigation tools are for the view cube, steering wheels, and show motion (discussed in Chapter 19). The view cube is displayed on the navigation bar when the normal display of the view cube is turned off.

The navigation bar can be placed at any location around the edge of the screen. In addition, the bar can be linked to the location of the view cube. The navigation bar can be quickly turned on or off by picking **Navigation Bar** in the **Viewport Controls** flyout in the viewport controls, or by picking the **Navigation Bar** button in the **Viewport Tools** panel of the **View** ribbon tab.

Repositioning the Navigation Bar

The navigation bar can be moved around the screen by picking the small customize button at the lower-right corner of the bar, and then selecting **Docking positions** to display the cascading menu. See **Figure 3-31**. The default setting is for the bar to be linked to the location of the view cube. This means that if the position of the view cube is changed, the navigation bar will follow. Test this by first confirming that **Link to View Cube** is checked in the menu. Next, move the view cube to one of the other locations, such as top left, using the **View Cube Settings** dialog box. Notice that the view cube and navigation bar move to the new location. Picking the **Undo** button on the **Quick Access** toolbar can quickly restore the previous location.

You can also move the navigation bar to a position of your choosing without moving the view cube. Simply uncheck the **Link to View Cube** option. Notice that a band appears at the top of the bar. When the pointer is moved into this band and you click, it becomes a move cursor. See **Figure 3-32**. Drag the bar to any edge location on the screen. The bar will dock at the closest screen edge. You cannot place it in a floating position in the middle of the screen.

Customizing the Navigation Bar

The tools that are displayed in the navigation bar can be changed by picking the customize button at the lower-right corner of the bar. By default, all of the tools are checked. If you wish to hide a tool, just select it to uncheck it in the list. All of the tools in the navigation bar can be removed or redisplayed in this manner. If the view cube is displayed on the screen, it will be grayed out in the list. If the view cube is currently not displayed, an option for it appears in the menu. In addition, a button for the view cube appears on the navigation bar. See **Figure 3-33**. The view cube can then be turned on using the navigation bar.

Figure 3-31.
The navigation
bar can be moved
around the screen by
selecting the small
customize button
at the lower-right
corner of the bar,
and then selecting
Docking positions in
the menu.

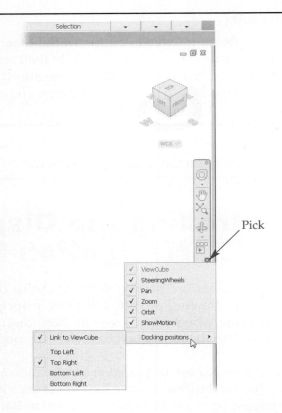

Pick

Figure 3-32.
Click and hold to
move the navigation
bar to any position
along the edge of the
screen.

Move
cursor

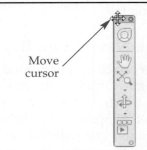

Figure 3-33.
If the view cube
is not currently
displayed, a button
for displaying it
appears on the
navigation bar.

Pick to
display
the view
cube

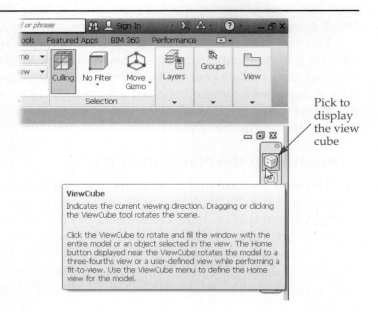

Chapter Review

Answer the following questions using the information in this chapter.

1. New named views are created in which dialog box?
2. Where is a thumbnail image of a new view displayed?
3. How do you select a standard isometric preset view using the view cube?
4. How is a standard orthographic view displayed using the view cube?
5. What is the difference between *parallel projection* and *perspective projection*?
6. What happens when one of the four view cube compass letters is picked?
7. Which command generates a continuous 3D orbit?
8. Which command can be used to produce a view that is parallel to the XY plane of the current UCS?
9. What is a *steering wheel*?
10. Briefly describe how to use a steering wheel.
11. The principal tool in the Tour Building wheel is the **Forward** tool. What is the purpose of this tool?
12. What are the three principal navigation tools found in the navigation bar?

Drawing Problems

1. General Open one of your 3D drawings from Chapter 2. Do the following.
   A. Use the view cube to create a pictorial view of the drawing.
   B. Set the Conceptual visual style current.
   C. Using a steering wheel, display different views of the model.
   D. Use the **Rewind** tool in the steering wheel to restore a previous view.
   E. Save the drawing as P3-1.

2. General Open one of your 3D drawings from Chapter 2. Do the following.
   A. Toggle the projection from parallel to perspective and turn off the view cube compass.
   B. Pick an edge on the view cube. Set the parallel projection current.
   C. Display three additional views based on view cube edges. Alternate between parallel and perspective projection.
   D. Put the model into a continuous orbit.
   E. Save the drawing as P3-2.

3. General Open one of your 3D drawings from Chapter 2. Do the following.
   A. Use a steering wheel to change the view of the model. Do this at least four times.
   B. Use the **Rewind** tool to restore a previous view.
   C. Repeat this for each of the previous views recorded with the steering wheel.
   D. Save the drawing as P3-3.

A user coordinate system (UCS) can be located on any surface and at any orientation for modeling purposes. In this example, a UCS is located on the inclined surface to draw the profile for the slot. The profile is extruded and then subtracted to create the final bracket model.

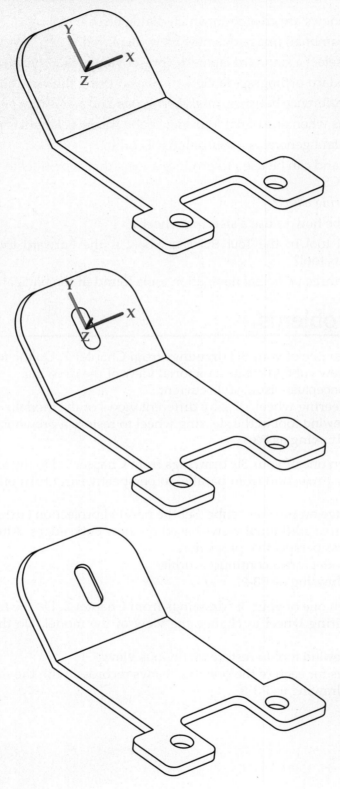

CHAPTER 4

Understanding Three-Dimensional Coordinates and User Coordinate Systems

Learning Objectives

After completing this chapter, you will be able to:

✓ Describe rectangular, spherical, and cylindrical methods of coordinate entry.
✓ Draw 3D polylines.
✓ Describe the function of the world and user coordinate systems.
✓ Move the user coordinate system to any surface.
✓ Rotate the user coordinate system to any angle.
✓ Use the UCS icon grips to move and rotate the UCS.
✓ Use a dynamic UCS.
✓ Save and manage user coordinate systems.
✓ Restore and use named user coordinate systems.
✓ Control UCS icon visibility in viewports.

As you learned in Chapter 1, any point in space can be located using X, Y, and Z coordinates. This type of coordinate entry is called *rectangular coordinate entry*. Rectangular coordinates are most commonly used for coordinate entry. However, there are actually three ways in which to locate a point in space. The other two methods of coordinate entry are spherical coordinate entry and cylindrical coordinate entry. These two coordinate entry methods are discussed in the following sections. In addition, this chapter introduces working with user coordinate systems (UCSs).

Introduction to Spherical Coordinates

Locating a point in 3D space with *spherical coordinates* is similar to locating a point on Earth using longitudinal and latitudinal values, with the center of Earth representing the origin. Lines of longitude connect the North and South Poles and provide an east-west measurement on Earth's surface. Lines of latitude horizontally extend around Earth and provide a north-south measurement. The origin (Earth's center) can be that of the default world coordinate system (WCS) or the current user coordinate system (UCS). See **Figure 4-1A**.

Figure 4-1.
A—Lines of longitude, representing the highlighted latitudinal segments in the illustration, run from north to south. Lines of latitude, representing the highlighted longitudinal segments, run from east to west. B—Spherical coordinates require a distance, an angle in the XY plane, and an angle from the XY plane.

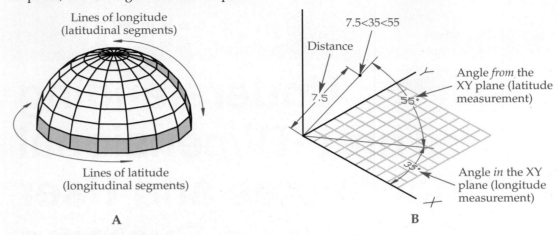

When entering spherical coordinates, the longitude measurement is expressed as the angle *in* the XY plane and the latitude measurement is expressed as the angle *from* the XY plane. See **Figure 4-1B**. A distance from the origin is also provided. The coordinates represent a measurement from the equator toward either the North or South Pole on Earth's surface. The following spherical coordinate entry is shown in **Figure 4-1B**.

7.5<35<55

This coordinate represents an ***absolute*** spherical coordinate, which is measured from the origin of the current UCS. Spherical coordinates can also be entered as ***relative*** coordinates. For example, a point drawn with the relative spherical coordinate @2<35<45 is located two units from the last point, at an angle of 35° *in* the XY plane, and at a 45° angle *from* the XY plane.

If dynamic input is turned on, the "second" or "next" coordinate entry is automatically a *relative* entry (by default). The @ symbol is not entered. To enter *absolute* coordinates with dynamic input turned on, enter an asterisk (*) before the first coordinate.

PROFESSIONAL TIP

Spherical coordinates are useful for locating features on a spherical surface. For example, they can be used to specify the location of a hole drilled into a sphere or a feature located from a specific point on a sphere. If you are working on such a spherical object, you might consider locating a UCS at the center of the sphere, then creating several different user coordinate systems rotated at different angles on the surface of the sphere. Any time a location is required, spherical coordinates can be used. Working with UCSs is introduced later in this chapter.

Using Spherical Coordinates

Spherical coordinates are well suited for locating points on the surface of a sphere. In this section, you will draw a solid sphere and then locate a second solid sphere with its center on the surface of the first sphere.

To draw the first sphere, select the **SPHERE** command. Specify the center point as 7,5 and the radius as 1.5 units. Display a southeast isometric pictorial view of the sphere. Alternately, you can use the view cube to create a different pictorial view. Also, set the Wireframe visual style current and switch to a parallel projection. Your drawing should look similar to **Figure 4-2A**.

Since you know the radius of the sphere, but the center of the sphere is not at the origin of the current UCS (the WCS), a relative spherical coordinate will be used to draw the second sphere. The sphere you drew is a solid and, as such, you can snap to its center using object snap. Set the center running object snap and then enter the **SPHERE** command again to draw the second sphere:

> Specify center point or [3P/2P/Ttr]: **FROM.**↵
> Base point: *(use the* **Center** *object snap to select the center of the existing sphere)*
> <Offset>: **@1.5<30<60**↵ *(1.5 is the radius of the first sphere)*
> Specify radius or [Diameter]: **.4.**↵

The objects should now appear as shown in **Figure 4-2B**. The center of the new sphere is located on the surface of the original sphere. This is clear after setting the Conceptual visual style current, **Figure 4-2C**. If you want the surfaces of the spheres to be tangent, add the radius value of each sphere (1.5 + .4) and enter this value when prompted for the offset from the center of the first sphere:

> <Offset>: **@1.9<30<60**↵

Notice in **Figure 4-2B** that the polar axes of the two spheres are parallel. This is because both objects were drawn using the same UCS, which can be misleading unless you understand how objects are constructed based on the current UCS. Test this by locating a cone on the surface of the large sphere, just below the small sphere. First, display a 3D wireframe view of the objects. Then, select the **CONE** command and continue as follows:

> Specify center point of base or [3P/2P/Ttr/Elliptical]: **FROM.**↵
> Base point: **CEN.**↵
> of *(pick the large sphere)*
> <Offset>: **@1.5<30<30.**↵
> Specify base radius or [Diameter]: **.25.**↵
> Specify height or [2Point/Axis endpoint/Top radius]: **1.**↵ *(before entering this value, make sure to move the crosshairs above the base of the cone to apply the value in the positive Z direction)*

The result of this construction with the Conceptual visual style set current is shown in **Figure 4-3**. Notice how the axis of the cone is parallel to the polar axis of the sphere.

Figure 4-2.
A—A three-unit diameter sphere shown from the southeast isometric viewpoint. B—A .8-unit diameter sphere is drawn with its center located on the surface of the original sphere. Also, lines have been drawn between the poles of the spheres. Notice how the polar axes are parallel. C—The objects after the Conceptual visual style is set current.

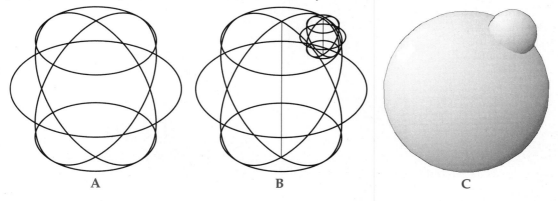

A   B   C

Figure 4-3.
The axis lines of objects drawn in the same user coordinate system are parallel. Notice that the cone does not project from the center of the large sphere.

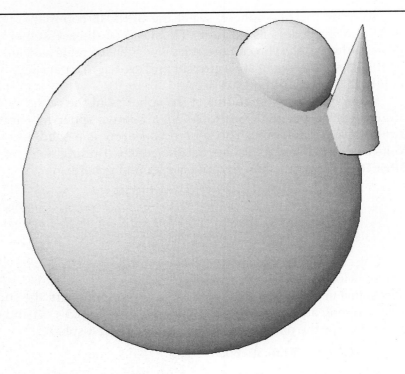

To draw the cone so that its axis projects from the center of the sphere, you will need to change the UCS. This is discussed later in the chapter.

Introduction to Cylindrical Coordinates

Locating a point in space with *cylindrical coordinates* is similar to locating a point on an imaginary cylinder. Cylindrical coordinates have three values. The first value represents the horizontal distance from the origin, which can be thought of as the radius of a cylinder. The second value represents the angle in the XY plane, or the rotation of the cylinder. The third value represents a vertical dimension measured up from the polar coordinate in the XY plane, or the height of the cylinder. See **Figure 4-4**. The absolute cylindrical coordinate shown in the figure is:

7.5<35,6

Figure 4-4.
Cylindrical coordinates require a horizontal distance from the origin, an angle in the XY plane, and a Z dimension.

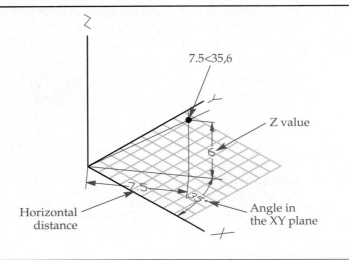

Like spherical coordinates, cylindrical coordinates can also be entered as relative coordinates. For example, a point drawn with the relative cylindrical coordinate @1.5<30,4 is located 1.5 units from the last point, at an angle of 30° in the XY plane of the previous point, and at a distance of four units up from the XY plane of the previous point.

NOTE

Turn off dynamic input before entering cylindrical coordinates. Dynamic input does not correctly interpret the entry when cylindrical coordinates are typed.

Using Cylindrical Coordinates

Cylindrical coordinates work well for attaching new objects to a cylindrical shape. An example of this is specifying coordinates for a pipe that must be attached to an existing pipe, tank, or vessel. In **Figure 4-5**, a pipe must be attached to a 12′ diameter tank at a 30° angle from horizontal and 2′-6″ above the floor. In order to draw the pipe properly as a cylinder, you will have to change the UCS, which you will learn how to do later in this chapter. An attachment point for the pipe can be drawn using the **POINT** command and cylindrical coordinates. First, set the drawing units to architectural. This can be done by selecting **Drawing Utilities>Units** from the **Application Menu** and specifying the length units as Architectural in the **Drawing Units** dialog box. Next, set the **PDMODE** system variable to 3. Enter the **POINT** command and continue as follows:

```
Current point modes: PDMODE=3 PDSIZE=0'-0"
Specify a point: FROM↵
Base point: CEN↵
of (pick the base of the cylinder)
<Offset>: @6'<30,2'6"↵ (The radius of the tank is 6'.)
```

The point can now be used as the center of the pipe (cylinder), **Figure 4-5B**. However, if you draw the pipe now, it will be parallel to the tank (large cylinder). By changing the UCS, as shown in **Figure 4-5C**, the pipe can be correctly drawn. You will learn how to do this later in this chapter.

Figure 4-5.
A—A plan view of a tank shows the angle of the pipe attachment. B—A 3D view from the southeast quadrant shows the pipe attachment point located with cylindrical coordinates. C—By creating a new UCS, the pipe can be drawn as a cylinder and correctly located without editing.

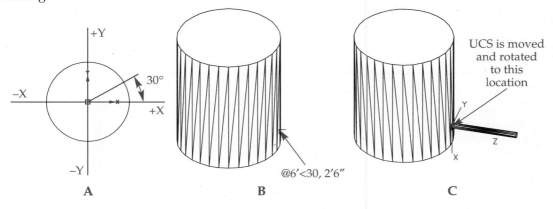

Complete the exercise on the companion website.

3D Polylines

3DPOLY

Ribbon

Home
> Draw

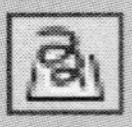

3D Polyline

Type

3DPOLY
3P

A polyline drawn with the **PLINE** command is a 2D object. All segments of the polyline must be drawn parallel to the XY plane of the current UCS. A *3D polyline*, on the other hand, can be drawn in 3D space. The Z coordinate value can vary from point to point in the polyline.

The **3DPOLY** command is used to draw 3D polylines. Any form of coordinate entry is valid for drawing 3D polylines. If polar tracking is on when using the **3DPOLY** command, you can pick points in the Z direction if the polar tracking alignment path is parallel to the Z axis.

The **Close** option can be used to draw the final segment and create a closed shape. There must be at least two segments in the polyline to use the **Close** option. The **Undo** option removes the last segment without canceling the command.

The **PEDIT** command can be used to edit 3D polylines. The **Spline curve** option of the **PEDIT** command is used to turn the 3D polyline into a B-spline curve based on the vertices of the polyline. A regular 3D polyline and the same polyline turned into a B-spline curve are shown in **Figure 4-6**.

Complete the exercise on the companion website.

Introduction to Working with User Coordinate Systems

All points in a drawing or on an object are defined with XYZ coordinate values (rectangular coordinates) measured from the 0,0,0 origin. Since this system of coordinates is fixed and universal, AutoCAD refers to it as the *world coordinate system (WCS)*. A *user coordinate system (UCS)*, on the other hand, can be defined with its origin at any location and with its three axes in any orientation desired, while remaining at 90° to each other. The **UCS** command is used to change the origin, position, and rotation of the coordinate system to match the surfaces and features of an object under construction. The symbol that identifies the orientation of the coordinate system is called the *UCS icon*. When set up to do so, the UCS icon reflects the changes in the orientation of the UCS and placement of the origin.

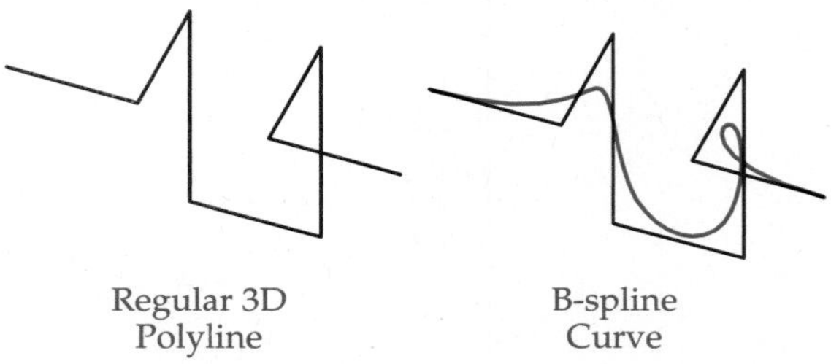

Figure 4-6.
A regular 3D polyline and the B-spline curve version after using the **PEDIT** command.

Regular 3D Polyline

B-spline Curve

The **UCS** command is used to create and manage UCSs. The quickest method of working with the UCS icon and the **UCS** command is to use the UCS icon shortcut menu and grip menus. Using this method will save you several steps and is described in this section. The **UCS** command and its options can be accessed on the **Coordinates** panel of the **Home** and **Visualize** tabs on the ribbon or by typing the command. Three selections in the ribbon provide access to all UCS options. These selections are introduced here and discussed in detail later in this chapter.

- **UCS.** Picking this button executes the **UCS** command. This command allows you to create a named UCS and provides access to all of the command options. The **UCS** command can also be selected by picking on the UCS drop-down list below the view cube. All of the command options are covered later in this chapter.
- **Named UCS Combo Control.** This drop-down list contains the six orthographic UCS options, which are covered later in this chapter. Any saved UCSs will be listed here, too.
- **UCS, Named UCS.** Picking this button displays the **UCS** dialog box. The three tabs in the dialog box contain a variety of UCS and UCS icon options and settings. These options and settings are described as you progress through the chapter.

The UCS Icon Shortcut Menu and Grip Menus

Creating and managing user coordinate systems is described in detail later in this chapter, but this is an introduction to a quick and dynamic method of working with the UCS and the UCS icon.

Many of the functions associated with managing the UCS and the UCS icon can be accomplished using the UCS icon shortcut menu. Grip menus are also available when the UCS icon is selected. The origin grip menu is useful for moving the UCS origin or realigning its orientation. For example, to move the UCS to the base of the wedge in **Figure 4-7A**, first select the UCS icon, then place the pointer over the UCS icon origin grip. Hover over the grip, but do not select it. The grip menu displays three options. Select **Move Origin Only**, then use an object snap to move the UCS origin to the base of the wedge. See **Figure 4-7B**.

To align a new UCS with the face of an object, select the icon and pick **Move and Align**. See **Figure 4-8A**. Then drag the icon to the face of the object. The alignment of the UCS icon depends on which edge of the face the pointer is moved over. In **Figure 4-8B**, the pointer is first moved over the pointed end of the wedge.

Figure 4-7.
A—The **Move Origin Only** option can be selected from the UCS icon origin grip menu. B—Use an object snap to relocate the UCS origin.

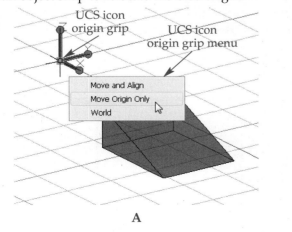

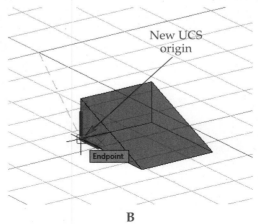

A   B

Figure 4-8.
A—To align the UCS with the face of an object, select the UCS icon and pick **Move and Align** from the UCS icon origin grip menu. B—The UCS icon is dragged over an edge and a point is selected using object snaps.

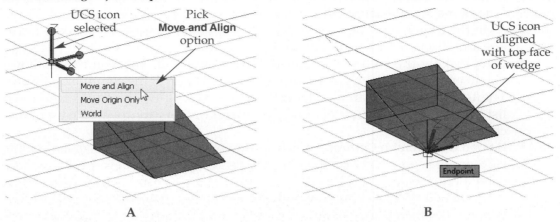

A   B

All of the options of the UCS command can be selected from the UCS icon shortcut menu, which is accessed by right-clicking on the UCS icon. See **Figure 4-9.** The options are discussed in this chapter as they apply.

Moving and Rotating the UCS

Earlier in this chapter, you used spherical coordinates to locate a small sphere on the surface of a larger sphere. You also drew a cone with the center of its base on the surface of the large sphere. However, the axis of the cone, which is a line from the center of the base to the tip of the cone, is not pointing to the center of the sphere. Refer to **Figure 4-3.** This is because the Z axes of the large sphere and cone are parallel to the world coordinate system (WCS) Z axis. The WCS is the default coordinate system of AutoCAD.

In order for the axis of the cone to project from the sphere's center point, the UCS must be changed. Working with different UCSs is discussed in more detail in the next section. However, the following is a quick overview and describes how to draw a cone with its axis projecting from the center of the sphere.

Figure 4-9.
The UCS icon shortcut menu provides access to the options of the **UCS** command and UCS icon settings.

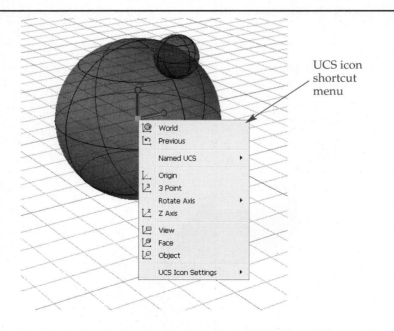

First, draw a three-unit diameter sphere with its center at 7,5. Display the drawing from the southeast isometric preset. To help see how the UCS is changing, make sure the UCS icon is displayed at the origin of the current UCS. The UCS icon drop-down list is found in the **Coordinates** panel of the **Home** and **Visualize** tabs of the ribbon and provides access to three options of the **UCSICON** command. See **Figure 4-10**. Pick the **Show UCS Icon at Origin** button to ensure the icon is displayed at the origin. This option can also be selected from the UCS icon shortcut menu. Also, set the X-ray visual style current, and set the opacity to 25%. In the **Visual Styles** panel of the **Visualize** tab, use the **Opacity** slider to adjust the value. You can also type a numerical value to the right of the slider.

Now, the sphere is drawn and the UCS icon is displayed at the origin of the current UCS (or at the lower-left corner of the screen, depending on the zoom level). However, the WCS is still the current user coordinate system. You are ready to start changing the UCS to meet your needs.

Pick the UCS icon and move the pointer over the origin grip. Select **Move Origin Only**. Now move the UCS origin to the center of the sphere using the center object snap. Notice that the UCS icon is now displayed at the center of the sphere, and its orientation has remained the same, **Figure 4-11**. Also, if the grid is displayed, the red X and green Y axes of the grid intersect at the center of the sphere.

Study **Figure 4-12** and continue as follows. Keep in mind that the point you are locating—the center of the cone on the sphere's surface—is 30° from the X axis and 30° from the XY plane. For ease of visualization, zoom in so the sphere fills the screen.

Right-click on the UCS icon to display the UCS icon shortcut menu. Then, select **Rotate Axis>Z**. Enter 30 for the rotation angle or select it dynamically. Watch the position of the UCS icon change when you press [Enter]. See **Figure 4-12B**.

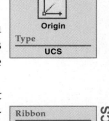

Figure 4-10.
The UCS icon drop-down list provides access to the **Origin**, **Off**, and **On** options of the **UCSICON** command.

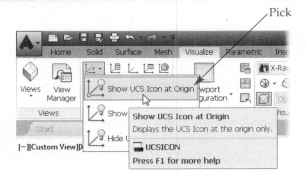

Figure 4-11.
The UCS origin is moved to the center of the sphere.

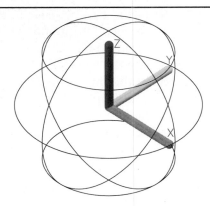

Figure 4-12.
A—The world coordinate system. B—The new UCS is rotated 30° in the XY plane about the Z axis. C—A line rotated up 30° from the XY plane represents the axis of the cone. D—The UCS is rotated 60° about the Y axis. The centerline of the cone coincides with the Z axis of this UCS.

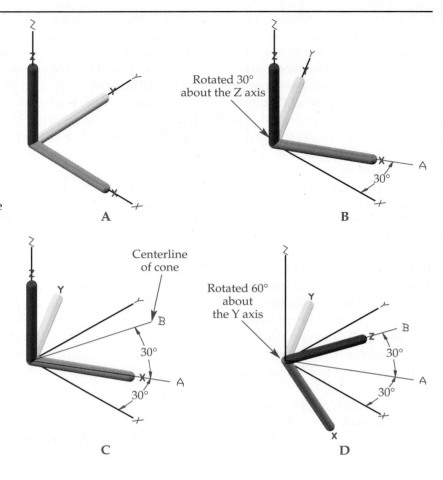

Next, right-click on the UCS icon and select **Rotate Axis>Y** in the UCS icon shortcut menu. Enter 60 for the rotation angle. Watch the position of the UCS icon change when you press [Enter]. See **Figure 4-12D**.

If the view cube is set to be oriented to the current UCS, then it rotates when the UCS is changed. Additionally, the grid rotates to match the new UCS. If the grid is not on, turn it on by pressing [Ctrl]+[G]. Remember, the grid is displayed on the XY plane of the current UCS.

The new UCS can be used to construct a cone with its axis projecting from the center of the sphere. **Figure 4-13A** shows the new UCS located at the center of the sphere. With the UCS rotated, rectangular coordinates can be used to draw the cone. Enter the **CONE** command and specify the center of the base as 0,0,1.5 (the radius of the sphere is 1.5 units). Enter a radius of .25 and a height of 1. The completed cone is shown in **Figure 4-13B**. You can see that the axis projects from the center of the sphere. **Figure 4-13C** shows the objects after setting the Conceptual visual style current.

This same basic procedure can be used in the tank and pipe example presented earlier in this chapter. To correctly locate the pipe (cylinder), first move the UCS origin to the point previously located on the surface of the cylinder. Next, rotate the UCS 30° about the Z axis. Then, rotate the UCS 90° about the Y axis. The Z axis of this new UCS aligns with the long axis of the pipe. Finally, use rectangular coordinates to draw the cylinder with its center at the point drawn in **Figure 4-5B**.

Once you have changed to a new UCS, you can quickly return to the WCS by picking **WCS** in the view cube drop-down list or by picking **World** in the UCS icon shortcut menu. The WCS provides a common "starting place" for creating new UCSs.

UCS

Ribbon
Home
> **Coordinates**
Visualize
> **Coordinates**

Y

Type
UCS

Figure 4-13.
A—A new UCS is created with the Z axis projecting from the center of the sphere. B—A cone is drawn using the new UCS. The axis of the cone projects from the center of the sphere. C—The objects after the Conceptual visual style is set current.

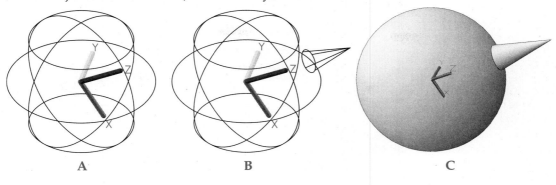

A   B   C

Exercise 4-3

www.g-wlearning.com/CAD/

Complete the exercise on the companion website.

Working with User Coordinate Systems

Once you understand a few of the basic options of user coordinate systems, creating 3D models becomes an easy and quick process. The following sections show how to display the UCS icon, change the UCS in order to work on different surfaces of a model, and name and save a UCS. As you saw in the previous section, working with UCSs is easy.

Displaying the UCS Icon

When a new drawing is started based on the acad3D.dwt template, the UCS icon is located at the WCS origin in the middle of the viewport. The display of the UCS icon is controlled by the **UCSICON** command. Turn the UCS icon on and off using the **Show UCS Icon** and **Hide UCS Icon** buttons on the **Coordinates** panel in the **Home** or **Visualize** tab of the ribbon. Refer to **Figure 4-10**.

If your drawing does not require viewports and altered coordinate systems, you may want to turn the icon off. The icon disappears until you turn it on again. If you redisplay the UCS icon and it does not appear at the origin, simply pick the **Show UCS Icon at Origin** button on the **Coordinates** panel in the ribbon.

You can also turn the icon on or off and set the icon to display at the UCS origin point using the options in the **Settings** tab of the **UCS** dialog box. See **Figure 4-14**. This dialog box is displayed by picking the **UCS, Named UCS...** button on the **Coordinates** panel in the ribbon or typing the **UCSMAN** command.

Notice that the **Allow Selecting UCS icon** option is checked by default in the **UCS Icon settings** area of the **Settings** tab. When this option is checked, it enables the display of the UCS icon grips. The UCS icon grips provide for dynamic and intuitive moving and rotating of the UCS icon. The **UCSSELECTMODE** system variable controls the **Allow Selecting UCS icon** setting and is set to 1 (on) by default.

Figure 4-14.
Setting UCS and
UCS icon options in
the **UCS** dialog box.

UCS icon
options

UCS
options

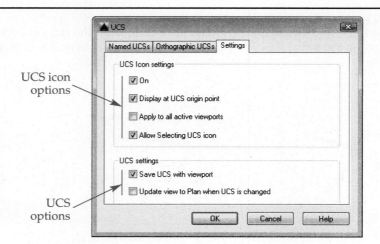

PROFESSIONAL TIP

It is recommended that you have the UCS icon turned on at all times when working in 3D drawings. It provides a quick indication of the current UCS.

Changing the Coordinate System

To construct a three-dimensional object, you must draw shapes at many different angles. Different planes are needed to draw features on angled surfaces. To construct these features, it is easiest to rotate the UCS to match any surface on an object. The following example illustrates this process.

The object in **Figure 4-15** has a cylinder on the angled surface. The **EXTRUDE** command, which is discussed in Chapter 6, is used to create the base of the object. The cylinder is then drawn on the angled feature. In Chapter 13, you will learn how to dimension the object as shown in **Figure 4-15**.

Figure 4-15.
This object can
be constructed
by changing the
orientation of the
coordinate system.

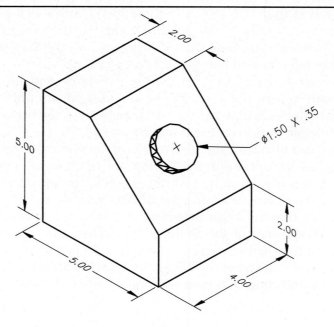

The first step in creating this model is to draw the side view of the base as a wireframe. You could determine the X, Y, and Z coordinates of each point on the side view and enter the coordinates. However, a lot of typing can be saved if all points share a Z value of 0. By rotating the UCS, you can draw the side view entering only X and Y coordinates. Start a new drawing and display the southeast isometric view. If the UCS icon is off, turn it on and display it at the origin.

Now, rotate the UCS 90° around the X axis. The quickest way to do this is to click directly on the UCS icon. Notice that as the pointer is moved over the UCS icon, the color of the icon changes to a light yellow. See **Figure 4-16A**. When you click on the icon, it displays grips. A square origin grip is displayed at the origin, and circular axis grips are displayed at the axis endpoints. See **Figure 4-16B**.

Once the UCS icon grips are displayed, move the pointer over the Y axis grip. The Y grip menu is displayed, **Figure 4-16C**. Select the **Rotate Around X Axis** option. The default value is 90. This is the desired rotation angle, so press [Enter]. The new UCS is now oriented parallel to the side of the new object. The UCS icon is displayed at the origin of the UCS. If needed, pan the view so the UCS icon is near the center of the view.

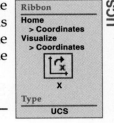

Ribbon

Home
> **Coordinates**
Visualize
> **Coordinates**

Type

UCS

Figure 4-16.
Using the UCS icon to rotate the UCS. A—When the pointer is moved over the UCS icon, the color of the icon changes to light yellow. B—Selecting the UCS icon displays a grip at the origin and grips at all three axis endpoints. C—The menu displayed when the pointer is moved over the Y axis grip to highlight the grip. To rotate the UCS 90° about the X axis, select the **Rotate Around X Axis** option.

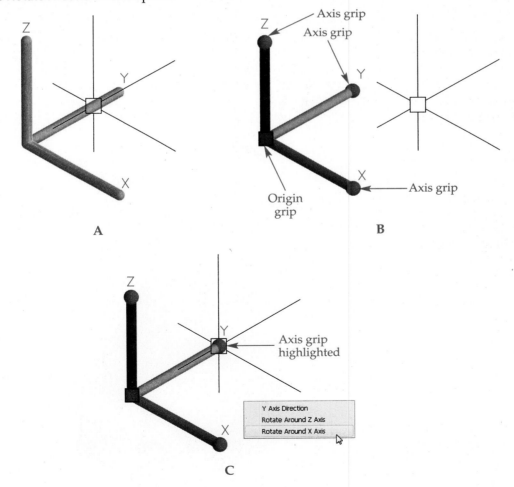

Next, use the **PLINE** command to draw the outline of the side view. Refer to the coordinates shown in **Figure 4-17**. When entering coordinates, you may want to turn off dynamic input. As an alternative, type a pound sign (#) before each coordinate with dynamic input turned on. This temporarily overrides the dynamic input of relative coordinates, which is controlled by the **DYNPICOORDS** system variable. The **PLINE** command is used instead of the **LINE** command for this model because a closed polyline can be extruded into a solid. Be sure to use the **Close** option to draw the final segment. A wireframe of one side of the object is created. Notice the orientation of the UCS icon.

Now, the **EXTRUDE** command is used to create the base as a solid. This command is covered in detail in Chapter 6. Make sure the same UCS used to create the wireframe side is current. Then, select the **EXTRUDE** command by picking the **Extrude** button on the **Modeling** panel of the **Home** tab in the ribbon. When prompted to select objects, pick the polyline and then press [Enter]. Next, move the mouse so the preview extends to the right and enter an extrusion height of 4. If dynamic input is off, you must enter –4.

The base is created as a solid. See **Figure 4-18**. You may want to switch to a parallel projection, as shown in the figure.

Saving a Named UCS

Once you have created a new UCS that may be used again, it is best to save it for future use. For example, you just created a UCS used to draw the wireframe of one side

---

Figure 4-17.
A wireframe of one side of the base is created. Notice the orientation of the UCS.

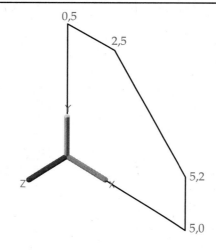

Figure 4-18.
The wireframe is extruded to create the base as a solid.

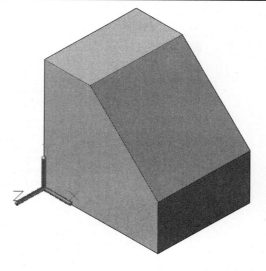

To see how to work with a dynamic UCS, recreate the cylinder on the angled face of the model completed in the previous discussion. First, delete the existing cylinder by picking on the cylinder and pressing the [Delete] key. Make sure that the UCS is oriented parallel to the side of the object. Refer to **Figure 4-18**. Next, select the **CYLINDER** command. Make sure the dynamic UCS function is on. Then, move the pointer over one of the surfaces of the object. Notice that the 3D crosshairs change when they are moved over a new surface. The red (X) and green (Y) crosshairs are flat on the face. For ease of visualizing the 3D crosshairs as they are moved across different surfaces, you can display the XYZ labels on the crosshairs. Open the **Options** dialog box, select the **3D Modeling** tab, and activate the **Show labels for dynamic UCS** option in the **3D Crosshairs** area.

The **CYLINDER** command is currently prompting to select a center point of the base. If you pick a point, this sets the center of the cylinder base and temporarily relocates the UCS so its XY plane lies on the selected face. When the command is completed, the UCS reverts to its previous orientation. Using the **CYLINDER** command options and the steps presented earlier, draw a 1.5″ diameter cylinder in the center of the angled face.

When using the dynamic UCS function, experiment with the behavior of the crosshairs as they are moved over different surfaces. The orientation of the crosshairs is related to the edge of the face that they are moved over. Can you determine the pattern by which the crosshairs are turned? The X axis of the crosshairs is always aligned with the edge that is crossed.

If you want to temporarily turn off the dynamic UCS function while working in a command, press and hold the [Shift]+[Z] key combination while moving the pointer over a face. As soon as you release the keys, the dynamic UCS function is reinstated.

Additional Ways to Change the UCS

As you have seen, the UCS can be moved to any location and rotated to any angle desired using the UCS icon grips. Location and alignment are controlled using the origin grip, and exact rotation of the three different UCS axes is controlled using the axis grips. The same UCS options are available when using the **UCS** command, but the UCS icon provides a quick alternative to entering the command or making ribbon or menu selections. The following sections discuss the UCS options available when using the UCS icon grips or the **UCS** command.

Moving and Aligning the UCS

The UCS icon origin grip is used to quickly move and align the UCS. Once the UCS icon is selected to display grips, you can pick on the origin grip and drag the icon dynamically. For example, drag the UCS icon to a model surface and orient the UCS parallel to that surface. If you are using dynamic input, you can enter a distance and angle defining a new location. You can also move the pointer over the grip to display the grip menu, as previously discussed. Pick **Move Origin Only** to relocate the UCS in its current orientation to a different origin point. Pick **Move and Align** to relocate the UCS to a new location and align it with an object face. Use object snaps for accuracy.

When the UCS icon is moved along a curved surface or solid, it will always remain normal to that object. In other words, the Z axis of the UCS will project from the center point of the radius curve. See **Figure 4-23**. In the example shown, the UCS is moved to a point on the outer surface of the model using the **Move and Align** option. Notice the direction of the Z axis in **Figure 4-23C**. If dynamic input is on, a tooltip is displayed when using the **Move and Align** option. You can press the [Ctrl] key to cycle through the options. The options are displayed on the command line to indicate the current mode.

Figure 4-23.
Using the UCS icon to move the UCS to a curved surface. In this example, the UCS icon origin grip is used to access the **Move and Align** option. A—The current UCS is located at the center point of the left curve of the model. This is apparent in the top view. B—A top view of the model. C—A new UCS is established on the outer surface by moving the UCS icon to the surface and picking. Note that the Z axis is normal to the surface.

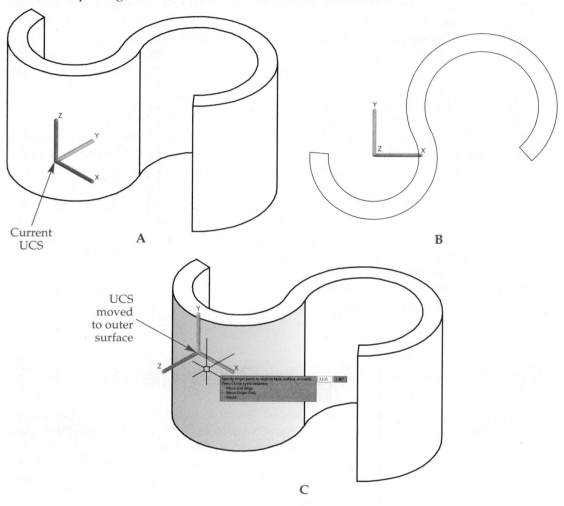

Current
UCS

A

B

UCS
moved
to outer
surface

C

Dynamically moving and aligning the UCS to an object surface can also be achieved by using the **Object** option of the **UCS** command. This option is also available in the UCS icon shortcut menu and is discussed later in this chapter.

Rotating the UCS on an Axis

The axis grips of the UCS icon enable you to rotate the icon in any direction, and to cycle between rotation options. For example, pick the UCS icon to display the grips, and then move the pointer over the Z axis grip (but do not select it). See **Figure 4-24A**. The first item in the Z grip menu is **Z Axis Direction**. Picking this option is the same as picking the Z axis grip. Select this option and notice that it enables you to select the direction of the Z axis. Use object snaps to assist, or turn ortho on to limit selections to 90°. In addition, if dynamic input is on, you can press the [Ctrl] key to cycle through the three options in the grip menu. See **Figure 4-24B**.

The other two options in the grip menu allow the Z axis to be moved in relation to one of the other two axes. Test each of these options to see how they function. The **Rotate Around X Axis** option limits rotation of the Z axis to around the X axis, and the **Rotate Around Y Axis** option limits rotation to around the Y axis.

Figure 4-24.
Using the UCS icon axis rotation options. A—The Z axis options appear after moving the pointer over the Z axis grip. Pick the **Z Axis Direction** option to set the direction of the Z axis. B—The new UCS after rotating the Z axis. Note that the other Z axis options can be accessed by pressing the [Ctrl] key.

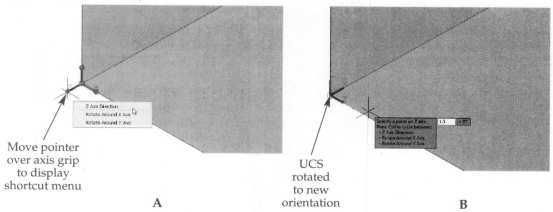

Move pointer over axis grip to display shortcut menu

A

UCS rotated to new orientation

B

Move the pointer to each of the other two UCS icon axis grips and note that a similar menu is displayed. Each axis of the UCS can be rotated in the same manner using the appropriate axis grip.

Additional UCS Options

Additional options for changing the UCS are discussed in the following sections. Some of these options can be selected from the UCS icon shortcut menu displayed by right-clicking on the UCS icon, shown in **Figure 4-19**, or by accessing the **Coordinates** panel in the **Home** or **Visualize** tab on the ribbon. All of the options discussed are available when entering the UCS command on the command line.

NOTE

Keep in mind that many of the procedures discussed in the following sections can be accomplished using the dynamic moving and rotating functions of the UCS icon grips.

Selecting Three Points to Create a New UCS

The **3 Point** option of the **UCS** command can be used to change the UCS to any flat surface. This option requires that you first locate a new origin, then a point on the positive X axis, and finally a point on the XY plane that has a positive Y value. See **Figure 4-25**. Use object snaps to select points that are not on the current XY plane. After you pick the third point—the point on the XY plane—the UCS icon changes its orientation to align with the plane defined by the three points. The **3 Point** option is available in the UCS icon shortcut menu.

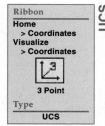

Ribbon

Home
> **Coordinates**
Visualize
> **Coordinates**

3 Point

Type

UCS

UCS

PROFESSIONAL TIP

When typing the **UCS** command, enter 3 at the Specify origin of UCS or [Face/NAmed/OBject/Previous/View/World/X/Y/Z/ZAxis] <World>: prompt. Notice that the option is not listed in the prompt.

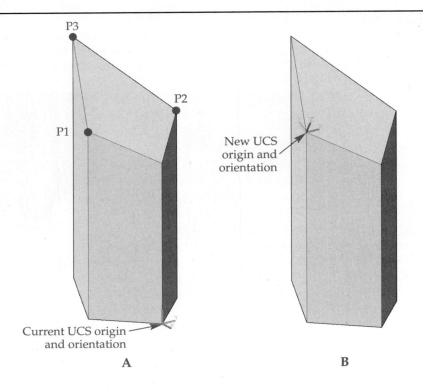

Figure 4-25.
A—A new UCS can be established by picking three points. P1 is the origin, P2 is on the positive X axis, and P3 is on the XY plane and has a positive Y value. B—The new UCS is created.

P3

P2

P1

New UCS origin and orientation

Current UCS origin and orientation

A

B

Selecting a New Z Axis

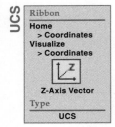

UCS

Ribbon
Home
> Coordinates
Visualize
> Coordinates

Z-Axis Vector
Type
UCS

The **ZAxis** option of the UCS command allows you to select the origin point and a point on the positive Z axis. Once the new Z axis is defined, AutoCAD sets the new X and Y axes. The **ZAxis** option is available in the UCS icon shortcut menu.

You will now add a cylinder to the lower face of the base created earlier. The cylinder extends into the base. Earlier, you moved the UCS to the angled face of the object, drew the cylinder on the angled face, and then restored the UCS to its previous position on the side face. If this is the current orientation, the Z axis does not project perpendicular to the lower face. Therefore, a new UCS must be created on the lower-right face. Change the UCS after entering the **ZAxis** option as follows.

1. Pick the origin of the new UCS. See **Figure 4-26A**. You may have to use an object snap to select the origin.
2. Pick a point on the positive portion of the new Z axis.
3. The new UCS is established and it can be saved if necessary.

Now, use 3D object snaps, AutoTrack, or 2D object snaps to draw a Ø.5″ cylinder centered on the lower face and extending 3″ into the base. Then, subtract the cylinder from the base part to create the hole, as shown in **Figure 4-26B**.

Setting the UCS to an Existing Object

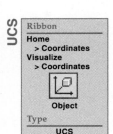

UCS

Ribbon
Home
> Coordinates
Visualize
> Coordinates

Object
Type
UCS

The **Object** option of the **UCS** command can be used to define a new UCS on an object. This option cannot be used with 3D polylines or xlines. There are also certain rules that control the orientation of the UCS. For example, if you select a circle, the center point becomes the origin of the new UCS. The pick point on the circle determines the direction of the X axis. The Y axis is relative to X and the UCS Z axis may or may not be the same as the Z axis of the selected object. The **Object** option is available in the UCS icon shortcut menu.

Look at **Figure 4-27A**. The circle is rotated an unknown number of degrees from the XY plane of the WCS. However, you need to create a UCS in which the circle is lying on the XY plane. Select the **Object** option of the **UCS** command and then pick the circle. The UCS icon may look like the one shown in **Figure 4-27B**. Notice how the

Figure 4-26.
A—Using the **ZAxis** option to establish a new UCS. B—The new UCS is used to create a cylinder, which is then subtracted from the base to create a hole.

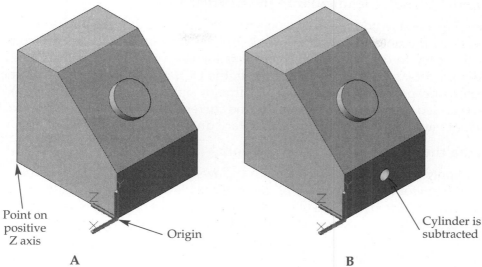

Point on positive Z axis

Origin

Cylinder is subtracted

A

B

X and Y axes are not aligned with the quadrants of the circle, as indicated by the grip locations. This may not be what you expected. The X axis orientation is determined by the pick point on the circle. Notice how the X axis is pointing at the pick point.

To rotate the UCS in the current plane so the X and Y axes of the UCS are aligned with the quadrants of the circle, use the **ZAxis** option of the **UCS** command. Select the center of the circle as the origin and then enter the absolute coordinate 0,0,1. This uses the current Z axis location, which also forces the X and Y axes to align with the object. Refer to **Figure 4-27C**. This method may not work with all objects.

Setting the UCS to the Face of a 3D Solid

The **Face** option of the **UCS** command allows you to orient the UCS to any face on a 3D solid, surface, or mesh object. The **Face** option is available in the UCS icon shortcut menu. Select the option and then pick a face on the model. After you have selected a face, you have the options of moving the UCS to the adjacent face or flipping the UCS 180° on the X axis, Y axis, or both axes. Use the **Next**, **Xflip**, or **Yflip** option to move or rotate the UCS as needed. Once you achieve the UCS orientation you want,

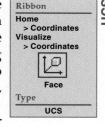

Ribbon

Home
 > **Coordinates**
Visualize
 > **Coordinates**

Face

Type
UCS

UCS

Figure 4-27.
A—This circle is rotated off of the WCS XY plane by an unknown number of degrees. It will be used to establish a new UCS. B—The circle is on the XY plane of the new UCS. However, the X and Y axes do not align with the circle's quadrants. C—The **ZAxis** option of the **UCS** command is used to align the UCS with the quadrants of the circle.

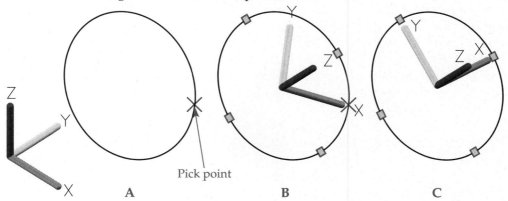

Pick point

A

B

C

Chapter 4 Understanding Three-Dimensional Coordinates and User Coordinate Systems   **101**

press [Enter] to accept. Notice in **Figure 4-28** how many different UCS orientations can be selected for a single face.

Setting the UCS Perpendicular to the Current View

You may need to add notes or labels to a 3D drawing that are plan to the current view, such as the note shown in **Figure 4-29**. The **View** option of the **UCS** command makes this easy to do. The **View** option is available in the UCS icon shortcut menu. Immediately after selecting the **View** option, the UCS rotates to a position so the new XY plane is perpendicular to the current line of sight (parallel to the screen). Now, anything added to the drawing is plan to the current view. The **View** option works on the current viewport only; other viewports are unaffected.

Applying the Current UCS to a Viewport

The **Apply** option of the **UCS** command allows you to apply the UCS in the current viewport to any or all model space or paper space viewports. Using the **Apply** option, you can have a different UCS displayed in every viewport or you can apply one UCS to all viewports. With the viewport that contains the UCS to apply active, enter the **Apply** option. This option is only available on the command line. However, the option does not appear in the command prompt. Enter either **A** or **APPLY** to select the option. Then, pick a viewport to which the current UCS will be applied and press [Enter]. To apply

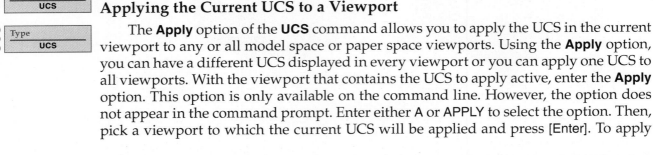

UCS

Ribbon
Home
  > Coordinates
Visualize
  > Coordinates

View

Type
  UCS

UCS

Type
  UCS

Figure 4-28.
Several different UCSs can be selected from a single pick point using the **Face** option of the **UCS** command. Given the pick point, five of the eight possibilities are shown here.

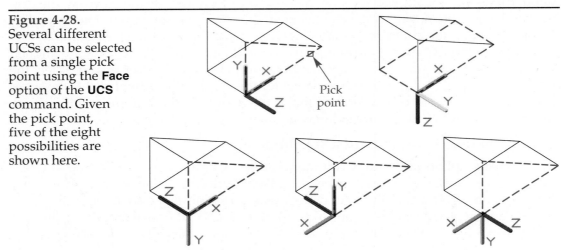

Pick point

Figure 4-29.
The **View** option of the **UCS** command allows you to place text plan to the current view.

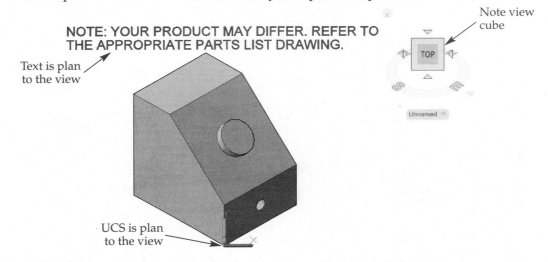

NOTE: YOUR PRODUCT MAY DIFFER. REFER TO THE APPROPRIATE PARTS LIST DRAWING.

Text is plan to the view

UCS is plan to the view

Note view cube

TOP

Unnamed

the current UCS to all viewports, enter the **All** option. See Chapter 5 for a complete discussion of model space viewports.

Preset UCS Orientations

AutoCAD has six preset orthographic UCSs that match the six standard orthographic views. With the current UCS as the top view (plan), all other views are arranged as shown in **Figure 4-30**. These orientations can be selected by using the **Named UCS Combo Control** drop-down list in the **Coordinates** panel of the **Home** or **Visualize** tab in the ribbon, entering the command on the command line, or using the **Orthographic UCSs** tab of the **UCS** dialog box. When using the command line, type the name of the UCS (FRONT, BACK, RIGHT, etc.) at the first prompt:

> Specify origin of UCS or [Face/NAmed/OBject/Previous/View/World/X/Y/Z/ZAxis]
> <World>: **FRONT** *or* **FR.**⏎

Note that there is no orthographic option listed for the command.

The **Relative to:** drop-down list at the bottom of the **Orthographic UCSs** tab of the **UCS** dialog box specifies whether each orthographic UCS is relative to a named UCS or absolute to the WCS. For example, suppose you have a saved UCS named Front Corner that is rotated 30° about the Y axis of the WCS. If you set current the top UCS relative to the WCS, the new UCS is perpendicular to the WCS, **Figure 4-31A**. However, if the top UCS is set current relative to the named UCS Front Corner, the new UCS is also rotated from the WCS, **Figure 4-31B**.

The Z value, or depth, of a preset UCS can be changed in the **Orthographic UCSs** tab of the **UCS** dialog box. First, right-click on the name of the UCS you wish to change. Then, pick **Depth** from the shortcut menu, **Figure 4-32A**. This displays the **Orthographic**

Figure 4-30.
The standard orthographic UCSs coincide with the six basic orthographic views.

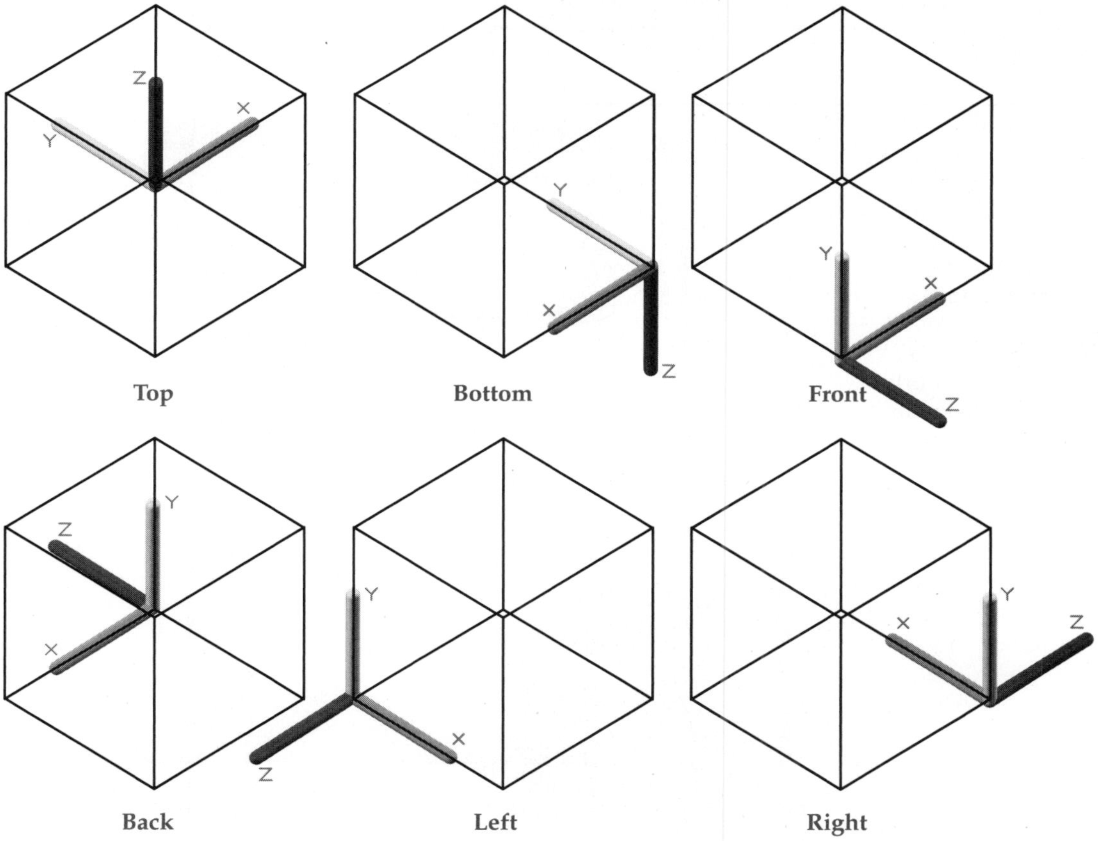

Top	Bottom	Front

Back	Left	Right

Figure 4-31.
The **Relative to:** drop-down list entry in the **Orthographic UCSs** tab of the **UCS** dialog box determines whether the orthographic UCS is based on a named UCS or the WCS. The UCS icon here represents the named UCS. A—Relative to the WCS. B—Relative to the named UCS.

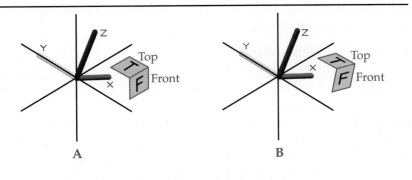

UCS depth dialog box. See **Figure 4-32B**. You can either enter a new depth value or specify the new location on screen by picking the **Select new origin** button. Once the new depth has been selected, it is reflected in the preset UCS list.

CAUTION

Changing the **Relative to:** setting affects *all* preset UCSs and *all* preset viewpoints! Therefore, leave this set to World unless absolutely necessary to change it.

Exercise 4-5

www.g-wlearning.com/CAD/

Complete the exercise on the companion website.

Figure 4-32.
A—The Z value, or depth, of a preset UCS can be changed by right-clicking on its name and selecting **Depth**. B—Enter a new depth value or pick the **Select new origin** button to pick a new location on screen.

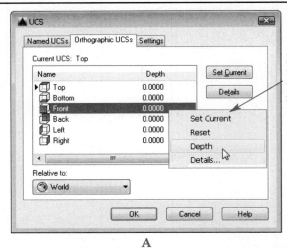

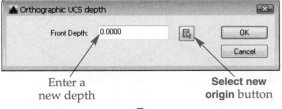

Managing User Coordinate Systems and Displays

You can create, name, and use as many user coordinate systems as needed to construct your model or drawing. As you saw earlier, AutoCAD allows you to name (save) coordinate systems for future use. User coordinate systems can be saved, renamed, set current, and deleted by using the **Named UCSs** tab of the **UCS** dialog box, **Figure 4-33**.

The **Named UCSs** tab contains the **Current UCS:** list box. This list box contains the names of all saved coordinate systems plus World. If other coordinate systems have been used in the current drawing session, Previous appears in the list. Unnamed appears if the current coordinate system has not been named (saved). The current UCS is indicated by a small triangle to the left of its name in the list and by the label at the top of the list. To make any of the listed coordinate systems active, highlight the name and pick the **Set Current** button.

A list of coordinate and axis values of the highlighted UCS can be displayed by picking the **Details** button. This displays the **UCS Details** dialog box shown in **Figure 4-34**.

If you right-click on the name of a UCS in the list in the **Named UCSs** tab, a shortcut menu is displayed. Using this menu, you can rename the UCS. You can also set the UCS current or delete it using the shortcut menu. The Unnamed UCS cannot be deleted, nor can the World UCS.

PROFESSIONAL TIP

You can also manage UCSs on the command line using the **UCS** command. In addition, by right-clicking on the UCS icon, you can select a UCS from the UCS icon shortcut menu to make it current.

Figure 4-33.
The **UCS** dialog box allows you to rename, list, delete, and set current an existing UCS.

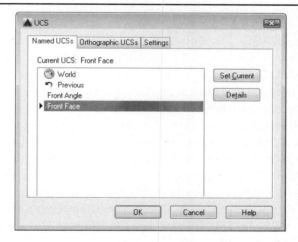

Figure 4-34.
The **UCS Details** dialog box displays the coordinate values of the selected UCS.

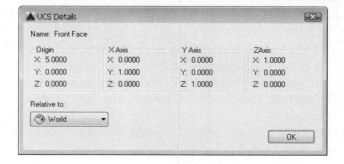

Setting an Automatic Plan Display

After changing the UCS, a plan view is often needed to give you a better feel for the XYZ directions. While you should try to draw in a pictorial view when possible as you construct a 3D object, some constructions may be much easier in a plan view. AutoCAD can be set to automatically make your view of the drawing plan to the current UCS. This is especially useful if you will be changing the UCS often, but want to work in a plan view.

The **UCSFOLLOW** system variable is used to automatically display a plan view of the current UCS. When it is set to 1, a plan view is automatically created in the current viewport when the UCS is changed. Viewports are discussed in Chapter 5. The default setting of **UCSFOLLOW** is 0 (off). After setting the variable to 1, a plan view will be automatically generated the next time the UCS is changed. The **UCSFOLLOW** variable generates the plan view only after the UCS is changed, not immediately after the variable is changed. However, if you select a different viewport, the previous viewport is set plan to the UCS if **UCSFOLLOW** has been set to 1 in that viewport. The **UCSFOLLOW** variable can be individually set for each viewport.

PROFESSIONAL TIP

To get the plan view displayed without changing the UCS, use the **PLAN** command, which is discussed in Chapter 3.

UCS Settings and Variables

As discussed in the previous section, the **UCSFOLLOW** system variable allows you to change how an object is displayed in relation to the UCS. There are also system variables that display information about the current UCS and other system variables that control UCS functions. These variables include:

- **UCSAXISANG.** (stored value) The default rotation angle for the **X**, **Y**, or **Z** option of the **UCS** command.
- **UCSBASE.** The name of the UCS used to define the origin and orientation of the orthographic UCS settings. It can be any named UCS.
- **UCSDETECT.** (on or off) Turns the dynamic UCS function on and off. The **Dynamic UCS** button on the status bar controls this variable, as does the [Ctrl]+[D] key combination.
- **UCSNAME.** (read only) Displays the name of the current UCS.
- **UCSORG.** (read only) Displays the XYZ origin value of the current UCS.
- **UCSORTHO.** (on or off) If set to 1 (on), the related orthographic UCS setting is automatically restored when an orthographic view is restored. If turned off, the current UCS is retained when an orthographic view is restored. Depending on your modeling preferences, you may wish to set this variable to 0.
- **UCSSELECTMODE.** (on or off) Enables the selection and manipulation of the UCS icon with grips. The default setting is 1 (on).
- **UCSVIEW.** (on or off) If this variable is set to 1 (on), the current UCS is saved with the view when a view is saved. Otherwise, the UCS is not saved with the view.
- **UCSVP.** Controls which UCS is displayed in viewports. The default value is 1, which means that the UCS configuration in the viewport is independent from all other UCS configurations. If the setting is 0, the UCS configuration in the current viewport is displayed. Each viewport can be set to either 0 or 1.
- **UCSXDIR.** (read only) Displays the XYZ value of the X axis direction of the current UCS.
- **UCSYDIR.** (read only) Displays the XYZ value of the Y axis direction of the current UCS.

UCS options and variables can also be managed in the **Settings** tab of the **UCS** dialog box. Refer to **Figure 4-14**. The options in the **UCS settings** area are:

- **Save UCS with viewport.** If checked, the current UCS settings are saved with the viewport and the **UCSVP** system variable is set to 1. This variable can be set for each viewport in the drawing. Viewports in which this setting is turned off, or unchecked, will always display the UCS settings of the current active viewport.
- **Update view to Plan when UCS is changed.** This setting controls the **UCSFOLLOW** system variable. When checked, the variable is set to 1. When unchecked, the variable is set to 0.

Chapter Review

Answer the following questions using the information in this chapter.

1. Explain *spherical coordinate entry.*
2. Explain *cylindrical coordinate entry.*
3. A new point is to be drawn 4.5″ from the last point. It is to be located at a 63° angle in the XY plane, and at a 35° angle from the XY plane. Write the proper spherical coordinate notation.
4. Write the proper cylindrical coordinate notation for locating a point 4.5″ in the horizontal direction from the origin, 3.6″ along the Z axis, and at a 63° angle in the XY plane.
5. Name the command that is used to draw 3D polylines.
6. Why is the command in question 5 needed?
7. Which command option is used to change a 3D polyline into a B-spline curve?
8. What is the *world coordinate system (WCS)?*
9. What is a *user coordinate system (UCS)?*
10. What effect does the **Show UCS Icon at Origin** option have on the UCS icon display?
11. Describe how to rotate the UCS so that the Z axis is tilted 30° toward the WCS X axis.
12. How do you return to the WCS from any UCS?
13. Which command controls the display of the user coordinate system icon?
14. What system variable controls the display of grips on the UCS icon? In which dialog box can it be set?
15. Briefly describe how to access the **Move and Align** option using the UCS icon grips.
16. How can you return to a previous UCS configuration using the UCS icon?
17. What is a *dynamic UCS* and how is one activated?
18. What is the function of the **3 Point** option of the **UCS** command?
19. How do you automatically create a display that is plan to a new UCS?
20. What is the function of the **Object** option of the **UCS** command?
21. What is the function of the **Apply** option of the **UCS** command?
22. In which dialog box is the **Orthographic UCSs** tab located?
23. Which command displays the **UCS** dialog box?
24. What appears in the **Named UCSs** tab of the **UCS** dialog box if the current UCS has not been saved?

Drawing Problems

For Problems 1–4, draw each object using solid primitives and Boolean commands to create composite solids. Measure the objects directly to obtain the necessary dimensions. If appropriate, apply geometric constraints to the base 2D drawing. Plot the drawings at a 3:1 scale using display methods specified by your instructor. Save the drawings as **P4-**(problem number).

1. General

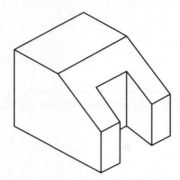

2. General

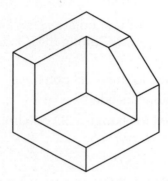

3. General

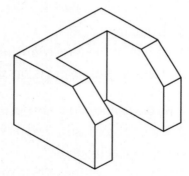

4. General

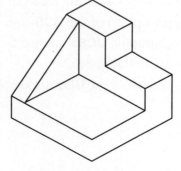

5. **Mechanical** Create the mounting bracket shown. Save the file as P4-5.

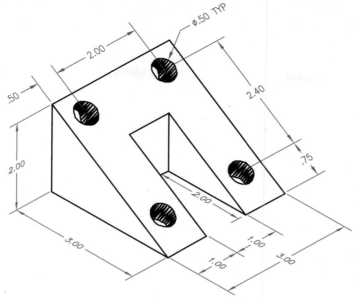

6. **Mechanical** Create the computer speaker as shown below. The large-radius, arched surface is created by drawing a three-point arc. The second point of the arc passes through the point located by the .26 and 2.30 dimensions. Save the file as P4-6.

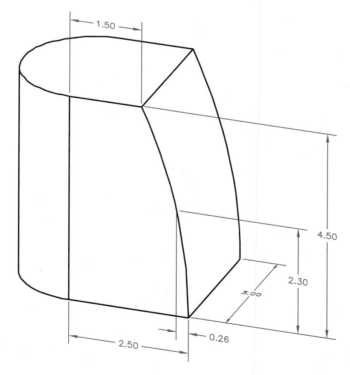

For Problems 7–9, draw each object using solid primitives and Boolean commands to create composite solids. Use the dimensions provided. Save the drawings as P4-*(problem number).*

7. Mechanical

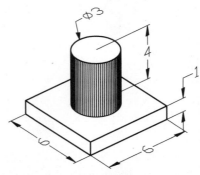

Pedestal #1

8. Mechanical

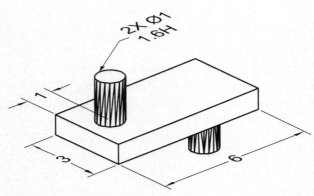

Locking Plate

9. Mechanical

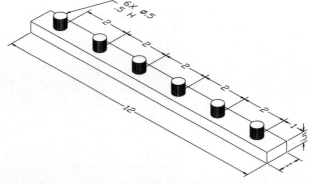

Pin Bar

10. Architectural Draw the Ø8" pedestal shown. It is .5" thick. The four feet are centered on a Ø7" circle and are .5" high. Save the drawing as P4-10.

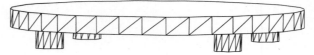

Pedestal #2

11. **Architectural** Four legs (cones), each 3″ high with a Ø1″ base, support this Ø10″ globe. Each leg tilts at an angle of 15° from vertical. The base is Ø12″ and .5″ thick. The bottom surface of the base is 8″ below the center of the globe. Save the drawing as P4-11.

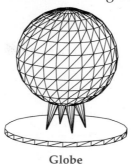

Globe

12. **Architectural** The table legs (A) are 2″ square and 17″ tall. They are 2″ in from each edge. The tabletop (B) is 24″ × 36″ × 1″. Save the drawing as P4-12.

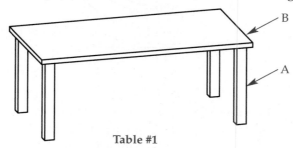

Table #1

13. **Architectural** The table legs (A) for the large table are Ø2″ and 17″ tall. The tabletop (B) is 24″ × 36″ × 1″. The table legs (C) for the small table are Ø2″ and 11″ tall. The tabletop (D) is 24″ × 14″ × 1″. All legs are 1″ in from the edges of the table. Save the drawing as P4-13.

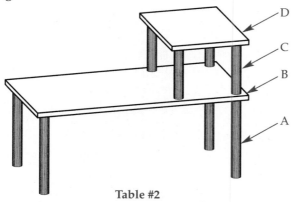

Table #2

14. **General** The spherical objects (A) are Ø4″. Object B is 6″ long and Ø1.5″. Save the drawing as P4-14.

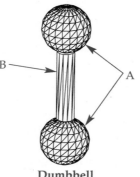

Dumbbell

15. **Architectural** Create the model of the globe using the dimensions shown. Save the file as P4-15.

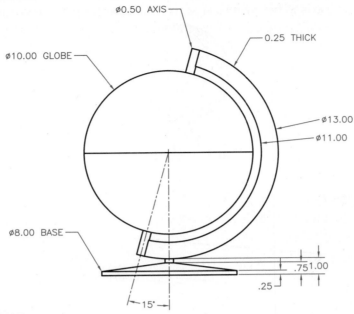

16. **Architectural** Object A is a ∅8″ cylinder that is 1″ tall. Object B is a ∅5″ cylinder that is 7″ tall. Object C is a ∅2″ cylinder that is 6″ tall. Object D is a .5″ × 8″ × .125″ box, and there are four pieces. The top surface of each piece is flush with the top surface of Object C. Object E is a ∅18″ cone that is 12″ tall. Create a smaller cone and hollow out Object E. Save the drawing as P4-16.

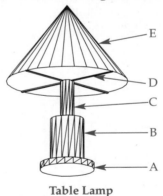

Table Lamp

17. **Architectural** Objects A and B are brick walls that are 5′ high. The walls are two bricks thick. Research the dimensions of standard brick and draw accordingly. Wall B is 7′ long and Wall A is 5′ long. Lamps are placed at each end of the walls. Object C is ∅2″ and 8″ tall. The center is offset from the end of the wall by a distance equal to the width of one brick. Object D is ∅10″. Save the drawing as P4-17.

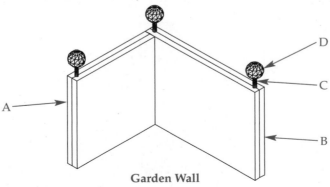

Garden Wall

18. **Architectural** Object A is ⌀18" and 1" tall. Object B is ⌀1.5" and 6' tall. Object C is ⌀6" and .5" tall. Object D is a ⌀10" sphere. Object E is a U-shaped bracket to support the shade (Object F). There are two items; draw them an appropriate size. Object F has a ⌀22" base and is 12" tall. Save the drawing as P4-18.

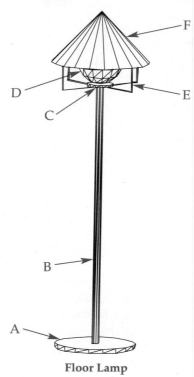

Floor Lamp

19. **General** This is a concept sketch of a desk organizer. Create a solid model using the dimensions given. Create and save new UCSs as needed. Inside dimensions of compartments can vary, but the thickness between compartments should be consistent. Do not add dimensions to the drawing. Plot your drawing on a B-size sheet of paper in a visual style specified by your instructor. Save the drawing as P4-19.

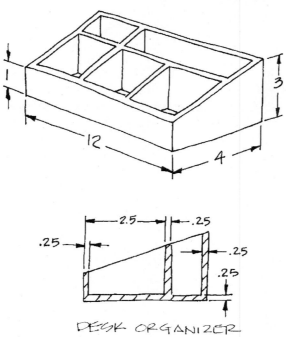

20. **General** This is a concept sketch of a pencil holder. Create a solid model using the dimensions given. Create and save new UCSs as needed. Do not add dimensions to the drawing. Plot your drawing on a B-size sheet of paper in a visual style specified by your instructor. Save the drawing as P4-20.

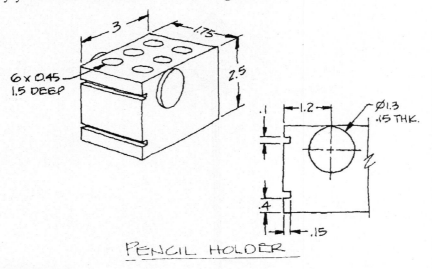

PENCIL HOLDER

21. **Mechanical** This is an engineering sketch of a window blind mounting bracket. Create a solid model using the dimensions given. Create and save new UCSs as needed. Do not add dimensions to the drawing. Create two plots, each of a different view, on B-size paper in the visual styles specified by your instructor. Save the drawing as P4-21.

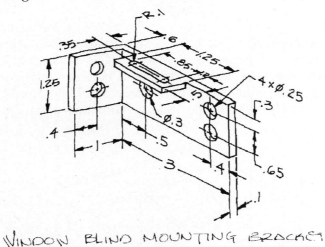

WINDOW BLIND MOUNTING BRACKET

CHAPTER

Using Model Space Viewports

Learning Objectives

After completing this chapter, you will be able to:

✓ Describe the function of model space viewports.
✓ Create and save viewport configurations.
✓ Alter the current viewport configuration.
✓ Use multiple viewports to construct a drawing.

A variety of views can be displayed in a drawing at one time using model space viewports. This is especially useful when constructing 3D models. Using the **VPORTS** command, you can divide the drawing area into two or more smaller areas. These areas are called *viewports*. Each viewport can be configured to display a different 2D or 3D view of the model.

The *active viewport* is the viewport in which a command will be applied. Any viewport can be made active, but only one can be active at a time. As objects are added or edited, the results are shown in all viewports. A variety of viewport configurations can be saved and recalled as needed. This chapter discusses the use of viewports and shows how they can be used for 3D constructions.

Understanding Viewports

The AutoCAD drawing area can be divided into a maximum of 64 viewports. However, this is impractical due to the small size of each viewport. Usually, the maximum number of viewports practical to display at one time is four. The number of viewports you need depends on the model you are creating. Each viewport can show a different view of an object. This makes it easier to construct 3D objects.

NOTE

The **MAXACTVP** (maximum active viewports) system variable sets the number of viewports that can be used at one time. The initial value is 64, which is the highest setting.

There are two types of viewports used in AutoCAD. The type of viewport created depends on whether it is defined in model space or paper space. *Model space* is the space, or mode, where the model or drawing is constructed. *Paper space*, or layout space, is the space where a drawing is laid out to be plotted. Viewports created in model space are called *model space viewports* or simply *model viewports*, and are also known as *tiled viewports*. Viewports created in paper space are called *layout viewports*, also known as *floating viewports*.

Model space is active by default when you start a new drawing. Model space viewports are created with the **VPORTS** command. Model space viewport configurations are for display purposes only and cannot be plotted. If you plot from model space, the content of the active viewport is plotted.

Model space viewports are described as *tiled viewports*. They are referred to as *tiled* because the edges of each viewport are placed side to side, as with floor tile, and they cannot overlap. Model space viewports are not AutoCAD objects and cannot be edited.

Floating (paper space) viewports are used to lay out the views of a drawing before plotting. They are described as *floating* because they can be moved around and overlapped. Paper space viewports are objects and can be edited. These viewports can be thought of as "windows" cut into a sheet of paper to "see into" model space. You can then display different scaled drawings (views) in these windows. For example, architectural details or sections and details of complex mechanical parts may be displayed in paper space viewports at different scales. Detailed discussions of paper space viewports are provided in *AutoCAD and Its Applications—Basics*.

The **VPORTS** command can be used to create viewports in a paper space layout. The process is very similar to that used to create model space viewports, which is discussed next. You can also use the **MVIEW** command to create paper space viewports.

Creating Viewports

Creating model space viewports allows you to work with multiple views of the same model. To work on a different view, simply pick with your pointing device in the viewport in which you wish to work. The picked viewport becomes active. Using viewports is a good way to construct 3D models because all views are updated as you draw. However, viewports are also useful when creating 2D drawings.

The project on which you are working determines the number of viewports needed. Keep in mind that the more viewports you display on your screen, the smaller the view in each viewport. Small viewports may not be useful to you. Four different viewport configurations are shown in **Figure 5-1**. As you can see, when 16 viewports are displayed, the viewports are very small. Normally, two to four viewports are used.

Quick Viewport Layout

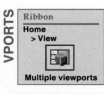

A four-view layout can be instantly displayed by picking the **Multiple viewports** button on the **View** panel in the **Home** tab of the ribbon. See **Figure 5-2**. This selection automatically creates a top, front, side, and pictorial view based on the current UCS. See **Figure 5-3**. This layout may be a good one to start from when working on models that require two or more views for construction purposes.

The screen display can be quickly returned to a single view by picking the **Single viewport** button on the **View** panel in the **Home** tab of the ribbon. Keep in mind, the resulting view will be the viewport that is current, which is the one surrounded by a highlighted frame.

Figure 5-1.
A—Two vertical viewports. B—Two horizontal viewports. C—Three viewports, with the largest viewport positioned at the right. D—Sixteen viewports.

Viewport controls

Crosshairs, UCS icon, and navigation bar appear in the active viewport

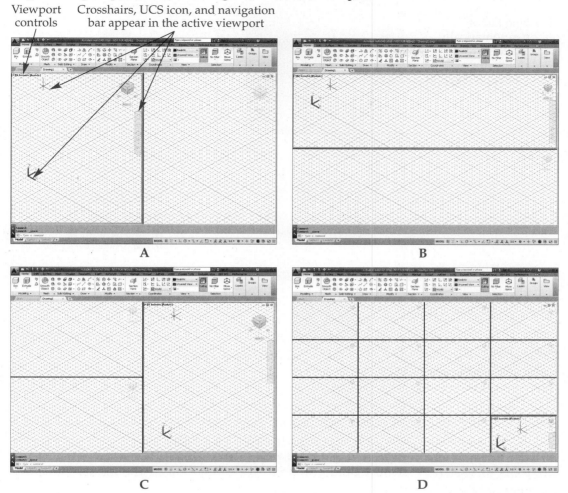

A

B

C

D

Figure 5-2.
A four-viewport layout can be instantly displayed by picking the **Multiple viewports** button in the **View** panel on the **Home** tab of the ribbon.

Pick to create a four-viewport configuration

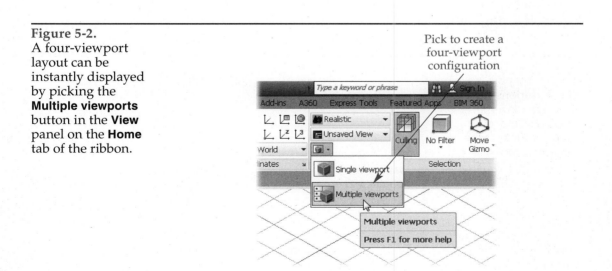

Figure 5-3.
Picking the **Multiple viewports** button in the **View** panel on the **Home** tab of the ribbon automatically creates a top, front, side, and pictorial view based on the current UCS.

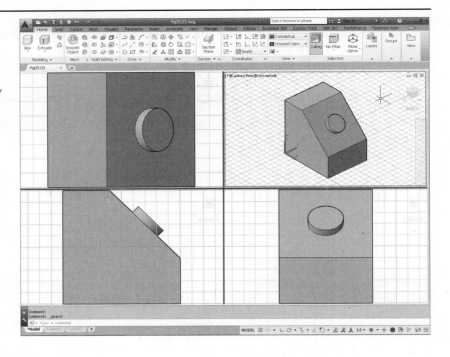

Viewport Configurations

A layout of viewport configurations can be quickly created by using the **Viewport Controls** flyout in the viewport controls located in the upper-left corner of the drawing window. See **Figure 5-4**. The viewport controls provide quick access to all viewport configurations and settings. Selecting **Viewport Configuration List** in the **Viewport Controls** flyout displays a menu with 12 different preset viewport configurations. Selecting **Configure...** in this menu displays the **Viewports** dialog box, **Figure 5-5**. This dialog box, which is also accessed with the **VPORTS** command, is used to save and manage viewport configurations. Another way to access the preset viewport configurations is to use the **Viewport Configuration** drop-down list in the **Model Viewports** panel on the **Visualize** tab of the ribbon. See **Figure 5-6**.

As shown in **Figure 5-5**, there are two tabs in the **Viewports** dialog box. The preset viewport configurations are available in the **New Viewports** tab. The preset viewport configurations include six different options for three-viewport configurations.

VPORTS

Type

VPORTS

Figure 5-4.
Preset viewport configurations can be accessed in the **Viewport Controls** flyout located in the viewport controls.

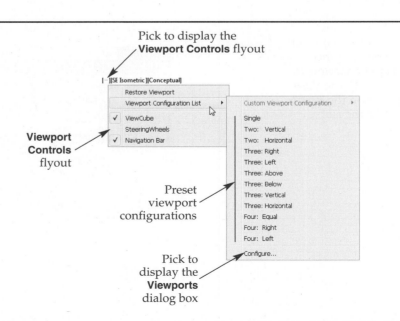

Figure 5-5.
Viewport configurations are created using the **New Viewports** tab of the **Viewports** dialog box.

Enter a name to save a configuration

Preset viewport configurations

Image preview of the selected configuration

Specify a visual style

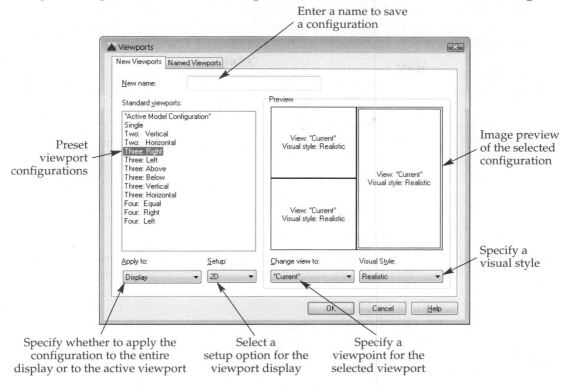

Specify whether to apply the configuration to the entire display or to the active viewport

Select a setup option for the viewport display

Specify a viewpoint for the selected viewport

Figure 5-6.
The **Viewport Configuration** drop-down list in the **Model Viewports** panel on the **Visualize** tab of the ribbon offers a quick way to recall a preset viewport configuration.

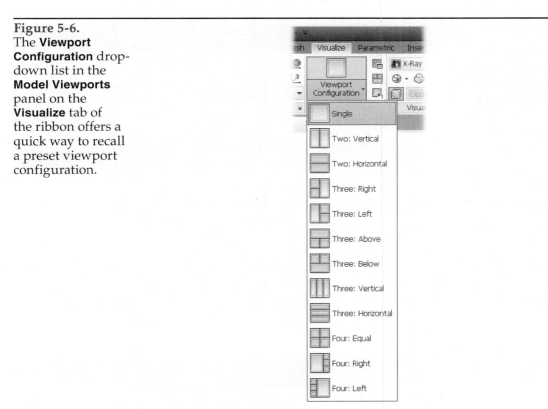

See **Figure 5-7.** When you pick the name of a configuration in the **Standard viewports:** list, the viewport arrangement is displayed in the **Preview** area. After you have selected, you can save the configuration by entering a name in the **New name:** text box and then picking **OK** to close the dialog box. When the **Viewports** dialog box closes, the configuration is displayed on screen.

NOTE

Notice in **Figure 5-1** that the UCS icon, viewport controls, and navigation bar are only displayed in the current viewport. The navigation bar may also be shortened because of the viewport size. In this case, simply pick the drop-down arrow at the bottom of the bar for a list of additional buttons.

Making a Viewport Active

After a viewport configuration has been created, a thick blue line surrounds the active viewport. When the screen cursor is moved inside of the active viewport, it appears as crosshairs. When moved into an inactive viewport, the standard Windows cursor appears.

Any viewport can be made active by moving the cursor into the desired viewport and pressing the pick button. You can also press the [Ctrl]+[R] key combination to switch viewports, or use the **CVPORT** (current viewport) system variable. Only one viewport can be active at a time.

Figure 5-7. Twelve preset tiled viewport configurations are provided in the **Viewports** dialog box.

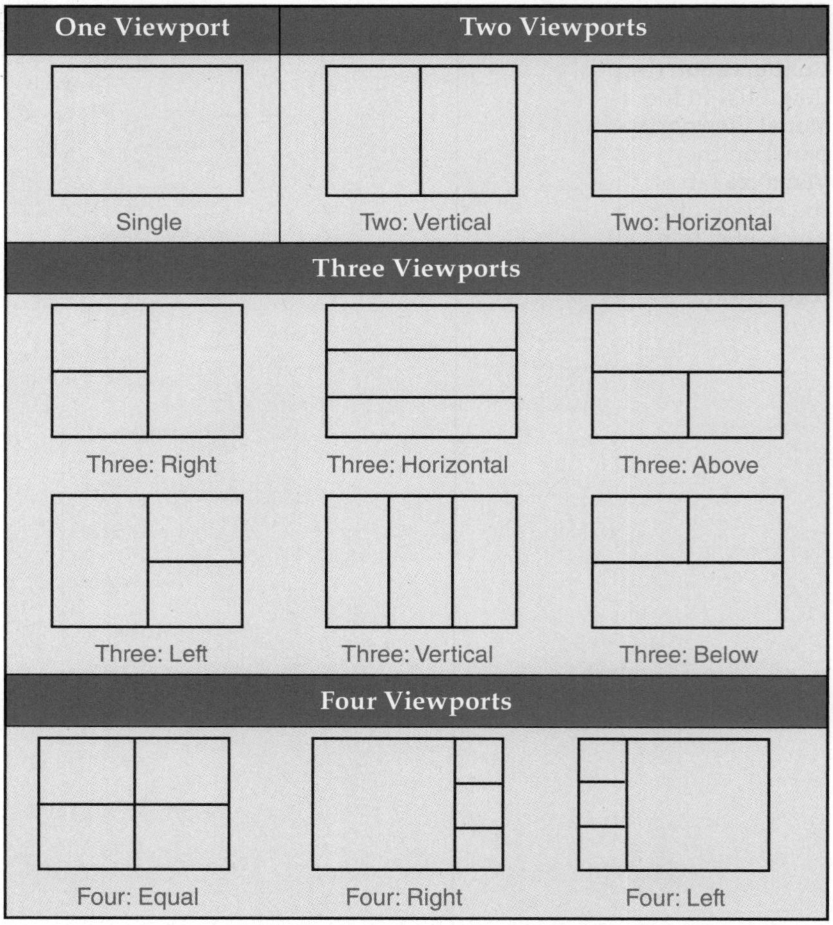

One Viewport	Two Viewports	
Single	Two: Vertical	Two: Horizontal

Three Viewports		
Three: Right	Three: Horizontal	Three: Above
Three: Left	Three: Vertical	Three: Below

Four Viewports		
Four: Equal	Four: Right	Four: Left

Command: **CVPORT**↵
Enter new value for CVPORT <*current*>: **3**.↵

The current value given is the ID number of the active viewport. The ID number is automatically assigned by AutoCAD, starting with 2. To change viewports with the **CVPORT** system variable, simply enter a different ID number. This technique may be used in custom programming for AutoCAD. Using the **CVPORT** system variable is also a good way to determine the ID number of a viewport. The number 1 is not a valid viewport ID number.

PROFESSIONAL TIP

Each viewport can have its own view, viewpoint, UCS, zoom scale, limits, grid spacing, and snap setting. Specify the drawing aids in all viewports before saving the configuration. When a viewport is restored, all settings are restored as well.

Managing Defined Viewports

If you are working with several different viewport configurations, it is easy to restore, rename, or delete existing viewports. You can do so using the **Viewports** dialog box. To access a list of named viewports, open the dialog box and select the **Named Viewports** tab. See **Figure 5-8**. To display a viewport configuration, select its name from the **Named viewports:** list. The selected configuration is displayed in the **Preview** area. If this is the arrangement you want, pick the **OK** button.

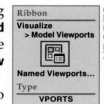

Assume you have saved the current viewport configuration. Now, you want to work in a specific viewport, but do not need other viewports displayed on screen. First, pick the viewport you wish to work in to make it active. Next, select **Viewport Configuration List** in the **Viewport Controls** flyout in the viewport controls and select **Single**. You can also use the **Viewport Configuration** drop-down list in the **Model Viewports** panel on the **Visualize** tab of the ribbon. The active viewport is displayed as the only viewport. To restore the original viewport configuration, select **Viewport Configuration List** in the **Viewport Controls** flyout in the viewport controls, and then

---

Figure 5-8.
The **Named Viewports** tab of the **Viewports** dialog box lists all named viewports and displays the selected configuration in the **Preview** area.

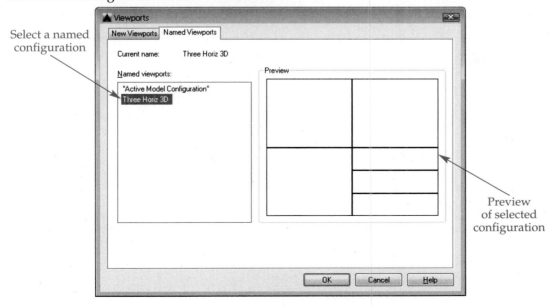

select **Custom Viewport Configuration**. This displays a list of named viewport configurations. Select the desired name and that configuration is displayed.

Viewport configurations can be renamed and deleted using the **Named Viewports** tab of the **Viewports** dialog box. To rename a viewport configuration, right-click on the name and pick **Rename** from the shortcut menu. To delete a viewport configuration, pick **Delete** from the shortcut menu. You can also press the [Delete] key to delete the highlighted viewport configuration. Pick **OK** to exit the dialog box.

PROFESSIONAL TIP

The **-VPORTS** command can be used to manage viewports on the command line. This may be required for some LISP programs where the dialog box cannot be used.

Using the Model Viewports Panel

The **Model Viewports** panel in the **Visualize** tab of the ribbon is shown in **Figure 5-9**. Picking the **Named Viewports...** button on the panel displays the **Viewports** dialog box with the **Named Viewports** tab active. The **Join** button is used to combine two viewports into a single viewport, as described in the next section. Picking the **Restore Viewports** button switches the display between a single viewport and the last multiple viewport configuration displayed. If no previous multiple viewport configuration has been displayed, picking this button displays a configuration of four viewports. Expanding the **Viewport Configuration** drop-down list displays the 12 preset viewport configurations. Selecting a configuration sets it current.

Joining Two Viewports

You can join two adjacent viewports in an existing configuration to form a single viewport. This process is often quicker than creating an entirely new configuration. However, the two viewports must form a rectangle when joined, **Figure 5-10**.

When you enter the **Join** option, AutoCAD first prompts you for the *dominant viewport*. All aspects of the dominant viewport are used in the new (joined) viewport. These aspects include the limits, grid, UCS, and snap settings.

> Select dominant viewport <current viewport>: *(select the viewport by picking in it or press [Enter] to set the current viewport as the dominant viewport)*
> Select viewport to join: *(select the other viewport)*

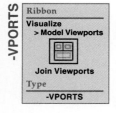

-VPORTS

Ribbon
Visualize
> Model Viewports

Join Viewports

Type
-VPORTS

Figure 5-9.
The **Model Viewports** panel in the **Visualize** tab of the ribbon.

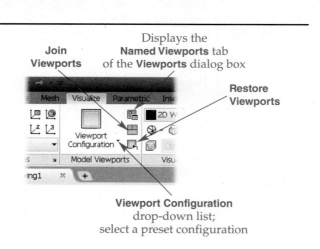

Join Viewports

Displays the **Named Viewports** tab of the **Viewports** dialog box

Restore Viewports

Viewport Configuration drop-down list; select a preset configuration

---

Figure 5-10.
Two viewports can be joined if they will form a rectangle. If the two viewports will not form a rectangle, they cannot be joined.

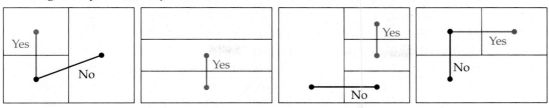

The two selected viewports are joined into a single viewport. If you select two viewports that do not form a rectangle, AutoCAD returns the message:

The selected viewports do not form a rectangle.

NOTE

You can adjust a viewport configuration dynamically by picking a viewport boundary and dragging the cursor. Viewports can be resized, added, and joined in this manner. To resize a viewport, pick a viewport boundary and drag it to the desired location. To add a new viewport, pick the small plus sign (+) on a viewport boundary and drag the green splitter bar to locate the new boundary. To join two viewports, pick a viewport boundary and drag it to another viewport boundary.

PROFESSIONAL TIP

Create only the number of viewports and viewport configurations needed to construct your drawing. Using too many viewports reduces the size of the image in each viewport and may confuse you. Also, it helps to zoom each view so that the objects fill the viewport.

Exercise 5-1

www.g-wlearning.com/CAD/

Complete the exercise on the companion website.

Applying Viewports to Existing Configurations and Displaying Different Views

You have total control over what is displayed in model space viewports. In addition to displaying various viewport configurations, you can divide an existing viewport into additional viewports or assign a different viewpoint to each viewport. The options for these functions are provided in the **New Viewports** tab of the **Viewports** dialog box. These options are located along the bottom of the **Viewports** dialog box, as shown in **Figure 5-5**:

- **Apply to.**
- **Setup.**
- **Change view to.**
- **Visual style.**

Apply To

When a preset viewport configuration is selected from the **Standard viewports:** list, it can be applied to either the entire display or the current viewport. The previous examples have shown how to create viewports that replace the entire display. Applying a configuration to the active viewport rather than the entire display can be useful when you need to display additional viewports.

For example, first create a configuration of three viewports using the Three: Right configuration option. Then, with the right (large) viewport active, open the **Viewports** dialog box again. Notice that the drop-down list under **Apply to:** is grayed out. Now, pick one of the standard configurations. This enables the **Apply to:** drop-down list. The default option is Display, which means the selected viewport configuration will replace the current display. Pick the drop-down list arrow to reveal the second option, Current Viewport. Pick this option and then pick the **OK** button. Notice that the selected viewport configuration has been applied to only the active (right) viewport. See **Figure 5-11**.

Setup

Viewports can be set up to display views in 2D or 3D. The 2D and 3D options are provided in the **Setup:** drop-down list. Displaying different views while working on a drawing allows you to see the results of your work on each view, since changes are reflected in each viewport as you draw. The selected viewport **Setup:** option controls the types of views available in the **Change view to:** drop-down list.

Change View To

The views that can be displayed in a selected viewport are listed in the **Change view to:** drop-down list. If the **Setup:** drop-down list is set to 2D, the views available to be displayed are limited to the current view and any named views. If 3D is active, the options include all of the standard orthographic and isometric views along with named views. When an orthographic or isometric view is selected for a viewport, the resulting orientation is shown in the **Preview** area. To assign a different viewpoint to a viewport, simply pick within a viewport in the **Preview** area to make it active and then

Figure 5-11.
The selected viewport configuration has been applied to the active viewport within the original configuration.

New configuration is applied to the viewport

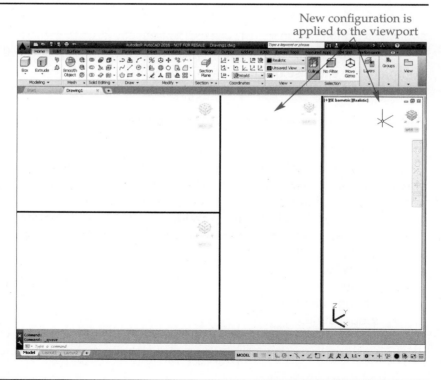

pick a viewpoint from the **Change view to:** drop-down list. Important: note that if you set a viewport to one of the orthographic preset views, the UCS is also changed (by default) in that viewport to the corresponding preset.

Visual Style

A visual style can be specified for a viewport. Pick within a viewport in the **Preview** area to make it active and then select a visual style from the **Visual Style:** drop-down list. All preset and saved visual styles are available in the drop-down list.

PROFESSIONAL TIP

If a standard naming convention is used to create viewport configurations, these named configurations can be saved with template drawings. This creates a consistent platform that all students or employees can use.

Exercise 5-2

www.g-wlearning.com/CAD/

Complete the exercise on the companion website.

Drawing in Multiple Viewports

When used with 2D drawings, viewports allow you to display a view of the entire drawing, plus views showing portions of the drawing. This is similar to using the **VIEW** command, except you can have several views on screen at once. You can also adjust the zoom magnification in each viewport to suit different areas of the drawing.

Viewports are also a powerful aid when constructing 3D models. You can specify different viewpoints in each viewport and see the model take shape as you draw. A model can be quickly constructed because you can switch from one viewport to another while drawing and editing. For example, you can draw a line from a point in one viewport to a point in another viewport simply by changing viewports while inside of the **LINE** command. The result is shown in each viewport.

In Chapter 4, you constructed a solid object. It was a base that had an angled surface from which a cylinder projected. See **Figure 5-12**. Now, you will construct the

Figure 5-12.
You will construct the object from Chapter 4 using multiple viewports.

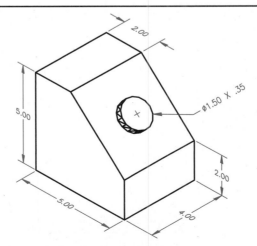

same object, but using two viewports. First, create a vertical configuration of two viewports. In the **Viewports** dialog box, set the right-hand viewport to display the southeast isometric view. Also, select the Conceptual visual style. Set the left-hand viewport to display the front view. Remember, this will also set the UCS to the front preset orthographic UCS in that viewport. Also, select the Wireframe visual style for the left-hand viewport. Close the dialog box and make the left-hand viewport active. Next, set the parallel projection current in both viewports.

Next, draw a polyline using the coordinates shown in **Figure 5-13**. Remember, to enter absolute coordinates with dynamic input turned on, type a pound sign (#) before the coordinate. This temporarily overrides the dynamic input of relative coordinate values. Be sure to use the **Close** option for the last segment. As you construct the side view, you can clearly see its true size and shape in the left-hand viewport. At the same time, you can see the construction in 3D in the right-hand viewport.

Notice in **Figure 5-13** that each view has a different UCS, as indicated by the UCS icon. Pick in each viewport in your drawing to view the different UCS orientations.

The next step is to extrude the shape to create the base. The **EXTRUDE** command is used the same way as in Chapter 4. With the left-hand viewport current, enter the **EXTRUDE** command, pick the polyline, and enter an extrusion height of −4 units. The base of the object is now complete, **Figure 5-14**.

Now, the cylinder needs to be created on the angled face. First, split the left-hand viewport into two horizontal viewports (top and bottom) using the **New Viewports**

Figure 5-13.
The screen is divided into two viewports. A front view of the object appears in the left-hand viewport and a 3D view appears in the right-hand viewport. Activate each viewport to view the UCS icon orientation as shown here.

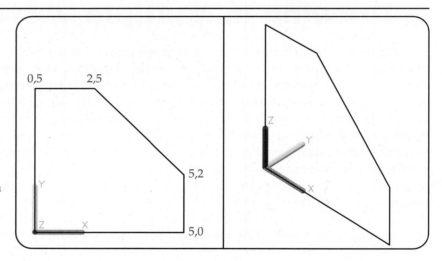

Figure 5-14.
The base of the object is now complete. A new UCS will be created on the angled face. If you are using the **3 Point** option of the UCS command, use the pick points indicated.

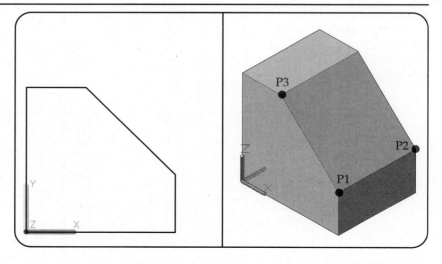

tab of the **Viewports** dialog box. Set both of the new viewports to display the current view. Pick the **OK** button to close the dialog box. Then, make the upper-left viewport current and set it up to always display a plan view of the current UCS by setting the **UCSFOLLOW** system variable to 1. In addition, set the **UCSVP** system variable to 0 so that the UCS configuration in the current viewport is displayed.

Using the UCS icon grips, as discussed in Chapter 4, create a new UCS on the angled face in the right-hand viewport. You can move the UCS icon dynamically using the origin grip, or you can use the **3 Point** option of the **UCS** command. If you are using the **3 Point** option, use the pick points shown in **Figure 5-14**. To quickly access the **3 Point** option, right-click on the UCS icon and select **3 Point** from the UCS icon shortcut menu. Notice in your drawing that after making the right-hand viewport active, the view in the upper-left viewport automatically changes to a plan view of the new current UCS. The view changes again after creating the new UCS on the angled face in the right-hand viewport.

With the right-hand viewport current, turn on 3D object snap. Turn off the dynamic UCS function, if it is on. Then, enter the **CYLINDER** command. When locating the center of the cylinder base, move the pointer to the angled face until the **Center of face** object snap is displayed. Then, pick the point. Next, enter a diameter of 1.5 units and a height of .35 units.

The object is now complete, **Figure 5-15**. Notice how the viewports have different UCSs. Each viewport can have its own UCS. The view in the upper-left viewport is the plan view of the current UCS. This is because of the **UCSFOLLOW** system variable setting of 1 in the upper-left viewport. In addition, the UCS orientation in the upper-left viewport is the same as the orientation in the right-hand viewport. The UCS orientation in one viewport is not affected by a change to the UCS in another viewport unless the **UCSVP** system variable is set to 0 in a viewport. If a viewport arrangement is saved with several different UCS configurations, then every named UCS remains intact and is displayed when the viewport configuration is restored.

To see the effect of the **UCSVP** system variable in the upper-left viewport, make that viewport current after completing the model in the right-hand viewport. The resulting display is shown in **Figure 5-15**. Next, make the lower-left viewport current, and then make the upper-left viewport current again. Now, notice that the front view is displayed in the upper-left viewport, and the UCS orientation is the same as the orientation in the lower-left viewport.

Figure 5-15.
The cylinder is drawn to complete the object. Notice the plan view in the upper-left viewport. Activate the upper-left viewport after completing the model in the lower-right viewport to view the UCS icon orientation as shown.

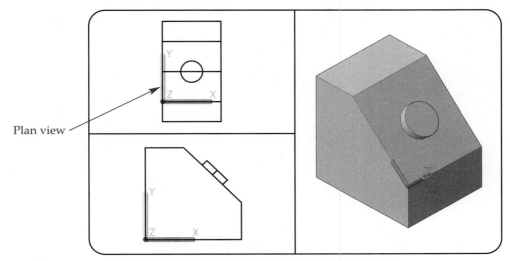

Plan view

The **REGEN** command affects only the current viewport. To regenerate all viewports at the same time, use the **REGENALL** command. This command can be entered by typing REGENALL.

The Quick Text mode is controlled by the **REGEN** command. Therefore, if you are working with text displayed with the Quick Text mode in viewports, be sure to use the **REGENALL** command in order for the text to be regenerated in all viewports.

NOTE

The **UCSFOLLOW** system variable and the **UCSVP** system variable can be set independently for each viewport. As discussed in the previous example, the UCS configuration in each viewport is controlled by the **UCSVP** system variable. When **UCSVP** is set to 1 in a viewport, the UCS is independent from all other UCSs, which is the default. If **UCSVP** is set to 0 in a viewport, its UCS will change to reflect any changes to the UCS in the current viewport.

Exercise 5-3

www.g-wlearning.com/CAD/

Complete the exercise on the companion website.

Chapter Review

Answer the following questions using the information in this chapter.

1. What is the purpose of *viewports*?
2. How do you name a configuration of viewports?
3. What is the purpose of saving a configuration of viewports?
4. Explain the difference between *tiled* and *floating* viewports.
5. Name the system variable controlling the maximum number of viewports that can be displayed at one time.
6. How can a named viewport configuration be redisplayed on screen?
7. How can a list of preset viewport configurations be accessed?
8. What relationship must two viewports have before they can be joined?
9. What is the significance of the dominant viewport when two viewports are joined?
10. When creating a new viewport configuration, how can you set a visual style in a viewport?

Drawing Problems

1. **General** Construct seven template drawings, each with a preset viewport configuration. Use the following configurations and names. Save each template under the same name as the viewport configuration.

Number of Viewports	Configuration	Name
2	Horizontal	TWO-H
2	Vertical	TWO-V
3	Right	THREE-R
3	Left	THREE-L
3	Above	THREE-A
3	Below	THREE-B
3	Vertical	THREE-V

2. **General** Construct one of the problems from Chapter 4 using viewports. Use one of your template drawings from Problem 5-1. Save the drawing as P5-2.

3. **Mechanical** This is an orthographic drawing of a light fixture bracket. Create it as a solid model. If appropriate, use geometric constraints on the base 2D shapes. Use solid primitives and Boolean commands as needed. Use the dimensions given. Similar holes have the same offset dimensions. Use multiple viewports to construct the drawing. Begin with a four-viewport layout. Then switch viewports as needed to work on specific areas of the model. Create new UCSs as needed. Display an appropriate pictorial view of the drawing in the upper-right viewport. Plot the 3D view of the drawing to scale on a B- or C-size sheet of paper. Save the drawing as P5-3.

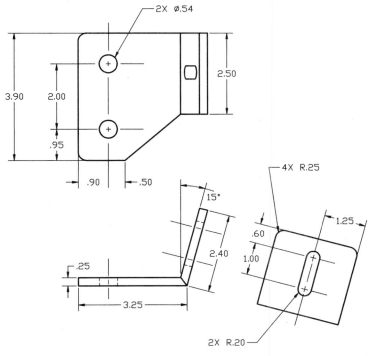

Light Fixture Bracket

Drawing Problems - Chapter 5

4. **Mechanical** This is an orthographic drawing of an angle bracket. Create it as a solid model. If appropriate, use geometric constraints on the base 2D shapes. Use solid primitives and Boolean commands as needed. Use the dimensions given. Similar holes have the same offset dimensions. Use multiple viewports to construct the drawing. Begin with a four-viewport layout. Then switch viewports as needed to work on specific areas of the model. Create new UCSs as needed. Display an appropriate pictorial view of the drawing in the upper-right viewport. Plot the 3D view of the drawing to scale on a B- or C-size sheet of paper. Save the drawing as P5-4.

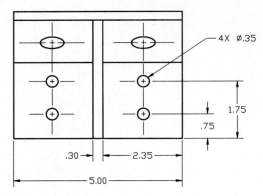

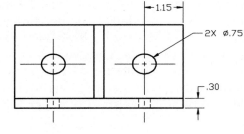

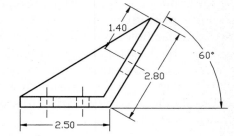

Angle Bracket

CHAPTER

Model Extrusions and Revolutions

Learning Objectives

After completing this chapter, you will be able to:

✓ Create solids and surfaces by extruding profiles.
✓ Extrude regions.
✓ Extrude surfaces.
✓ Create symmetrical 3D solids and surfaces by revolving profiles.
✓ Revolve regions.
✓ Revolve surfaces.
✓ Extrude and revolve objects using mathematical expressions and constraints.
✓ Use solid extrusions and revolutions as construction tools.

A complex shape can be created by *extruding* a profile into a 3D solid or surface model. You have been introduced to the operation in previous chapters. Symmetrical objects can be created by revolving a profile about an axis to create a new shape. This chapter discusses the **EXTRUDE** and **REVOLVE** commands and using these commands as construction tools.

Creating Extruded Models

An extrusion is a profile that has been "extruded" into a 3D solid or surface. The **EXTRUDE** command allows you to create extrusions from lines, arcs, elliptical arcs, 2D and 3D polylines, 2D and 3D splines, circles, ellipses, 2D solids, 3D solid faces and edges, regions, planar surfaces, surface edges, and donuts.

By default, closed objects, such as circles, polygons, closed polylines, and donuts, are converted to a *solid extrusion* when they are extruded. A solid extrusion represents a solid object and has mass properties. Open-ended objects, such as lines, arcs, polylines, elliptical arcs, and splines, are converted to a *surface extrusion* when they are extruded. Surface extrusions have no mass properties.

Extrusions can be created along a straight line or along a path curve. A taper angle can also be applied as you extrude an object. **Figure 6-1** illustrates a polygon extruded into a solid.

When the **EXTRUDE** command is selected, you are prompted to select the objects to extrude. Select the objects and press [Enter]. You are then prompted for the extrusion

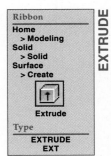

Ribbon

Home
> Modeling
Solid
> Solid
Surface
> Create

Extrude

Type
EXTRUDE
EXT

EXTRUDE

Figure 6-1.
The **EXTRUDE** command creates a 3D solid or surface from a profile object. A—The initial, closed profile. B—The extruded solid object shown with hidden lines removed.

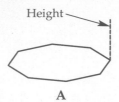

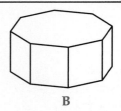

A

B

height. If a pictorial view is displayed, you can move the cursor to establish a specific height. The height is always applied along the Z axis of the selected object, not the current UCS. A positive value extrudes the height above the XY plane of the object. A negative height value extrudes the height below the XY plane. If a pictorial view is displayed, you can drag the mouse to set the extrusion above or below the XY plane and then enter the height value.

The **Direction** option provides an alternate way to create the extrusion. After selecting this option, specify a start and end point to define the extrusion direction and height. This is similar to extruding along a path, discussed in the next section.

Before entering a height, you can specify a taper angle. The taper angle can be any value *between* +90° and –90°. A positive angle tapers to the inside of the object from the base. A negative angle tapers to the outside of the object from the base. See **Figure 6-2**. However, the taper angle cannot result in edges that "fold into" the extruded object.

If the command is selected from the **Solid** tab in the ribbon, the mode is automatically set to solid. However, if an open profile is selected, a surface is still created. If the command is selected from the **Surface** tab in the ribbon, the mode is automatically set to surface. In either case, you can use the **Mode** option to change the output type to solid or surface.

Keep in mind, objects that are not closed, such as lines and polylines, *always* result in extruded surfaces. A solid can only be created by the **EXTRUDE** command if the original object is closed.

NOTE

If a surface is created using the **EXTRUDE** command, the type of surface created is controlled by the **SURFACEMODELINGMODE** system variable. The settings available are discussed later in this chapter.

PROFESSIONAL TIP

Objects such as polylines, lines, and arcs that have a thickness can be converted to surfaces using the **CONVTOSURFACE** command. Circles and closed polylines with a thickness can be converted to solids using the **CONVTOSOLID** command. The thickness property for an object is controlled by the Thickness setting in the **Properties** palette.

Figure 6-2.
A—A positive angle tapers to the inside of the object from the base. B—A negative angle tapers to the outside of the object.

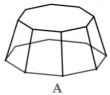

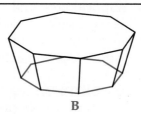

A

B

Extrusions along a Path

A profile shape can be extruded along a path to create a 3D solid or surface. The path can be a line, circle, arc, ellipse, elliptical arc, 2D or 3D polyline, helix, 3D solid edge, surface edge, or 2D or 3D spline. Multiple line segments and other objects can be first joined to form a polyline path. The corners of angled segments on the extruded object are mitered, while curved segments are smooth. See **Figure 6-3**.

When open objects, such as lines, arcs, polylines, elliptical arcs, and splines, are used as the profile, they are converted to a swept surface when extruded along a path. A *sweep* is a solid or surface that is created when an open or closed curve is pulled, or swept, along a 2D or 3D path. An extrusion is really a form of a sweep. Sweeps are discussed in detail in Chapter 7.

To extrude along a path, enter the **EXTRUDE** command and select the objects to extrude. When prompted for the height of the extrusion, select the **Path** option. If needed, first enter a taper angle. Then, pick the object to be used as the extrusion path.

Objects can also be extruded along a line at an angle to the base object, **Figure 6-4**. Notice that the plane at the end of the extruded object is parallel to the original object. Also notice that the length of the extrusion is the same as that of the path. The path does not need to be perpendicular to the object.

If the path begins perpendicular to the profile, the cross section of the resulting extrusion is perpendicular to the path, regardless if the path is a straight line, curve, or spline. See **Figure 6-5**. If the path is a spline or curve that does not begin perpendicular to the profile, the profile may not remain perpendicular to the path as it is extruded.

If one of the endpoints of the path is not on the plane of the object to be extruded, the path is temporarily moved to the center of the profile. The extrusion is then created as if the path were connected to the original object, as shown in **Figure 6-4**.

Figure 6-3.
A—Angled segments are mitered when an object is extruded. B—Curved segments are smoothed when an object is extruded.

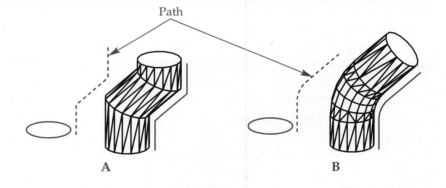

Path

A

B

Figure 6-4.
A—An object extruded along a path. B—The end of an object extruded along an angled path is parallel to the original object.

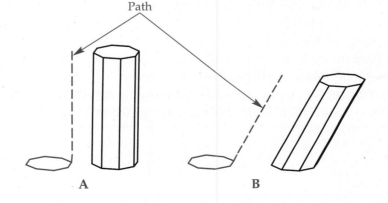

Path

A

B

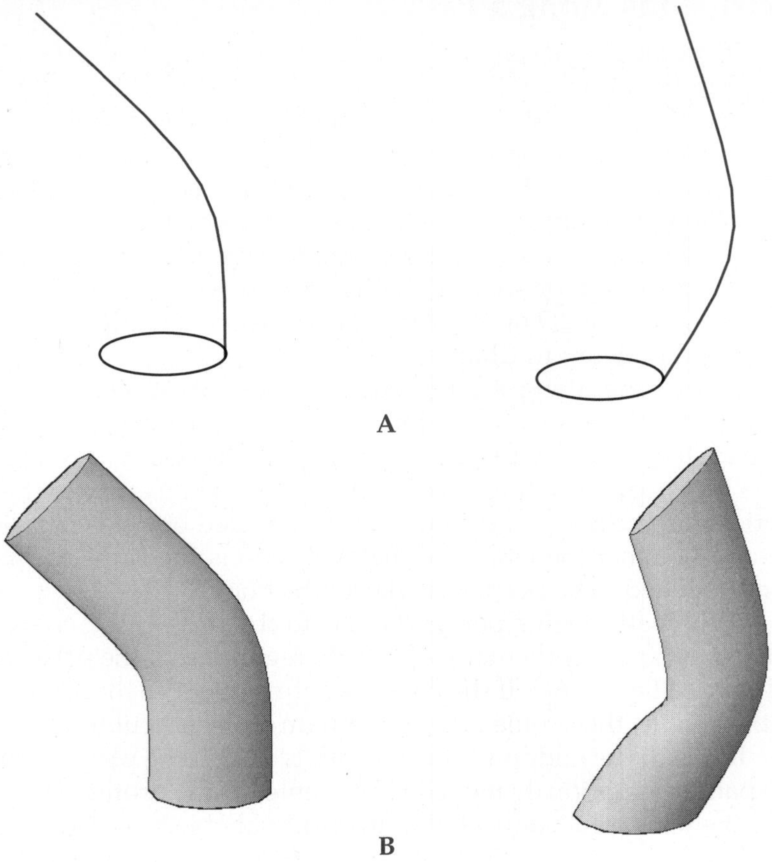

Figure 6-5.
A—Splines can be used as extrusion paths. Notice that the profile on the right is not perpendicular to the start of the path. B—The resulting extrusions.

A

B

NOTE

The **DELOBJ** system variable allows you to delete or retain the original extruded objects and path definitions. This system variable has no effect when a surface extrusion is created and the **SURFACEASSOCIATIVITY** system variable is set to 1 (the default setting), so as to maintain associativity with the original defining geometry. The **DELOBJ** system variable settings are:

0   All original profile geometry (objects that were extruded) and path geometry is retained.

1   Profile geometry is deleted. Path geometry is retained.

2   Profile and path geometry is deleted.

3   Profile and path geometry is deleted when the extruded entity is a solid. (This is the default setting.)

–1 You are prompted to delete profile geometry. Path geometry is retained.

–2 You are prompted to delete all profile and path geometry.

–3 You are prompted to delete all profile and path geometry when the extruded entity is a solid. Defining geometry is deleted for certain modeling operations that result in a solid. (This setting is necessary to maintain compatibility with older versions of AutoCAD.)

The **DELOBJ** system variable also affects the **REVOLVE**, **SWEEP**, **LOFT**, **CONVTOSOLID**, **CONVTOSURFACE**, **CONVTONURBS**, and **CONVTOMESH** commands.

Extruding Regions

In Chapter 2, you learned how to create 2D regions. As an example, you created the top view of the base shown in **Figure 6-6A** as a region. Regions can be extruded to create 3D solids. The base you created in Chapter 2 can be extruded to create the final solid shown in **Figure 6-6B**. Any features of the region, such as holes, are extruded the same height as the rest of the object. If the profile was created as polylines, the holes must be separately extruded and then subtracted from the solid. Using this method, you can construct a fairly complex 2D region that includes curved profiles, holes, slots, etc. See **Figure 6-6C**. Then, a complex 3D solid can be quickly created. See **Figure 6-6D**. Additional features can be added using editing commands or Boolean operations.

Exercise 6-1

www.g-wlearning.com/CAD/

Complete the exercise on the companion website.

Figure 6-6.
A—The 2D region that will be extruded. B—The solid object created by extruding the region, shown in a 3D wireframe display. C—A 2D region created from a variety of objects. The objects forming the outer area include polylines, splines, arcs, and lines. The internal circles form two separate regions and are subtracted using a Boolean operation to create a single region. D—The solid object after extruding the region with the Conceptual visual style set current.

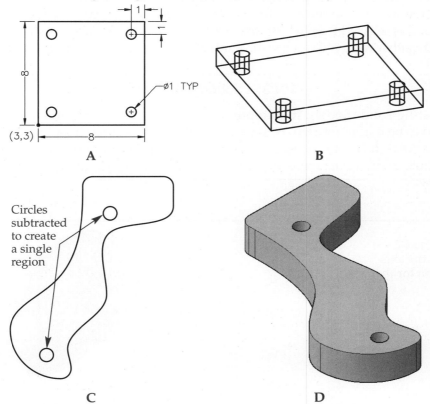

Extruding a Surface

A planar surface can be extruded into a solid object in the same manner as a region. Nonplanar (curved) surfaces can be extruded, but only into extruded surfaces. A planar surface can be quickly converted to a solid using the **EXTRUDE** command. Simply select the surface when prompted to select objects. The surface can be extruded in a specific direction, along a path, or at a taper angle.

Any closed object, such as a circle, rectangle, polygon, or polyline, can be converted into a surface with the **Object** option of the **PLANESURF** command. After accessing the **PLANESURF** command, enter the **Object** option and select the closed object. The resulting surface can then be extruded into a 3D solid. As previously discussed, the **Mode** option of the **EXTRUDE** command allows you to set the output type to solid or surface.

PLANESURF

Ribbon
Surface
> Create

Planar Surface

Type
PLANESURF

REVOLVE

Ribbon
Home
> Modeling
Solid
> Solid
Surface
> Create

Revolve

Type
REVOLVE
REV

Creating Revolved Models

The **REVOLVE** command allows you to create solids and surfaces by revolving a shape about an axis. Shapes that can be revolved include lines, arcs, circles, ellipses, polygons, 2D and 3D polylines, 2D and 3D splines, regions, planar surfaces, and donuts. The selected object can be revolved at any angle up to 360°. Open curves and line segments create revolved surfaces. Closed curves can be used to create either solids or surfaces. Surface revolutions have no mass properties.

When the command is selected, you are prompted to pick the objects to revolve. Then, you must define the axis of revolution. The default option is to pick the two endpoints of an axis of revolution. This is shown in **Figure 6-7**. You can also revolve about an object or the X, Y, or Z axis of the current UCS. Once the axis is defined, you are prompted to enter the angle through which the profile will be revolved. When the angle is specified by either moving the cursor or entering a value, the revolution is created.

If the command is selected from the **Solid** tab in the ribbon, the mode is automatically set to solid. However, if an open profile is selected, a surface is created instead of a solid. If the command is selected from the **Surface** tab in the ribbon, the mode is automatically set to surface. In either case, you can use the **Mode** option to change the output type.

PROFESSIONAL TIP

When creating solid models, it is important to consider that the final part will most likely need to be manufactured. Be aware of manufacturing processes and methods as you design parts. It is easy to create a part in AutoCAD with internal features that may be impossible to manufacture, especially when revolving a profile.

Figure 6-7.
Points P1 and P2 are selected as the axis of revolution for the profile.

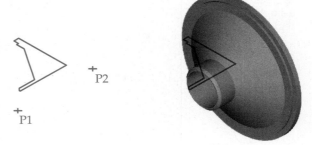

Revolving about an Object

You can select an object, such as a line, as the axis of revolution. **Figure 6-8** shows a solid created using the **Object** option of the **REVOLVE** command. Both a full circle (360°) revolution and a 270° revolution are shown. Enter the **Object** option when prompted for the axis of revolution. Then, pick the axis object and enter the angle through which the profile will be rotated. You can use the **Start Angle** option before entering an angle of revolution. This allows you to specify the point at which the revolution starts and then the angle of revolution. The **Reverse** option allows you to reverse the initial direction in which the profile is revolved.

Revolving about the X, Y, or Z Axis

The X axis of the current UCS can be used as the axis of revolution by selecting the **X** option of the **REVOLVE** command. The origin of the current UCS is used as one end of the X axis line. Notice in **Figure 6-9** that two different shapes can be created from the same profile by changing the UCS origin. No hole appears in the object in **Figure 6-9B** because the profile was revolved about an edge that coincides with the X axis. The Y or Z axis can also be used as the axis of revolution. See **Figure 6-10**.

Exercise 6-2

www.g-wlearning.com/CAD/

Complete the exercise on the companion website.

Figure 6-8.
An axis of revolution can be selected using the **Object** option of the **REVOLVE** command. Here, the line is selected as the axis.

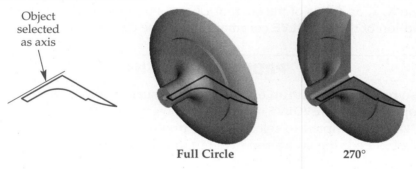

Full Circle   270°

Figure 6-9.
A—A solid is created using the X axis as the axis of revolution. B—A different object is created with the same profile by changing the UCS origin.

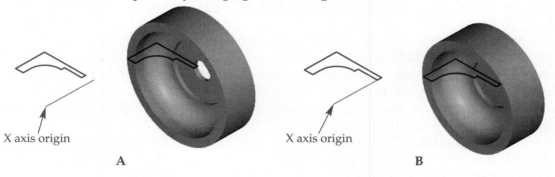

A   B

Figure 6-10.
A—A solid is created using the Y axis as the axis of revolution. B—A different object is created by changing the UCS origin.

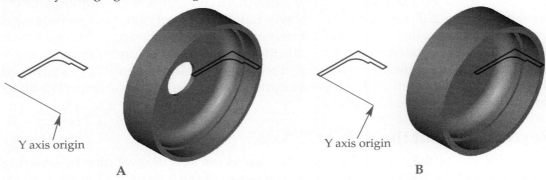

Y axis origin

A

Y axis origin

B

Revolving Regions

Earlier in this chapter, you learned that regions can be extruded. In this manner, holes, slots, keyways, etc., can be created. Regions can also be revolved. A complex 2D shape can be created using Boolean operations on regions. Then, the region can be revolved. One advantage of this method is it may be easier to create a region than trying to create a complex 2D profile as a single, closed polyline. As previously discussed, you can use the **Mode** option of the **REVOLVE** command to set the output type to solid or surface.

Revolving Surfaces

Just as planar and nonplanar surfaces can be extruded, they can also be revolved. When the **REVOLVE** command is selected, simply pick the surface when prompted to select objects. The surface can be revolved about an axis defined by two pick points, an object, or the X, Y, or Z axis of the current UCS. As previously discussed, you can use the **Mode** option of the **REVOLVE** command to set the output type to solid or surface.

PROFESSIONAL TIP

You can also revolve a face on an existing solid or surface into a new solid or surface. The **Mode** option of the **REVOLVE** command allows you to set the output type to solid or surface. When prompted to select objects, press the [Ctrl] key and pick the face to revolve. Subobject editing is covered in detail in Chapter 11.

Using Extrude and Revolve to Create Surfaces

When creating an extruded or revolved model as a surface, the **SURFACEMODELINGMODE** system variable controls the type of surface created. The default setting of 0 creates a *procedural surface*. This is a standard surface composed of multiple flat polygons, but it has no control vertices. By default, this type of surface is associated with the object used to create it. The **SURFACEASSOCIATIVITY** system variable controls this setting.

A NURBS surface can be created by setting the **SURFACEMODELINGMODE** system variable to a value of 1. A *NURBS surface* is composed of splines and contains control vertices that enable the control of the curve shape with great precision.

Surface modeling variables can be set on the **Create** panel in the **Surface** tab of the ribbon. See **Figure 6-11**. When the **Surface Associativity** button is on, the **SURFACEASSOCIATIVITY** system variable is set to 1. When the **NURBS Creation** button is on, the **SURFACEMODELINGMODE** system variable is set to 1.

Extruding Surfaces from Objects with Dimensional Constraints

When the **Mode** option is set to create a surface, you can use the **Expression** option to create a parametric relationship between a dimensional constraint and the extruded surface. For example, the circle in **Figure 6-12** has been given a dimensional constraint named dia1. To create an extruded surface cylinder with a height that is always one-half of the diameter, enter the **Expression** option. Then, at the **Enter expression:** prompt, enter:

dia1/2

In this manner, using appropriate formulaic expressions, you can create an extruded surface that has a mathematical relationship to a 2D object that has a dimensional constraint.

CAUTION

When you enter an expression while creating a solid, the result is calculated, but the dimensional constraint is removed. Therefore, the parametric aspect of using an expression is lost.

Figure 6-11.
Surface modeling variables can be set on the **Create** panel in the **Surface** tab of the ribbon.

Pick to associate the surface with the original object

Pick to create a NURBS surface

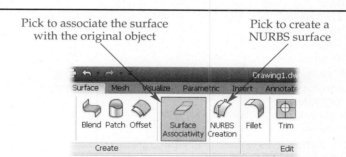

Figure 6-12.
A—The circle has been given a dimensional constraint named dia1. B—The circle is extruded using an expression to create a cylinder surface with a height that is one-half the distance of dia1. When the diameter is changed, the height changes to maintain the relationship.

dia1=3.0000

A

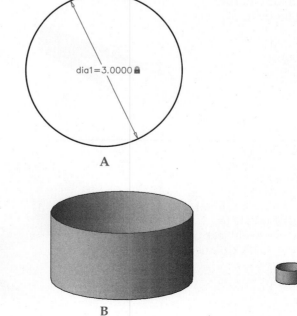

B

Revolving Surfaces from Objects with Dimensional Constraints

When constructing a revolved surface, the **Expression** option of the **REVOLVE** command can be used with dimensional constraints in the same manner as when using the **EXTRUDE** command. Examples of mathematical expressions used with constraints are shown in **Figures 6-13** and **6-14**. These examples illustrate how linear and angular dimensional constraints can be used to calculate an angular value for a revolved surface.

The profile in **Figure 6-13A** is to be revolved at an angle that is one-half that of the angular dimension (ang2). This is achieved by selecting the **Expression** option and then entering:

ang2/2

As shown in **Figure 6-13B**, the height of the model changes when the angular dimension is changed.

In **Figure 6-14**, the same 2D geometry is used. However, the aligned dimension d2 is used to create a negative value to revolve the object below the XY plane. The following formula is entered:

d2−31

The result of subtracting 31 from the value of d2 is processed as the angular value for the revolution. The result is shown in **Figure 6-14**.

Figure 6-13.
A—The 2D profile is to be revolved at an angle that is one-half of the angular dimension ang2. B—The revolved surface is created using an expression that references the angular dimension ang2. When the angle is changed, the height changes to maintain the relationship.

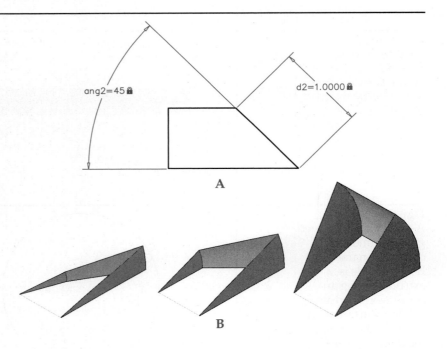

Figure 6-14.
The surface is created using a negative value based on the dimensional constraint d2 to revolve the object below the XY plane.

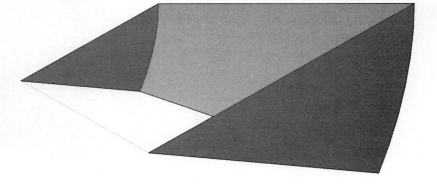

Using Extrude and Revolve as Construction Tools

It is unlikely that an extrusion or revolution will result in a finished object. Rather, these operations will be used with other model construction methods, such as Boolean operations in solid modeling, to create the final composite 3D object. The next sections discuss how to use **EXTRUDE** and **REVOLVE** with other construction methods to create a finished solid object.

Creating Features with Extrude

You can create a wide variety of features with the **EXTRUDE** command. Study the shapes shown in Figure 6-15. These detailed solid objects were created by drawing a profile and then using the **EXTRUDE** command. The objects in Figures 6-15C and 6-15D must be constructed as regions before they are extruded. For example, the five holes (circles) in Figure 6-15D must be removed from the base region using the **SUBTRACT** command.

Look at Figure 6-16. This is part of a clamping device used to hold parts on a mill table. There is a T-slot milled through the block to receive a T-bolt and one side is stair-stepped, under which parts are clamped. If you look closely at the end of the object, most of the detail can be drawn as a 2D region and then extruded. However, there are also two holes in the top of the block to allow for bolting the clamp to the mill table. These features must be added to the extruded solid.

First, change the UCS to the front preset orthographic UCS. Display a plan view of the UCS. Then, draw the profile shown in Figure 6-17 using the **PLINE** command. You can draw it in stages, if you like, and then use the **PEDIT** command to join all segments into a single polyline.

Next, use the **EXTRUDE** command to create the 3D solid. Extrude the profile a distance of −6 units with a 0° taper. This will extrude the object away from you. Display the object from the southeast isometric preset viewpoint or use the view cube to display a pictorial view. The object should look similar to Figure 6-16 without the holes in the top. Set the Conceptual visual style current, if you like.

The two holes are ⌀.5 units and evenly spaced on the surface through which they pass. Change to the WCS and draw a construction line from midpoint to midpoint, as shown in Figure 6-18. Then, set **PDMODE** to an appropriate value, such as 3, and use the **DIVIDE** command to divide the construction line into three parts. The two points created by the **DIVIDE** command are equally spaced on the surface and can be used to locate the two holes.

Figure 6-15.
Detailed solids can be created by extruding the profile of an object. The profiles are shown here in color.

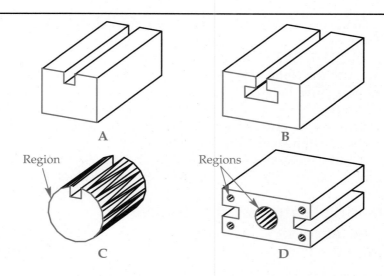

Figure 6-16.
Most of this object can be created by extruding a profile. However, the holes must be added after the extruded solid is created.

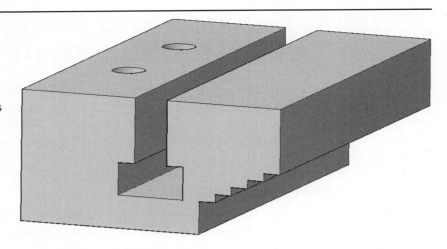

Figure 6-17.
This is the profile that will be extruded for the clamping block. Notice the dimensional constraints that have been applied.

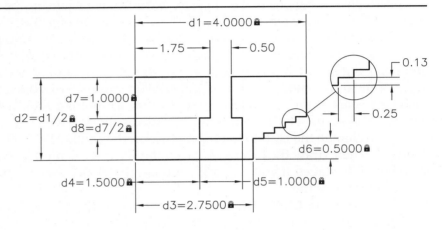

Figure 6-18.
Draw a construction line (shown here in color) and divide it into three parts.

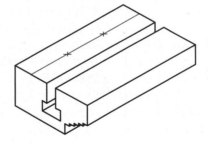

There are two ways to create a hole. You can draw a circle and extrude it to create a cylinder or you can draw a solid cylinder. Either way, you need to subtract the cylinder to create the hole. Drawing a solid cylinder is probably easiest. When prompted for a center, use the **Node** object snap to select the point. Then, enter the diameter. Finally, enter a negative height so that the cylinder extends into the solid or drag the cylinder down in the 3D view so it extends all of the way through the block. The actual height is not critical, as long as it extends through the block.

You can either copy the first cylinder to the second point or draw another cylinder. When both cylinders are located, use the **SUBTRACT** command to remove them from the solid. The object is now complete and should look like **Figure 6-16.**

Creating Features with Revolve

The **REVOLVE** command is very useful for creating symmetrical, round objects. Many times, however, the object you are creating is not completely symmetrical. For example, look at the camshaft in **Figure 6-19**. For the most part, this is a symmetrical, round object. However, the cam lobes are not symmetrical in relation to the shaft and bearings. The **REVOLVE** command can be used to create the shaft and bearings. Then, the cam lobes can be created and added.

Start a new drawing and make sure the WCS is the current UCS. Using the **PLINE** command, draw the profile shown in **Figure 6-20A**. This profile will be revolved through 360°, so you only need to draw half of the true plan view of the cam profile. The profile represents the shaft and three bearings.

Next, display the drawing from the southwest isometric preset viewpoint. Then, use the **REVOLVE** command to create the base camshaft as a 3D solid. Pick the endpoints shown in **Figure 6-20A** as the axis of revolution. Revolve the profile through 360°. Perform a zoom extents and set the Conceptual visual style current to clearly see the object.

Now, you need to create one cam lobe. Change the UCS to the left orthographic preset. Then, draw a construction point in the center of the left end of the camshaft. Use the **Center** object snap and an appropriate **PDMODE** setting. Next, draw the profile shown in **Figure 6-20B**. Use the construction point as the center of the large radius. You may want to create a new layer and turn off the display of the base camshaft.

Once the cam lobe profile is created, use the **REGION** command to create a region. Then, use the **EXTRUDE** command to extrude the region to a height of −.5 units (into the camshaft). The extrusion should have a 0° taper. If you turned off the display of the base camshaft, turn it back on.

Figure 6-19.
For the most part, this object is symmetrical about its center axis. However, the cam lobes are not symmetrical about the axis.

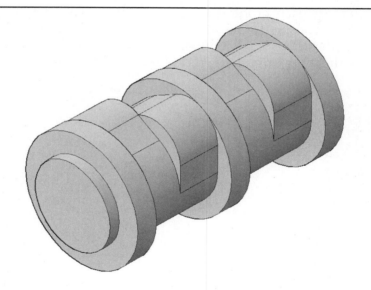

Figure 6-20.
A—This profile will be revolved to create the shaft and bearings. Notice the parallel constraints that have been applied. B—This is the profile of one cam lobe, which will be extruded.

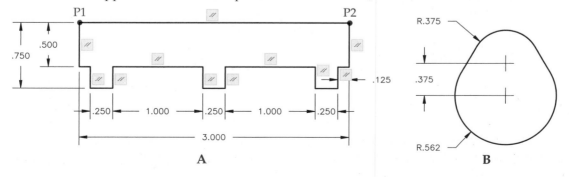

One cam lobe is created, but it is not in the proper position. With the left UCS current, move the cam lobe –.375 units on the Z axis. If a different UCS is current, the axis of movement will be different. This places the front surface of the cam lobe on the back surface of the first bearing. Now, make a copy of the lobe that is located –.5 units on the Z axis. Finally, copy the first two cam lobes –1.25 units on the Z axis.

You now need to rotate the four cam lobes to their correct orientations. Make sure the left UCS is still current. Then, rotate the first and third cam lobes 30°. If a different UCS is current, you can use the **3DROTATE** command. The center of rotation should be the center of the shaft. There are many points on the shaft to which the **Center** object snap can snap; they are all acceptable. You can also use the construction point as the center of rotation. Rotate the second and fourth cam lobes –30° about the same center.

Finally, use the **UNION** command to join all objects. The final object should appear as shown in **Figure 6-19**. Use the view cube to see all sides of the object. You can also create a rotating display using the **3DORBIT** command.

Multiple Intersecting Extrusions

Many solid objects have complex curves and profiles. These can often be constructed from the intersection of two or more extrusions. The resulting solid is a combination of only the intersecting volumes of the extrusions. The following example shows the construction of a coat hook.

1. Construct the first profile, **Figure 6-21A**.
2. Construct the second profile located on a common point with the first, **Figure 6-21B**.
3. Construct the third profile located on the common point, **Figure 6-21C**.
4. Extrude each profile the required dimension into the same area. Be careful to specify positive or negative heights for each extrusion, **Figures 6-21D** and **6-21E**.
5. Use the **INTERSECT** command to create a composite solid from the volume shared by the three extrusions, **Figure 6-21F**.

INTERSECT

Ribbon
Home
> **Solid Editing**
Solid
> **Boolean**

Solid, Intersect

Type
INTERSECT
IN

Figure 6-21.
Constructing a coat hook. A—Draw the first profile. B—Draw the second profile. C—Draw the third profile. All three profiles should have a common origin. D—Extrude each profile so that the extruded objects intersect. E—The extruded objects after the Conceptual visual style is set current. F—Use the **INTERSECT** command to create the composite solid. The final solid is shown here with the Conceptual visual style set current.

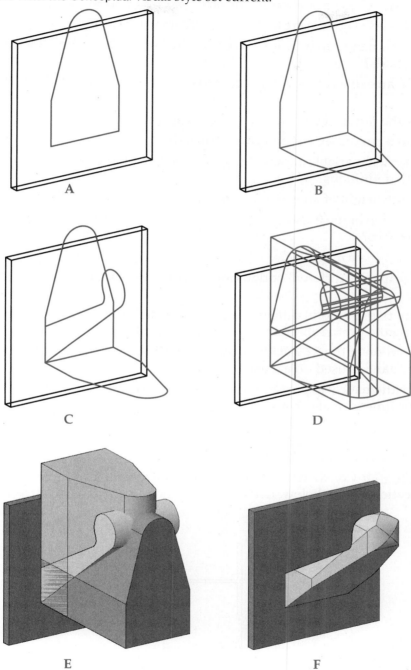

Chapter Review

Answer the following questions using the information in this chapter.

1. What is an *extrusion*?
2. How do you create a surface extrusion?
3. Briefly describe how to create a solid extrusion.
4. Which command can be used to convert circles and closed polylines with a thickness to solids?
5. How can an extrusion be constructed to extend below the XY plane of the current UCS?
6. What is the range in which a taper angle can vary?
7. How can a curved extrusion be constructed?
8. Which system variable allows you to delete or retain the original extruded objects and path definitions?
9. How is the height of an extrusion applied in relation to the original object?
10. Which option of the **PLANESURF** command can be used to convert a closed object into a surface?
11. What is a *surface revolution*?
12. What are the five different options for selecting the axis of revolution for a revolved solid?
13. How can two (or more) different solids be created from the same 2D profile when revolving the profile about the X, Y, or Z axis of the current UCS?
14. Which option of the **REVOLVE** command controls the type of object created when the command is used on a circle?
15. What mathematical expression would be used to revolve a profile 90° less than the angular dimensional constraint named ang4?

Drawing Problems

1. Architectural Construct a 12′ long section of wide flange structural steel with the cross section shown below. Use the dimensions given. Save the drawing as P6-1.

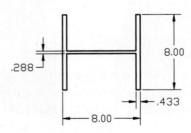

*Problems 2–7. These problems require you to use a variety of solid modeling methods to construct the objects. Use **EXTRUDE**, **REVOLVE**, solid primitives, new UCSs, Boolean commands, and editing tools such as extrusions and revolutions to assist in construction. If appropriate, apply geometric constraints to the base 2D drawing. Do not create section views. Do not add dimensions to the models. For objects that have dimensions with tolerances, use the base dimension from which the upper and lower limits are calculated to create the model. Save each drawing as P6-(problem number).*

2. Mechanical

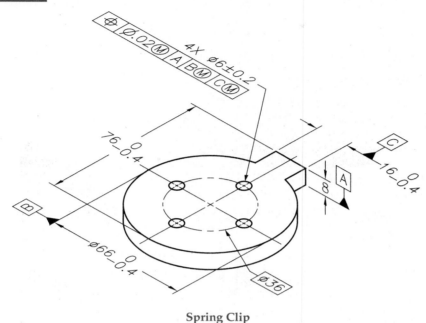

Spring Clip

3. Mechanical

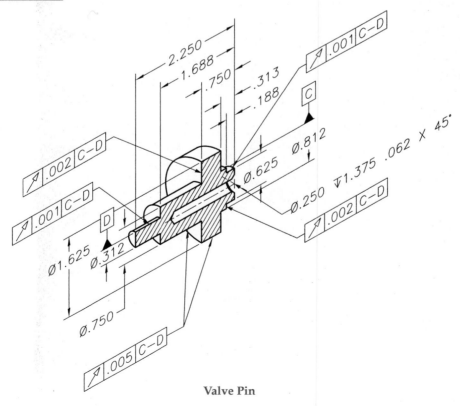

Valve Pin

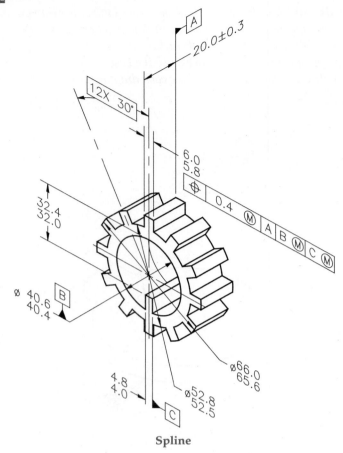

Spline

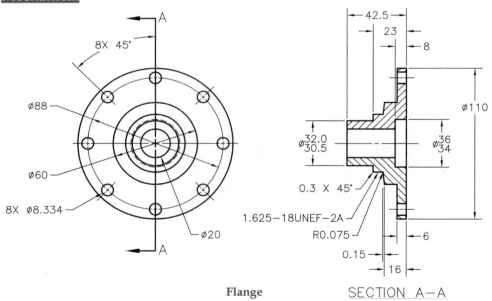

Flange   SECTION A—A

6. **Mechanical**

$6 \times \varnothing 6 \, {}^{0.2}_{0}$

$6 \times 60°$

$\varnothing \, {}^{28.1}_{28.0}$

100

20

$\varnothing \, {}^{60.25}_{60.0}$

$2 \times \varnothing \, {}^{40.2}_{40.0}$

10

30°

30

80

$4 \times \varnothing 4 \, {}^{+0.2}_{0}$

SECTION A-A

Nozzle

7. **Mechanical**

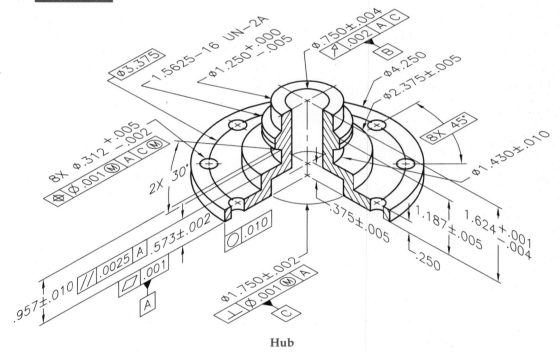

$\varnothing 3.375$

1.5625−16 UN−2A

$\varnothing 1.250 \, {}^{+.000}_{-.005}$

$\varnothing .750 \pm .004$

$\nearrow \, | .002 | A | C |$

B

$\varnothing 4.250$

$\varnothing 2.375 \pm .005$

8X 45°

$\varnothing 1.430 \pm .010$

8X $\varnothing .312 \, {}^{+.005}_{-.002}$

$\oplus | \varnothing .001 (M) | A | C (M) |$

2X 30°

$| \bigcirc | .010 |$

$.573 \pm .002$

$| // | .0025 | A |$

$.957 \pm .010$

$| \angle | .001 |$

A

$.375 \pm .005$

$\varnothing 1.750 \pm .002$

$| \perp | \varnothing .001 (M) | A |$

C

$1.187 \pm .005$

$1.624 \, {}^{+.001}_{-.004}$

.250

Hub

Chapter 6 Model Extrusions and Revolutions **149**

Drawing Problems - Chapter 6

8. **Architectural** Create the stairway shown below using the following parameters. Save the file as P6-8.
   A. Use the detail for the riser and tread dimensions.
   B. There are 13 risers.
   C. The stairs are 42″ wide.
   D. The landing at the top of the stairs is 48″ long from the face of the last riser.
   E. The vertical wall is 15′-1″ high on the inside and 17′-7″ long.
   F. The floor is 8′ wide on the inside and 17′-7″ long.
   G. Draw the floor and the wall as 1″ thick.
   H. The center of the banister is 3″ away from the wall and 34″ above the steps.
   I. The ends of the banister are directly above the face of the first and last riser.
   J. Use the detail for the profile of the banister. Use **EXTRUDE** as needed.

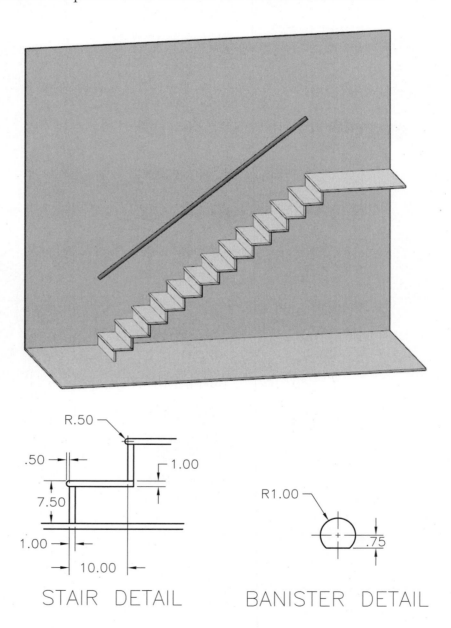

STAIR DETAIL

BANISTER DETAIL

9. **Architectural** In this problem, you will refine the seat of the kitchen chair that you started in Chapter 2. You will use an extrusion following a path to create a curved, receding edge under the seat.

A. Open P2-11 from Chapter 2. If you have not yet completed this model, do so now.

B. Create a path for the extrusion by drawing a polyline that exactly matches either the upper or lower edge of the seat. Refer to the drawing shown.

C. Change the view and UCS as needed to display a plan view of the edge of the seat. Draw the profile shown.

D. Extrude the profile along the path and then subtract the extrusion from the seat.

E. Save the drawing as P6-9.

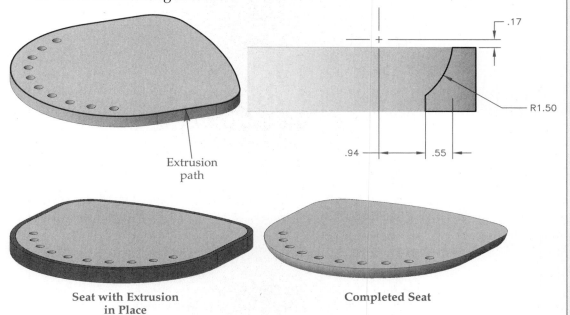

Extrusion
path

.17

R1.50

.94   .55

Seat with Extrusion
in Place

Completed Seat

10. **Architectural** In this problem, you will be taking a manually drawn layout from an archive. You are to create a 3D model of the garage to update the archive.

A. Review the manually drawn layout. Make note of the construction details shown.

B. Research any additional details needed to construct the model. For example, the thickness of the doors is not listed as these are purchased items. However, you need to know these dimensions to draw the 3D model.

C. Using what you have learned, create the garage as a solid model. Be sure to create all components, including the studs in the walls, the footings, and the anchor bolts.

D. Create layers as needed. For example, you may wish to place the wall sheathing on a layer so it can be hidden to show the studs.

E. Save the drawing as P6-10.

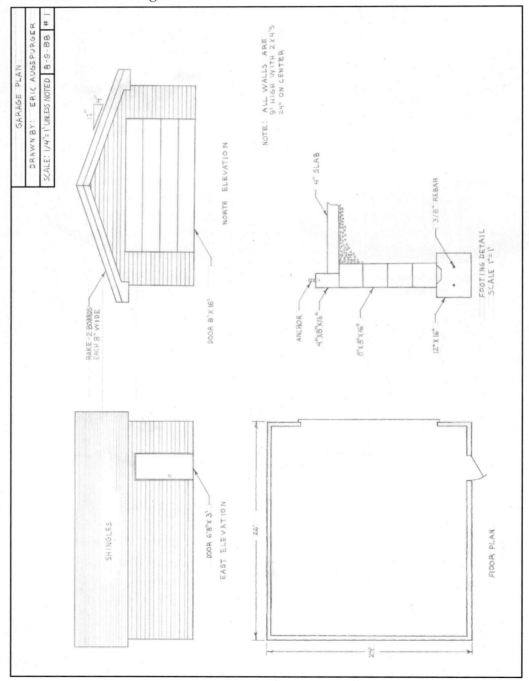

Sweeps and Lofts

Learning Objectives

After completing this chapter, you will be able to:

✓ Sweep shapes along a 2D or 3D path to create a solid or surface object.
✓ Create 3D solid or surface objects by lofting a series of cross sections.

In the previous chapter, you learned about extruded solids and surfaces. Sweeps and lofts are similar to extrusions. In fact, an extrusion is really just a type of sweep. A *sweep* is an object created by extruding a single profile along a path object. Sweeping an open shape along the path results in a surface object. If a closed shape is swept, a solid or surface object can be created. A *loft* is an object created by extruding between two or more 2D profiles. The shape of the loft object blends from one cross-sectional profile to the next. The profiles can control the loft, or the loft can be controlled by one path or multiple guide curves. As with a sweep, open shapes result in surfaces and closed shapes give you solids. Open and closed shapes cannot be used together in the same loft.

Creating Swept Surfaces and Solids

The **SWEEP** command is used to create swept surfaces and solids. The command requires at least two objects:

- Shape to be swept.
- 2D or 3D shape to be used as the sweep path.

The profile can be aligned with the path, you can specify the base point, a scale factor can be applied, and the profile can be twisted as it is swept. The command procedure and options are the same for both swept solids and surfaces.

Sweeping an open shape creates a surface. See **Figure 7-1**. The objects that can be swept to create surfaces include lines, arcs, elliptical arcs, 2D polylines, 2D or 3D splines, and 3D edge subobjects. Sweeping a closed shape creates a solid by default, but surfaces may be created as well. See **Figure 7-2**. Closed shapes that can be swept include circles, ellipses, closed 2D polylines, closed 2D splines, regions, and 3D solid face subobjects. The sweep path can be a line, arc, circle, ellipse, elliptical arc, 2D or 3D polyline, 2D or 3D spline, helix, or 3D edge subobject.

Ribbon

Home
 > Modeling
Solid
 > Solid
Surface
 > Create

Sweep

Type

SWEEP

SWEEP

Figure 7-1.
A—This open shape will be swept along the path (shown in color). B—The resulting surface.

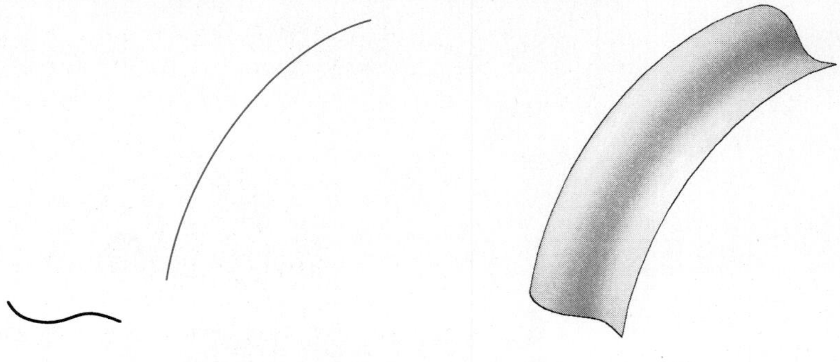

A

B

Figure 7-2.
A—This closed shape will be swept along the path (shown in color). B—The resulting solid.

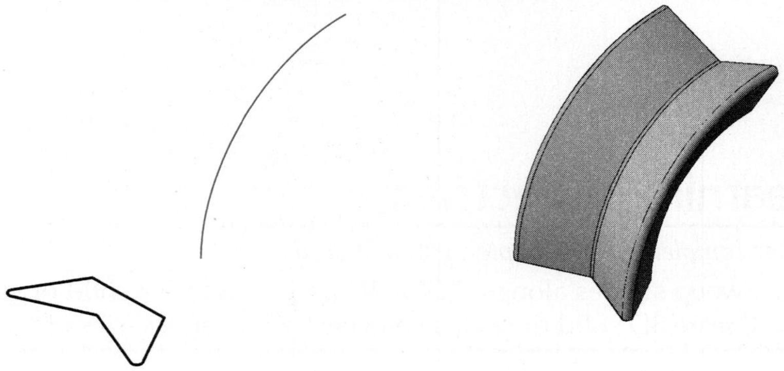

A

B

When the command is initiated, you are prompted to select the objects to sweep. Select the profile(s) and press [Enter]. Planar faces of solids may be selected by holding the [Ctrl] key as you select. Multiple profiles can be selected. They are swept along the same path, but separate objects are created.

Next, you are prompted to select the path. The path and profile can lie on the same plane. Select the object to be used as the sweep path. To select the edge of a surface or solid as the path, press the [Ctrl] key and then select the edge. Once you select the path, the profile is moved to be perpendicular to the path and extruded along it. The sweep starts at the endpoint of the path nearest to where you select it.

The **Mode** option that is available when the **SWEEP** command is initiated controls the closed profiles creation mode. This allows you to change the way in which closed shapes are handled. Normally, the **SWEEP** command produces a solid object when a closed shape is swept, but surfaces may be created by changing this mode. For example, a circle swept with a line as the path will normally create a solid cylinder. Setting the **Mode** option to **Surface** will result in a tube created as a surface model.

Exercise 7-1

www.g-wlearning.com/CAD/

Complete the exercise on the companion website.

Changing the Alignment of the Profile

By default, the profile is aligned perpendicular to the sweep path. However, you can create a sweep where the profile is not perpendicular to the path. See **Figure 7-3**. After the **SWEEP** command is initiated, select the profile and press [Enter]. Before selecting the path, enter the **Alignment** option. The default setting of **Yes** means that

Figure 7-3.
A—The profile and path for the sweep. B—By default, the profile is aligned perpendicular to the path when swept. C—Using the **Alignment** option, the profile can be swept so it is not perpendicular to the path.

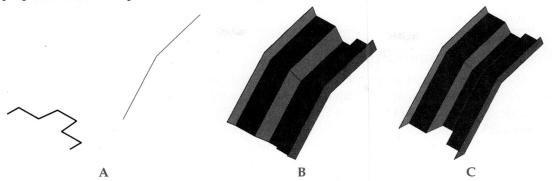

A   B   C

the profile will be moved so it is perpendicular to the path. If you select **No**, the profile is kept in the same position relative to the path as it is swept. The position of the shape determines the alignment. This will not work if the profile and path are coplanar.

Changing the Base Point

The base point is the location on the shape that will be moved along the path to create the sweep. By default, if the shape intersects the path, the profile is swept along the path at the point of intersection. If the shape does not intersect the path, the default base point depends on the type of object being swept. When lines and arcs are swept, the default base point is their midpoint. Open polylines have a default base point at the midpoint of their total length.

The base point can be any point on the shape or anywhere in the drawing. See **Figure 7-4**. To change the base point, use the **Base point** option of the **SWEEP** command. When the command is initiated, select the profile and press [Enter]. Before selecting the path, enter the **Base point** option. Next, pick the new base point. It does not have to be on an existing object. Once the new base point is selected, pick the path to create the sweep.

Figure 7-4.
A—The profile and path for the sweep. B—The sweep is created with the default base point. C—The end of the path is selected as the base point. Notice the difference in this sweep and the one shown in B.

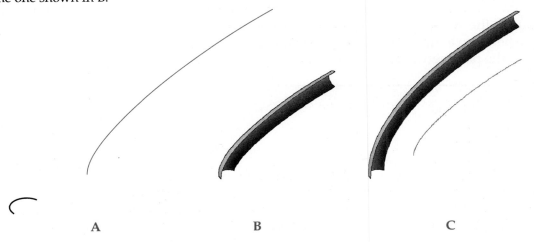

A   B   C

If the location of the base point in relationship to the path is important, line up the shape with the path before starting the **SWEEP** command. Turn off the **Alignment** option in this situation.

Scaling the Sweep Profile

By default, the size of the profile remains uniform from the beginning of the path to the end. However, using the **Scale** option of the **SWEEP** command, you can change the scale of the profile at the end of the path. This, in effect, tapers the sweep. **Figure 7-5** shows a .25 scale applied to a sweep object. A 2D polyline with multiple segments must be edited using the **Fit** or **Spline** option in order to be used as a path. Sharp corners will not work with the **Scale** option. A 3D polyline path must be a spline.

Once the **SWEEP** command is initiated, select the profile and press [Enter]. Before selecting the path, enter the **Scale** option. You are prompted for the scale. Enter the scale value and press [Enter]. The scale value must be greater than zero. You can also enter the **Reference** option. With this option, pick two points for the first reference line and then two points for the second reference line. The difference in scale between the two distances is the scale value. Once the scale is set, pick the path to create the sweep. The **Expression** option allows you to enter a mathematical expression to specify the scaling of the sweep. When used with a surface sweep, it creates a parametric relationship between a dimensional constraint and the swept surface. If the sweep results in a solid, the parametric relationship is not maintained.

Twisting the Sweep

The profile can be rotated as it is swept along the length of the path by using the **Twist** option of the **SWEEP** command. The angle that you enter indicates the rotation of the shape along the path of the sweep. The higher the number, the more twists in the sweep. **Figure 7-6** shows how a simple, closed profile and a straight line can be used to create a milling tool. The profile was swept with a 270° twist.

Once the **SWEEP** command is initiated, select the profile and press [Enter]. Then, before selecting the path, enter the **Twist** option. You are prompted for the twist angle or to enter the **Bank** option.

Banking is the natural rotation of the profile on a 3D sweep path, similar to a banked curve on a racetrack. See **Figure 7-7**. The path must be 3D (nonplanar) to set banking. The banking option is disabled for a 2D path, although you can go through the process of turning it on when creating the sweep. Once you use the **Bank** option to turn banking on, it is on by default the next time the **SWEEP** command is used. To turn it off, enter a twist angle of zero (or the twist angle you wish to use). The **Expression** option allows you to enter a mathematical expression to specify the number of rotations in the sweep. When used with a surface sweep, it creates a parametric relationship between a dimensional constraint and the swept surface. If the sweep results in a solid, the parametric relationship is not maintained.

Figure 7-5.
A—The profile and path for the sweep. B—The resulting sweep. Notice how the .25 scale results in a tapered sweep.

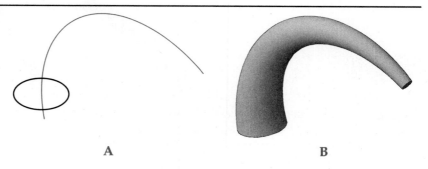

A   B

Figure 7-6.
A—The profile and path for creating the end mill. B—The resulting end mill model. Notice how the profile is twisted (rotated) as it is swept.

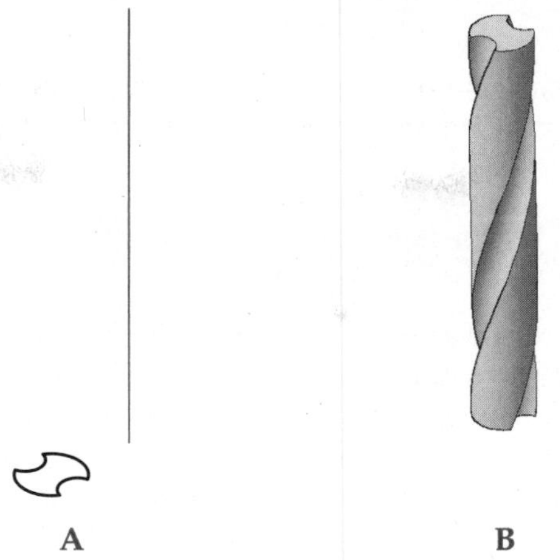

Figure 7-7.
A—The profile and path for the sweep are shown in color. B—Banking is off for this sweep. When viewed from the side, you can see that the profile does not bank through the curve. Look at the upper-right corner. C—Banking is on for this sweep. Notice how the profile banks or leans through the curve. Compare this to B.

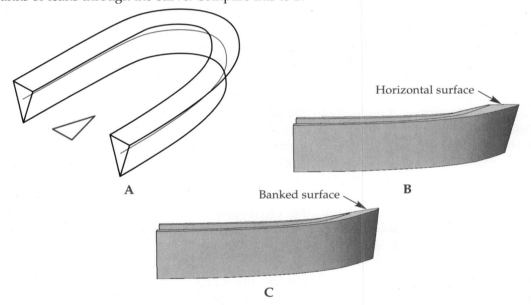

PROFESSIONAL TIP

The **Expression** option available for scaling and twisting a sweep can be used as a simple calculation tool for both swept solids and surfaces. For example, if you want to twist the sweep to an angle one and one half times 45°, select the **Expression** option and enter 1.5*45. The sweep will twist 67.5°.

NOTE

The sweep options can be changed after the sweep is created using the **Properties** palette. In the **Geometry** section, you will find Profile rotation (alignment), Bank (banking), Twist along path (twist angle), and Scale along path (scale) settings.

Creating Lofted Objects

The **LOFT** command is used to create lofted surfaces and solids based on a series of cross-sectional profiles. **Figure 7-8** shows an example of a loft formed from a rectangle, circle, and polygon. The loft may be guided by only the cross sections, as shown in the figure, by a path, or by guide curves. Lofting open shapes results in a surface object, while lofting closed shapes creates a solid. Open and closed shapes cannot be combined in the same loft.

Objects that can be used as cross sections include lines, circles, arcs, points, ellipses, elliptical arcs, 2D polylines, 2D splines, regions, edge subobjects, surfaces, face subobjects, and helices. Points may be used for the first and last cross sections only. The loft path may be a line, circle, arc, ellipse, elliptical arc, spline, helix, 2D or 3D polyline, or edge subobject. Guide curves may be composed of lines, arcs, elliptical arcs, 2D or 3D splines, 2D or 3D polylines, and edge subobjects. However, 2D polylines are limited to only one segment.

Once the command is initiated, you are prompted to select the cross-sectional profiles. Pick each profile in the order in which it should appear in the loft and press [Enter]. Be sure to individually select the cross sections in the order of the loft creation. You may not get the desired loft if you randomly select them or use a window selection. As the profiles are selected, the loft is previewed in a semitransparent state, allowing the user to make adjustments.

The **Mode** option that is available when you start the command controls the closed profiles creation mode and allows you to change the way that closed shapes are handled. It behaves exactly as it does in the **SWEEP** command.

To avoid having to draw point objects for cross sections, you can use the **Point** option to pick any point as either the start point or the end point of the loft. If you pick a point first, it is the start point and you can pick as many shapes as you want as the other cross sections. If you pick the other cross sections first, then the point must be the endpoint of the loft. The other cross sections must be closed shapes. Open shapes will not work with the **Point** option.

If you need to use the edges of existing 3D objects as cross sections, the **Join multiple edges** option works well. The edges must be touching at their end points and form a cross section. See **Figure 7-9**. This option is used to define a single cross section. When you press [Enter], you are again prompted to select cross sections. The option can be used again to select additional cross sections.

Figure 7-8.
A—The three profiles will be lofted to create a solid. B—The resulting loft with the default settings.

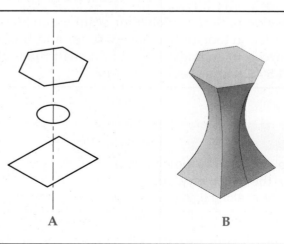

A   B

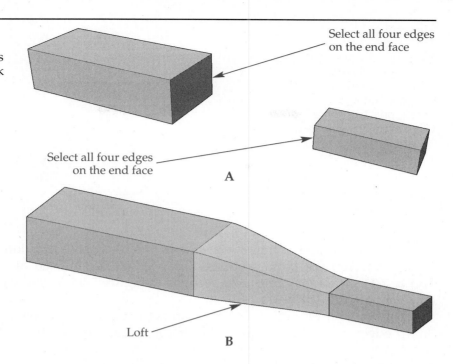

Figure 7-9.
The **Join multiple edges** option allows you to make a quick transition from the edges of one 3D object to another. A—Use the option twice to select two cross sections. B—The resulting loft is a transition between the two objects.

Select all four edges on the end face

Select all four edges on the end face

A

Loft

B

After selecting cross sections, you are prompted to select how the loft is to be controlled. As previously discussed, you can control the loft by the cross sections, a path, or guide curves. These options are discussed in the next sections.

Controlling the Loft with Cross Sections

The **Cross sections only** option of the **LOFT** command is useful when the 2D cross sections are drawn in their proper locations in space. The command determines the transition from one cross section to the next. The cross sections are not moved by the command.

When you select the **Settings** option, the **Loft Settings** dialog box appears, **Figure 7-10.** The settings in this dialog box control the transition or contour between cross sections. As settings are changed, the preview is updated. When all settings have been made, pick the **OK** button to close the dialog box and create the loft.

Figure 7-10.
The **Loft Settings** dialog box is used to control the transition between profiles.

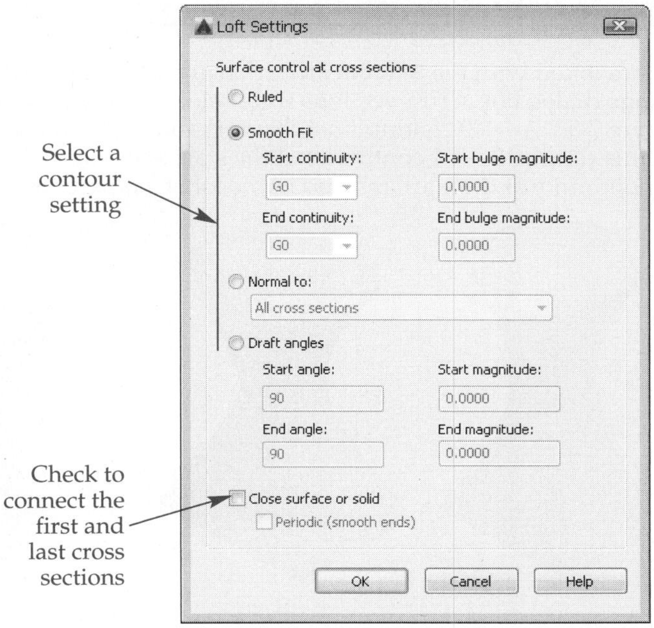

Select a contour setting

Check to connect the first and last cross sections

Loft Settings

Surface control at cross sections

○ Ruled
● Smooth Fit

Start continuity: G0
Start bulge magnitude: 0.0000
End continuity: G0
End bulge magnitude: 0.0000

○ Normal to:
All cross sections

○ Draft angles

Start angle: 90
Start magnitude: 0.0000
End angle: 90
End magnitude: 0.0000

☐ Close surface or solid
☐ Periodic (smooth ends)

OK   Cancel   Help

Figure 7-11.
The profiles in Figure 7-8A are lofted with the **Ruled** option selected in the **Loft Settings** dialog box. Compare this to Figure 7-8B.

When the **Ruled** option is selected in the dialog box, the loft has straight transitions between the cross sections. Sharp edges are created at each cross section. **Figure 7-11** shows the same cross sections in **Figure 7-8A** lofted with the **Ruled** option on. Compare this to **Figure 7-8B**.

The **Smooth Fit** option creates a smooth transition between the cross sections. Sharp edges are only created at the first and last cross sections. This is the default setting and the one used to create the loft shown in **Figure 7-8B**. The **Start continuity:** and **End continuity:** settings control the tangency and curvature of the first and last cross sections. The **Start bulge magnitude:** and **End bulge magnitude:** settings control the size of the curve at the first and last cross sections. These options only apply if the first and/or last cross sections are regions. See **Figure 7-12**. The start and end continuity can be set to G0, G1, or G2. The G0 (positional continuity) setting creates a sharp edge. The loft transitions precisely at the location of the profile. The G1 (tangential continuity) setting is the default. It creates the largest bulge radius. The G2 (curvature or continuous continuity) setting creates a smaller bulge radius. For all three settings, bulge magnitude is initially set to 0.5. Altering the value will make the bulge either larger or smaller but has no effect when continuity is set to G0.

When the **Normal to:** option is selected in the **Loft Settings** dialog box, you can choose how the normal of the transition is treated at the cross sections. A *normal* is a vector extending perpendicular to the cross section. When the transition is normal to a cross section, it is perpendicular to the cross section. You can set the transition normal

Figure 7-12.
Region profiles are lofted with the **Smooth Fit** option and different continuity settings selected in the **Loft Settings** dialog box. Cross sections were selected from bottom to top. Compare these results with Figure 7-8B. A—G0 continuity at the start and end of the loft. Bulge magnitude has no effect at this setting. B—G1 continuity at the start and end of the loft. This is the default setting. C—G2 continuity at the start and end of the loft. Compare the bulge radius to B.

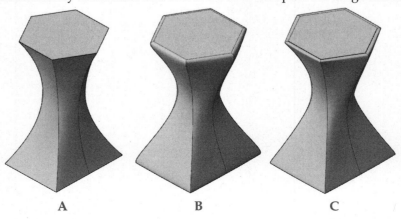

A   B   C

Figure 7-13.
The profiles in Figure 7-8A are lofted with the **Normal to:** option on in the **Loft Settings** dialog box. Cross sections were selected from bottom to top. Compare these results with Figure 7-8B and Figure 7-12. A—Start cross section. B—End cross section. C—Start and End cross sections. D—All cross sections.

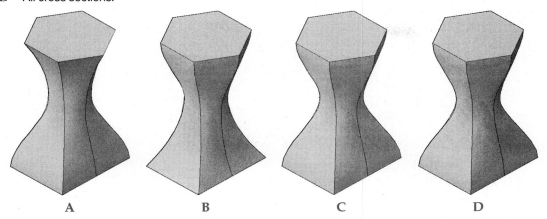

to the first cross section, last cross section, both first and last cross sections, or all cross sections. See **Figure 7-13**. Select the normal setting in the drop-down list. You will have to experiment with these settings to get the desired loft shape.

In manufacturing, plastic or metal parts are sometimes formed in a two-part mold. A slight angle is designed into the parts on the inside and outside surfaces to make removing the part from the mold easier. This taper is called a *draft angle*. The **Draft angles** option allows you to add a taper to the beginning and end of the loft.

When setting the draft angle, you can set the angle and magnitude. See **Figure 7-14**. The default draft angle is 90°, which means the transition is perpendicular to the cross section. The magnitude represents the relative distance from the cross section, in the same direction as the draft angle, before the transition starts to curve toward the next cross section. Magnitude settings depend on the size of the cross sections, the draft angle values, and the distance between the cross sections. You may have to experiment with different magnitude and angle settings to get the desired loft shape.

The **Close surface or solid** option is used to connect the last cross section to the first cross section. See **Figure 7-15**. This option "closes" the loft, similar to the **Close** option of the **LINE** or **PLINE** command. The shapes from **Figure 7-8** are shown in **Figure 7-16** with the **Close surface or solid** option on. This option is only available when the **Ruled** or **Smooth Fit** option is selected.

Figure 7-14.
When setting the draft angle, you can set the angle and the magnitude. A—Draft angle of 90° and a magnitude of zero. B—Draft angle of 30° and a magnitude of 180. C—Draft angle of 60° and a magnitude of 180.

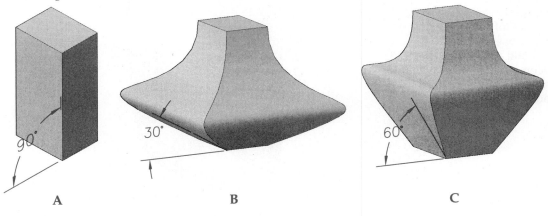

Figure 7-15.
A—These profiles will be used to create a sealing ring. They should be selected in a counterclockwise direction starting with the first cross section. B—The resulting loft with the default settings. Notice the gap between the first and last cross sections. C—By checking the **Close surface or solid** check box in the **Loft Settings** dialog box, the loft continues from the last cross section to the first cross section.

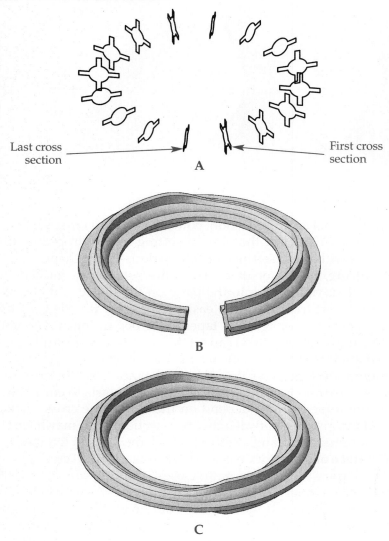

Figure 7-16.
The same cross sections from Figure 7-8 are used in this loft. However, the **Close surface or solid** option has been applied. Notice how the loft is inside out.

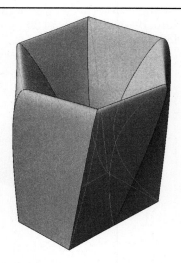

If the **Smooth Fit** option is selected and the **Close surface or solid** check box is checked, the **Periodic (smooth ends)** check box is available. If a closed-loop loft is created similar to the loft in **Figure 7-15**, the seam may kink if the loft is reshaped in some way. Checking the **Periodic (smooth ends)** check box will help alleviate this problem.

PROFESSIONAL TIP

It may be easier to experiment with the loft settings after the loft is created. Select the loft and change the settings in the **Properties** palette. In addition, special grips and handles appear at the locations of the profiles. Picking on a grip will open a shortcut menu and dragging a handle will alter the loft dynamically.

Exercise 7-3

www.g-wlearning.com/CAD/

Complete the exercise on the companion website.

Controlling the Loft with Guide Curves

Guide curves are lines that control the shape of the transition between cross sections. They do not have to be *curves*. They can be lines, arcs, elliptical arcs, splines (2D or 3D), polylines (2D or 3D), or edge subobjects. There are four rules to follow when using guide curves:
- The guide curve should start on the first cross section.
- The guide curve should end on the last cross section.
- The guide curve should intersect all other cross sections.
- The surface control in the **Loft Settings** dialog box must be set to **Smooth Fit** (**LOFTNORMALS** = 1).

When the **Guides** option of the **LOFT** command is entered, you are prompted to select the guide curves. Select all of the guide curves and press [Enter]. The loft is created. The order in which guide curves are selected is not important.

For example, **Figure 7-17A** shows two circles that will be lofted. If the **Cross sections only** option is used, a cylinder is created, **Figure 7-17B**. However, if the **Guides** option is used and the two guide curves shown in **Figure 7-17A** are selected, one side of the cylinder is deformed similar to a handle or grip. See **Figure 7-17C**.

Figure 7-17.
A—These two circles will be lofted. The objects shown in color will be used as guide curves. B—When the circles are lofted using the **Cross sections only** option, a cylinder is created. C—When the **Guides** option is used and the guide curves shown in A are selected, the resulting loft is shaped like a handle or grip.

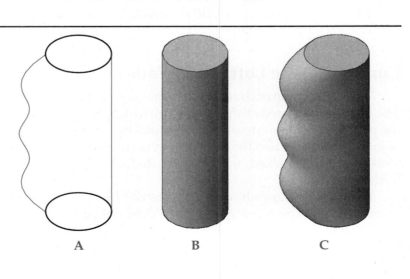

A   B   C

Figure 7-18.
A—The open profiles shown in black and the guide curve shown in color will be used to create a fabric covering for the three solid objects. B—The resulting fabric covering. This is a surface because the profiles were open.

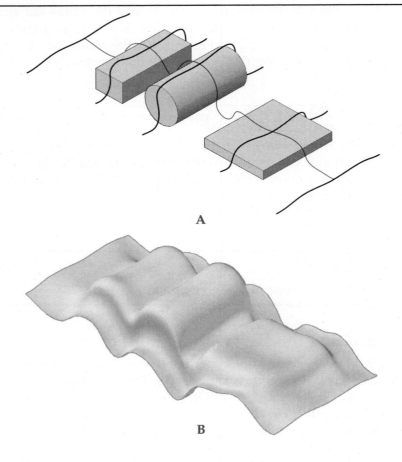

A

B

Lofting is used to create open-contour shapes such as fenders, automobile interior parts, fabrics, and other ergonomic consumer products. **Figure 7-18** shows the use of open 2D splines in the construction of a fabric covering. Notice how each cross section is intersected by the guide curve. There is a cross section at the beginning of the guide curve and one at the end. These conditions fulfill the rules outlined earlier.

CAUTION

Guide curves only work well when the surface control is set to **Smooth Fit** (**LOFTNORMALS** = 1). If you get an error message when using guide curves or the curves are not reflected in the end result, make sure **LOFTNORMALS** is set to 1 and try it again.

Controlling the Loft with a Path

The **Path** option of the **LOFT** command places the cross sections along a single path. The path must intersect the planes on which each of the cross sections lie. However, the path does *not* have to physically touch the edge of each cross section, as is required of guide curves. When the **Path** option is entered, you are prompted to select the path. Once the path is picked, the loft is created. The cross sections remain in their original positions.

Figure 7-19 shows how 2D shapes can be positioned at various points on a path to create a loft. The rectangular shape does not cross the path. However, as long as the path intersects the plane of the rectangle, which it does, the shape will be included in the loft definition. The last shape at the top of the helix is a point object, causing the loft to taper.

Figure 7-19.
A—The profiles shown in black will be lofted along the path shown in color. Notice how the rectangular profile is not intersected by the path, but the path does intersect the plane on which the rectangle lies. B—The resulting loft.

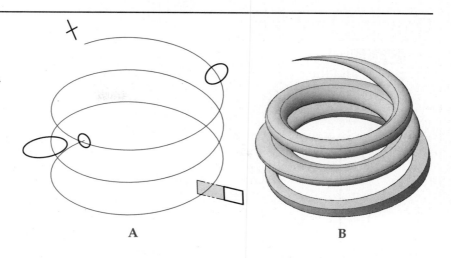

A

B

Exercise 7-4

Complete the exercise on the companion website.

www.g-wlearning.com/CAD/

Chapter Review

Answer the following questions using the information in this chapter.

1. What is a *loft*?
2. What option of the **SWEEP** command determines whether the sweep will be a solid or a surface?
3. When using the **SWEEP** command, on which endpoint of the path does the sweep start?
4. What is the purpose of the **Base point** option of the **SWEEP** command?
5. After the sweep or loft is created, how may the creation options be changed?
6. Which objects may be used as a sweep path?
7. How is the alignment of a sweep set to be perpendicular to the start of the path?
8. Which **SWEEP** command option is used to taper the sweep?
9. What does the **Bank** option of the **SWEEP** command do?
10. What is the difference between the **Ruled** and **Smooth Fit** options in the **LOFT** command?
11. How can you close a loft?
12. List five objects that may be used as guide curves in a loft.
13. What are the four rules that must be followed when using guide curves?
14. When using the **Path** option of the **LOFT** command, what must the path intersect?
15. How can a loft be created so it tapers to a point at its end?

Drawing Problems

1. **General** Create as a sweep the wedding ring shown. Use the half ellipse as the profile and the circle as the sweep path. Save the drawing as P7-1.

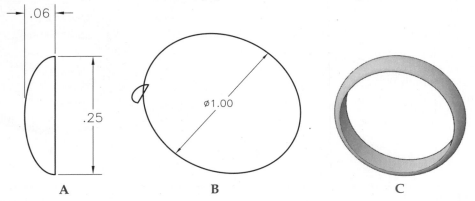

A   B   C

2. **General** Create the aluminum can shown by lofting the circular cross sections. There are nine circular cross sections centered along the Z axis. Select the cross sections in order, starting with the Ø2 circle at the bottom and proceeding to the top. The four Ø2 and Ø1.94 circles at the top create the folded lip. Select the four circles in the appropriate order. The last circle to select is the Ø1.75 circle, which is below the four Ø2 and Ø1.94 circles and creates the recess in the top. Save the drawing as P7-2.

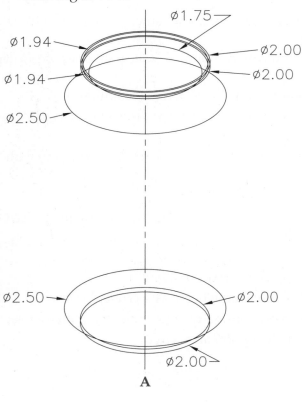

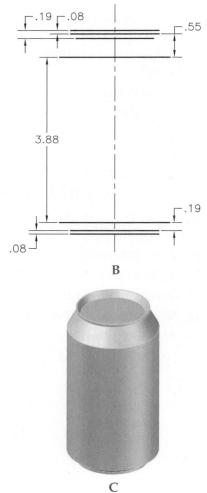

3. **General** Create the two shampoo bottles shown. One design uses cross sections only and the other uses a guide curve. Each bottle is made up of two loft objects. Join the pieces so each bottle is one solid. Save the drawing as P7-3.

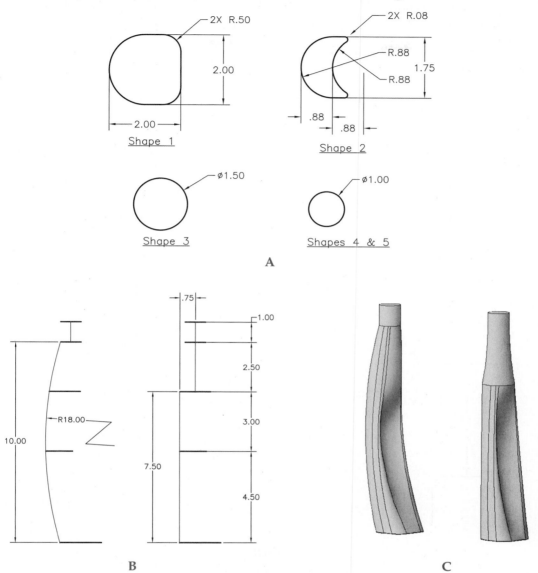

4. **Mechanical** Create as a loft the automobile fender shown below. Use the object shown in color as a guide curve or a path. Save the drawing as P7-4.

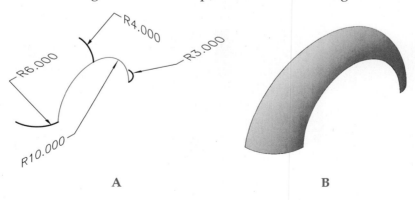

5. **Mechanical** Draw as a loft the C-clamp shown. Use the shapes (A, B, C, and D) as the cross sections and the polyline (in color) as the guide curve. Add ∅1 unit cylinders to the ends. Make one cylinder .125H and the other 1.125H. The cylinders should be centered on profile D and located at the ends of the loft as shown. Make a ∅.625 hole through the larger cylinder. Save the drawing as P7-5.

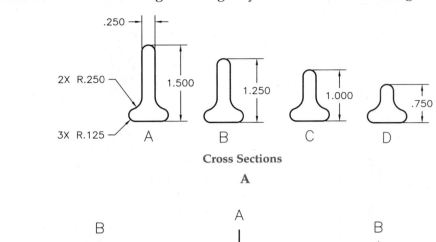

Cross Sections

A

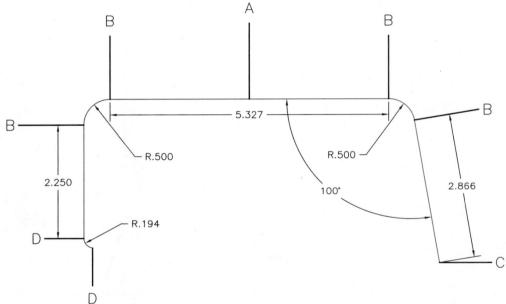

Layout

B

C

6. **Architectural** Create the lamp shade shown. Create two separate loft objects for the top and the bottom. Then, union the two pieces. Finally, scale a copy and hollow out the lamp shade. Save the drawing as P7-6.

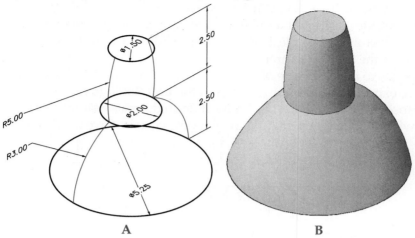

A B

7. **General** In this problem, you will draw a racetrack for toy cars by sweeping a 2D shape along a polyline path.
   A. Draw the polyline path shown with the coordinates given. Turn it into a spline.
   B. Draw the 2D profile shown using the dimensions given. Turn it into a region or a polyline.
   C. Use the **SWEEP** command to create the racetrack, as shown in the shaded view.
   D. You may have to use the **Properties** palette to adjust the sweep after it is drawn.
   E. Save the drawing as P7-7.

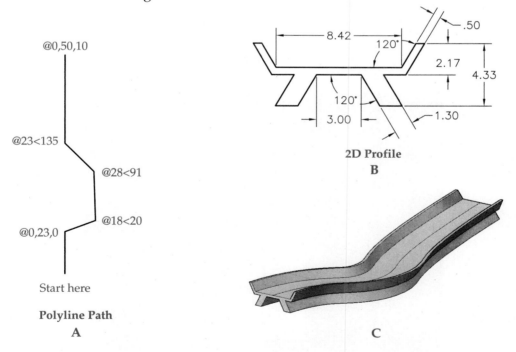

2D Profile
B

Polyline Path
A

C

8. **Architectural** Use the **LOFT** command to create a window curtain. Create a spline or a polyline converted into a spline similar to the top profile object (A). Position the spline at the top of the window on the left side. Copy the spline and position it about halfway down the window. Use the **SCALE** command to make the copied spline smaller but keep the outside edge lined up with the top spline. You may want to use grips to edit the spline and make the pleats deeper after scaling. Copy the smaller spline to the bottom near the floor. The three splines will be used for the first loft object. Mirror the three splines to the other side of the window for the second loft object. When creating each loft, experiment with the settings in the **Loft Settings** dialog box to change the appearance of the curtain to your liking. If desired, create a simple 3D model of a room with a window opening including the window curtain. Save the drawing as P7-8.

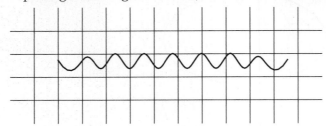

**Top Profile
Original Spline**

A

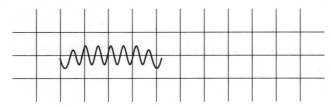

**Middle Profile and
Bottom Profile**

B

**Completed Room
with Window Curtain**

C

9. **Architectural** Create the furniture leg shown. Either the **LOFT** command or the **SWEEP** command may be used. However, one command may work better than the other. The profile dimensions refer to the bottom end (small end) of the leg. The top is twice the size of the bottom. Note that the top and the bottom of the leg are parallel. Save the drawing as P7-9.

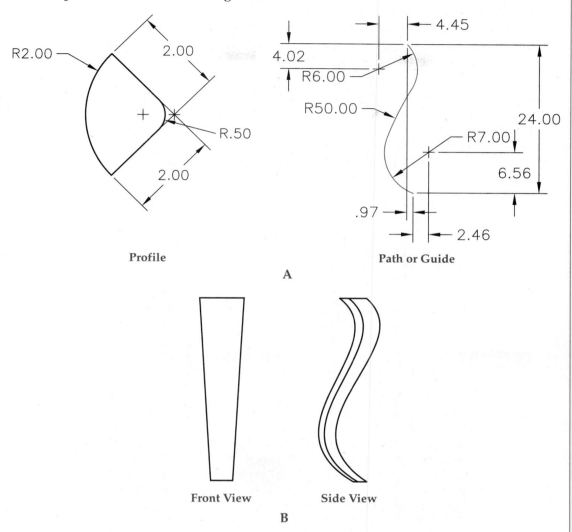

Profile

Path or Guide

A

Front View   Side View

B

C

10. **Mechanical** In this problem, you will cut a UNC thread in a cylinder by sweeping a 2D shape around a helix and subtracting it.
A. Draw a ∅.25 cylinder that is 1.00 in height.
B. Draw the thread cutter profile shown below. The long edge of the cutter should be aligned with the vertical edge of the cylinder.
C. Draw a helix centered on the cylinder with base and top radii of .125, a turn height of .050, and a total height of 1.000.
D. Sweep the 2D shape along the helix. Then, subtract the resulting solid from the cylinder. Refer to the shaded view.
E. If time allows, create another cutter profile to cut a .0313 × 45° chamfer on the end of the thread. Use a circle as a sweep path or revolve the profile about the center of the cylinder.
F. Save the drawing as P7-10.

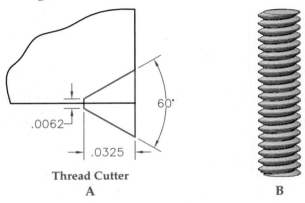

Thread Cutter
A

B

11. **Architectural** In this problem, you will add a seatback to the kitchen chair you started modeling in Chapter 2. In Chapter 6, you refined the seat.
A. Open P6-9 from Chapter 6.
B. Draw an arc for the top of the bow. Using a 14.25″ length of the arc, divide it into seven equal segments.
C. Position eight ∅.50 circles at the division points.
D. Draw circles at the top of each hole in the seat.
E. Create a loft between each lower circle and each upper circle.
F. Using the information in the drawings, create the outer bow for the seatback.
G. Save the drawing as P7-11.

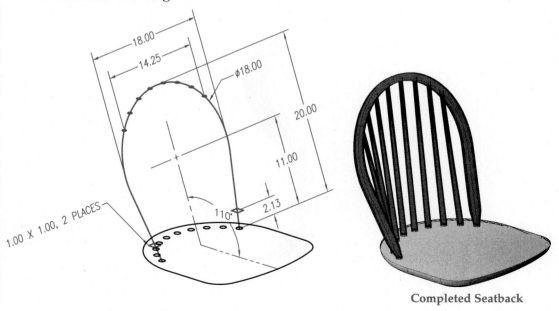

Completed Seatback

CHAPTER

Creating and Working with Solid Model Features

Learning Objectives

After completing this chapter, you will be able to:

✓ Change properties on solids.
✓ Align objects.
✓ Rotate objects in three dimensions.
✓ Mirror objects in three dimensions.
✓ Create 3D arrays.
✓ Fillet solid objects.
✓ Chamfer solid objects.
✓ Slice a solid using various methods.
✓ Construct features on solid models.
✓ Remove features from solid models.

Changing Properties Using the Properties Palette

Properties of 3D objects can be modified using the **Properties** palette, which is thoroughly discussed in *AutoCAD and Its Applications—Basics*. This palette is displayed using the **PROPERTIES** command. You can also select a solid object, right-click, and select **Properties** from the shortcut menu.

The **Properties** palette lists the properties of the currently selected object. For example, **Figure 8-1** lists the properties of a selected solid sphere. Some of its properties are parameters, which is why solid modeling in AutoCAD can be considered parametric. You can change the sphere's radius; diameter; X, Y, Z coordinates; linetype; linetype scale; color; layer; lineweight; and visual settings. The categories and properties available in the **Properties** palette depend on the selected object.

To modify an object property, select the property. Then, enter a new value in the right-hand column. The drawing is updated to reflect the changes. You can leave the **Properties** palette open as you continue with your work.

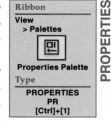

PROPERTIES

Ribbon
View
  > Palettes

Properties Palette
Type
  PROPERTIES
  PR
  [Ctrl]+[1]

Figure 8-1.
The **Properties** palette can be used to change many of the properties of a solid.

Type of object selected

Category

Properties within the category

Selected property to modify

History settings

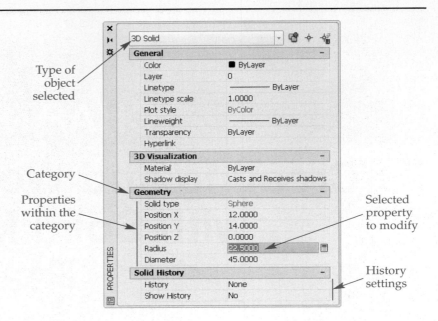

Solid Model History

AutoCAD can automatically record a history of a composite solid model's construction. A *composite solid* is created by a Boolean operation or by using the **SOLIDEDIT** command. The retention of a composite solid's history is controlled by the History property setting in the **Solid History** category of the **Properties** palette. By default, the History property is set to None. Refer to **Figure 8-1**. This means that the history is not saved and that edges, vertices, and faces of solids can be directly edited. If the History property in the **Properties** palette is set to Record, the solid history is "recorded," or retained. This allows you to work with the geometry used in modeling operations to create the composite solid. Depending on your modeling preferences, it is generally a good idea to have the history recorded. Then, at any time, you can graphically display all of the geometry that was used to create the model.

The retention of solid model history for newly created composite solids is controlled by the **SOLIDHIST** system variable. By default, the **SOLIDHIST** system variable is set to 0. This means that all new solids have their History property set to None and solid history is not recorded. If the **SOLIDHIST** system variable is set to a value of 1, all new solids have their History property set to Record and solid history is preserved. With either setting of the system variable, 0 (None) or 1 (Record), the **Properties** palette can be used to change the setting for individual solids. The current setting of the **SOLIDHIST** system variable is indicated by the **Solid History** button located in the **Primitive** panel of the **Solid** tab on the ribbon. See **Figure 8-2**. With a default setting of 0 (None), the **Solid History** button has a white background. Picking this button when it has a white background changes the background color to blue and sets the **SOLIDHIST** system variable to 1 (Record).

To view the graphic history of a composite solid, set the Show History property in the **Properties** palette to Yes. All of the geometry used to construct the model is displayed. If the **SHOWHIST** system variable is set to 0, the Show History property is set to No for all solids and cannot be changed. If this system variable is set to 2, the Show History property is set to Yes for all solids and cannot be changed. A **SHOWHIST** setting of 1 allows the Show History property to be individually set for each solid. This is the default setting.

An example of showing the history on a composite solid is provided in **Figure 8-3**. In **Figure 8-3A**, the model appears in its current edited format. The History property setting

Figure 8-2.
The **Solid History** button, located in the **Primitive** panel of the **Solid** tab on the ribbon, is used to control the **SOLIDHIST** system variable setting. This button has a white background when the **SOLIDHIST** system variable is set to 0 (the default value). If picked, the button has a blue background.

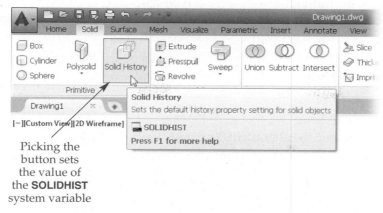

Picking the button sets the value of the **SOLIDHIST** system variable

Figure 8-3.
A—The object appears in its current state with the Show History property turned off. The History property was initially set to Record before subtracting two solid primitives. B—The Show History property is set to Yes and the display of isolines has been turned on. C—When the sphere is selected while pressing the [Ctrl] key, grips are displayed to indicate the sphere primitive subobject is selected. The move grip tool can be used to edit the original sphere primitive.

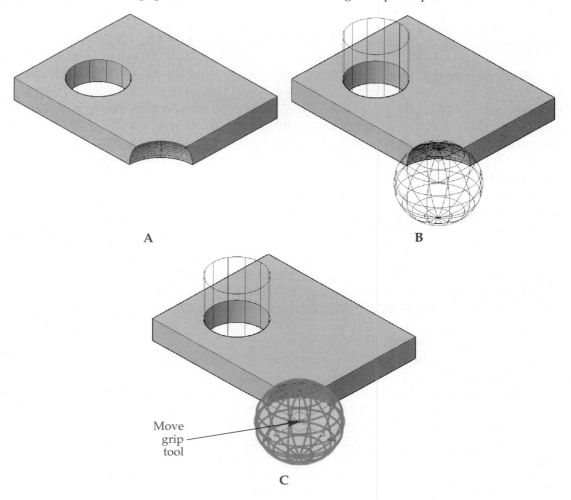

A

B

Move grip tool

C

was initially set to Record before performing two Boolean subtraction operations. The Show History property is currently set to No and the Conceptual visual style is set current. In **Figure 8-3B**, the Show History property is set to Yes. Isolines have also been turned on. You can see the geometry that was used in the Boolean subtraction operations. Using subobject editing techniques, the individual geometry can be selected and edited. In **Figure 8-3C**, the sphere (the subtracted object) is selected while pressing the [Ctrl] key. Selecting the sphere in this manner selects the solid primitive *subobject* and displays a grip tool for editing the subobject. Using grip tools for editing 3D objects is introduced later in this chapter. Subobject editing is discussed in detail in Chapter 11.

NOTE

It is good practice to set the **SOLIDHIST** system variable to 1 (Record) before starting all 3D modeling work. This ensures that you will be able to edit your 3D solid models if the need arises.

PROFESSIONAL TIP

If the Show History property is set to Yes to display the components of the composite solid, as seen in **Figure 8-3**, the components will appear when the drawing is plotted. Be sure to set the Show History property to No before you print or plot.

Selection Cycling

When selecting objects that are on top of each other or occupy the same space, *selection cycling* is the preferred method to select one of the objects. When editing, you may need to erase, move, or copy one of the objects that overlap. Press the [Ctrl]+[W] key combination to turn on selection cycling. You can then select the object needed from the **Selection** dialog box. When you need to cycle through objects at a pick point:

1. At the "select objects" prompt, press the [Ctrl]+[W] key combination. AutoCAD displays the message <Selection Cycling on>. Pick to select the object you want.
2. When the **Selection** dialog box appears, select the desired object from the list.
3. Press the [Enter] key.

You can use the **SELECTIONCYCLING** system variable to turn on selection cycling instead of using the [Ctrl]+[W] key combination. The **Selection Cycling** button on the status bar is used for toggling selection cycling. The **Selection Cycling** button does not appear on the status bar by default. Select the **Selection Cycling** option in the status bar **Customization** flyout to add the **Selection Cycling** button to the status bar. There are three settings for the **SELECTIONCYCLING** system variable:

- Off (0).
- On, but the list dialog box does not display (1).
- On and the list dialog box displays the selected objects that can be cycled through (2).

It is recommended that you turn on selection cycling. With selection cycling turned on, you can cycle and select the faces that may overlap one another.

In **Figure 8-4**, the two 3D objects occupy the same space. The tapered 3D object needs to be moved up using the **3DMOVE** command. Turn on selection cycling. When you select the tapered object, the **Selection** dialog box lists the objects overlapping at the pick point. Select which object you want to work with by picking it in the dialog box. As your cursor is over an object in the list, the object is highlighted in the drawing area.

Type
SELECTIONCYCLING
Toolbar
Status Bar
Selection Cycling

SELECTIONCYCLING

Figure 8-4.
Using selection cycling to choose which object to select.

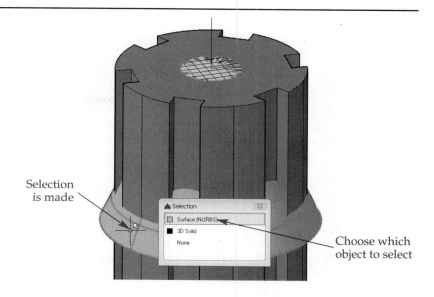

Selection is made

Selection
☐ Surface (NURBS)
■ 3D Solid
None

Choose which object to select

NOTE

You can also press the [Shift] key and space bar at the same time to cycle through overlapping objects.

PROFESSIONAL TIP

If multiple 3D objects occupy the same space, selection cycling gives you the ability to select the correct object needed to complete the edit.

Aligning Objects in 3D

AutoCAD provides two different methods with which to move and rotate objects in a single command. This is called *aligning* objects. The simplest method is to align 3D objects by picking source points on the first object and then picking destination points on the object to which the first one is to be aligned. This is accomplished with the **3DALIGN** command, which allows you to both relocate and rotate the object. The second method is possible with the **ALIGN** command. The **ALIGN** command aligns 2D or 3D objects according to three sets of alignment pairs. You first select the source object, then the first source point, and finally the first destination point on the destination object. This technique is repeated two more times to align one object to another. This method of aligning allows you to not only move and rotate an object, but scale the object being aligned.

PROFESSIONAL TIP

Use the **3DALIGN** command to create assemblies.

Move and Rotate Objects in 3D Space

3DALIGN

Ribbon

Home
> **Modify**

3D Align

Type

3DALIGN
3AL

The basic operation of moving and rotating an object relative to a second object or set of points is done with the **3DALIGN** command. It allows you to reorient an object in 3D space. Using this command, you can correct errors of 3D construction and quickly manipulate 3D objects. The **3DALIGN** command requires existing points (source) and the new location of those existing points (destination).

For example, refer to **Figure 8-5**. The wedge in **Figure 8-5A** is aligned in its new position in **Figure 8-5B** as follows. Set the **Intersection** or **Endpoint** running object snap to make point selection easier. Refer to the figure for the pick points.

Select objects: *(pick the wedge)*
1 found
Select objects: ↵
   Specify source plane and orientation…
Specify base point or [Copy]: *(pick P1)*
Specify second point or [Continue] <C>: *(pick P2)*
Specify third point or [Continue] <C>: *(pick P3)*
   Specify destination plane and orientation…
Specify first destination point: *(pick P4)*
Specify second destination point or [eXit] <X>: *(pick P5)*
Specify third destination point or [eXit] <X>: *(pick P6)*

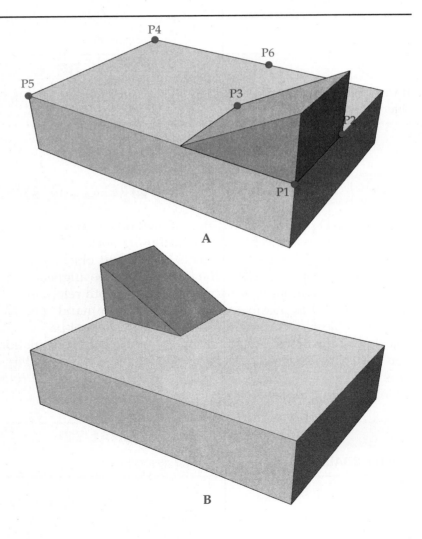

Figure 8-5.
The **3DALIGN** command can be used to properly orient 3D objects. A—Before aligning. Note the pick points. B—After aligning.

You can also use the **3DALIGN** command to align cylindrical 3D objects. The procedure is similar to aligning planar objects, which require three pick points. However, you only need two pick points per object.

For example, the socket head cap screw in **Figure 8-6** is aligned to a new position using the **3DALIGN** command. Set the **Center** running object snap to make point selection easier. Refer to the figure for the pick points.

> Select objects: *(pick the socket head cap screw)*
> 1 found
> Select objects: ↵
>   Specify source plane and orientation…
> Specify base point or [Copy]: *(pick P1)*
> Specify second point or [Continue] <C>: *(pick P2)*
> Specify third point or [Continue] <C>: ↵
>   Specify destination plane and orientation…
> Specify first destination point: *(pick P3)*
> Specify second destination point or [eXit] <X>: *(pick P4)*
> Specify third destination point or [eXit] <X>: ↵

The idea with this command is you are aligning a plane defined by three points with another plane defined by three points. The planes do not need to correspond to actual planar faces. In the example shown in **Figure 8-5**, the alignment planes coincide with planar faces. However, in the example shown in **Figure 8-6**, the alignment planes define planes on which axes lie.

PROFESSIONAL TIP

You can also use the **3DALIGN** command to copy an object, rather than move it, and realign it at the same time. Just select the **Copy** option at the Specify base point or [Copy]: prompt. Then, continue selecting the points as previously discussed.

Figure 8-6.
Using the **3DALIGN** command to align cylindrical objects. A—Before aligning. Note the pick points. B—After aligning.

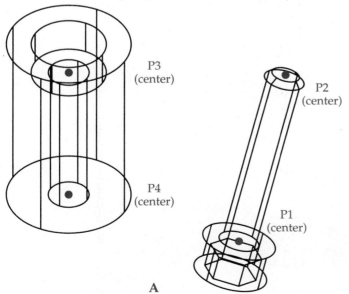

Complete the exercise on the companion website.

Move, Rotate, and Scale Objects in 3D Space

ALIGN

Ribbon

Home
> Modify

Align

Type

ALIGN
AL

The **ALIGN** command has the same functions of the **3DALIGN** command, but adds the ability to scale an object. See **Figure 8-7**. The 90° bend must be rotated and scaled to fit onto the end of the HVAC assembly. Two source points and two destination points are required, **Figure 8-7A**. Then, you can choose to scale the object.

Select objects: (*pick the 90° bend*)
1 found
Select objects: ↵
Specify first source point: (*pick P1*)
Specify first destination point: (*pick P2; a line is drawn between the two points*)
Specify second source point: (*pick P3*)
Specify second destination point: (*pick P4; a line is drawn between the two points*)
Specify third source point or <continue>: ↵
Scale objects based on alignment points? [Yes/No] <N>: Y↵

The 90° bend is aligned and uniformly scaled to meet the existing ductwork object. See **Figure 8-7B**. You can also align using three source and three destination points. However, when doing so, you cannot scale the object.

PROFESSIONAL TIP

Before using 3D editing commands, set running object snaps to enhance your accuracy and speed.

Figure 8-7.
Using the **ALIGN** command. A—Two source points and two destination points are required. Notice how the bend is not at the proper scale. B—You can choose to scale the object during the operation. Notice how the aligned bend is also properly scaled.

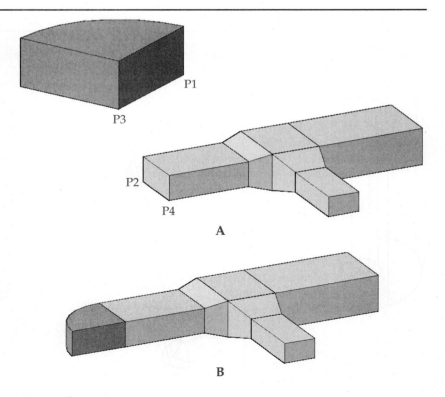

A

B

Exercise 8-2

www.g-wlearning.com/CAD/

Complete the exercise on the companion website.

Move, Rotate, and Scale Gizmos

A *gizmo*, also called a *grip tool*, appears when an object or a subobject is selected. This tool is used to specify how the transformation (movement, rotation, or scaling) is applied. Using gizmos (grip tools) is discussed in detail in Chapter 11.

The move gizmo allows movement of an object or a subobject along the X, Y, or Z axis or on the XY, XZ, or YZ plane. The rotate gizmo allows rotation about the X, Y, or Z axis. The scale gizmo allows scaling along the X, Y, or Z axis or XY, XZ, or YZ plane. You can set the default gizmo displayed when an object is selected by using the gizmo flyout in the **Selection** panel of the **Solid** ribbon tab. You can also use the **Show gizmos** flyout on the status bar. The **Show gizmos** flyout does not appear on the status bar by default. Select the **Gizmo** option in the status bar **Customization** flyout to add the **Show gizmos** flyout to the status bar.

The move gizmo is displayed by default. To use the move gizmo, pick any axis to move along that axis. See **Figure 8-8**. To move along a plane, pick the rectangular area at the intersection of the axes. To use the rotate gizmo, pick the circle with the center about which you wish to rotate the selection. To use the scale gizmo, pick an axis to scale along that axis. Pick the triangular area at the intersection of the axes to uniformly scale the selection. Pick the area between the inner and outer triangular areas to nonuniformly scale along that plane. Using gizmos to move and rotate objects is discussed in the following sections.

Ribbon

Solid
> Selection

Move Gizmo

Rotate Gizmo

Scale Gizmo

No Gizmo

Type
DEFAULTGIZMO

DEFAULTGIZMO

NOTE

Mesh objects, discussed in Chapter 9, can be nonuniformly scaled along an axis or plane. The scale gizmo cannot be used to nonuniformly scale solid or surface objects.

Figure 8-8.
Gizmos are used to transform (move, rotate, or scale) objects or subobjects. A—Move gizmo. B—Rotate gizmo. C—Scale gizmo.

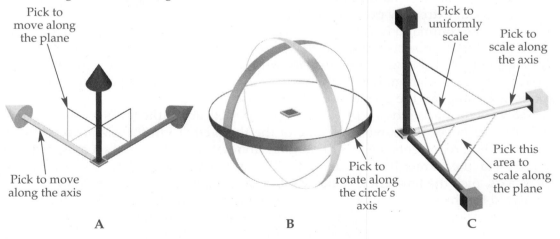

Pick to move along the plane

Pick to move along the axis

Pick to rotate along the circle's axis

Pick to uniformly scale

Pick to scale along the axis

Pick this area to scale along the plane

A

B

C

3D Moving

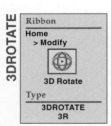

3DMOVE

Ribbon
Home
> Modify

3D Move

Type
3DMOVE
3M

The **3DMOVE** command allows you to quickly move an object along any axis or plane of the current UCS. When the command is initiated, you are prompted to select the objects to move. If the 2D Wireframe visual style is current, the visual style is temporarily changed to the Wireframe visual style because the gizmo is not displayed in 2D mode. After selecting the objects, press [Enter]. The move gizmo is displayed in the center of the selection set. By default, the move gizmo is also displayed when a solid is selected with no command active.

The move gizmo is a tripod that appears similar to the shaded UCS icon. Refer to **Figure 8-8A**. You can relocate the tool by right-clicking on the gizmo and selecting **Relocate Gizmo** from the shortcut menu. Then, move the gizmo to a new location and pick. You can also realign the gizmo using the shortcut menu.

If you move the pointer over the X, Y, or Z axis of the gizmo, the axis changes to yellow. To restrict movement along that axis, pick the axis. If you move the pointer over one of the right angles at the origin of the gizmo, the corresponding two axes turn yellow. Pick to restrict the movement to that plane. You can complete the movement by either picking a new point or by direct distance entry.

NOTE

If the **GTAUTO** system variable is set to 1, the move gizmo is displayed when a solid is selected with no command active. If the **GTLOCATION** system variable is set to 0, the gizmo is placed at the same location as the UCS icon. Both variables are set to 1 by default.

3D Rotating

3DROTATE

Ribbon
Home
> Modify

3D Rotate

Type
3DROTATE
3R

As you have seen in earlier chapters, the **ROTATE** command can be used to rotate 3D objects. However, the command can only rotate objects in the XY plane of the current UCS. This is why you had to change UCSs to properly rotate objects. The **3DROTATE** command, on the other hand, can rotate objects on any axis regardless of the current UCS. This is an extremely powerful editing and design tool.

When the command is initiated, you are prompted to select the object(s) to rotate. After selecting the objects, press [Enter]. The rotate gizmo is displayed in the center of the selection set. The gizmo provides you with a dynamic, graphic representation of the three axes of rotation. Refer to **Figure 8-8B**. After selecting the objects, you must specify a location for the gizmo, which is the base point for rotation.

Now, you can use the gizmo to rotate the objects about the tool's local X, Y, or Z axis. As you hover the cursor over one of the three circles in the gizmo, a vector is displayed that represents the axis of rotation. To rotate about the tool's X axis, pick the red circle on the gizmo. To rotate about the Y axis, pick the green circle. To rotate about the Z axis, pick the blue circle. Once you select a circle, it turns yellow and you are prompted for the start point of the rotation angle. You can enter a direct angle at this prompt or pick the first of two points defining the angle of rotation. When the rotation angle is defined, the object is rotated about the selected axis.

The following example rotates the bend in the HVAC assembly shown in **Figure 8-9A**. Set the **Center of face** 3D object snap. Then, select the command and continue:

Current positive angle in UCS: ANGDIR=*(current)* ANGBASE=*(current)*
Select objects: *(pick the bend)*
1 found
Select objects: ↵
Specify base point: *(acquire the center of the face and pick)*
Pick a rotation axis: *(pick the circle indicated in Figure 8-9A)*
Specify angle start point or type an angle: **180**↵

Note that the rotate gizmo remains visible through the base point and the angle of rotation selections. The rotated object is shown in **Figure 8-9B**.

If you need to rotate an object on an axis that is not parallel to the current X, Y, or Z axes, use a dynamic UCS with the **3DROTATE** command. Chapter 4 discussed the benefits of using a dynamic UCS when creating objects that need to be parallel to a surface other than the XY plane. With the object selected for rotation and the dynamic UCS option active (press the [Ctrl]+[D] key combination), right-click and select **Relocate Gizmo**. Make sure the **Respect Dynamic UCS** option is checked. Then, move the rotate gizmo over a face of the object. The gizmo aligns itself with the surface so that the Z axis is perpendicular to the face. Carefully place the gizmo over the point of rotation using object snaps. Make sure the tool is correctly positioned before picking to locate it. Then, enter an angle or use polar tracking to rotate the object about the appropriate axis on the gizmo.

PROFESSIONAL TIP

By default, the move gizmo is displayed when an object is selected with no command active. To toggle between the move gizmo and the rotate gizmo, select the grip at the tool's origin and press the space bar. Then, pick a location for the tool's origin. You can toggle back to the move gizmo using the same procedure.

Figure 8-9.
A—Use object snap tracking or object snaps to place the grip tool in the middle of the rectangular face. Then, select the axis of rotation. B—The completed rotation.

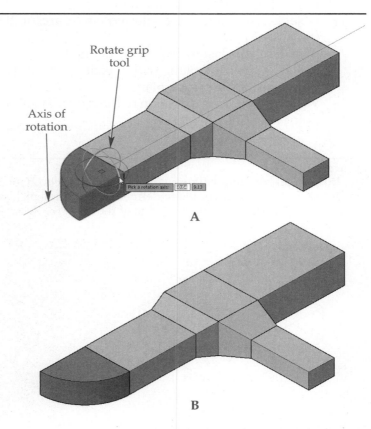

Complete the exercise on the companion website.

3D Mirroring

The **MIRROR** command can be used to rotate 3D objects. However, like the **ROTATE** command, the **MIRROR** command can only work in the XY plane of the current UCS. Often, to properly mirror objects with this command, you have to change UCSs. The **MIRROR3D** command, on the other hand, allows you to mirror objects about any plane regardless of the current UCS.

The default option of the command is to define a mirror plane by picking three points on that plane, **Figure 8-10A**. Object snaps should be used to accurately define the mirror plane. To mirror the wedge in **Figure 8-10A**, set the **Midpoint** object snap, select the command, and use the following sequence. The resulting drawing is shown in **Figure 8-10B**.

Select objects: *(pick the wedge)*
1 found
Select objects: ↵
Specify first point of mirror plane (3 points) or
[Object/Last/Zaxis/View/XY/YZ/ZX/3points] <3points>: *(pick P1, which is the mid-point of the box's top edge)*
Specify second point on mirror plane: *(pick P2)*
Specify third point on mirror plane: *(pick P3)*
Delete source objects? [Yes/No] <N>: ↵

There are several different ways to define a mirror plane with the **MIRROR3D** command. The options include the following:

- **Object.** The plane of the selected circle, arc, or 2D polyline segment is used as the mirror plane.
- **Last.** Uses the last mirror plane defined.
- **Zaxis.** Defines the plane with a pick point on the mirror plane and a point on the Z axis of the mirror plane.
- **View.** The viewing direction of the current viewpoint is aligned with a selected point to define the plane.

Figure 8-10.
The **MIRROR3D** command allows you to mirror objects about any plane regardless of the current UCS. A—The mirror plane defined by the three pick points is shown here in color. Point P1 is the midpoint of the top edge of the base. B—A copy of the original is mirrored.

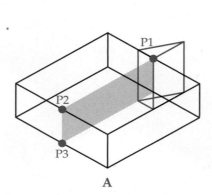
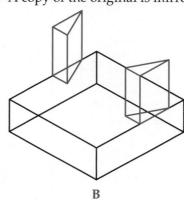

A

B

- **XY, YZ, ZX.** The mirror plane is placed parallel to one of the three basic planes of the current UCS and passes through a selected point.
- **3points.** Allows you to pick three points to define the mirror plane, as shown in **Figure 8-10**.

Exercise 8-4

www.g-wlearning.com/CAD/

Complete the exercise on the companion website.

Creating 3D Arrays

An *array* is an arrangement of objects in a 2D or 3D pattern. An array can be created as a rectangular, polar, or path array. You probably used arrays to complete some of the problems in previous chapters. A 2D array is created on the XY plane of the current UCS. A 3D array is an arrangement of objects in 3D space. The **ARRAYRECT**, **ARRAYPOLAR**, and **ARRAYPATH** commands can be used to create both 2D and 3D arrays. These commands provide the same functions as the **Rectangular**, **Polar**, and **Path** options of the **ARRAY** command and are available in the drop-down menu located in the **Modify** panel of the **Home** tab on the ribbon. See **Figure 8-11**.

An array can be created as an associative or non-associative array. Creating an associative array creates an *array object*, which can be modified as a single entity. For example, you can edit the source object to change all of the items in the array at once. You can also perform other modifications, such as deleting one or more items in the array, while maintaining the associativity of the arrayed items.

When creating a 3D array, the information you specify depends on the type of array being created. Many of the options are similar to those used when creating a 2D array. The following sections discuss 3D rectangular, polar, and path arrays.

NOTE

The legacy **3DARRAY** command can also be used to create 3D arrays, but it is limited to creating rectangular and polar arrays and cannot create associative arrays. Using the **ARRAYRECT**, **ARRAYPOLAR**, or **ARRAYPATH** command is the preferred method to create a 3D array.

3D Rectangular Arrays

In a *3D rectangular array,* as with a 2D rectangular array, you must enter the number of rows and columns. However, you must also specify the number of *levels,*

Figure 8-11.
The **ARRAYRECT**, **ARRAYPOLAR**, and **ARRAYPATH** commands can be accessed from the drop-down menu located in the **Modify** panel of the **Home** tab on the ribbon.

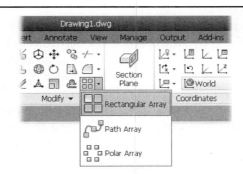

which represents the third (Z) dimension. The command sequence is similar to that used when creating a 2D array.

An example of where a 3D rectangular array may be created is the layout of structural columns on multiple floors of a commercial building. In **Figure 8-12A**, you can see two concrete floor slabs of a building and a single structural column. It is now a simple matter of arraying the column in rows, columns, and levels.

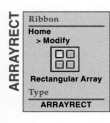

To draw a 3D rectangular array, select the **ARRAYRECT** command. Select the object to array and press [Enter]. An initial pattern of three rows, four columns, and one level forms. You can select one of the array grips and drag to increase or decrease the number of rows, number of columns, or spacing dynamically. See **Figure 8-12B**. Without exiting the command, you can then adjust the row, column, level, and spacing values by using the **Array Creation** ribbon tab. See **Figure 8-12C**. In **Figure 8-12D**, there are three rows, five columns, and two levels. To set the number of levels and spacing between levels, use the **Array Creation** ribbon tab, right-click and select **Levels** from the shortcut menu, or use the prompts on the command line. Note that for the following command sequence, the drawing units have been set to architectural.

> Select grip to edit array or [ASsociative/Base point/COUnt/Spacing/COLumns/Rows/Levels/eXit] <eXit>: **R**↵
> Enter the number of rows or [Expression] <*current*>: **3**↵
> Specify the distance between rows or [Total/Expression] <*current*>: **10'**↵
> Specify the incrementing elevation between rows or [Expression] <0">: ↵
> Select grip to edit array or [ASsociative/Base point/COUnt/Spacing/COLumns/Rows/Levels/eXit] <eXit>: **COL**↵
> Enter the number of columns or [Expression] <*current*>: **5**↵
> Specify the distance between columns or [Total/Expression] <*current*>: **10'**↵
> Select grip to edit array or [ASsociative/Base point/COUnt/Spacing/COLumns/Rows/Levels/eXit] <eXit>: **L**↵
> Enter the number of levels or [Expression] <1>: **2**↵
> Specify the distance between levels or [Total/Expression] <*current*>: **12'8"**↵
> Select grip to edit array or [ASsociative/Base point/COUnt/Spacing/COLumns/Rows/Levels/eXit] <eXit>: ↵

The result is shown in **Figure 8-12D**. By default, an associative array is created. You can create a non-associative array by selecting the **Associative** option. The **Associative** option setting is maintained by AutoCAD the next time an array command is accessed. You can verify the current setting of this option after entering the command and selecting objects. The Associative = Yes or Associative = No prompt appears. If the Associative = No prompt appears and you want to create an associative array, select the **Associative** option.

As shown in the previous sequence, when setting the distance between rows, you can also define an elevation increment between rows when the Specify the incrementing elevation between rows or [Expression] <0">: prompt appears. The elevation increment is different from the distance between levels. The elevation increment sets the spacing between rows along the Z axis so that each successive row is drawn on a higher or lower plane. This option can also be used when creating a 3D polar array and is discussed in the next section.

The **Base point** option is used to define the base point for the array. By default, the base point defined by AutoCAD is the centroid of the object(s) selected. You may want to select a more logical base point, such as an endpoint of an edge or the center point of a circular face. In **Figure 8-12B**, the center of the cylindrical base of the column has been selected as the base point of the array.

If any of the properties of an associative array require changes, use the **ARRAYEDIT** command or the **Properties** palette to edit the array. The arrayed items will update based on the changes. In **Figure 8-12E**, the array is shown after changing the number of columns to 7 and the column spacing to 6'-8".

Figure 8-12.
A—Two floors and one structural column are drawn. Creating a 3D rectangular array will place all of the required columns on both floors at the same time. B—After entering the **ARRAYRECT** command, you can pick a grip and drag to set the number of rows or columns or the spacing between rows or columns dynamically. The floor objects are removed from the view for illustration purposes only. C—The **Array Creation** ribbon tab can be used to set values for rows, columns, and levels. D—An associative 3D rectangular array made up of three rows and five columns on two levels. E—An arrangement of three rows and seven columns after editing the properties of the associative array.

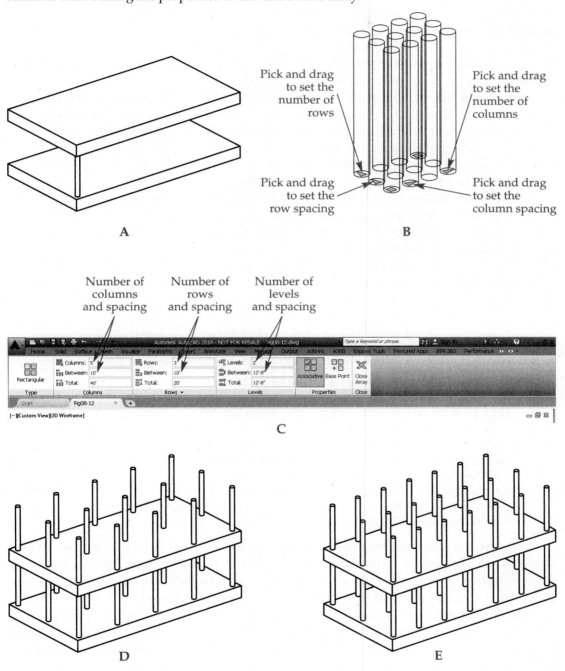

3D Polar Arrays

A *3D polar array* is similar to a 2D polar array. However, the axis of rotation in a 2D polar array is parallel to the Z axis of the current UCS. In a 3D polar array, you can define a centerline axis of rotation that is not parallel to the Z axis of the current UCS. In other words, you can array an object in a UCS different from the current one. In addition, as with a 3D rectangular array, you can array the object in multiple "levels" along the Z axis. The **ARRAYPOLAR** command can be used to create a 3D arrangement in rows, levels, or both rows and levels.

ARRAYPOLAR

Ribbon
Home
> Modify

Polar Array

Type
ARRAYPOLAR

To create a 3D polar array, select the **ARRAYPOLAR** command. Select the object to array and press [Enter]. Then, pick the center point of the array or use the **Axis of rotation** option to select a centerline axis. The **Axis of rotation** option allows you to select a centerline axis that is different from the Z axis of the current UCS. Using this option requires you to pick two points to define the axis. In **Figure 8-13A**, the leg attached to the hub must be arrayed about the center axis. The center axis is drawn as a centerline. Use the **Axis of rotation** option to pick the two endpoints of the axis. Once you define the axis, an initial pattern of six items forms. You can use the array grips to adjust the pattern dynamically or you can use the **Array Creation** ribbon tab to make settings. The settings available include the number of items, angle between items, fill angle, number of rows and spacing between rows, and number of levels and spacing between levels. The same settings can be made using the prompts on the command line.

Specify center point of array or [Base point/Axis of rotation]: **A**↵
Specify first point on axis of rotation: *(pick one endpoint of the axis centerline)*
Specify second point on axis of rotation: *(pick the other endpoint of the axis centerline)*
Select grip to edit array or [ASsociative/Base point/Items/Angle between/Fill angle/ ROWs/Levels/ROTate items/eXit] <eXit>: **I**↵
Enter number of items in array or [Expression] <6>: **6**↵
Select grip to edit array or [ASsociative/Base point/Items/Angle between/Fill angle/ ROWs/Levels/ROTate items/eXit] <eXit>: **F**↵
Specify the angle to fill (+=ccw, −=cw) or [EXpression] <360>: ↵
Select grip to edit array or [ASsociative/Base point/Items/Angle between/Fill angle/ ROWs/Levels/ROTate items/eXit] <eXit>: **AS**↵
Create associative array [Yes/No] <Yes>: **N**↵
Select grip to edit array or [ASsociative/Base point/Items/Angle between/Fill angle/ ROWs/Levels/ROTate items/eXit] <eXit>: ↵

The result is shown in **Figure 8-13B**. In this case, a non-associative array is created. The legs are arrayed as individual objects and the resulting arrayed objects can be selected individually. As previously discussed, the **Associative** option setting is maintained by AutoCAD the next time an array command is accessed.

The **Rows** option is used to create multiple rows of arrayed objects. After selecting this option, enter the number of rows and the distance between rows. Then, set an elevation increment value to control the spacing along the Z axis between each successive row. An array of seats in a theater can be created in this manner. **Figure 8-14** shows an example of arraying a single seat to create multiple rows of seats. To create this array, select the **ARRAYPOLAR** command, select the first seat, and use the following command sequence. If needed, use the **Associative** option to create an associative array. For this example, the default **Center point** option is used to set the center point of the array. Using this option is sufficient because the Z axis of the current UCS is

Figure 8-13.
A— Six new legs need to be arrayed about the center axis of the existing part. B—The model after using the **ARRAYPOLAR** command.

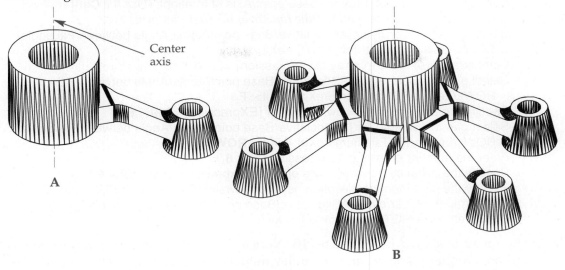

Figure 8-14.
Using the **Rows** option of the **ARRAYPOLAR** command to create a polar array of seats arranged in multiple rows. A—The first seat is modeled on the bottom platform. B—The result after using the **ARRAYPOLAR** command.

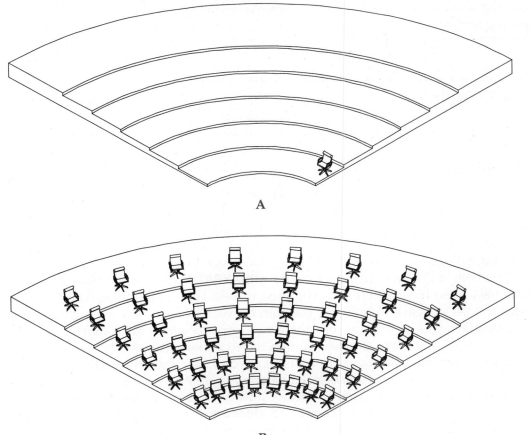

parallel to the required axis of rotation. Once you define the center axis, you can adjust the array using grips, the **Array Creation** ribbon tab, or the command line prompts.

Specify center point of array or [Base point/Axis of rotation]: *(Use the* **Center** *object snap to pick the center point of the top arc of the first platform)*
Select grip to edit array or [ASsociative/Base point/Items/Angle between/Fill angle/ ROWs/Levels/ROTate items/eXit] <eXit>: **I⏎**
Enter number of items in array or [Expression] <6>: **8⏎**
Select grip to edit array or [ASsociative/Base point/Items/Angle between/Fill angle/ ROWs/Levels/ROTate items/eXit] <eXit>: **F⏎**
Specify the angle to fill (+=ccw, −=cw) or [EXpression] <360>: **80⏎**
Select grip to edit array or [ASsociative/Base point/Items/Angle between/Fill angle/ ROWs/Levels/ROTate items/eXit] <eXit>: **ROWS⏎**
Enter the number of rows or [Expression] <1>: **6⏎**
Specify the distance between rows or [Total/Expression] *<current>*: **96⏎**
Specify the incrementing elevation between rows or [Expression] <0.0000>: **7⏎**
Select grip to edit array or [ASsociative/Base point/Items/Angle between/Fill angle/ ROWs/Levels/ROTate items/eXit] <eXit>: **⏎**

The result is shown in **Figure 8-14B**. Notice that each successive row of seats is situated on a higher plane. In more complex models, the **Levels** option can be used to create multiple levels of rows.

NOTE

The **Expression** option allows you to enter a mathematical expression to calculate one of the array parameters. For example, you can enter an expression to calculate the number of items, rows, or levels. The **Expression** option can also be used to define the spacing between rows and the incrementing elevation.

Exercise 8-6

www.g-wlearning.com/CAD/

Complete the exercise on the companion website.

3D Path Arrays

A *3D path array* is similar to a 2D path array. Objects can be arrayed along a path or a segment of a path. The path, also called a *path curve*, can be a line, circle, arc, ellipse, spline, polyline, helix, or 3D polyline. See **Figure 8-15**. As with a 3D polar array, you can create a 3D arrangement in rows, levels, or both rows and levels. A 3D path array can be created as an associative or non-associative array.

To create a 3D path array, select the **ARRAYPATH** command. Select the object to array and press [Enter]. You are prompted to select the path curve. The object to be arrayed does not have to intersect the path curve. Once you select the path curve, an initial pattern forms. You can use the array grips to adjust the pattern dynamically or you can use the **Array Creation** ribbon tab to make settings. The settings can also be made using the command prompts. If needed, use the **Associative** option to set whether an associative or non-associative array is created.

The **Method** option is used to specify how objects are distributed along the path. The two options available are **Divide** and **Measure**. The **Divide** option distributes the specified number of objects evenly along the path. The **Measure** option distributes the objects at specific distances along the path. The **Items** option works in conjunction with the **Method** option and allows you to specify the number of objects. When using the **Measure** option, you can specify the spacing between objects or the total distance

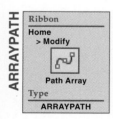

ARRAYPATH

Ribbon
Home
> Modify
Path Array
Type
ARRAYPATH

Figure 8-15.
Examples of 3D path arrays used to array posts.

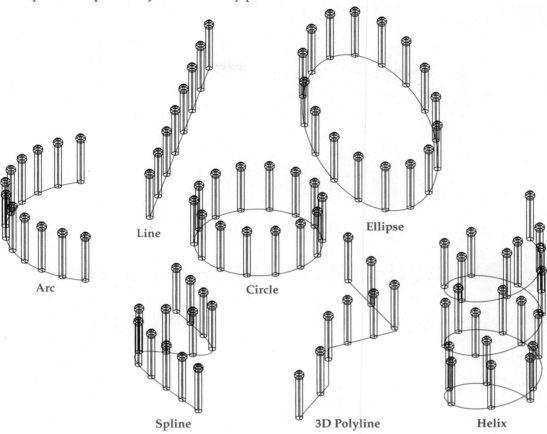

Arc

Line

Circle

Ellipse

Spline

3D Polyline

Helix

between the first and last objects. If the path curve changes in length, the number of objects will increase or decrease automatically.

The **Base point** option is used to set the array's base point. The default base point of the array is the endpoint of the path curve closest to where you select it. This point serves as the start point of the array. Depending on the result you want, you can select a different base point (start point), such as a point on the object.

The default orientation of the array is the current orientation of the object. The **Tangent direction** option allows you to pick two points to define a different orientation. The **Normal** option is used to align the object "normal" to the path. Using this option aligns the Z axis of the object perpendicular to the path.

After specifying the base point and orientation of the array or using the defaults, continue as follows. When using the **Divide** option, specify the number of items. When using the **Measure** option, specify the distance between each object or the total distance between the first and last objects. Distances can be set using the **Array Creation** ribbon tab. In **Figure 8-16A**, the **Divide** option is used:

> Select grip to edit array or [ASsociative/Method/Base point/Tangent direction/Items/
>   Rows/Levels/Align items/Z direction/eXit] <eXit>: **M**↵
> Enter path method [Divide/Measure] <*current*>: **D**↵
> Select grip to edit array or [ASsociative/Method/Base point/Tangent direction/Items/
>   Rows/Levels/Align items/Z direction/eXit] <eXit>: **I**↵
> Enter number of items along path or [Expression] <*current*>: **10**↵
> Select grip to edit array or [ASsociative/Method/Base point/Tangent direction/Items/
>   Rows/Levels/Align items/Z direction/eXit] <eXit>: ↵

Using the **ARRAYPATH** command. A—Specifying 10 items to array and using the **Divide** option. B—Using the **Measure** option to specify 8 items to array and a distance of 25 units between objects. C—The resulting array.

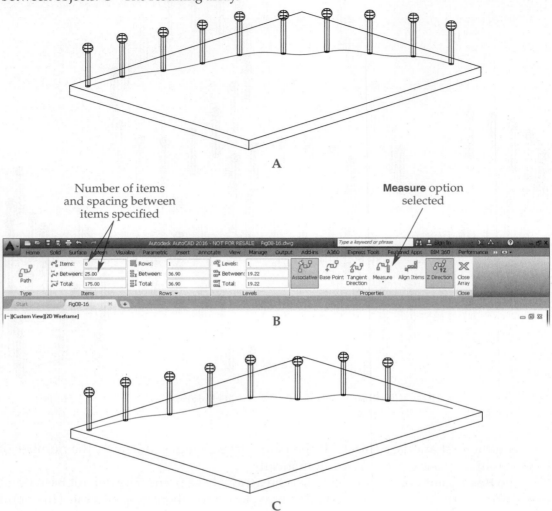

In **Figure 8-16B**, the **Array Creation** ribbon tab is shown after selecting the **Measure** option. The number of items is set to 8 and the distance between items is set to 25 using the settings in the **Items** panel. The result is shown in **Figure 8-16C**.

The **Align items** option is used to align the arrayed objects tangent to the direction of the array path. The **Z direction** option is used to change the Z axis direction of the arrayed objects. By default, the Z axis of each object is aligned in the same orientation used by the original object. If the **Z direction** option is set to **No**, the Z axis of the object changes direction to follow the path as the path changes direction.

The **Rows** and **Levels** options allow you to create an arrangement in rows and levels. The options are similar to those used with the **ARRAYPOLAR** command. In **Figure 8-17**, an oval table is created using the **Levels** option. The table feet and middle supports are created as a 3D path array. One of the feet is arrayed along an elliptical path to create an array consisting of two levels. In this example, a non-associative array is created. This allows the middle supports to be edited as individual objects after creating the array. The middle supports have a smaller diameter than the feet and a longer length. To complete the table, the bottom shelf is copied along the Z axis to create the top.

Complete the exercise on the companion website.

Figure 8-24.
Slicing a solid with a planar object. A—The circle is drawn at the proper orientation and in the correct location. B—The completed slice.

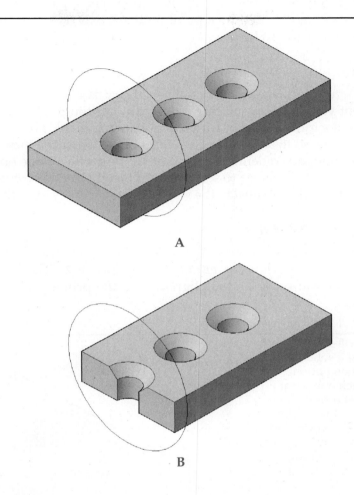

A

B

Figure 8-25.
Slicing a solid with a surface. A—Draw the surface and locate it within the solid to be sliced. The solid is represented here by the wireframe. B—The completed slice with both sides retained. The top can now be moved and rotated as shown here.

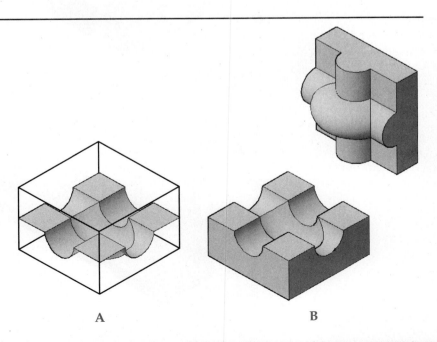

A

B

Z Axis

You can specify one point on the cutting plane and one point on the Z axis of the plane. See **Figure 8-26**. This allows you to have a cutting plane that is not parallel to the current UCS XY plane. First, select the **SLICE** command, pick the object to slice, and press [Enter]. Next, enter the **Zaxis** option. Then, pick a point on the XY plane of the cutting plane followed by a point on the Z axis of the cutting plane. Finally, pick the side of the object to keep.

View

A cutting plane can be established that is aligned with the viewing plane of the current viewport. The cutting plane passes through a point you select, which sets the depth along the Z axis of the current viewing plane. First, select the **SLICE** command, pick the object to slice, and press [Enter]. Next, enter the **View** option. Then, pick a point in the viewport to define the location of the cutting plane on the Z axis of the viewing plane. Use object snaps to select a point on an object. The cutting plane passes through this point and is parallel to the viewing plane. Finally, pick the side of the object to keep.

XY, YZ, and ZX

You can slice an object using a cutting plane that is parallel to any of the three primary planes of the current UCS. See **Figure 8-27**. The cutting plane passes through the point you select and is aligned with the primary plane of the current UCS that

Figure 8-26.
Slicing a solid using the **Zaxis** option.
A—Pick one point on the cutting plane and a second point on the Z axis of the cutting plane.
B—The resulting slice.

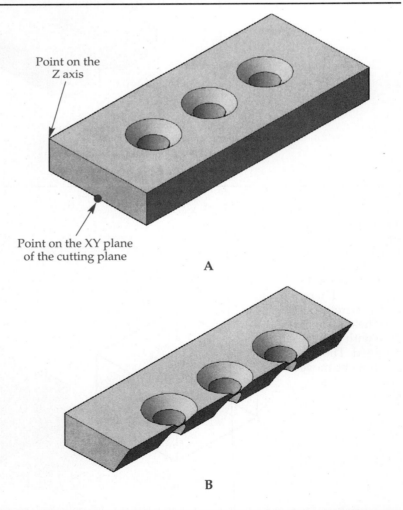

Point on the Z axis

Point on the XY plane of the cutting plane

A

B

you specify. First, select the **SLICE** command, pick the object to slice, and press [Enter]. Next, enter the **XY**, **YZ**, or **ZX** option, depending on the primary plane to which the cutting plane will be parallel. Then, pick a point on the cutting plane. Finally, pick the side of the object to keep.

Three Points

Three points can be used to define the cutting plane. This allows the cutting plane to be aligned at any angle, similar to using the **Zaxis** option. See **Figure 8-28**. First, select the **SLICE** command, pick the object to slice, and press [Enter]. Then, enter the **3points** option. Pick three points on the cutting plane and then select the side of the object to keep.

Exercise 8-10

www.g-wlearning.com/CAD/

Complete the exercise on the companion website.

Removing Features

Sometimes, it may be necessary to remove a feature that has been constructed. For example, suppose you placed a R.5 fillet on an object based on an engineering sketch. Then, the design is changed to a R.25 fillet. Subobject editing is the best technique to accomplish this. Subobject editing is covered in Chapter 11.

Figure 8-27.
Slicing a solid using the **XY**, **YZ**, and **ZX** options. A—The object before slicing. The UCS origin is in the center of the first hole and at the midpoint of the height. B—The resulting slice using the **XY** option. C—The resulting slice using the **YZ** option. D—The resulting slice using the **ZX** option.

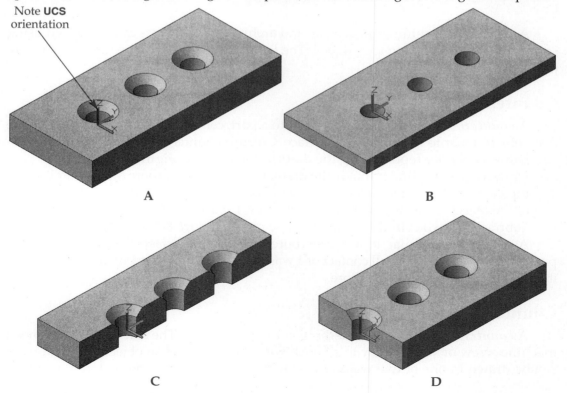

Note **UCS** orientation

A

B

C

D

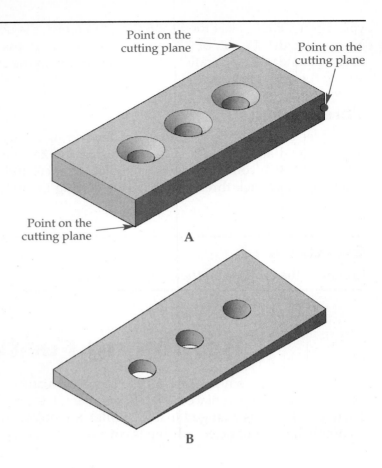

Figure 8-28.
Slicing a solid using the **3points** option. A—Specify three points to define the cutting plane. B—The resulting slice.

Point on the cutting plane

Point on the cutting plane

Point on the cutting plane

A

B

Constructing Features on Solid Models

A variety of machining, structural, and architectural features can be created using some basic solid modeling techniques. The features discussed in the next sections are just a few of the possibilities.

Counterbore and Spotface

A *counterbore* is a recess machined into a part, centered on a hole, that allows the head of a fastener to rest below the surface. Create a counterbore as follows.
1. Draw a cylinder representing the diameter of the hole, **Figure 8-29A**.
2. Draw a second cylinder that is the diameter of the counterbore and center it at the top of the first cylinder. Move the second cylinder so it extends below the surface of the object to the depth of the counterbore, **Figure 8-29B**.
3. Subtract the two cylinders from the base object, **Figure 8-29C**.

A *spotface* is similar to a counterbore, but is not as deep. See **Figure 8-30**. It provides a flat surface for full contact of a washer or underside of a bolt head. Construct it in the same way as a counterbore.

Countersink

A *countersink* is like a counterbore with angled sides. The sides allow a flat-head machine screw or wood screw to sit flush with the surface of an object. A countersink can be drawn in one of two ways. You can draw an inverted cone centered on a hole

Figure 8-29.
Constructing a counterbore.
A—Draw a cylinder to represent a hole.
B—Draw a second cylinder to represent the counterbore.
C—Subtract the two cylinders from the base object.

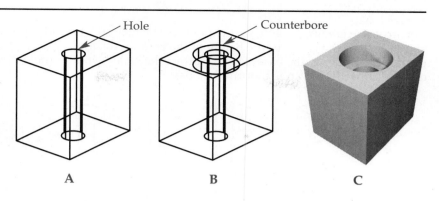

A   B   C

Figure 8-30.
Constructing a spotface.
A—The bottom of the second, larger-diameter cylinder should be located at the exact depth of the spotface. However, the height may extend above the surface of the base. Then, subtract the two cylinders from the base.
B—The finished solid.

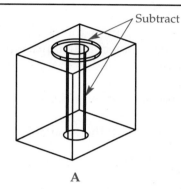

A   B

and subtract it from the base or you can chamfer the top edge of a hole. Chamfering is the quickest method.

1. Draw a cylinder representing the diameter of the hole, **Figure 8-31A**.
2. Subtract the cylinder from the base object.
3. Select the **CHAMFER** or **CHAMFEREDGE** command.
4. If using the **CHAMFER** command, select the top edge of the base object, enter the chamfer distance(s), and pick the top edge of the hole, **Figure 8-31B**.
5. If using the **CHAMFEREDGE** command, pick the top edge of the hole and enter the chamfer distance(s). Preview the result before completing the operation.

Boss

A **boss** serves the same function as a spotface. However, it is an area raised above the surface of an object. Draw a boss as follows.

1. Draw a cylinder representing the diameter of the hole. Extend it above the base object higher than the boss is to be, **Figure 8-32A**.

Figure 8-31.
Constructing a countersink. A—Subtract the cylinder from the base to create the hole. B—Chamfer the top of the hole to create a countersink.

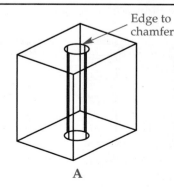

A   B

Figure 8-32.
Constructing a boss. A—Draw a cylinder for the hole so it extends above the surface of the object. B—Draw a cylinder the height of the boss on the top surface of the object. C—Union the large cylinder to the base. Then, subtract the small cylinder (hole) from the unioned objects. D—Fillet the edge to form the boss.

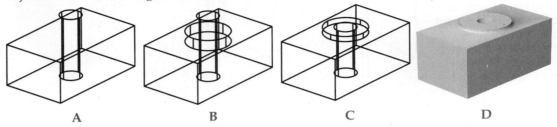

A   B   C   D

2. Draw a second cylinder the diameter of the boss. Place the base of this cylinder above the top surface of the base object a distance equal to the height of the boss. Give the cylinder a negative height value so that it extends inside of the base object, **Figure 8-32B**.
3. Union the base object and the second cylinder (boss). Subtract the hole from the unioned object, **Figure 8-32C**.
4. Fillet the intersection of the boss with the base object, **Figure 8-32D**.

O-Ring Groove

An *O-ring* is a circular seal that resembles a torus. It sits inside of a groove constructed so that part of the O-ring is above the surface. An *O-ring groove* can be constructed by placing the center of a circle on the outside surface of a cylinder. Then, revolve the circle around the cylinder. Finally, subtract the revolved solid from the cylinder.

1. Construct the cylinder to the required dimensions, **Figure 8-33A**.
2. Rotate the UCS on the X axis (or appropriate axis).
3. Draw a circle with a center point on the surface of the cylinder, **Figure 8-33B**.
4. Revolve the circle 360° about the center of the cylinder, **Figure 8-33C**.
5. Subtract the revolved object from the cylinder, **Figure 8-33D**.

PROFESSIONAL TIP

In many cases, you will draw the O-ring as a torus. A copy of the torus can be used to create the O-ring groove instead of revolving a circle as described in the previous section.

Figure 8-33.
Constructing an O-ring groove. A—Construct a cylinder; this one has a round placed on one end. B—Draw a circle centered on the surface of the cylinder. C—Revolve the circle 360° about the center of the cylinder. D—Subtract the revolved object from the cylinder. E—The completed O-ring groove.

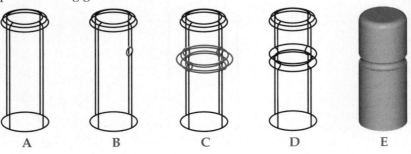

A   B   C   D   E

Architectural Molding

   Architectural molding features can be quickly constructed using extrusions. First, construct the profile of the molding as a closed shape, **Figure 8-34A**. Then, extrude the profile the desired length, **Figure 8-34B**.

   Corner intersections of molding can be quickly created by extruding the same shape in two different directions and then joining the two objects. First, draw the molding profile. Then, copy and rotate the profile to orient the local Z axis in the desired direction, **Figure 8-35A**. Next, extrude the two profiles the desired lengths, **Figure 8-35B**. Finally, union the two extrusions to create the mitered corner molding, **Figure 8-35C**.

Exercise 8-11

www.g-wlearning.com/CAD/

Complete the exercise on the companion website.

Figure 8-34.
A—The molding profile. B—The profile extruded to the desired length.

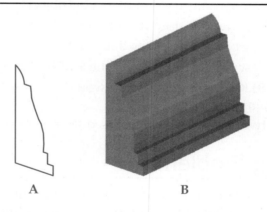

A   B

Figure 8-35.
Constructing corner molding. A—Copy and rotate the molding profile. B—Extrude the profiles to the desired lengths. C—Union the two extrusions to create the mitered corner. Note: The view has been rotated. D—The completed corner.

Copy and rotate the profile

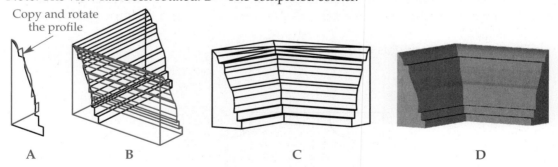

A   B   C   D

Chapter Review

Answer the following questions using the information in this chapter.

1. Which properties of a solid can be changed in the **Properties** palette?
2. What does the History property control?
3. What is the preferred command for aligning objects to create an assembly of parts?
4. How does the **3DALIGN** command differ from the **ALIGN** command?
5. How does the **3DROTATE** command differ from the **ROTATE** command?
6. How does the **MIRROR3D** command differ from the **MIRROR** command?
7. Which command allows you to create a rectangular array by defining rows, columns, and levels?
8. How does a 3D polar array differ from a 2D polar array?
9. Which option of the **ARRAYPATH** command can be used to evenly distribute the arrayed object along the path?
10. Which command is used to fillet a 3D object?
11. Which command is used to chamfer a 3D object?
12. Name four types of surfaces that can be used to slice objects.
13. Briefly describe the function of the **SLICE** command.
14. Which **SLICE** command option would be used to create a contoured solid object like a computer mouse?
15. Which two commands can be used to create a countersink?

Drawing Problems

1. **Piping** Construct an 8″ diameter tee pipe fitting using the dimensions shown below. Hint: Extrude and union two solid cylinders before subtracting the cylinders for the inside diameters.
   A. Use **EXTRUDE** to create two sections of pipe at 90° to each other, then use **UNION** to union the two pieces together.
   B. Fillet and chamfer the object to finish it. The chamfer distance is .25″ × .25″.
   C. The outside diameter of all three openings is 8.63″ and the pipe wall thickness is .322″.
   D. Save the drawing as P8-1.

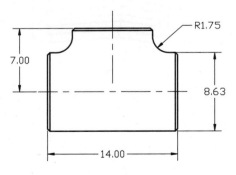

2. **Piping** Construct an 8″ diameter, 90° elbow pipe fitting using the dimensions shown below.
   A. Use **EXTRUDE** or **SWEEP** to create the elbow.
   B. Chamfer the object. The chamfer distance is .25″ × .25″. Note: You cannot use the **CHAMFER** command.
   C. The outside diameter is 8.63″ and the pipe wall thickness is .322″.
   D. Save the drawing as P8-2.

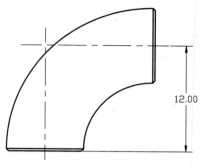

12.00

3. **Architectural** Construct picture frame moldings using the profiles shown below.
   A. Draw each of the closed profiles shown. Use your own dimensions for the details of the moldings.
   B. The length and width of A and B should be no larger than 1.5″ × 1″.
   C. The length and width of C and D should be no larger than 3″ × 1.5″.
   D. Construct an 8″ × 12″ picture frame using moldings A and B.
   E. Construct a 12″ × 24″ picture frame using moldings C and D.
   F. Save the drawing as P8-3.

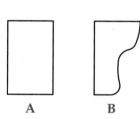

A   B

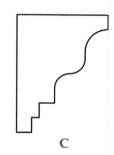

C

D

Problems 4–7. These problems require you to use a variety of solid modeling functions to construct the objects. Use all of the solid modeling and editing commands you have learned so far to assist in construction. Create new UCSs as needed and use a dynamic UCS when practical. Use **SOLIDHIST** *and* **SHOWHIST** *to record and view the steps used to create the solid models. Create copies of the completed models and split them as required to show the internal features visible in the section views. Do not add dimensions to the models. For objects that have dimensions with tolerances, use the base dimension from which the upper and lower limits are calculated to create the model. Save each drawing as* P8-*(problem number).*

4. Mechanical

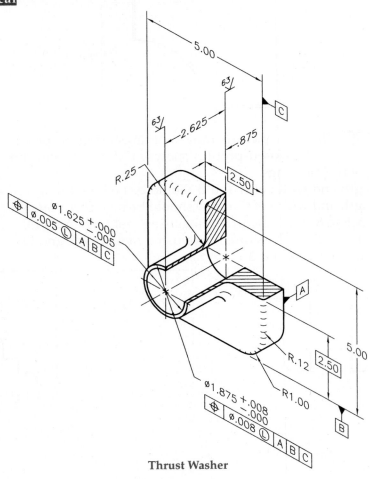

Thrust Washer

5. Mechanical

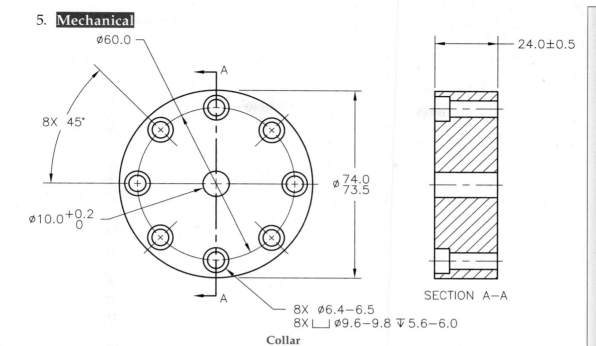

Collar

8X ⌀6.4−6.5
8X ⌴ ⌀9.6−9.8 ▽5.6−6.0

⌀60.0

8X 45°

⌀10.0 +0.2 −0

⌀ 74.0 / 73.5

24.0±0.5

SECTION A−A

6. Mechanical

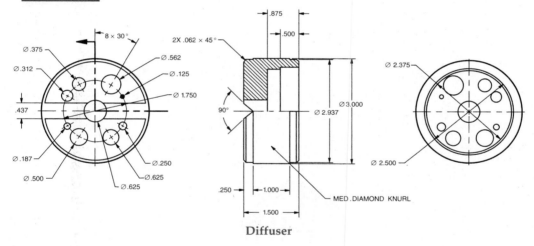

Diffuser

8 × 30°

2X .062 × 45°

⌀ .375
⌀ .312
⌀ .125
⌀ .562
⌀ .187
⌀ 1.750
⌀ .500
⌀ .625
⌀ .625
⌀ .250

.437

90°

.875
.500
⌀ 3.000
⌀ 2.937
.250
1.000
1.500
MED. DIAMOND KNURL

⌀ 2.375
⌀ 2.500

7. Mechanical

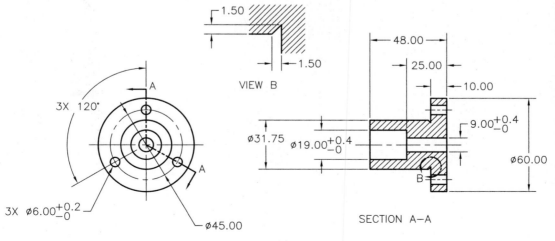

Bushing

1.50
1.50
VIEW B

3X 120°

3X ⌀6.00 +0.2 −0

⌀45.00

48.00
25.00
10.00
9.00 +0.4 −0
⌀60.00
⌀31.75
⌀19.00 +0.4 −0

SECTION A−A

Chapter 8 Creating and Working with Solid Model Features **211**

8. **Architectural** In this problem, you will add legs to the kitchen chair you started in Chapter 2. In Chapter 6, you refined the seat, and in Chapter 7, you added the seatback. In this chapter, you complete the model. In Chapter 17, you will add materials to the model and render it.

A. Open P7-11 from Chapter 7.

B. Using the **LINE** command, create the framework for lofting the legs and cross-bars as shown. The crossbars are at the midpoints of the legs. Position the lines for the double crossbar about 4.5" apart.

C. Using a combination of the **3DROTATE** and **3DMOVE** commands in conjunction with new UCSs, draw and position circles as shapes for lofting the legs and cross-bars. The legs transition from ⌀1.00 at the ends to ⌀1.25 at the midpoints. The crossbars transition from ⌀.75 at the ends to ⌀1.00 at a position 2" from each end.

D. Create the legs and crossbars. The completed model is shown.

E. Save the drawing as P8-8.

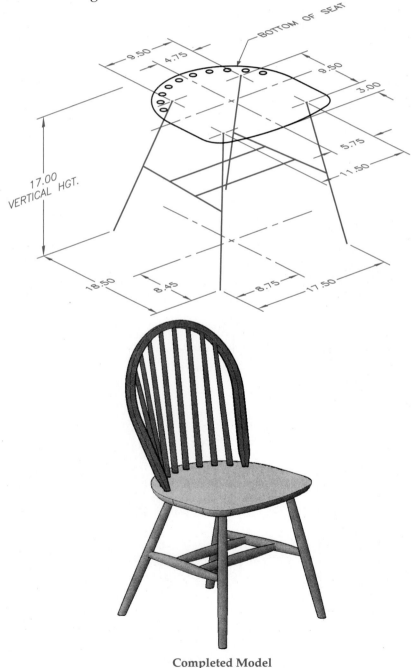

Completed Model

Mesh Modeling

Learning Objectives

After completing this chapter, you will be able to:

✓ Explain tessellation divisions and values.
✓ Create mesh primitives.
✓ Create a smoothed mesh object.
✓ Create a refined mesh object.
✓ Generate a mesh by converting a solid or surface.
✓ Generate a solid or surface by converting a mesh.
✓ Execute editing techniques on mesh objects.
✓ Create a split face on a mesh.
✓ Produce an extruded mesh face.
✓ Apply a crease to mesh subobjects.
✓ Create and close mesh object gaps.
✓ Create a new mesh face by collapsing a mesh face or edge.
✓ Merge mesh faces to form a single mesh face.
✓ Construct a new mesh face by spinning a triangular mesh face.

Overview of Mesh Modeling

Mesh primitives can be used to create freeform designs. See **Figure 9-1**. The tools for creating and editing meshes extend the capability of AutoCAD's 3D modeling tools. There are two key workflows that the designer considers:

- The creation of 3D models, which can be solids, surfaces, or meshes.
- Editing the 3D models to create unique shapes.

Mesh models can be created as mesh primitives, mesh forms, or freeform mesh shapes. A *mesh model* consists of vertices, edges, and faces. You can modify a mesh by adjusting the smoothness or by adding creases, extrusions, splits, and gaps. You can also distort a mesh to create unique freeform shapes.

A mesh model is a type of surface model. *Subdivision surfaces* is another term for mesh models. Mesh models do *not* have volume or mass. Rather, mesh models only define the shape of the design.

Figure 9-1.
Constructing an ergonomic mouse as a mesh model. A—The basic mesh primitive. B—Using editing tools, the mesh is reformed. C—The completed mesh model.

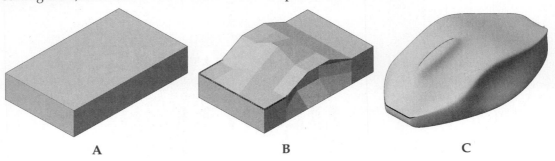

A   B   C

Mesh objects can be created using one of these methods:
- Construct mesh primitives (**MESH** command).
- Convert an existing solid or surface into a mesh object (**MESHSMOOTH** command).
- Construct mesh forms that are ruled, revolved, tabulated, or edge-defined objects (**RULESURF**, **REVSURF**, **TABSURF**, and **EDGESURF** commands).
- Convert legacy surface objects into mesh objects using commands such as **3DFACE**, **3DMESH**, and **PFACE**.

The tools used to create and modify meshes are found on the **Mesh** tab in the ribbon. The commands used to create mesh primitives are similar to those used to create solid primitives, which are discussed in Chapter 2. However, mesh primitives have face mesh objects that are divided into smaller faces. These divisions are based on tessellation division values (smoothness), as discussed in this chapter.

NOTE

A mesh model is a unique type of object in AutoCAD. A mesh model is a type of surface model, but it is different from a procedural surface model or a NURBS surface model. Unique construction and editing methods are used with mesh models. AutoCAD procedural surfaces and NURBS surfaces are precision surfaces, which makes them relatives to solids. Procedural surfaces and NURBS surfaces are discussed in Chapter 10.

PROFESSIONAL TIP

The **RULESURF**, **REVSURF**, **TABSURF**, and **EDGESURF** commands are available for creating mesh forms. These commands are not discussed in this chapter. Other surface modeling methods are typically more useful and provide more flexibility in editing. Refer to Chapter 10 for detailed coverage on surface modeling.

Tessellation Division Values

Tessellation divisions are the basic foundation for the smoothness of a mesh object. Tessellation divisions on a mesh object consist of planar shapes (faces) that fit together to form the surface. See **Figure 9-2A**. These divisions display the edges of a mesh face that can then be edited.

Figure 9-2.
A—Tessellation divisions are key to mesh modeling. They define the smoothness of the mesh model. B—Facets on a mesh face.

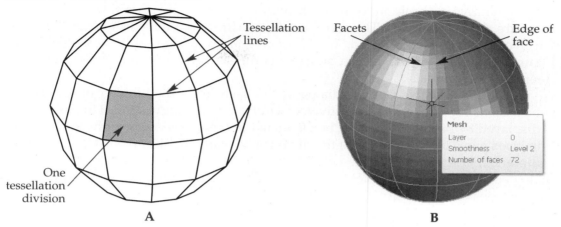

When creating mesh primitives, set the mesh tessellation divisions *before* creating a mesh primitive shape. Setting the proper values for the mesh tessellation divisions ensures the model has enough faces, edges, and vertices for editing. The default tessellation divisions are listed in the **Mesh Primitive Options** dialog box, which is discussed in the next section.

Within the tessellation division, the face consists of structures known as *facets*. See **Figure 9-2B**. The facets create a grid pattern that is related to the smoothness of the mesh. As the smoothness level of the mesh object increases, the facets increase in number, resulting in a smoother, more rounded surface. As the smoothness level decreases, the facets decrease in number, resulting in a rougher, less rounded surface.

You can change the default smoothness in the **Mesh Primitive Options** dialog box or, if the primitive is being drawn using the command line, by entering the **Settings** option of the **MESH** command. **Figure 9-3** shows an example of a box mesh primitive created using the default tessellation divisions. The default settings create a box with no smoothness, length divisions of three, width divisions of three, and height divisions of three.

NOTE

Any mesh object or subobject that has a level of smoothness of 1 or higher can be refined by converting facets to editable faces. Refining a mesh is discussed later in this chapter. You can also adjust the appearance of the facets by setting the **VSLIGHTINGQUALITY** system variable. Graphic card and monitor display *may* affect visibility.

Figure 9-3.
This box mesh primitive is drawn with the default settings. A—Displayed with the 2D Wireframe visual style current. B—After the **HIDE** command is used.

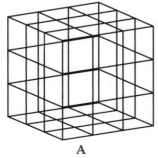
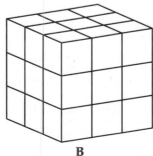

Drawing Mesh Primitives

MESH

Ribbon

Mesh
> Primitives

Mesh Box

Mesh Cone

Mesh Cylinder

Mesh Pyramid

Mesh Sphere

Mesh Wedge

Mesh Torus

Type

MESH

A primitive is a basic building block. Just as there are solid primitives in AutoCAD, there are mesh primitives. The seven *mesh primitives* are the mesh box, mesh cone, mesh cylinder, mesh pyramid, mesh sphere, mesh wedge, and mesh torus. See **Figure 9-4**. These primitives can be used as the starting point for creating complex freeform mesh models.

The **Mesh Primitive Options** dialog box is used to set the number of tessellation subdivisions for each mesh primitive object created. See **Figure 9-5**. This dialog box is displayed by picking the dialog box launcher button at the lower-right corner of the **Primitives** panel in the **Mesh** tab of the ribbon or by using the **MESHPRIMITIVEOPTIONS** command. Set the number of tessellation subdivisions in the **Mesh Primitive Options** dialog box *before* creating a mesh primitive. There is no way to change the number of subdivisions after the primitive is created.

The tessellation subdivisions for each primitive are based on the dimensions required to create the primitive. For example, a box has length, width, and height subdivisions. On the other hand, a mesh cylinder has axis, height, and base subdivisions. To set the subdivisions, select the primitive in the tree on the left-hand side of the **Mesh Primitive Options** dialog box. The subdivision properties are then displayed below the tree. Enter the number of subdivisions as required and then close the dialog box. All new primitives of that type will have this number of subdivisions until the setting is changed in the dialog box. Existing primitives are *not* affected.

Figure 9-4.
An overview of AutoCAD's mesh primitives and the dimensions required to draw them.

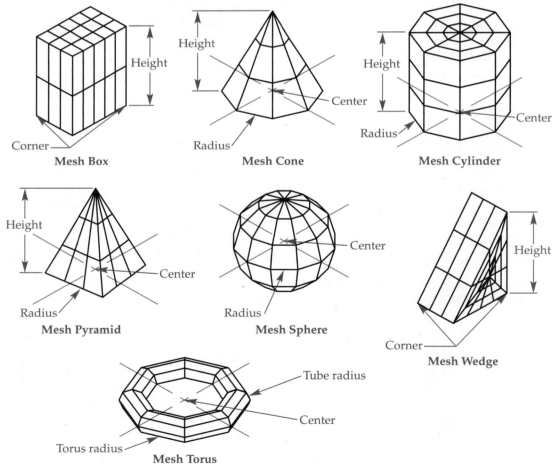

Figure 9-5.
The **Mesh Primitive Options** dialog box.

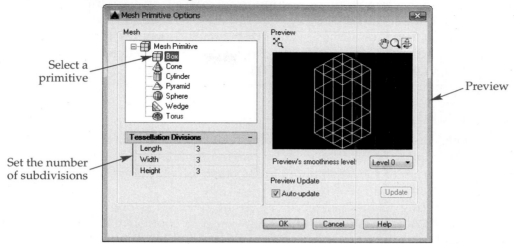

Select a primitive

Set the number of subdivisions

Preview

The following is an example of how to create a mesh box primitive. The mesh box is the mesh primitive used for most base shapes. Refer to **Figure 9-4** for the information required to draw mesh primitives.

1. Open the **Mesh Primitive Options** dialog box.
2. Select the box primitive in the tree.
3. Enter 3 in each of the Length, Width, and Height property boxes to set the subdivisions.
4. Close the dialog box.
5. Select the command for drawing a mesh box.
6. Specify the first corner of the base of the mesh box.
7. Specify the opposite corner of the base of the mesh box.
8. Specify the height of the mesh box.

PROFESSIONAL TIP

Creating mesh primitives is similar to creating solid primitives. Creating solid primitives is discussed in Chapter 2.

Exercise 9-1

www.g-wlearning.com/CAD/

Complete the exercise on the companion website.

Converting Between Mesh and Surface or Solid Objects

AutoCAD offers the flexibility of converting between solid, surface, and mesh objects. This allows you to select the type of modeling that offers the best tools for the task at hand, then convert the model into a form appropriate for the end result. The next sections discuss converting between solid, surface, and mesh objects and the settings that control the conversion.

Mesh Tessellation Options

The **Mesh Tessellation Options** dialog box contains many mesh options, **Figure 9-6**. This dialog box is displayed by picking the dialog box launcher button at the lower-right corner of the **Mesh** panel in the **Mesh** tab of the ribbon or by using the **MESHOPTIONS** command.

When converting to mesh objects, the resulting mesh is one of three different mesh types. The **FACETERMESHTYPE** system variable controls which type of mesh is created. Selecting Smooth Mesh Optimized in the **Mesh type:** drop-down list converts objects to the optimized mesh type (**FACETERMESHTYPE** = 0). This is the default and recommended setting. At this setting, the smoothness level can be set for the resulting mesh using the options in the **Smooth Mesh After Tessellation** area. Selecting Mostly Quads in the **Mesh type:** drop-down list creates faces that are mostly quadrilateral (**FACETERMESHTYPE** = 1). Selecting Triangle in the **Mesh type:** drop-down list creates faces that are mostly triangular (**FACETERMESHTYPE** = 2).

The **Mesh distance from original faces:** setting is the maximum deviation of the mesh faces (**FACETERDEVSURFACE**). Simply put, this setting determines how closely a converted mesh shape matches the original solid or surface shape.

The **Maximum angle between new faces:** setting is the maximum angle of a surface normal of two adjoining faces (**FACETERDEVNORMAL** system variable). The higher the value, the more faces created in very curved areas and the less faces created in flat areas. Increasing the value is good for objects that have curved areas, such as fillets, rounds, holes, or other tightly curved areas.

The **Maximum aspect ratio for new faces:** setting is the upper limit for the ratio of height to width for new faces (**FACETERGRIDRATIO** system variable). By adjusting this value, long faces can be avoided, such as those that would be created from a cylindrical

Figure 9-6.
The **Mesh Tessellation Options** dialog box.

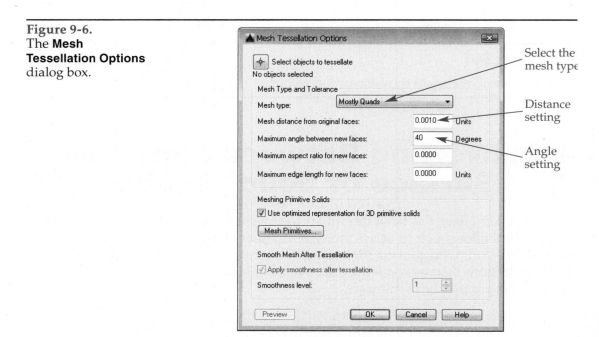

Select the mesh type

Distance setting

Angle setting

object during the conversion process. A low value will create a cleaner look in the formed faces. The default setting is 0, which specifies that no limitation is applied to the aspect ratio. A setting of 1 specifies that the height must be equal to the width. A setting greater than 1 specifies that the height may exceed the width. A setting between 0 and 1 specifies that the width may exceed the height.

The **Maximum edge length for new faces:** setting is the maximum length any edge can be (**FACETERMAXEDGELENGTH** system variable). The default setting is 0, which allows the size of the mesh to be determined by the size of the 3D model. A setting of 1 or higher results in a reduced number of faces and less accuracy compared to the original model.

When the **Use optimized representation for 3D primitive solids** check box is checked, the mesh settings in the **Mesh Primitive Options** dialog box are applied when converting a solid primitive. This is the default setting. If the check box is unchecked, the settings in the **Mesh Tessellation Options** dialog box are applied.

NOTE

Converting swept solids and surfaces, regions, closed polylines, 3D face objects, and legacy polygon and polyface mesh objects may produce unexpected results. If this happens, undo the operation and try making setting adjustments in the **Mesh Tessellation Options** dialog box for better results.

Converting from a Solid or Surface to a Mesh

The **MESHSMOOTH** command is used to convert a solid or surface object into a mesh object. See **Figure 9-7**. The command is easy to use. First, enter the command. Then, select the solids or surfaces to be converted. Finally, press [Enter] and the objects are converted into mesh objects. If you select an object that is not a primitive, you may receive a message indicating that the command works best on primitives. If you receive this message, choose to create the mesh.

Ribbon

Home
> Mesh
Mesh
> Mesh

Smooth Object

Type

MESHSMOOTH
SMOOTH
CONVTOMESH

MESHSMOOTH

PROFESSIONAL TIP

With a good understanding of solid modeling, create a solid model first. Then, convert it to a mesh object using the **MESHSMOOTH** command and edit the mesh to create a freeform design.

Figure 9-7.
Converting an existing solid or surface into a mesh object. A—The existing objects shown are a surface (left) and solid (right). B—After converting each object into a mesh using the **MESHSMOOTH** command.

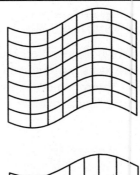
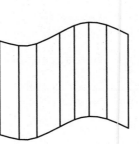

A

B

Converting From a Mesh to a Solid or Surface

Mesh objects can be converted into solids or surfaces using the **CONVTOSOLID** and **CONVTOSURFACE** commands. The faces on the resulting solid or surface can be smoothed or faceted and optimized or not. This is controlled by the **SMOOTHMESHCONVERT** system variable. Select one of the four possible settings *before* converting a mesh to a solid or surface. The settings are described as follows:

- **Smoothed and optimized.** Coplanar faces are merged into a single face. The overall shape of some faces can change. Edges of faces that are not coplanar are rounded. (**SMOOTHMESHCONVERT** = 0)
- **Smoothed and not optimized.** Each original mesh face is retained in the converted object. Edges of faces that are not coplanar are rounded. (**SMOOTHMESHCONVERT** = 1)
- **Faceted and optimized.** Coplanar faces are merged into a single, flat face. The overall shape of some faces can change. Edges of faces that are not coplanar are creased or angular. (**SMOOTHMESHCONVERT** = 2)
- **Faceted and not optimized.** Each original mesh face is converted to a flat face. Edges of faces that are not coplanar are creased or angular. (**SMOOTHMESHCONVERT** = 3)

To convert a mesh to a solid, first set the smoothing option, as described above. Then, select the **CONVTOSOLID** command. Next, select the mesh objects to convert and press [Enter]. The objects are converted from mesh objects to solid objects based on the selected smoothing option.

To convert a mesh to a surface, first set the smoothing option, as described above. Then, select the **CONVTOSURFACE** command. Next, select the mesh objects to convert and press [Enter]. The objects are converted from mesh objects to surface objects based on the selected smoothing option.

The examples shown in **Figure 9-8A** are simple mesh objects. In **Figure 9-8B**, the mesh objects have been converted into solid objects with the **Faceted, Optimized** button selected in the **Convert Mesh** panel on the **Mesh** tab in the ribbon. In **Figure 9-8C**, the mesh objects have been converted into surface objects with the **Smooth, Optimized** button selected.

PROFESSIONAL TIP

There are some mesh shapes that cannot be converted to a 3D solid. If using grips to edit the mesh, gaps or holes between the faces may be created. Smooth the mesh object to close the gaps or holes. Also, during the mesh editing process, mesh faces may be created that intersect each other and cannot be converted to a 3D solid. Converting this type of a mesh into a 3D solid will result in the following error message:

Mesh not converted because it is not closed or it self-intersects.
Object cannot be converted.

In some cases, there may be a mesh shape that cannot be converted to a solid object, but can be converted to a surface.

Figure 9-8.
A—Three basic mesh objects.
B—The mesh objects converted into faceted solids.
C—The mesh objects converted into smoothed surfaces.

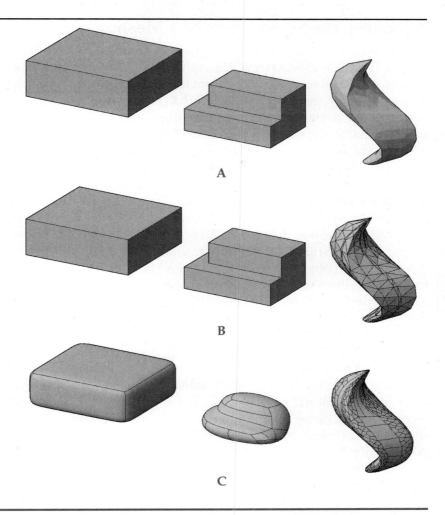

A

B

C

Exercise 9-3

www.g-wlearning.com/CAD/

Complete the exercise on the companion website.

Smoothing and Refining a Mesh Object

The roundness of a mesh object is increased by increasing the smoothness level. The smoothness can be set before creating the mesh, as described earlier in this chapter. The mesh object can also be refined or have its smoothness increased after it is created. This is described in the following sections. When smoothing or refining a mesh object, the number of tessellation subdivisions is either increased or decreased. The lowest smoothness level is 0 and the highest smoothness level is 4. The default smoothness level is 0.

PROFESSIONAL TIP

When creating primitives, begin with the least amount of faces possible. You can always refine the mesh model to create more faces.

Adjusting the Smoothness of a Mesh

MESHSMOOTHMORE

Ribbon

Home
> Mesh
> Smooth More
Mesh
> Mesh

Smooth More

Type

MESHSMOOTHMORE
MORE

When a mesh is smoothed, it changes form to more closely represent a rounded shape. The **MESHSMOOTHMORE** command is used to increase the level of smoothness of a mesh object. The maximum level of smoothness attained with this command is controlled by the **SMOOTHMESHMAXLEV** system variable. The default setting is 4. For example, a mesh object with a level of smoothness of 0 is considered to have no smoothness. If the value of the **SMOOTHMESHMAXLEV** system variable is 4, you can increase the smoothness of the mesh object up to level 4.

Once the **MESHSMOOTHMORE** command is selected, pick the mesh objects for which to increase the smoothness. Then, press [Enter] and the level is increased by one. See **Figure 9-9**. If the selection set includes objects that are not meshes, you have the opportunity to either filter out the non-mesh objects or convert them to meshes.

MESHSMOOTHLESS

Ribbon

Home
> Mesh
> Smooth Less
Mesh
> Mesh

Smooth Less

Type

MESHSMOOTHLESS
LESS

When a mesh is desmoothed, it changes form to more closely represent a boxed shape. The **MESHSMOOTHLESS** command is used to decrease the level of smoothness of a mesh object. Once the command is selected, pick the mesh objects for which to decrease the smoothness. Then, press [Enter] and the level is decreased by one. If the selection set includes objects that are not meshes, AutoCAD displays a message informing you that the non-mesh objects will be filtered out.

NOTE

You can change the **SMOOTHMESHMAXLEV** system variable setting to create a higher level of mesh model smoothness. The setting can range from 1 to 255. However, a recommended range of smoothness levels is 1–5. Using a level of 5 as the upper limit should be sufficient enough to create a very smooth model. Using a higher level may generate too many faces (a dense mesh) and affect system performance. In this case, you may receive an error message stating that the operation cannot be completed because your system does not have enough physical memory.

PROFESSIONAL TIP

The **MESHSMOOTHMORE** and **MESHSMOOTHLESS** commands change the smoothing one level at a time. You can use the **Properties** palette to change the smoothness to any level in one step.

Figure 9-9.
Using the **MESHSMOOTHMORE** command to increase the level of smoothness from level 0 to level 4. From left to right, smoothness levels 0, 1, 2, 3, 4.

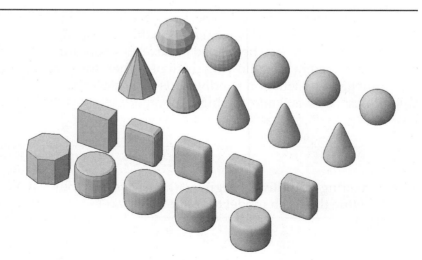

Exercise 9-4

www.g-wlearning.com/CAD/

Complete the exercise on the companion website.

Refining a Mesh

Refining a mesh increases the number of subdivisions in a mesh object. Refining adds detail to the mesh to make it more realistic with smooth, flowing lines. This also gives a greater selection for editing the mesh faces, vertices, or edges. The **MESHREFINE** command is used to refine a mesh. When the command is used, the number of subdivisions is quadrupled. See **Figure 9-10**. The smoothness level of the original object must be level 1 or higher.

Ribbon

Home
> Mesh
Mesh
> Mesh

Refine Mesh

Type
MESHREFINE
REFINE

MESHREFINE

The entire mesh can be refined (all faces at once) or a selected face can be refined. When you refine a mesh *model*, the increased amount of face subdivisions becomes a new smoothness level of 0. However, if you refine an individual *face*, the level of smoothness is not reset.

Be careful not to create too dense of a mesh. Creating a mesh that is too dense may result in face, edge, and vertex subobjects that are very small. This may make it difficult to select subobjects and edit the mesh.

Exercise 9-5

www.g-wlearning.com/CAD/

Complete the exercise on the companion website.

Editing Meshes

As discussed earlier, the second key workflow of mesh modeling is the ability to edit the mesh. The tools for editing a mesh are found in the **Mesh Edit** and **Selection** panels on the **Mesh** tab of the ribbon. See **Figure 9-11**. The face, edge, and vertex subobjects can be edited to change the shape of the mesh. These subobjects can be moved, rotated, or scaled. A face on the mesh can be split or extruded.

Subobject filters are used to assist in the selection of a mesh face, edge, or vertex *before* it is moved, rotated, or scaled. This is especially true for a very dense mesh. First, right-click in the drawing window and select **Subobject Selection Filter** to display the cascading menu, **Figure 9-12A**. Then, select the filter you wish to use. You can also select the filter in the **Selection** panel of the **Mesh** tab in the ribbon, **Figure 9-12B**. A subobject filter can also be selected from the **Filters object selection** flyout on the status bar. This flyout does not appear on the status bar by default. Select the **Selection Filtering** option in the status bar **Customization** flyout to add the **Filters object selection** flyout to the status bar.

Figure 9-10.
Refining a mesh.
A—The original mesh with a smoothness level of 1.
B—The refined mesh, which now has more faces and a smoothness level of 0.

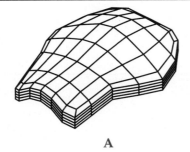

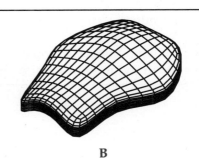

A

B

Figure 9-11.
The tools for editing a mesh are found in the **Mesh Edit** and **Selection** panels of the **Mesh** tab in the ribbon.

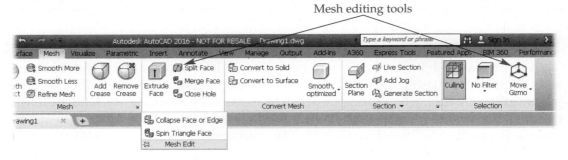

Figure 9-12.
Selecting a subobject filter. A—Using the shortcut menu. B—Using the drop-down list in the **Selection** panel of the **Mesh** ribbon tab.

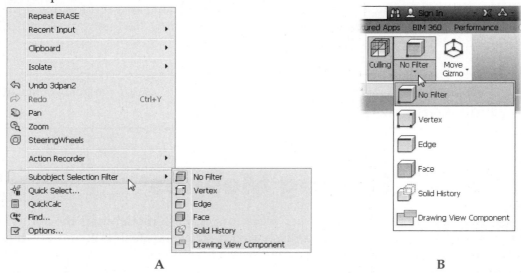

Subobject editing is discussed in detail in Chapter 11. The same procedures discussed in that chapter for solids can be applied to mesh models.

NOTE

Use the [Ctrl] key and left mouse button to select the subobjects or use the **SUBOBJSELECTIONMODE** system variable for subobject filtering. These settings apply:

- 0 = off
- 1 = vertices
- 2 = edges
- 3 = faces
- 4 = solid history subobjects
- 5 = drawing view components

Using Gizmos

A *gizmo*, also called a grip tool, appears when a face, edge, or vertex subobject on a mesh model is selected. This tool is used to specify how a movement, rotation, or scaling transformation is applied. A visual style other than 2D Wireframe must be current in order for the gizmo to appear. Using gizmos (grip tools) is introduced in Chapter 8 and covered in detail in Chapter 11.

You can switch between the move, rotate, and scale gizmos by using the drop-down list in the **Selection** panel on the **Mesh** tab of the ribbon. You can also use the **Show gizmos** flyout on the status bar. The **Show gizmos** flyout does not appear on the status bar by default. Select the **Gizmo** option in the status bar **Customization** flyout to add the **Show gizmos** flyout to the status bar.

When you make a subobject selection on a mesh model, the ribbon displays *context-sensitive panels* based on the selection. For example, if a face subobject is selected, the ribbon displays the **Crease** and **Edit Face** panels, **Figure 9-13**. Context-sensitive panels allow improved editing efficiency by displaying tools commonly used for the selected item. These panels are indicated by the green bar under the panel name. Once the selection is canceled, the context-sensitive panels are no longer displayed.

PROFESSIONAL TIP

Before selecting a face, edge, or vertex to perform a move, rotate or scale operation, set the appropriate selection filter and gizmo.

Extrude a Face

You can select a face subobject on a mesh and extrude it. Extruding a mesh face adds new features to the mesh. This creates new faces that can be edited. The **MESHEXTRUDE** command is used to extrude mesh faces. The command sequence to extrude a mesh face is similar to extruding to create a solid shape. The **EXTRUDE** command is covered in detail in Chapter 6.

To extrude a face, enter the command. The command automatically turns on the face subobject filter. This is indicated by the **Face** button on the **Selection** panel in the **Mesh** tab on the ribbon. Next, pick the face subobject(s) to extrude. Finally, enter an extrusion height. In **Figure 9-14**, a camera is being developed as a mesh model. The **MESHEXTRUDE** command is used to extrude an individual face as the first step in creating the lens tube.

> Ribbon
> **Mesh**
> **> Mesh Edit**
>
> **Extrude Face**
> Type
> **MESHEXTRUDE**

MESHEXTRUDE

PROFESSIONAL TIP

Extrude a face using the **MESHEXTRUDE** command instead of moving a face. This will give greater editing control over an individual face.

Figure 9-13.
A context-sensitive panel is only displayed when certain objects or subobjects are selected. In this case, the **Crease** and **Edit Face** panels are displayed when a face subobject is selected.

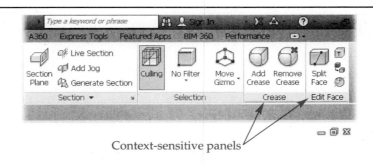

Context-sensitive panels

Figure 9-14.
Extruding a mesh face. A—Select the face to extrude. B—The process adds faces to the model.

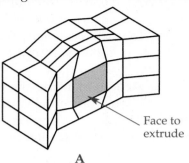

Face to extrude

A

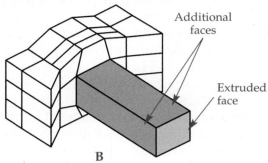

Additional faces

Extruded face

B

Split a Face

Splitting a mesh face is used to increase the number of faces on the model without refining a mesh. Splitting one face creates two faces. This is easier than refining the mesh. Think of these new split faces as subdivisions of an existing face.

The **MESHSPLIT** command is used to split a face. Once the command is entered, select the face to split. Then, specify the starting point and the ending point of a line defining the split. You can specify any two points on the mesh face. **Figure 9-15** shows a mesh model with three faces that have been split.

Once the faces are split, mesh editing options, such as extruding, moving, rotating, or scaling, can be used on the new faces. In **Figure 9-16**, the split faces from **Figure 9-15** have been extruded, and the model has been smoothed to level 3.

MESHSPLIT

Ribbon
Mesh
> Mesh Edit

Split Face

Type
MESHSPLIT
SPLIT

Exercise 9-6

www.g-wlearning.com/CAD/

Complete the exercise on the companion website.

Figure 9-15.
Three faces have been split on this mesh model. Each face has been split into two faces.

Split lines

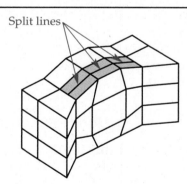

Figure 9-16.
Three faces have been extruded on this mesh model and smoothing has been applied. Notice how the model edges are rounded.

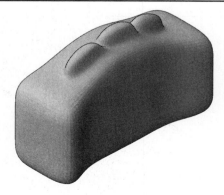

Applying a Crease to a Mesh Model

A *crease* is a sharpening of a mesh subobject, much like a crease in folded paper. A crease sharpens or squares off an edge or flattens a face. This prevents the subobject from being smoothed. Creases can be applied to faces, edges, and vertices. A smoothness level of 1 or higher must be assigned to the mesh object for the creases to have an effect, but they can be applied at any smoothness level. Once a crease is applied, any existing smoothing is removed from the subobject. If the mesh smoothness is increased, any creased subobjects are not smoothed. Also, creasing an edge *before* an object is smoothed will limit the mesh editing capabilities.

An example of where a crease might be applied is the bottom of a computer mouse. As the design of the computer mouse begins, the base of the mesh model is rounded when the model is smoothed. See **Figure 9-17A**. By creasing all bottom edges of the mouse, the bottom is squared off. See **Figure 9-17B**. A flat bottom is a requirement of the design intent because the mouse must sit flat on a desk.

In another example, the trigger pads on the game controller shown in **Figure 9-18** have been creased to create a flat surface. When the conceptual design is finished, it can be converted to a solid. This will give the designer a unique shape to which buttons and joysticks can be added. This unique shape would be difficult, if not impossible, to create from scratch as a solid model.

Once the command is entered, select the subobjects to crease. You do not need to press the [Ctrl] key to select subobjects, but filters should be used to make selection easier. Once the subobjects are selected, press [Enter]. This prompt appears:

Specify crease value [Always] <Always>:

If you enter a crease value, this is the highest smoothing level for which the crease is retained. If the smoothness level is set higher than the crease value, the creased subobject is smoothed. The Always option forces the crease to be retained for all smoothness levels. In most cases, this is the recommended option.

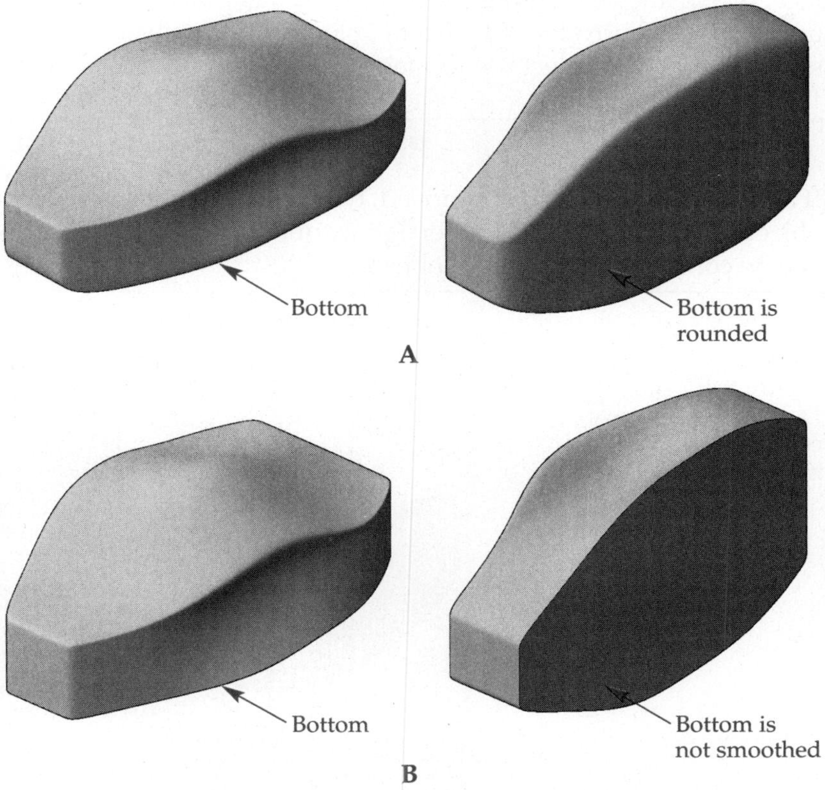

Ribbon
Mesh
> Mesh

Add Crease

Type
MESHCREASE
CREASE

MESHCREASE

Figure 9-17.
Adding creases to a mesh. A—Before the creases are applied, the bottom of the mouse is rounded. This is obvious if the object is rotated (right). B—After the creases are added, the bottom is flat.

Bottom

Bottom is rounded

A

Bottom

Bottom is not smoothed

B

Figure 9-18.
In this example, creases have been added to the mesh to create flat areas on the game pad for buttons.

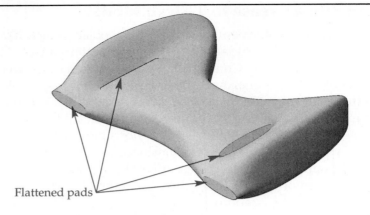

Flattened pads

Exercise 9-7

www.g-wlearning.com/CAD/

Complete the exercise on the companion website.

Removing a Crease

Removing a crease is a simple process. The **MESHUNCREASE** command is used to do this. Enter the command and then select the subobjects from which a crease is to be removed. Then, press [Enter]. The crease is removed and smoothing is applied as appropriate.

Creating a Mesh Gap

As the designer creates meshes, editing the mesh faces is the next step toward final design. There will be times when the designer needs to modify a mesh by deleting faces. The techniques of erasing or deleting a mesh face aid in the design. To delete a mesh face, use the **ERASE** command or press the [Delete] key after selecting the mesh face. By erasing or deleting the mesh face, a gap in the mesh occurs. **Figure 9-19** shows a game pad controller with *multiple* top faces removed. Use the [Ctrl] key or the face subobject filter to select the top faces and then press the [Delete] key.

PROFESSIONAL TIP

If a mesh face is erased or removed, the mesh object is not considered a *watertight* mesh object. A mesh that is not watertight cannot be converted to a solid object. However, it can be converted to a surface object.

Figure 9-19.
Deleting faces creates a gap on the mesh model.

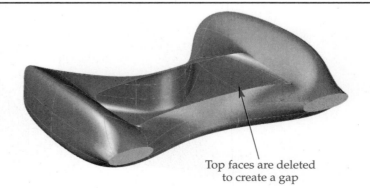

Top faces are deleted to create a gap

Close Gaps in Mesh Objects

After a mesh face has been erased or deleted, the designer has the ability to close the mesh gap. The **MESHCAP** command is used to close the gap, **Figure 9-20.** This is done by creating a new face between selected, continuous edges. Once the command is entered, select the edges of the surrounding mesh object faces. The selected edges do not need to form a closed loop. If a closed loop is not formed, AutoCAD automatically closes the loop. Then, press [Enter] to complete the command and close the mesh gap.

To select multiple edges at once, use the **Chain** option of the **MESHCAP** command. A quick way to access this option is to right-click and select **Chain** from the shortcut menu after entering the **MESHCAP** command. The **Chain** option allows you to select multiple continuous edges with a single pick.

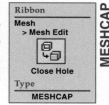

Ribbon
Mesh
> Mesh Edit

Close Hole

Type
MESHCAP

MESHCAP

PROFESSIONAL TIP

When using the **MESHCAP** command, the selected mesh face edges should be on the same plane whenever possible. Also, the **MESHCAP** command cannot be used with edges that are shared by adjacent faces, such as edges that form the boundary of a hole feature.

Exercise 9-8

www.g-wlearning.com/CAD/

Complete the exercise on the companion website.

Collapse a Mesh Face or Edge

After a mesh is created, the surrounding mesh faces may be converged to create a different and unique mesh face shape. This is called collapsing a mesh. The designer can collapse surrounding mesh faces at the *center* of a selected mesh edge or face. New mesh faces are created. This helps the designer create new mesh face shapes for further editing.

The **MESHCOLLAPSE** command is used to collapse a mesh. Once the command is selected, pick the face or edge to collapse. The command is immediately applied and then ends.

Figure 9-21 shows an in-progress conceptual design for a computer mouse. The side must be collapsed to a point and then moved inward. Enter the command and select the middle edge. Be sure to use filters to help select the edge. Once the edge is collapsed, the new vertex can be selected for subobject editing and moved inward.

Ribbon
Mesh
> Mesh Edit

Collapse Face or Edge

Type
MESHCOLLAPSE

MESHCOLLAPSE

Figure 9-20.
A gap on a mesh model can be closed with the **MESHCAP** command.

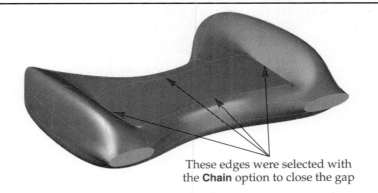

These edges were selected with the **Chain** option to close the gap

Figure 9-21.
Collapsing a mesh. A—The edge to be selected. B—The mesh is collapsed and the new vertex is edited.

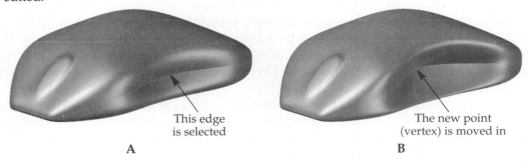

This edge
is selected

The new point
(vertex) is moved in

A

B

Exercise 9-9

www.g-wlearning.com/CAD/

Complete the exercise on the companion website.

Merge Mesh Faces

The designer can merge adjacent mesh faces into a new, single mesh face. The **MESHMERGE** command is used to do this. Two or more adjacent faces can be selected to merge to create a different or unique face. For best results, the faces should be on the same plane. Do *not* try to merge faces that are not adjacent or are on corners.

Figure 9-22 shows a tape dispenser. The three faces on the front need to be merged to create a new mesh shape. Enter the command and set the face subobject filter. Then, select the three adjoining faces to merge. Press [Enter] to complete the command.

Exercise 9-10

www.g-wlearning.com/CAD/

Complete the exercise on the companion website.

Figure 9-22.
Merging faces. A—The faces to be merged. B—The resulting model.

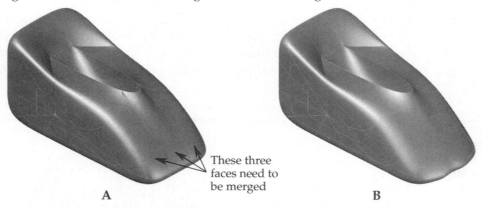

These three
faces need to
be merged

A

B

Spin Mesh Faces

Mesh faces have different and unique shapes when created. Additional unique shapes can be created by spinning a triangular mesh face. Spinning a mesh face spins the adjoining edge of *two* triangle mesh faces. Modifying a mesh face by spinning rotates the newly shared edge. The **MESHSPIN** command is used to spin a face.

PROFESSIONAL TIP

Use the **Vertex** option of the **MESHSPLIT** command to create triangular mesh faces.

Figure 9-23 shows a pocket camera. The camera's top button has been split to create two triangular faces. Before the button top can be extruded and smoothed, the top adjoining faces must be spun.

Select the command, then pick the two triangular faces. Press [Enter] to complete the command. Notice how the diagonal line of the triangular faces is now in a different orientation. You can then extrude one of the faces up or down to create a new button. Smooth the camera to create its final design shape.

Figure 9-23.
Spinning a face. A—The face has been split, but the orientation is wrong. B—The face is spun and the orientation is correct.

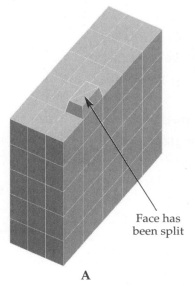
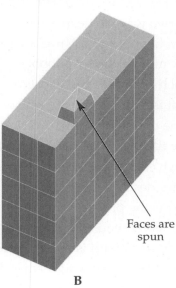

Face has been split

Faces are spun

A

B

Chapter Review

Answer the following questions using the information in this chapter.

1. Of what does a mesh model consist?
2. What is another term for *mesh models*?
3. What are *tessellation divisions*?
4. When creating a mesh primitive, when should mesh tessellation divisions be set?
5. What are *facets*?
6. For what is the **Mesh Primitive Options** dialog box used?
7. How is a mesh box created?
8. How is a mesh sphere created?
9. How is a mesh torus created?
10. What is the purpose of the **DELOBJ** system variable?
11. Which command converts a mesh object to a surface object?
12. Which command converts a mesh object to a solid object?
13. How is the roundness of a mesh object increased?
14. Which command is used to convert an existing solid or surface to a mesh object?
15. Name the system variable that controls the maximum level of smoothness attained with the **MESHSMOOTHMORE** command.
16. List two ways to decrease the smoothness of a mesh.
17. What happens to the mesh when you refine it?
18. How many types of subobjects does a mesh have? List them.
19. Which keyboard key is used to select subobjects for editing?
20. What is a *context-sensitive panel*?
21. Name the three operations that can be performed with a gizmo.
22. How do you cycle through the three different gizmos?
23. Which command is used to extrude a mesh face?
24. Briefly describe the process for extruding a mesh face.
25. What is the process for splitting a mesh face?
26. Why would you crease a mesh model?
27. Which command is used to remove a crease?
28. Explain why you would erase or delete a mesh face during the design process.
29. Which command is used to close gaps in a mesh object?
30. What is the purpose of collapsing a mesh face or edge?

CHAPTER 10

Advanced Surface Modeling

Learning Objectives

After completing this chapter, you will be able to:

✓ Understand and work with different types of surface models.
✓ Create procedural surfaces.
✓ Create NURBS surfaces.
✓ Create network surfaces.
✓ Create surface models from existing surfaces.
✓ Blend and patch surfaces.
✓ Offset, fillet, extend, and trim surfaces.
✓ Convert existing models to NURBS surfaces.
✓ Edit NURBS surface control vertices.
✓ Convert 2D objects to surfaces.
✓ Thicken a surface into a solid.
✓ Sculpt watertight surfaces into solids.
✓ Extract curves from existing surfaces.

Overview

This chapter describes advanced surface modeling techniques and workflows used in AutoCAD. Surface modeling provides the ability to create a more freeform shape with tools that solid modeling cannot provide. You have been introduced to basic surface modeling techniques in previous chapters. As you have learned, one way to create surface models is to extrude, revolve, sweep, or loft profiles. This chapter builds on those techniques. In this chapter, you will develop an understanding of surface modeling techniques that can stand alone in the design process or work in combination with other modeling techniques.

As discussed in previous chapters, a **solid model** is created with a closed and bounded profile and has mass and volume properties. A **mesh** consists of vertices, edges, and faces that define the 3D mesh shape. A mesh does not have mass or volume. A **surface model** can be thought of as a thin-walled object with no "Z" depth. A surface model does not have mass or volume.

There are a number of workflows in AutoCAD available to the 3D designer. The following approaches can be considered depending on the nature of the work or the requirements of a specific application:

- Creating 3D models as solids, meshes, procedural surfaces, or NURBS surfaces (procedural surfaces and NURBS surfaces are discussed in the next section).
- Using Boolean operations on solids to create composite solids.
- Slicing composite solids using surfaces.
- Converting solids to mesh models.
- Converting solids to surface models.
- Converting surface models to NURBS surfaces.

These are just a few of the possible workflows. Editing techniques are also available and often play a significant role in surface modeling.

Understanding Surface Model Types

There are two basic types of surface models in AutoCAD: procedural surfaces and NURBS surfaces. A *procedural surface* is a standard surface object without control vertices. By default, a procedural surface, when created, is an *associative surface*. This means that the surface maintains associativity to the defining geometry or to other surrounding surfaces. Editing the defining geometry of an associative surface, or an adjacent surface in a "chain" of associative surfaces, modifies the surface.

A *NURBS surface* is based on splines or curves. The acronym *NURBS* stands for *non-uniform rational B-spline*. NURBS surfaces are based on a mathematical model and are used to create organic, freeform shapes. NURBS surfaces have control vertices that can be manipulated to edit the shape of the surface with great precision. Unlike a procedural surface, a NURBS surface cannot be created as an associative surface.

A third type of surface in AutoCAD is a *generic surface*. A generic surface has no associative history and no control vertices.

The type of surface model created is controlled by the **SURFACEMODELINGMODE** system variable. The default setting, 0, creates procedural surfaces. If the **SURFACEMODELINGMODE** system variable is set to 1, NURBS surfaces are created.

When creating a procedural surface, the **SURFACEASSOCIATIVITY** system variable setting determines whether an associative surface is created. The default setting, 1, creates associative surfaces. This system variable has no effect when creating NURBS surfaces.

Surface models can be created from either closed and bounded geometry or open profile geometry. When using modeling commands such as **EXTRUDE**, **REVOLVE**, **SWEEP**, and **LOFT**, the **Mode** option determines whether a surface model or solid model is created. Open profile curves always create surfaces, regardless of the **Mode** option setting.

The advantage to a procedural surface is the ease with which the surface can be created based on common shapes. In addition, working with procedural surfaces allows the designer to take advantage of associative modeling. Based on the design intent, the designer can use profile curves such as lines, circles, arcs, ellipses, helices, points, polylines, 3D polylines, and splines as the basis for the model. A procedural surface model can then be created using commands such as **EXTRUDE**, **REVOLVE**, **SWEEP**, **LOFT**, or **PLANESURF** (as discussed in previous chapters). If created as an associative surface, the model is linked to the defining geometry and can be modified by editing the geometry.

The commands used to create surface models are located in the **Surface** tab of the ribbon. See **Figure 10-1**. As discussed in Chapter 6, selecting a command from the **Create** panel in the **Surface** tab automatically sets the model creation mode to

Figure 10-1.
The **Surface** tab of the ribbon.

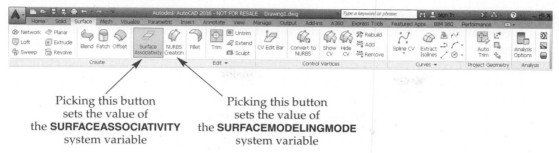

Picking this button
sets the value of
the **SURFACEASSOCIATIVITY**
system variable

Picking this button
sets the value of
the **SURFACEMODELINGMODE**
system variable

surface. Referring to **Figure 10-1**, notice the **Surface Associativity** and **NURBS Creation** buttons in the **Create** panel. The status of the **Surface Associativity** button indicates the **SURFACEASSOCIATIVITY** system variable setting. By default, the button has a blue background to indicate the system variable is turned on. The status of the **NURBS Creation** button indicates the **SURFACEMODELINGMODE** system variable setting. By default, this button does not have a blue background. This indicates that procedural surfaces are created by default.

PROFESSIONAL TIP

Procedural surfaces are also referred to as *explicit surfaces*. The terms *procedural surface* and *explicit surface* are interchangeable. This text uses procedural surfaces to illustrate various methods of surface creation. NURBS surfaces, as discussed later in this chapter, are used in creating models that represent sophisticated freeform shapes. Procedural surfaces can be converted to NURBS surfaces in order to model more sophisticated surface model shapes.

NOTE

Planar surface creation is discussed in Chapter 2. Extruded and revolved surfaces are discussed in Chapter 6. Swept and lofted surfaces are discussed in Chapter 7.

Exercise 10-1

www.g-wlearning.com/CAD/

Complete the exercise on the companion website.

Working with Associative Surfaces

Procedural surfaces, when created, are associative by default. An associative surface adjusts to the modifications made to the defining profile geometry (or other adjoining surfaces). This provides flexibility in the design. However, it is important to remember that when modifying the shape of an associative surface, you modify the profile geometry, *not* the surface. Modifying the profile geometry maintains the associative relationship. If you pick on the surface and then attempt to modify it, AutoCAD issues a warning that the surface will lose its associativity with the defining curve, surface, or parametric equation. If you choose to continue with the operation, the associativity is lost. You can cancel the operation to preserve the associativity.

Set the **SURFACEASSOCIATIVITY** system variable to 1 to create associative surfaces. If the system variable is set to 0, surfaces that are created have no associativity to defining profile curves or other surfaces.

An example of creating a procedural surface model is shown in **Figure 10-2**. In this example, an associative surface is created from a series of cross-sectional profiles using the **LOFT** command. As discussed in Chapter 7, a loft is created by selecting the cross-sectional profiles in order. In **Figure 10-2A**, four open profile curves are selected to create the loft. By default, this creates a surface model. In **Figure 10-2B**, the object is shown after lofting. In **Figure 10-2C**, the object is shown after moving and scaling the third open profile. Because the surface model is associative, the model updates and adjusts to conform to the new curve shape.

In **Figure 10-3**, a loft surface is created from three closed cross-sectional profiles (circles). The surface is created by using the **Mode** option or by selecting the command from the **Surface** tab in the ribbon. In **Figure 10-3B**, the object is shown after moving and scaling the third closed profile at the top. Because the surface is associative, the model updates and conforms to the new curve shape.

PROFESSIONAL TIP

The **DELOBJ** system variable is ignored if the **SURFACEASSOCIATIVITY** system variable is set to 1. Also, when creating a NURBS surface, the surface is a NURBS surface and is not associative.

Figure 10-2.
Creating a loft surface as an associative surface model. A—The original cross-sectional profiles. B—The model after using the **LOFT** command. C—The model after editing one of the cross-sectional profiles.

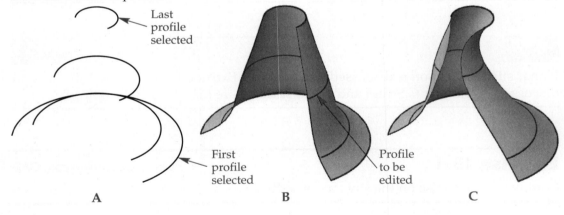

A   **B**   **C**

Figure 10-3.
Modifying a loft surface created from closed profiles (circles). A—The original loft surface. B—The model after editing the closed profile at the top.

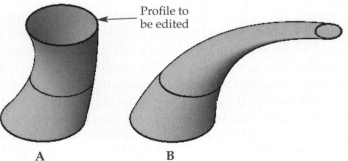

A   **B**

Removing Surface Associativity

The surface associativity can be removed from a surface once the surface has been created. This can be done by selecting the surface and opening the **Properties** palette. The Maintain Associativity property in the **Surface Associativity** category controls the associativity of the surface. See **Figure 10-4**. By default, the property is set to Yes. Selecting **Remove** from the drop-down list removes the associativity and changes the property to None. This converts the surface to a generic surface.

The **Show Associativity** property in the **Surface Associativity** category controls whether adjoining associative surfaces are highlighted when a surface is selected in order to indicate dependency. When this property is set to Yes and a surface is selected, AutoCAD highlights other surfaces to which the surface is dependent. This can be useful for identifying associative relationships in a chain of surfaces.

PROFESSIONAL TIP

When moving, scaling, or rotating an associative surface, be sure to select the underlying curve geometry defining the surface. Failure to select the underlying geometry will result in the loss of the associativity.

Determining Modeling Workflows for Procedural Surfaces and NURBS Surfaces

When the design of a 3D model requires a freeform shape that would be difficult to create using solids, start by creating a procedural surface. You can convert the surface as required. A practical application is creating a surface model of a car fender. Start with a lofted surface based on four guide curves. Finish creating the fender by creating several procedural surfaces or patches, as discussed later in this chapter. Then, convert the fender surfaces to NURBS surfaces as needed and add further editing techniques for a more freeform sculpted shape.

Different factors determine when to use procedural surface modeling and NURBS surface modeling. For example, create procedural surfaces when it is important to maintain associativity and you plan to edit the original geometry. On the other hand, NURBS surfaces have control vertices that typically permit greater flexibility when editing. NURBS surfaces are often very useful for modeling organic shapes. The extent to which the design will require further editing can serve as a guideline for determining the best modeling approach.

The following sections discuss surface modeling commands and techniques available in AutoCAD. Procedural surfaces are shown in examples as the default creation method. NURBS surface creation and editing techniques are covered later in this chapter.

Figure 10-4.
Associativity settings for a surface are accessed in the **Properties** palette.

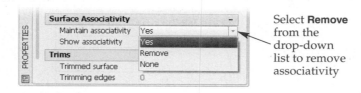

Select **Remove** from the drop-down list to remove associativity

---

Creating Network Surfaces

A *network surface* is a surface model created by a group or "network" of profile curves or edges. A network surface is similar to a loft surface. As with a loft, the defining profiles can be open or closed curves, such as splines. The defining profiles can also be the edges of existing objects, including region edges, surface edge subobjects, and solid edge subobjects. The curves or edges selected can intersect at coincident points, but do not have to intersect.

The **SURFNETWORK** command is used to create network surfaces. After selecting this command, select the curves or edges defining the first direction of the surface. Make sure to select the curves in the order of surface creation. Then, press [Enter]. Next, select the curves or edges defining the second direction of the surface. Press [Enter] when you are done selecting the profiles. This creates the surface and ends the command. See **Figure 10-5**. A network surface is created as an associative surface by default.

The curves selected for the two directions define the U and V directions of the surface. The U and V directions can be thought of as the local directions of the surface and can be defined in either order. The U and V directions define the "flow" of the surface.

Figure 10-6 shows examples of creating network surfaces from similar sets of profile curves. In **Figure 10-6A**, a series of connected profile curves defines the network surface. In **Figure 10-6B**, two of the curves do not intersect with other profiles. Notice the differences between the resulting surface models.

Creating a network surface from region edges, surface subobject edges, and solid subobject edges is shown in **Figure 10-7**. Creating a network surface in this manner may result in some unexpected surface shapes. To select the profile edges, press and hold the [Ctrl] key. You can also use the edge subobject filter by selecting **Edge** from the **Selection** panel in the **Home** tab of the ribbon. In **Figure 10-7**, the four objects are located at different "Z" heights. The network surface is created from the four nonintersecting edges.

Referring to **Figure 10-6**, when working in a wireframe display, isolines appear to represent the curved surfaces of the surface model. The **SURFU** and **SURFV** system variables control the number of isolines displayed in the U and V directions of the surface. The number of isolines does not include the lines defining the object's

Ribbon

Surface
> Create

Network Surface

Type

SURFNETWORK

SURFNETWORK

Figure 10-5.
Creating a network surface. Splines are used as the profiles in this example. A—The profiles used to define the surface are selected in the numbered order shown. Profiles 1–3 define the first direction and are shown in color. Profiles 4–8 define the second direction. B—The resulting surface model.

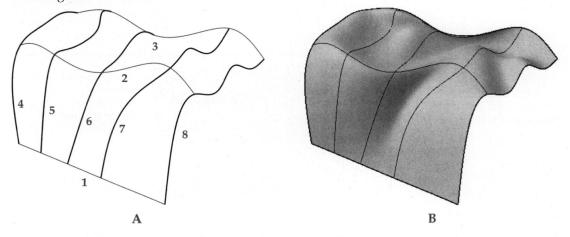

A

B

Figure 10-6.
Network surfaces. A—Profiles 1–3 define the first direction of the surface and are shown in color. Profiles 4 and 5 define the second direction of the surface. The resulting surface model is shown in wireframe and shaded form. B—Profiles 1–4 define the first direction of the surface and are shown in color. Profiles 5 and 6 define the second direction of the surface. The resulting surface model is shown in wireframe and shaded form.

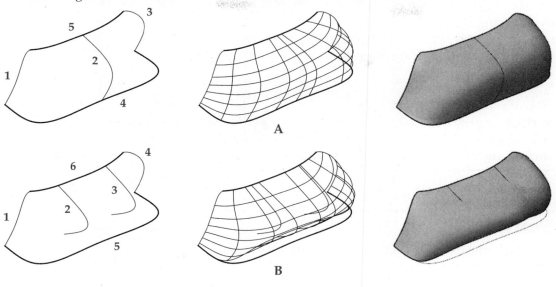

Figure 10-7.
Creating a network surface from the edges of existing objects. A—Two region edges (1 and 2) are selected to define the first direction of the surface. A surface subobject edge (3) and solid subobject edge (4) are selected to define the second direction of the surface. B—The resulting surface model.

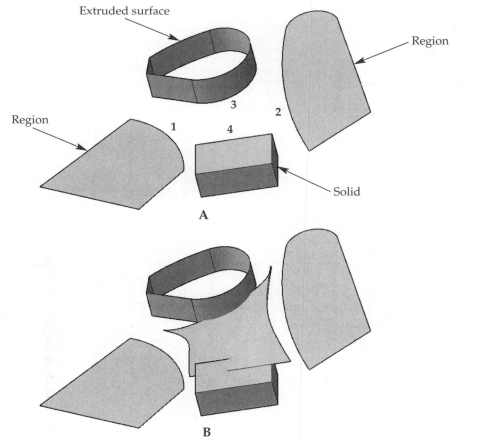

boundary. The default setting for both system variables is 6. These settings can be changed for a surface by selecting the surface and opening the **Properties** palette. The U isolines and V isolines properties in the **Geometry** category control the number of isolines displayed in the U and V directions. See **Figure 10-8**. Setting a higher value can give you a better understanding of the curvature of the model. When working in a shaded display instead of a wireframe display, you can view the isoline representation by hovering the cursor over the model. The default settings for the **SURFU** and **SURFV** system variables can be changed in the **3D Objects** area of the **3D Modeling** tab in the **Options** dialog box. The settings can range from 0 to 200.

PROFESSIONAL TIP

Surface associativity plays an important role in the creation of network surfaces. When editing a network surface with associativity, select a curve that forms the basis for the surface and modify it as needed. The result will be a new network surface shape.

Exercise 10-2

www.g-wlearning.com/CAD/

Complete the exercise on the companion website.

Figure 10-8.
Isolines defining the curvature of the surface model appear when working in a wireframe display. A—The **Properties** palette is used to set the number of isolines in the U and V directions of the surface. B—A network surface with the default number of isolines in the U and V directions. C—The surface after increasing the values of the U isolines and V isolines properties from 6 to 10.

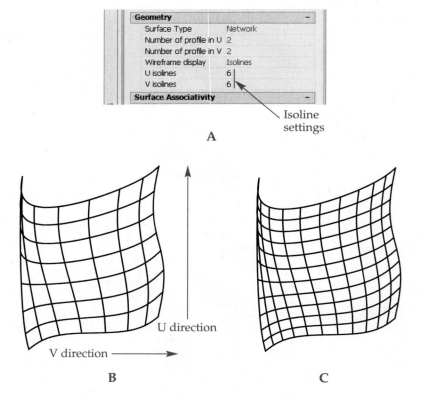

Copyright Goodheart-Willcox Co., Inc.

Creating Surfaces from Existing Surfaces

In addition to creating surfaces from profile curves, you can create surfaces from existing surfaces using the **SURFBLEND**, **SURFPATCH**, **SURFOFFSET**, **SURFFILLET**, and **SURFEXTEND** commands. These commands are discussed in the following sections. Surface models created with these commands are created as associative surfaces by default. This maintains associativity between the surfaces used to create the surface and the resulting surface.

Blend Surfaces

When working with surface models, there are situations when you need to "blend" together surfaces that do not meet or touch. The **SURFBLEND** command is used to create a *blend surface* between two surface edges or two solid edges. When blending surfaces, you select the surface edges to blend, *not* the surfaces.

Select the **SURFBLEND** command and then select the first edge to blend. Press and hold the [Ctrl] key to select the first edge or use the edge subobject filter. You can select multiple edges or use the **Chain** option to select a chain of continuous edges. Press [Enter] after defining the first edge. Next, select the second edge. Select a single edge or multiple edges, or use the **Chain** option to select a chain of continuous edges. When you press [Enter], a preview of the surface appears. You can press [Enter] to accept the default settings, as shown in **Figure 10-9**, or you can use the **Continuity** and **Bulge magnitude** options to specify the continuity and bulge magnitude settings at the edges. Different settings can be applied at each edge. The settings are similar to those when creating a loft, as discussed in Chapter 7.

Continuity defines how the surfaces blend together at the starting and ending edges. The following options are available:

- **G0 (positional continuity).** This option creates a sharp transition between surfaces. The position of the surfaces is maintained continuously at the surface edges. This option is used for creating flat surfaces.
- **G1 (tangential continuity).** This option forms surfaces so that the end tangents match at the edges. The two surfaces blend together tangentially. This is the default option, as shown in **Figure 10-9B**.
- **G2 (curvature).** This option creates a curvature blend between the surfaces. The surfaces share the same curvature.

Examples of blend surfaces with different continuity settings are shown in **Figure 10-10**.

Ribbon
Surface > Create
Surface Blend
Type
SURFBLEND

SURFBLEND

Figure 10-9.
A blend surface is created between two existing surfaces by selecting a starting and ending edge. A—Two existing loft surfaces. B—The model after creating a blend surface with the default settings of the **SURFBLEND** command.

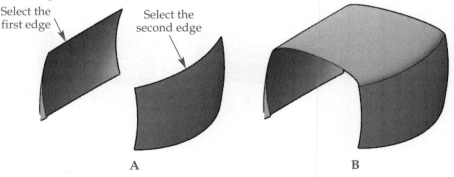

Select the first edge

Select the second edge

A

B

Bulge magnitude defines the size or "bulge" of the radial transition where the surfaces meet. See **Figure 10-11**. The default setting is 0.5. Valid values range from 0 to 1. A greater value is valid, but results in a larger roundness to the blend. Using different surface modeling techniques instead of entering a value greater than 1 is recommended. If the surface is set to G0 (positional continuity), changing the default bulge magnitude value has no effect.

Using different continuity and bulge magnitude settings modifies the surface and provides a way to create different blend surface shapes. You can change the continuity by using the grips that appear when creating the blend surface. Picking on a grip displays a menu with the continuity options. The same grips appear when selecting a blend surface after it has been created. You can also use the **Properties** palette to edit the settings of a blend surface.

Blend surfaces can also be created between the edges of regions or solid objects. An example of creating a blend surface between two solid subobject edges to form a cap is shown in **Figure 10-12**.

PROFESSIONAL TIP

Continuity settings are retained when exporting a 3D model to other 3D CAD modeling applications.

Figure 10-10.
Blend surfaces created with different continuity settings. A—G0 continuity. B—G1 continuity. C—G2 continuity.

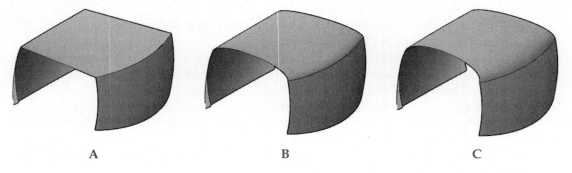

A   B   C

Figure 10-11.
Bulge magnitude determines the size of the radial transition at the edges where surfaces meet. A—The original model consists of two surfaces created from extruded arcs. B—The model after capping the ends with blend surfaces. For each surface, the continuity is set to G1 and the bulge magnitude is set to 0. C—For each surface, the continuity is set to G1. The bulge magnitude at the left edge is set to 0.5. The bulge magnitude at the right edge is set to 1.

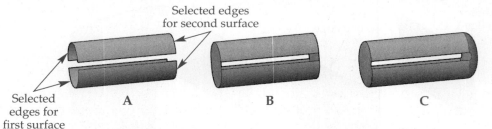

Selected edges
for second surface

Selected
edges for
first surface   A   B   C

Figure 10-12.
Creating a blend surface between two solid subobject edges. A—The original model consists of two solid boxes. B—A blend surface is created between two subobject edges. The continuity is set to G2 and the bulge magnitude is set to 1.

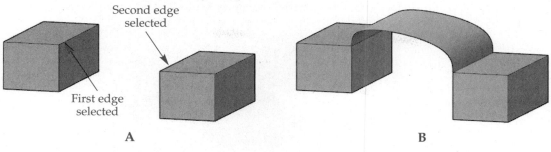

Second edge selected

First edge selected

A

B

Exercise 10-3

www.g-wlearning.com/CAD/

Complete the exercise on the companion website.

Patch Surfaces

A *patch surface* is used to create a "patch" over an opening in an existing surface. A patch surface is used when it is necessary to close an opening or gap in the model. You can think of a patch surface as one of the many squares making up a quilt.

Ribbon

Surface
> Create

Surface Patch
Type
SURFPATCH

SURFPATCH

The **SURFPATCH** command is used to create a surface patch based on one or more edges forming a closed loop. You can select one or more surface edges or a series of curves. As when using the **SURFBLEND** command, you can specify the continuity and bulge magnitude to define the curvature of the surface.

Select the **SURFPATCH** command and then select one or more surface edges defining a closed loop. You can use the **Chain** option to select a chain of continuous surface edges. You can also use the **Curves** option to select multiple curves forming a closed loop. After selecting the edges or curves, press [Enter]. A preview appears and you can press [Enter] to create the surface using the default settings. The **Continuity** and **Bulge magnitude** options can be used to change the default settings as previously discussed. The default continuity setting is G0. The default bulge magnitude setting is 0.5.

The **Guides** option allows you to use a guide curve to constrain the shape of the surface patch. You can select one or more curves to define the guide curve. You can also select points to define the guide curve. When selecting points, use object snaps as needed.

Examples of creating patch surfaces are shown in **Figure 10-13**. In **Figure 10-13A**, the top of the tent requires a patch. To create the patch, the single edge representing the opening in the model is selected with the default **Surface edges** option. In **Figure 10-13B**, the patch surface is created with the continuity set to G1. This is the appropriate setting for the patch surface. In this case, you would want the patch to be tangent to the existing surface. The default bulge magnitude (0.5) is used. In **Figure 10-13C**, the patch surface is created with the continuity set to G1, but the bulge magnitude is set to 1. Notice the different result.

When using the **Guides** option, draw a curve to serve as the guide curve prior to selecting the **SURFPATCH** command. In **Figure 10-14A**, the top of the tent requires a patch. A new middle post has been added to the model, and a spline is drawn to serve as a guide curve. After selecting the **SURFPATCH** command, select the surface edge. Then, select the **Guides** option, select the spline, and press [Enter]. Press [Enter] to create the patch surface, or adjust the continuity and bulge magnitude settings as needed. See **Figure 10-14B**.

Figure 10-13.
Creating a patch surface to close an opening in a surface model. A—The original model. The single surface edge indicated forms a closed loop and is selected to generate the patch surface. B—A patch surface created with the continuity set to G1 and the bulge magnitude set to 0.5. C—A patch surface created with the continuity set to G1 and the bulge magnitude set to 1.

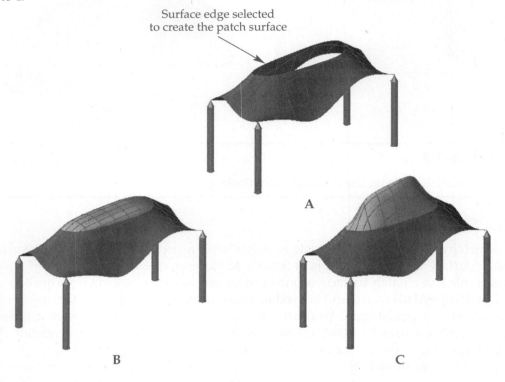

Surface edge selected
to create the patch surface

A

B

C

Figure 10-14.
Using a guide curve to constrain the shape of a patch surface. A—The original model. The middle post has been added. The spline is drawn for use with the **Guides** option. B—The patch surface created after selecting the spline as the guide curve. Notice the resulting shape.

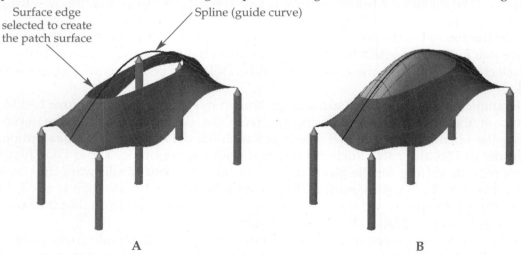

Surface edge
selected to create
the patch surface

Spline (guide curve)

A

B

In **Figure 10-15**, the game controller is to be redesigned with a new top shape. The model has been converted from a mesh model to a surface. In the original model (the mesh model), the top faces were deleted. The surface opening has eight continuous

Figure 10-15.
Using the **Chain** option to select multiple surface edges to define the patch surface. A—The original model is a surface converted from a mesh model. The top opening includes eight surface edges. B—The patch surface created after using the **Chain** option to select the edges. The continuity is set to G2 and the bulge magnitude is set to 0.7.

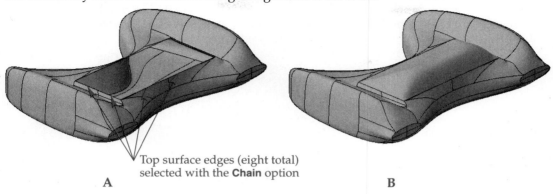

Top surface edges (eight total) selected with the **Chain** option

A

B

surface edges. The **Chain** option is used to assist in selecting edges to create the surface patch. After selecting the **SURFPATCH** command, select the **Chain** option and select one of the edges. The remaining edges are automatically selected. Next, press [Enter]. You can adjust the continuity and bulge magnitude settings or press [Enter] to create the patch surface. See **Figure 10-15B**.

Using the **Curves** option of the **SURFPATCH** command is shown in **Figure 10-16**. In **Figure 10-16A**, multiple curves have been created to design a sophisticated free-form shape. First, the **LOFT** command is used to create six loft surfaces defining the sides. Refer to the shaded surfaces shown in **Figure 10-16B**. Then, the **Curves** option of the **SURFPATCH** command is used to create three surface patches forming the top of the model. Refer to **Figure 10-16C**. To use the **Curves** option, enter it after selecting the **SURFPATCH** command. Then, select the curves defining the patch surface.

NOTE

Surfaces created using the **SURFPATCH** command may differ from those created with the **LOFT** command. In addition, you may come across design situations where curves used with the **LOFT** command will not create a surface, but will when used with the **SURFPATCH** command.

PROFESSIONAL TIP

The **PREVIEWCREATIONTRANSPARENCY** system variable controls the transparency of surface previews when using the **SURFBLEND**, **SURFPATCH**, and **SURFFILLET** commands. The default setting is 60. Setting a higher value increases the transparency of the surface preview.

Exercise 10-4

www.g-wlearning.com/CAD/

Complete the exercise on the companion website.

Figure 10-16.
Using the **LOFT** and **SURFPATCH** commands to create a freeform design from a series of curves. A—The original model. The **LOFT** command is used to create six separate loft surfaces to form the sides. Each loft surface consists of two cross-sectional curves (shown in color). B—The **Curves** option of the **SURFPATCH** command is used to create three separate surface patches forming the top of the model. Each surface is created from four curves (shown in color). C—The model after creating surface patches. For each surface patch, the continuity is set to G1 and the bulge magnitude is set to 0.

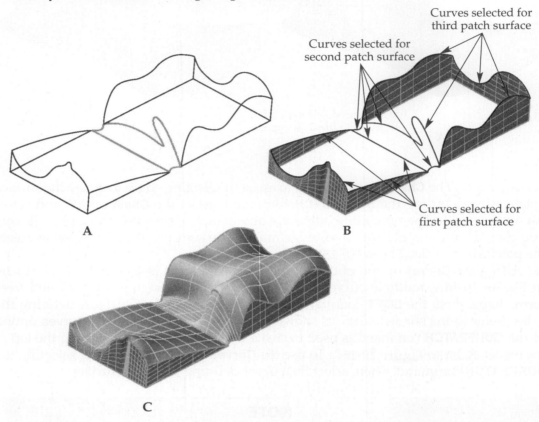

Curves selected for
third patch surface

Curves selected for
second patch surface

Curves selected for
first patch surface

A

B

C

Offsetting Surfaces

The **SURFOFFSET** command allows you to offset a surface to create a new, parallel surface at a specified distance. You can offset a surface in one direction or in both directions from an existing surface. You can also offset a region to create a new surface.

Select the **SURFOFFSET** command and select the surface or region to offset. Then, press [Enter]. A preview of the offset surface appears with offset arrows indicating the direction of the offset. See **Figure 10-17A**. Next, specify the offset distance. If the design calls for the offset surface to be located on the opposite side, select the **Flip direction** option. You can offset the surface to both sides by selecting the **Both sides** option. After specifying the offset distance, press [Enter] to create the offset surface. In **Figure 10-17B**, the hair dryer housing is offset to the inside of the existing surface at a distance of .125.

The **Solid** option allows you to create a new solid based on the specified offset distance. In **Figure 10-17C**, the **Solid** option is used to create a solid. This is similar to using the **THICKEN** command, as discussed later in this chapter.

The **Connect** option can be used when you have more than one surface to offset and you need to maintain connection between the surfaces. When using this option, the original surfaces must be connected. See **Figure 10-18**.

Figure 10-17.
Using the **SURFOFFSET** command. A—After selecting the surface to offset, offset arrows are displayed to indicate the direction of the offset. B—The surface is offset to the inside of the hair dryer housing at a distance of .125. C—The **Solid** option is used to create a new solid using the specified offset distance from the base surface.

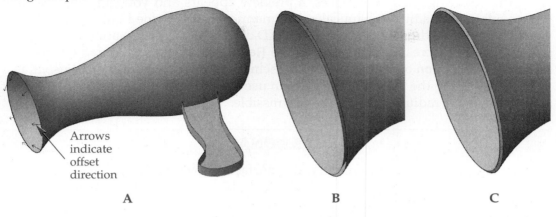

Arrows indicate offset direction

A   B   C

Figure 10-18.
The **Connect** option of the **SURFOFFSET** command is used to maintain the connection between surfaces when offsetting multiple surfaces. A—The original model. B—The vertical surfaces are offset using the **Connect** option.

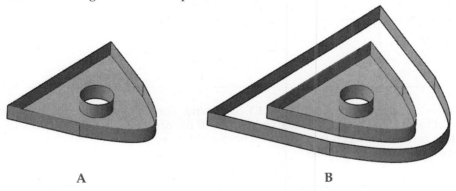

A   B

The **Expression** option allows you to enter an expression to constrain the offset distance. Surface associativity must be enabled in order to use this option.

Exercise 10-5

www.g-wlearning.com/CAD/

Complete the exercise on the companion website.

Creating Fillet Surfaces

The **SURFFILLET** command is used to create a fillet between two existing surfaces. The fillet created is a rounded surface that is tangent to the existing surfaces. You can create fillet surfaces from existing surfaces or regions. The existing objects do not have to intersect. Using the **SURFFILLET** command is similar to using other fillet commands in AutoCAD. Commands used to fillet solids are introduced in Chapter 8.

To create a fillet surface, select the **SURFFILLET** command. First, set a radius using the **Radius** option. Then, select two surfaces. AutoCAD stores the radius you specify as the setting for the **FILLETRAD3D** system variable. If you do not specify a radius, the current **FILLETRAD3D** system variable setting is used.

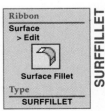

Ribbon
Surface
> Edit

Surface Fillet

Type
SURFFILLET

SURFFILLET

Creating a fillet surface between surfaces that do not meet is shown in **Figure 10-19A**. Creating a fillet surface between surfaces that share an edge or intersect is shown in **Figure 10-19B** and **Figure 10-19C**. By default, the existing surfaces are trimmed to form the new surface. The surface trimming mode can be set by using the **Trim surface** option.

After selecting the two surfaces, a preview appears and you can drag the fillet grip to adjust the radius dynamically. If dynamic input is turned on, you can type a value. If the fillet radius is too large, AutoCAD displays a message stating that the fillet surface cannot be created. When using the **Radius** option to set the radius, you can select the **Expression** option to enter a mathematical expression for the radius value.

After creating the fillet surface, you can use the **Properties** palette to edit the fillet surface radius. A radius of zero is not permissible.

PROFESSIONAL TIP

You can use the **UNION** command to union surfaces. However, it is not recommended. You will lose the surface associativity between the surfaces and the defining profile curves. Use surface editing commands instead.

Exercise 10-6

www.g-wlearning.com/CAD/

Complete the exercise on the companion website.

Figure 10-19.
Fillet surfaces created with the **SURFFILLET** command. For each example, the **Trim surface** option is set to **Yes**. A—A fillet surface created to fill the area between two surfaces. B—A fillet surface created between two surfaces meeting at an edge. C—A fillet surface created between two intersecting surfaces.

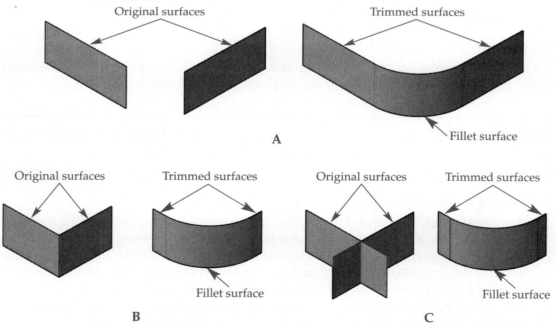

Extending Surfaces

You can add length to an existing surface using the **SURFEXTEND** command. When extending a surface, you can specify whether the new surface is created as a continuation of the existing surface or as a new surface. The surface extends to a new length using the specified distance.

Select the **SURFEXTEND** command and select one or more surface edges to extend. After selecting an edge, press [Enter]. A preview of the extended surface appears and you can drag the cursor dynamically to set the distance. You can also enter a distance by typing a value. The **Expression** option allows you to enter a mathematical expression for the extension distance. If you specify a distance and press [Enter], the surface is extended using the default settings. See **Figure 10-20**.

Before specifying the extension distance, you can use the **Modes** option to specify the extension mode. The two options are **Extend** and **Stretch**. The default **Extend** option is used to extend the surface in the same direction as the existing surface and attempt to maintain the surface shape based on the surface contour. The **Stretch** option is also used to extend the surface in the same direction as the existing surface. However, the resulting extension may not have the same surface contour.

After specifying the extension mode, you can use the **Creation type** option to set the type of surface created. The two options are **Append** and **Merge**. The default **Append** option is used to create a new surface extending from the original surface. This option results in two surfaces instead of one surface. The **Merge** option is used to extend the surface as one surface. After extending a surface, you can use the **Properties** palette to edit the extension distance.

Exercise 10-7

www.g-wlearning.com/CAD/

Complete the exercise on the companion website.

Trimming Surfaces

The **SURFTRIM** command can be used to trim surfaces or regions using other existing surfaces. You can trim any part of a surface where the surface intersects with another surface, region, or curve. In addition, you can project an existing object onto a surface to serve as a trimming boundary. The object to be trimmed and the cutting object do not have to intersect. When an associative surface is trimmed, it remains associative and retains the ability to be modified by editing the cutting object.

Figure 10-20.
Extending an existing surface using the **SURFEXTEND** command. A—The original model. B—The extended surface.

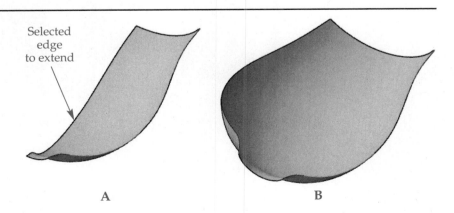

Selected edge to extend

A

B

SURFTRIM

Ribbon

Surface
> Edit

Surface Trim

Type

SURFTRIM

Select the **SURFTRIM** command, select one or more surfaces or regions to trim, and press [Enter]. Next, you are prompted to select the cutting objects. Select one or more curves, surfaces, or regions. Then, press [Enter]. You are then prompted to select the surface areas to be trimmed. Select one or more areas. As you select each area, it is trimmed by the cutting object(s). If you trim an area that you wish to restore, use the **Undo** option. When you are finished trimming, press [Enter] to end the command. See **Figure 10-21**.

The **Extend** and **Projection direction** options are available after selecting the **SURFTRIM** command. The **Extend** option determines whether a surface used as a cutting edge is extended to meet the surface to be trimmed. By default, this option is set to **Yes**. The **Projection direction** option specifies the projection method used for projected geometry, as discussed in the next section.

Exercise 10-8

www.g-wlearning.com/CAD/

Complete the exercise on the companion website.

Using Projected Geometry to Trim Surfaces

With the **SURFTRIM** command, you can trim surfaces using cutting objects other than existing surfaces. Objects used in this manner are referred to by AutoCAD as *curves*. Selecting a curve, such as an arc or circle, allows you to project the geometry onto the surface and use it as the cutting edge. See **Figure 10-22**. In the example shown, the cell phone case is selected as the surface to trim. The arcs located above the cell phone case are selected as the cutting edges. When selected, the arcs are projected onto the surface by AutoCAD. See **Figure 10-22B**. The areas to be trimmed are then selected to complete the sides. See **Figure 10-22C**.

To use a curve instead of an existing surface for trimming, select the curve after selecting the object to trim. You can select lines, arcs, circles, ellipses, polylines, splines, and helices. The **Projection direction** option of the **SURFTRIM** command can be used

Figure 10-21.
Trimming surfaces with the **SURFTRIM** command. A—The original model consists of three loft surfaces. Two of the loft surfaces are to be trimmed. B—The model after trimming.

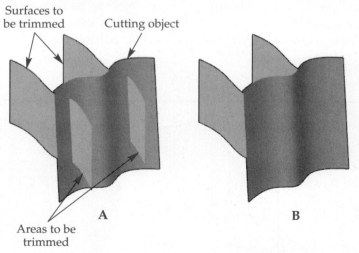

Surfaces to
be trimmed

Cutting object

Areas to be
trimmed

A

B

Figure 10-22.
Using projected geometry with the **SURFTRIM** command. A—The cell phone case model is a surface created from a mesh. The arcs drawn above the model are used for trimming the sides. B—Selecting the two curves and pressing [Enter] projects the curves to the model surface. The areas within the projected geometry on each side are selected as the areas to trim. C—The model after trimming.

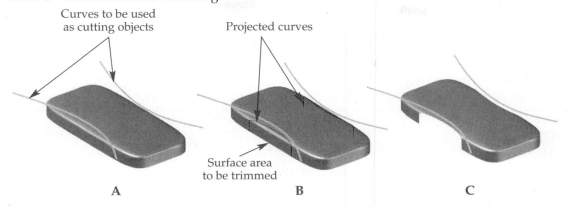

Curves to be used
as cutting objects

Projected curves

Surface area
to be trimmed

A   B   C

to set the projection method used by AutoCAD when projecting curves onto a surface. The following settings are available.

- **Automatic.** The cutting object is projected onto the surface to be trimmed. The projection is based on the current viewing direction. In a plan view, the projection of the cutting object is in the viewing direction. In a 3D view, the projection of a planar curve is normal to the curve, and the projection of a 3D curve is parallel to the direction of the Z axis of the current UCS. The **Automatic** option is set by default.
- **View.** The cutting object is projected in a direction based on the current view.
- **UCS.** The cutting object is projected in the positive or negative direction of the Z axis of the current UCS.
- **None.** The cutting object is not projected and must lie on the surface in order to perform the trim.

PROFESSIONAL TIP

The **SURFTRIM** command defaults to the **Automatic** option. Automatically projected geometry is used in most trim situations.

NOTE

When using the **PROJECTGEOMETRY** command, the **SURFACEAUTOTRIM** system variable controls automatic trimming. The default setting is 0. A setting of 1 enables automatic trimming of surfaces. The current setting is indicated by the **Auto Trim** button in the **Project Geometry** panel of the **Surface** tab on the ribbon.

Exercise 10-9

www.g-wlearning.com/CAD/

Complete the exercise on the companion website.

Ribbon
Surface
> Edit

⊕

Surface Untrim
Type
SURFUNTRIM

Untrimming Surfaces

If you need to restore a trimmed surface back to its original shape, use the **SURFUNTRIM** command. After selecting this command, select the edge of the surface area to untrim. If the surface has multiple trimmed edges, you can use the **Surface** option. The **SURFUNTRIM** command untrims surfaces trimmed by the **SURFTRIM** command. It does not untrim surfaces trimmed using the **PROJECTGEOMETRY** command.

PROFESSIONAL TIP

Open the **Properties** palette if you are unsure if a surface has been trimmed. The Trimmed surface property setting indicates the surface trim status.

NOTE

Exercise caution if you are trimming an associative surface using another surface as a cutting object. When using another surface as the cutting object, turn off surface associativity (set the **SURFACEASSOCIATIVITY** system variable to 0) before trimming to avoid potential problems in future edits.

NURBS Surfaces

When creating a NURBS surface, you use many of the same commands that you would use to create a procedural surface. You can use splines and various curve shapes to create NURBS surfaces. In addition, you can convert procedural surfaces into NURBS surfaces.

As discussed earlier in this chapter, a NURBS surface is created when the **SURFACEMODELINGMODE** system variable is set to 1. In addition, NURBS surfaces are non-associative. The setting of the **SURFACEASSOCIATIVITY** system variable has no effect when creating a NURBS surface.

The advantage of working with NURBS surfaces is that you use control vertices to control or influence the shape of the surface. The ability to edit control vertices provides significant flexibility in creating and sculpting freeform, organic shapes. For example, in computer animation work, NURBS surface modeling techniques are commonly used for modeling characters to produce the organic shape desired. In AutoCAD, you can create highly sophisticated, freeform shapes using NURBS surface models. This chapter introduces the NURBS surface modeling tools available in AutoCAD.

NURBS Surface Modeling Workflows

There are two common workflows used in NURBS surface modeling. You can begin by creating procedural surfaces and then convert them to NURBS surfaces, or you can create the initial surfaces as NURBS surfaces.

When you start the modeling process by working from procedural surfaces, the following workflow is common:

* Create procedural surfaces using commands such as **EXTRUDE, REVOLVE, SWEEP, LOFT, PLANESURF,** and **SURFNETWORK.** The **SURFACEMODELINGMODE** system variable should be set to 0.
* Create other surfaces, such as blend surfaces, patches, fillets, and offset surfaces. Use the commands presented in this chapter.
* Convert the surfaces into NURBS surfaces.
* Edit the NURBS surfaces as needed to create the desired sculpted shape.

13. How do you create a new solid when using the **SURFOFFSET** command?

14. What are the two creation type options available when using the **SURFEXTEND** command? What is the purpose of each option?

15. What are the three object types that can be used as cutting objects when trimming a surface?

16. What are the two basic ways to create a NURBS surface?

17. What command is used to display control vertices on a NURBS surface?

18. What command can be used to convert a 2D line or polyline into a surface model?

19. What is the purpose of the **THICKEN** command and which type of object does it create?

20. What is the preferred command to convert a watertight series of surfaces into a solid?

21. Name three objects you can use to extract isoline curves when using the **SURFEXTRACTCURVE** command.

22. What is the default direction used for extracting isoline curves with the **SURFEXTRACTCURVE** command?

Drawing Problems

1. General Create a loft surface from the three cross sections shown. The spacing between cross sections is 5 units. The circle cross section is a Ø6 circle. Use your own coordinates for the circle center point and the ellipse center points. To draw the two ellipse cross sections, refer to the dimensions given. The ellipses are centered on the same center point along the Z axis (refer to the top view). The major axis of the top ellipse is parallel to the minor axis of the lower ellipse. Use your own orientation for the ellipse axes relative to the circle (the exact orientation is not important). When creating the loft, create a procedural surface with associativity. After creating the loft, edit it by changing the dimensions of the ellipse cross sections. Refer to the dimensions given. Save the drawing as P10-1.

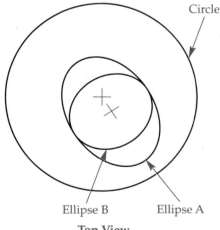

Top View

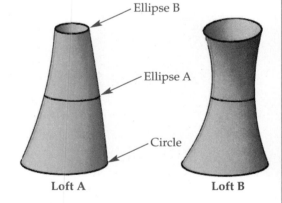

Loft A   Loft B

Cross Section	Dimensions (Loft A)	Edited Dimensions (Loft B)
Ellipse A	Major diameter = 3.80 Minor diameter = 2.70	Major diameter = 3.00 Minor diameter = 2.00
Ellipse B	Major diameter = 2.70 Minor diameter = 2.30	Major diameter = 6.00 Minor diameter = 4.00

2. **General** Create the cell phone case shown. Create the case as an extruded surface. Create a patch surface for the top surface. Then, use the arcs to trim the sides.
   A. Draw the base profile in the top view using the dimensions given.
   B. Draw the arcs in the top view. The arcs should extend past the perimeter of the cell phone case. Draw the first arc on the right side of the profile and then mirror it to the other side. Move the arcs along the Z axis so they are located .75 units above the bottom of the cell phone case.
   C. Extrude the base profile to a height of .5 units.
   D. Create the top of the cell phone case by creating a patch surface with C2 continuity and a bulge magnitude of .125.
   E. Trim the sides by projecting the arcs and selecting areas to trim.
   F. Save the drawing as P10-2.

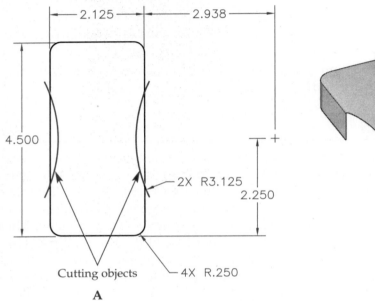

A

B

3. **General** Create the hair dryer handle shown. Create the base profile using a spline and a straight line segment. Use the profile to create a planar surface. Then, extrude the planar surface and fillet the bottom end. Finally, convert the model to a NURBS surface and edit the top surface.

　A.　Draw a spline and a line to create a profile similar to the profile shown. Use the **CV** option of the **SPLINE** command to create a CV spline. Dimensions are not important. Create the general shape by picking points to define the control vertices as indicated. Draw a line to form the straight segment at the top of the handle. Then, use the **JOIN** command to join the line to the spline. The resulting object should be a single, closed spline.

　B.　Create a planar surface from the profile.

　C.　Extrude the planar surface to a height of .85 units.

　D.　Use the **SURFFILLET** command to fillet the bottom end of the handle. Use a radius of .375 units. Set the trimming mode to **Yes**.

　E.　Using the **CONVTONURBS** command, convert the model to a NURBS surface. Edit the control vertices of the top surface to create a different shape.

　F.　Save the drawing as P10-3.

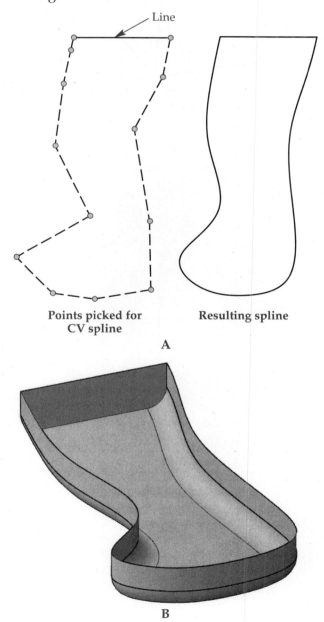

Line

Points picked for CV spline

Resulting spline

A

B

4. **Mechanical** Create the computer speaker using surface modeling commands. Use associative surfaces and edit the height from 4.5 units to 6 units, as shown. Save the drawing as P10-4.

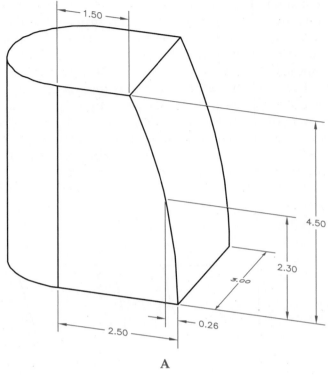

A

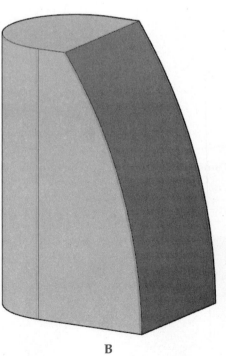

B

5. **Architectural** Create the kitchen chair shown. In earlier chapters, you created this model as a solid model. In this problem, use surface modeling commands to create the model. Use the drawings shown to create the seat, seatback, legs, and crossbars.

A. Create the top of the seat using the profile shown. Use the edge profile to create the curved, receding edge extending to the bottom of the seat. The bottom of the seat is 1" below the top.

B. Create the profile geometry for the seatback. Loft the circular cross sections to create the supports. Loft the square and circular cross sections to create the bow.

C. Create the framework for lofting the legs and crossbars as shown on the next page. The crossbars are at the midpoints of the legs. Position the lines for the double crossbar about 4.5" apart. The legs transition from ⌀1.00 at the ends to ⌀1.25 at the midpoints. The crossbars transition from ⌀.75 at the ends to ⌀1.00 at a position 2" from each end.

D. Save the drawing as P10-5.

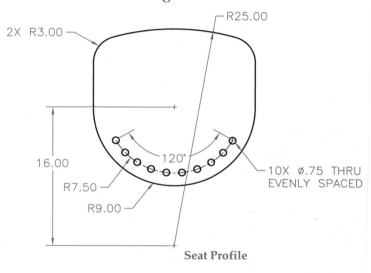

Seat Profile

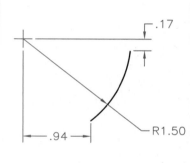

Edge Profile

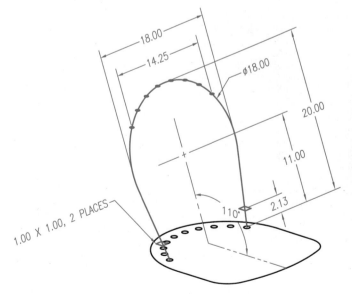

Seatback Profiles

Completed Seatback

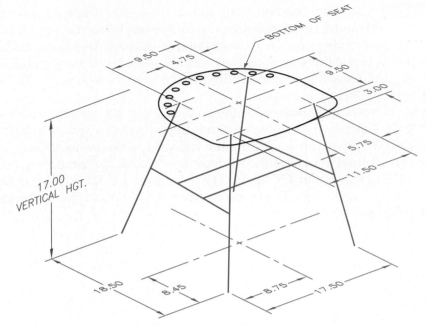

Profiles for Legs and Crossbars

Completed Model

CHAPTER 11

Subobject Editing

Learning Objectives

After completing this chapter, you will be able to:

✓ Select subobjects (faces, edges, and vertices).
✓ Edit solids using grips.
✓ Edit fillet and chamfer subobjects.
✓ Edit composite solid model subobjects.
✓ Edit face subobjects.
✓ Edit edge subobjects.
✓ Edit vertex subobjects.
✓ Extrude a 2D or 3D object using the **PRESSPULL** command.
✓ Offset planar face edges using the **OFFSETEDGE** command.

Grip Editing

There are three basic types of 3D solids in AutoCAD. The commands **BOX**, **WEDGE**, **PYRAMID**, **CYLINDER**, **CONE**, **SPHERE**, **TORUS**, and **POLYSOLID** create 3D solid *primitives*. *Swept objects* are 3D solids created from closed profiles using the **EXTRUDE**, **REVOLVE**, **SWEEP**, and **LOFT** commands. Finally, 3D solid *composites* are created by a Boolean operation (**UNION**, **SUBTRACT**, or **INTERSECT**) or by using the **SOLIDEDIT** command. The **SOLIDEDIT** command is discussed in Chapter 12. Smooth-edged solid primitives are achieved by converting mesh objects to solids.

There are two basic types of grips—base and parameter—that may be associated with a solid object. These grips provide an intuitive means of modifying solids. Base grips are square and parameter grips are typically arrows. The editing techniques that can be performed with these grips are discussed in the next sections.

3D Solid Primitives

The 3D solid primitives all have basically the same types of grips (base and parameter). However, not all grips are available on all primitives. All primitives have a base grip at the centroid of the base. This grip functions like a standard grip in 2D work. It can be used to stretch, move, rotate, scale, or mirror the solid.

Boxes, wedges, and pyramids have square base grips at the corners that allow the size of the base to be changed. See **Figure 11-1**. The object dynamically changes in the viewport as you select and move the grip or after you type the new coordinate location for the grip and press [Enter]. If ortho is off, the length and width can be changed at the same time by dragging the grip, except in the case of a pyramid. The triangular parameter grips on the base allow the length or width to be changed. The height can be changed using the top parameter grip. Each object has one parameter grip for changing the height of the apex and one for changing the height of the plane on which the base sits. A pyramid also has a parameter grip at the apex for changing the radius of the top.

Cylinders, cones, and spheres have four parameter grips for changing the radius of the base, or the cross section in the case of a sphere. See **Figure 11-2**. Cylinders and cones also have parameter grips for changing the height of the apex and the height of the plane on which the base sits. Additionally, a cone has a parameter grip at the apex for changing the radius of the top.

A torus has a parameter grip located at the center of the tube. See **Figure 11-3**. This grip is used to change the radius of the torus. Parameter grips at each quadrant of the tube are used to change the radius of the tube.

Figure 11-1.
Boxes, wedges, and pyramids have square base grips at the corners of the base, a parameter grip at the center of the base or bottom edge, and parameter grips on the sides of the base and center of the top face, edge, or vertex.

Figure 11-2.
Cylinders, cones, and spheres have four parameter grips for changing their radius. Cylinders and cones have parameter grips for changing their height. Cones also have a parameter grip for changing the radius of the top.

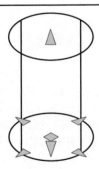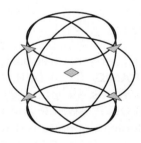

Figure 11-3.
A torus has a parameter grip located at the center of the tube for changing the radius of the torus. There are also parameter grips for changing the radius of the tube.

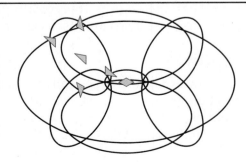

A polysolid does not have parameter grips. Instead, a base grip appears at each corner and edge midpoint of the starting face of the solid. See **Figure 11-4**. Use these grips to change the cross-sectional shape or height of the polysolid. The corners do not need to remain square. Base grips also appear at the endpoint and midpoint of each segment center-line. Use these grips to change the location of each segment's midpoint or endpoints.

Grips in AutoCAD are multifunctional. For example, if dynamic input is turned on, you can hover over the parameter grips of an object to display various design param-eters. In **Figure 11-5A**, the box primitive is selected to display grips. Hovering over the parameter grip at the side of the base displays the length dimension. Hovering over parameter grips on other types of primitives, such as a cylinder or sphere, displays the corresponding dimension of the shape. See **Figure 11-5B** and **Figure 11-5C**. You can quickly subobject edit a primitive by selecting a parameter grip and typing a delta value to add or remove geometry. For example, suppose a cylinder has a height of 10 units. The new height requirement is 15. With dynamic input turned on, select the cylinder, pick the parameter grip at the apex, and drag it so the dynamic input changes to delta entry. In the dynamic input box, type 5 for the new height and press [Enter].

Swept Solids

Extrusions, revolutions, sweeps, and lofts are considered swept solids and typi-cally have base grips located at the vertices of their profiles. These can be used to change the size of the profile and, thus, the solid. Other grips that appear include:

- A parameter grip appears on the upper face of an extrusion for changing the height.
- A base grip appears on the axis of a revolved solid for changing the location of the axis in relation to the profile.
- Base grips appear on the vertices of a sweep path for modifying the swept object.

Composite Solids

Composite solids are created by using one of the Boolean commands (**UNION**, **SUBTRACT**, or **INTERSECT**) on a solid. Solids that have been modified using any of the options of the **SOLIDEDIT** command also become composite solids, as do meshes converted into solids. The solid may still look like a primitive, sweep, loft, etc., but it is a composite. The grips available with the previous objects are no longer available, unless performing subobject editing on a composite created with a Boolean command (as discussed later in this chapter). Composite solids have a base grip located at the centroid of the base surface. This grip can be used to stretch, move, rotate, scale, or mirror the solid when the 2D Wireframe visual style is current. See **Figure 11-6A**. This grip also appears if no gizmo is selected to display when a 3D visual style is current. The gizmo drop-down list selection in the **Selection** panel on the **Home** tab of the

Figure 11-4.
A polysolid has a base grip at each corner and edge midpoint of the starting face of the solid and one at the endpoint and midpoint of each segment.

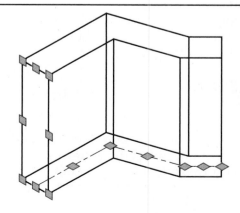

Figure 11-5.
Using parameter grips to display design data for a box, cylinder, and sphere. A—Hovering over the parameter grip on the side of the base displays the box length. B—Hovering over the parameter grip at the apex of the cylinder displays the cylinder height. C—Hovering over the parameter grip at the cross section of the sphere displays the sphere radius.

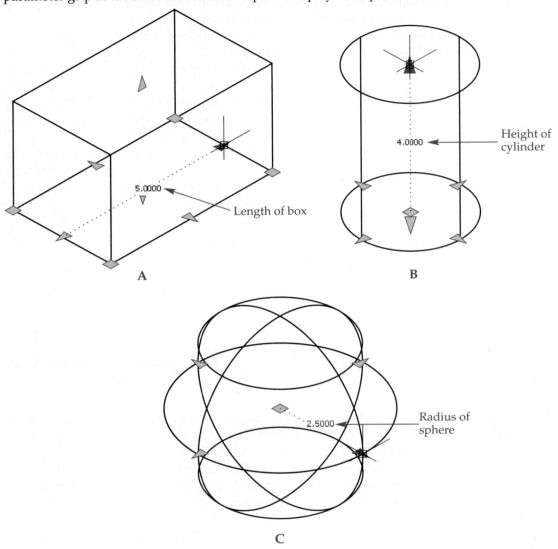

ribbon controls the display of the default gizmo in a 3D visual style. If the move gizmo is set to appear in a 3D visual style (the default option), it appears when you select a composite solid. See **Figure 11-6B**. You can use the move, rotate, or scale gizmo to make modifications.

The grip options for composite solids converted from meshes are similar to those for other composite solids. Composite solids converted from meshes have several grips, but these all act as base grips.

NOTE

The **Show gizmos** flyout on the status bar can also be used to select the default gizmo that appears when an object is selected. The **Show gizmos** flyout does not appear on the status bar by default. Select the **Gizmo** option in the status bar **Customization** flyout to add the **Show gizmos** flyout to the status bar.

Figure 11-6.
Grip editing options available for composite solids. A—A base grip appears on the composite solid when the object is selected if the 2D Wireframe visual style is current. Right-clicking displays a shortcut menu with the grip editing options. B—The move gizmo appears when the object is selected if a 3D visual style is current and the move gizmo is selected in the **Selection** panel in the **Home** tab of the ribbon. Right-clicking on the gizmo displays the shortcut menu.

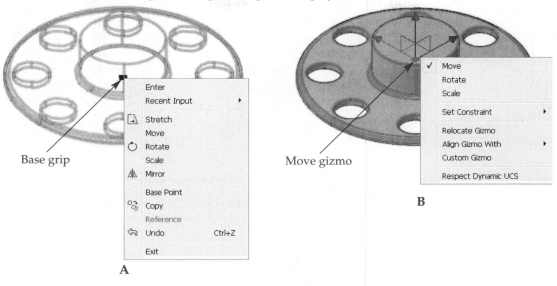

Exercise 11-1

www.g-wlearning.com/CAD/

Complete the exercise on the companion website.

Using Grips with Surfaces

Surfaces can be edited using grips in the same manner previously discussed with solids. A planar surface created with **PLANESURF** can be moved, rotated, scaled, and mirrored, but not stretched. Base grips are located at each corner.

As you learned in Chapter 6, a variety of AutoCAD objects can be extruded to create a surface. Three of these objects—arc, line, and polyline—are shown extruded into surfaces in **Figure 11-7**. Notice the location and type of grips on the surface extrusions. Base grips are located on the original profile that was extruded to make the surface. These grips enable you to alter the shape of the surface. You can use the parameter grips on the top of the surface to change the extrusion height, the extrusion direction, or the radius of the top.

Surfaces that have been extruded, or swept, along a path can be edited with grips. Also, the grips located on the path allow you to change the shape of the surface extrusion. See **Figure 11-8**.

Overview of Subobject Editing

AutoCAD solid primitives, such as cylinders, wedges, and boxes, are composed of three types of subobjects: faces, edges, and vertices. As discussed in Chapter 9, mesh models are also made up of subobjects. In addition, the objects that are used with Boolean commands to create a composite solid are considered subobjects, if the

Figure 11-7.
Surfaces extruded from an arc, line, and polyline. Notice the grips.

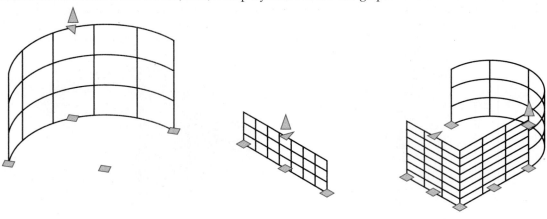

Figure 11-8.
A—Grips can be used to modify the path on this swept surface. B—The swept surface after grip editing.

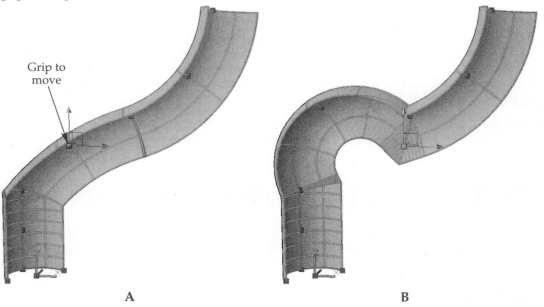

Grip to
move

A   B

solid history is recorded. The primitive subobjects can be edited. See **Figure 11-9**. Once selected, the primitive subobjects can be modified or deleted as needed from the composite solid. **Figure 11-10** illustrates the difference between a composite solid model, the solid primitives used to construct it, and an individual subobject of one of the primitives.

Subobjects can be easily edited using grips, which provide an intuitive and flexible method of solid model design. For example, suppose you need to rotate a face subobject in the current XY plane. You can select the subobject, pick its base grip, and then cycle through the editing functions to the rotate function. You can also use the **ROTATE** command on the selected subobject.

To select a subobject, press the [Ctrl] key and pick the subobject. You can select multiple subobjects and subobjects on multiple objects. To select a subobject that is hidden in the current view, first display the model as a wireframe. After creating a selection set, select a grip and edit the subobject as needed. Multiple objects can be

Figure 11-9.
A—If the solid history is recorded, selecting a subobject solid primitive within a composite solid displays its grips. B—The grips on the primitive can be used to edit the primitive.

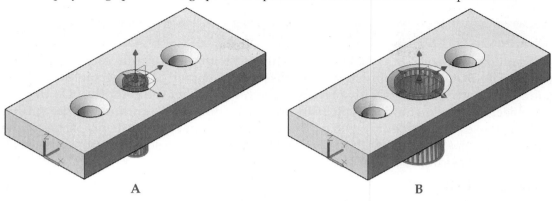

A   B

Figure 11-10.
A—The composite solid model is selected. Notice the single base grip. B—The wedge primitive subobject has been selected. Notice the grips associated with the primitive. C—An edge subobject within the primitive subobject is selected for editing.

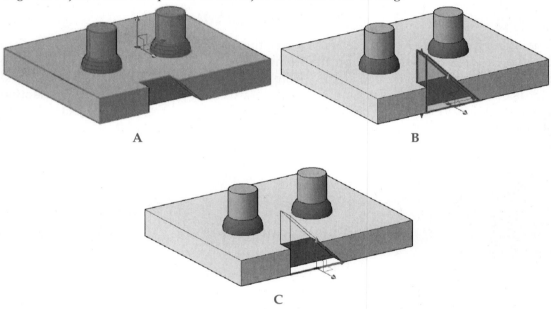

A   B

C

selected in this manner. To deselect objects, press the [Shift]+[Ctrl] key combination and pick the objects to be removed from the selection set.

If objects or subobjects are overlapping, use selection cycling to select the object you need. Press the [Ctrl]+[W] key combination to turn on selection cycling. This is the same as picking the **Selection Cycling** button on the status bar to turn it on. The **Selection Cycling** button does not appear on the status bar by default. Select the **Selection Cycling** option in the status bar **Customization** flyout to add the **Selection Cycling** button to the status bar. To use selection cycling, pick the desired object where it is near overlapping objects and use the **Selection** dialog box to select the object. Selection cycling is introduced in Chapter 8.

The [Ctrl] key method can be used to select subobjects for use with editing commands such as **MOVE**, **COPY**, **ROTATE**, **SCALE**, and **ERASE**. Some commands, like **STRETCH** and **MIRROR**, are applied to the entire solid. Other operations may not be applied at all, depending on which type of subobject is selected. You can also use

Type
SELECTIONCYCLING
Toolbar
Status Bar
Selection Cycling

SELECTIONCYCLING

the **Properties** palette to change the color of edge and face subobjects or the material assigned to a face. The color of a subobject primitive can also be changed, but not the color of its subobjects.

Subobject Editing Fillets and Chamfers

If you improperly create a fillet or chamfer, edit it by holding down the [Ctrl] key and selecting the fillet or chamfer. Then, use editing methods to change the fillet or chamfer. A fillet has a parameter grip that can be used to change the radius of the fillet.

On some occasions during the editing process of fillets or chamfers, the fillet radius or the chamfer distances will not appear in the **Properties** palette. If this is the case and you only see a single face grip on the surface of the fillet or chamfer, erase the fillet or chamfer and reapply it. Fillet and chamfer subobject editing is a form of composite solid editing.

Face Subobject Editing

Faces of 3D solids can be modified using commands such as **MOVE**, **ROTATE**, and **SCALE** or by using grips and gizmos (grip tools). To select a face on a 3D solid, press the [Ctrl] key and pick within the boundary of the face. Do not pick the edge of the face. The face subobject filter can also be used to limit the selection to a face.

Face grips are circular and located in the center of the face, as shown in **Figure 11-11**. In the case of a sphere, the grip is located in the center of the sphere since there is only one face. The same is true of the curved face on a cylinder or cone.

If you select a solid primitive, all of the grips associated with that primitive are displayed. See **Figure 11-12A**. If you edit a solid primitive face, the history of the primitive is deleted and the object becomes a composite solid. Then, when the object is selected, a single base grip is displayed. See **Figure 11-12B**.

The same holds true for composite solids, if the solid history has been recorded. Solid model history is discussed in Chapter 8. If you select a primitive subobject in a composite solid with a recorded solid history, all of the grips associated with the primitive are displayed. If you edit a solid primitive face of the composite solid, the history of the primitive is deleted and the object becomes a new composite solid. Then, when the subobject is selected, a single base grip is displayed.

While pressing the [Ctrl] key and selecting a face, it may be difficult to select the face you want or to deselect faces you do not need. Use the view cube to transparently change the viewpoint. You can also press and hold the [Shift] key and press and hold the mouse wheel button at the same time to activate the transparent **3DORBIT** command.

NOTE

The recorded history of a composite solid can be displayed by selecting the solid, opening the **Properties** palette, and changing the Show History property in the **Solid History** category to Yes.

Figure 11-11.
Face grips are located in the center of face subobjects.

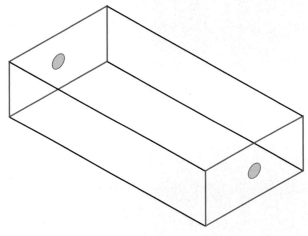
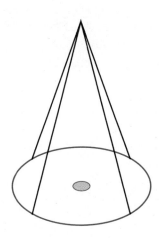

Figure 11-12.
A—This primitive is selected for editing. Notice the grips associated with the primitive.
B—If the primitive is edited, the history of that primitive is deleted and a single base grip is
displayed.

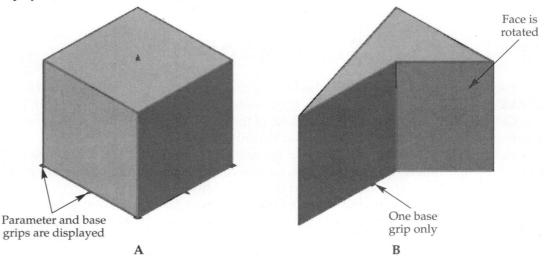

Face is
rotated

Parameter and base
grips are displayed

One base
grip only

A

B

Moving Faces

When a face of a 3D solid is moved, all adjacent faces are dragged and stretched
with it. The shape of the original 3D primitive or solid determines the manner in
which the face can be moved and how adjacent faces react. A face can be moved using
the **MOVE** command, **3DMOVE** command, or move gizmo, or by dragging the face's
base grip. When moving a face, use the gizmo, polar tracking, or direct distance entry.
Otherwise, the results may appear correct in the view in which the edit is made,
but, when the view is changed, the actual result may not be what you wanted. See
Figure 11-13.

Figure 11-13.
A—The original solid primitives. B—The box is dynamically edited without using exact
coordinates or distances. C—When the viewpoint is changed, you can see that dynamic
editing has produced unexpected results.

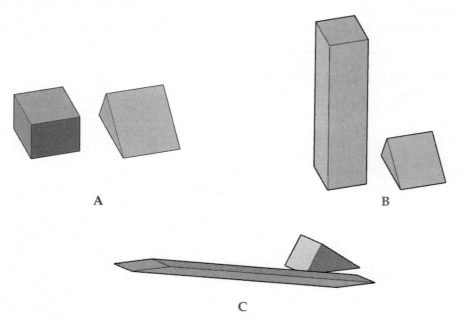

A

B

C

The move gizmo, as discussed earlier in this chapter, is displayed by default when the face is selected. To use this gizmo, move the pointer over the X, Y, or Z axis of the gizmo; the axis changes to yellow. To restrict movement along that axis, pick the axis. If you move the pointer over one of the right angles at the origin of the gizmo, the corresponding two axes turn yellow. Pick to restrict the movement to that plane. You can complete the movement by either picking a new point or by using direct distance entry.

The face base grip is used to access options when dynamically moving a face. First, select the face. Then, hover over the face base grip to display the base grip shortcut menu. See **Figure 11-14A**. The options in this menu determine the effects of the edit on the selected face and adjacent faces. When using the **Extend Adjacent Faces** option, the moved face maintains its shape and orientation. However, its size is modified because the planes of adjacent faces are maintained. See **Figure 11-14B**. If you select **Move Face**, the moved face maintains its size, shape, and orientation. The shape and plane of adjacent faces are changed. See **Figure 11-14C**. If you select **Allow Triangulation**, the moved face maintains its size, shape, and orientation. However, adjacent faces are subdivided into triangular faces, if needed.

During the edit, you can press the [Ctrl] key to cycle through the move options available in the base grip shortcut menu. For example, after selecting the face, pick the face grip or gizmo. Then, press and release the [Ctrl] key to cycle through the options.

Figure 11-14.
A—The original solid primitive. B—Using the **Extend Adjacent Faces** option keeps the adjacent faces in their original planes, but alters the modified face. C—When using the **Move Face** option, the face maintains its shape, size, and orientation.

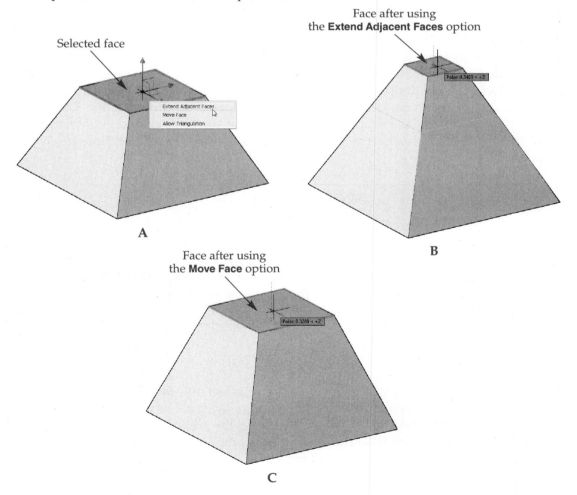

It is always important to keep the design intent of your solid model in mind. If you are creating a conceptual design, you may be able to use subobject grip editing and gizmos without entering precise coordinates. But, if you are working on a design for manufacturing or production, it is usually critical to use tools such as direct distance entry, gizmos with exact values, and polar tracking for greater accuracy.

Rotating Faces

Before rotating any primitive or subobject, you must know in which plane the rotation is to occur. The **ROTATE** command permits a rotation in the current XY plane. But, you can get around this limitation by using the **3DROTATE** command. This command allows you to select a rotation plane by means of the rotate gizmo discussed in Chapter 8.

The rotate gizmo provides a dynamic, graphic representation of the three axes of rotation. To rotate about the X axis of the tool, pick the red circle on the gizmo. To rotate about the Y axis, pick the green circle. To rotate about the Z axis, pick the blue circle. Once you select a circle, it turns yellow and you are prompted for the start point of the rotation angle. You can enter a direct angle at this prompt or pick a point to define the angle of rotation. When the rotation angle is defined, the face is rotated about the selected axis.

For example, in **Figure 11-15A**, the top face is selected and the rotate gizmo is placed on a corner of the face. After picking the rotation axis on the gizmo, specify the angle start point and then the angle end point. Notice in **Figure 11-15B** that dynamic input can be used to enter an exact angle value. The result is shown in **Figure 11-15C**.

The [Ctrl] key is used to access options when dynamically rotating a face. **Figure 11-16A** shows a rotation without pressing [Ctrl]. The shape and size of the face being rotated is maintained, while the adjacent faces change. **Figure 11-16B** shows a rotation after pressing [Ctrl] once. The shape and size of the face being rotated changes, while the plane and shape of adjacent faces are maintained. Pressing the [Ctrl] key a second time maintains the shape and orientation of the selected face, but triangular faces may be created on adjacent faces. Pressing the [Ctrl] key a third time resets the function.

When you are working in a 3D view, you may want to use the **3DMOVE** and **3DROTATE** commands exclusively during editing sessions. If this is the case, set the **GTDEFAULT** system variable to 1 (the default is 0). This automatically executes the **3DMOVE** or **3DROTATE** command in a 3D view when you select the **MOVE** or **ROTATE** command. If the 2D Wireframe visual style is current, the Wireframe visual style is set current for the duration of the command.

Scaling Faces

Scaling a face is a simple procedure. First, select the face to be scaled. Then, select a base point and dynamically pick to change the scale or use a scale factor. See **Figure 11-17**. Pressing the [Ctrl] key has no effect on the scaling process, except to turn it off or on, if the base point is on the same plane as the face. However, if the base point is not on the same plane as the selected face, then pressing the [Ctrl] key has the same effect as for a rotate face-editing operation.

Figure 11-15.
Using the rotate gizmo to rotate a face. A—The top face is selected and the gizmo is placed on a base point of the face. B—The axis of rotation is specified and a starting point for the angle is selected. C—The completed rotation.

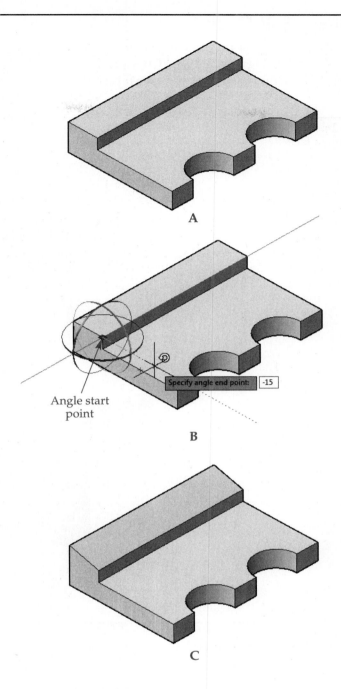

Angle start point

Specify angle end point: -15

A

B

C

Coloring Faces

To change the color of a face, use the [Ctrl] key selection method to select the face. Next, open the **Properties** palette. See **Figure 11-18**. In the **General** category, pick the drop-down list for the Color property. Select the desired color or pick Select Color... and select a color from the **Select Color** dialog box. To change the material applied to the face, pick the drop-down list for the Material property in the **3D Visualization** category. Select a material from the list. A material must be loaded into the drawing to be available in this drop-down list. Materials are discussed in detail in Chapter 16.

Extruding a Solid Face

Planar faces on 3D solids can be extruded into new solids. Refer to the HVAC duct assembly shown in **Figure 11-19A**. A new, reduced trunk needs to be created on the left end of the assembly. This requires two pieces: a reducer and the trunk.

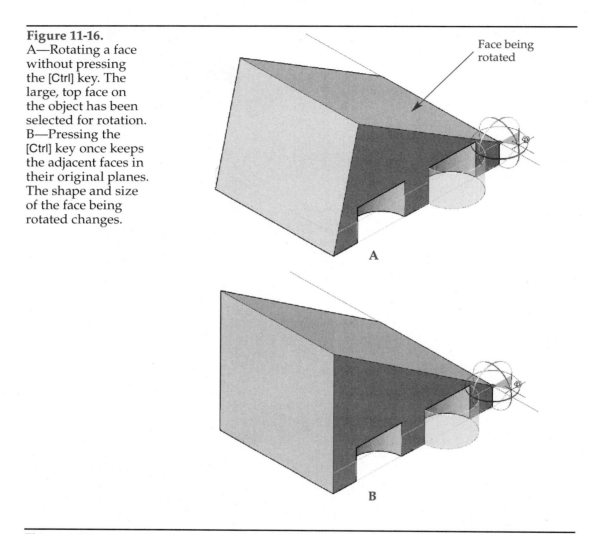

Figure 11-16.
A—Rotating a face without pressing the [Ctrl] key. The large, top face on the object has been selected for rotation. B—Pressing the [Ctrl] key once keeps the adjacent faces in their original planes. The shape and size of the face being rotated changes.

Face being rotated

A

B

Figure 11-17.
Scaling a face. A—The original solid. B—The dark face is scaled down. C—The dark face is scaled up.

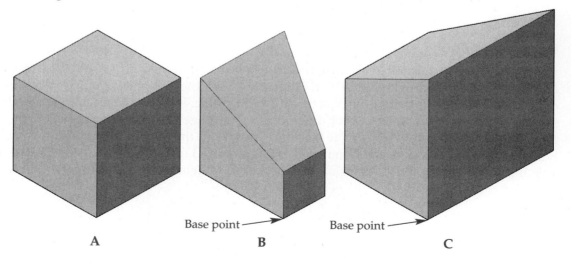

Base point

Base point

A

B

C

First, select the **EXTRUDE** command. At the "select objects" prompt, press the [Ctrl] key and pick the face subobject to be extruded. Next, since this is a reduced trunk, specify a taper angle. Enter the **Taper angle** option and specify the angle. In this case, a 15° angle is used. Finally, specify the extrusion height. The height of the reducer is 12". See **Figure 11-19B**.

Figure 11-18.
Changing the color of a face or the material assigned to it.

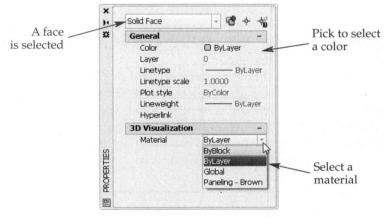

Figure 11-19.
A—A new, reduced trunk needs to be created on the left end of the HVAC assembly. The face shown in color will be extruded. B—The **Taper angle** option of the **EXTRUDE** command is used to create the reducer. The face shown in color is extruded to create the extension. C—The **EXTRUDE** command is used to create an extension from the reducer.

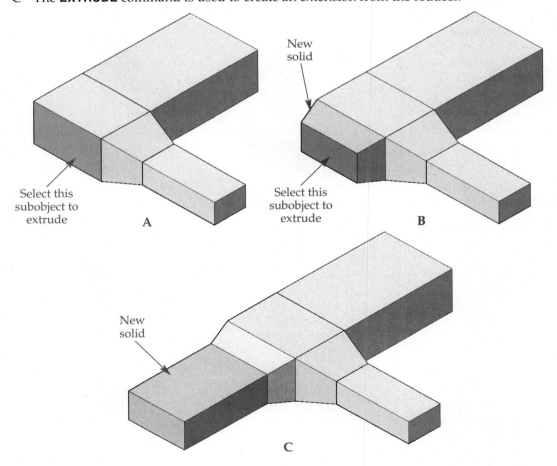

Now, the new trunk needs to be created. Select the **EXTRUDE** command. Press the [Ctrl] key and pick the face to extrude. Since this piece is not tapered, enter the extrusion height, which in this case is 44″. See **Figure 11-19C**. The two new pieces are separate solid objects. If the assembly is to be one solid, use the **UNION** command and join the two new solids to the assembly.

Revolving a Solid Face

Planar faces on 3D solids can be revolved in the same manner as other AutoCAD objects to create new solids. Refer to **Figure 11-20A**. The face on the left end of the HVAC duct created in the last section needs to be revolved to create a 90° bend. First, select the **REVOLVE** command. At the "select objects" prompt, press the [Ctrl] key and pick the face subobject to be revolved.

Next, the axis of revolution needs to be specified. You can pick the two endpoints of the vertical edge, but you can also pick the edge subobject. Enter the **Object** option of the command, press the [Ctrl] key, and select the edge subobject.

Finally, the 90° angle of revolution needs to be specified. **Figure 11-20B** shows the face revolved into a new solid. The bend is a new, separate solid. If necessary, use the **UNION** command to join the bend to the assembly.

Exercise 11-3

www.g-wlearning.com/CAD/

Complete the exercise on the companion website.

Figure 11-20.
A—The face on the left end of the HVAC duct (shown in color) needs to be revolved to create a 90° bend.
B—Use the **REVOLVE** command and pick the face subobject to be revolved.

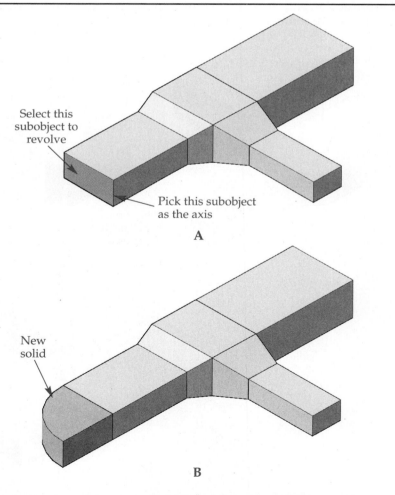

Select this subobject to revolve

Pick this subobject as the axis

A

New solid

B

Edge Subobject Editing

Individual edges of a solid can be edited using grips and gizmos in the same manner as faces. To select an edge subobject, press the [Ctrl] key and pick the edge. The edge subobject filter can also be used to limit the selection to an edge.

Grips on linear edges are rectangular and appear in the middle of the edge, **Figure 11-21**. In addition to solid edges, the edges of regions can be altered using **MOVE**, **ROTATE**, and **SCALE**, but grips are not displayed on regions as they are on solid subobjects.

Remember, editing subobjects of a primitive removes the primitive's history. This should always be a consideration if it is important to preserve the solid primitives that were used to construct a 3D solid model. Instead of editing the primitive subobjects at their subobject level, it may be better to add or remove material with a Boolean operation, thus preserving the solid's history.

NOTE

There are several ways to perform a 3D edit. Keep the following options in mind when working with subobject editing.

- Selecting the **MOVE**, **ROTATE**, or **SCALE** command and picking a subobject will not display a gizmo, regardless of the current gizmo button displayed in the **Selection** panel and the **Show gizmos** flyout.
- Selecting the **3DMOVE**, **3DROTATE**, or **3DSCALE** command and picking a subobject will display the appropriate gizmo, regardless of the current gizmo button displayed in the **Selection** panel and the **Show gizmos** flyout.
- Picking a subobject without having selected a command first will display the gizmo corresponding to the current gizmo button in the **Selection** panel and the **Show gizmos** flyout.

Figure 11-21.
Edge grips are rectangular and displayed in the middle of the edge.

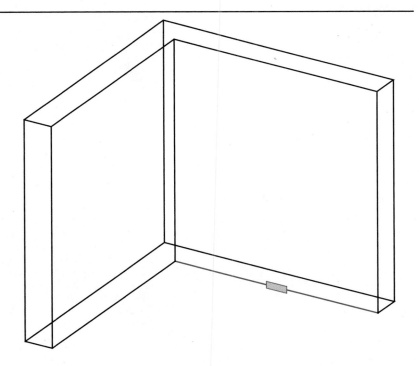

Moving Edges

To move an edge, select it using the [Ctrl] key, as previously discussed. See **Figure 11-22A**. By default, the gizmo corresponding to the gizmo button in the **Selection** panel and the **Show gizmos** flyout appears. If the move gizmo is not current, select the **Move Gizmo** button in the **Selection** panel or the **Show gizmos** flyout. Select the appropriate axis handle and dynamically move the edge or use direct distance entry, **Figure 11-22B**. The move gizmo remains active until the [Esc] key is pressed to deselect the edge.

If you pick an edge grip to turn it hot, the gizmo is bypassed. This places you in the standard grip editing mode. You can stretch, move, rotate, scale, and mirror the edge. In this case, the stretch function works in the same manner as the move gizmo, but less reliably. You must be careful to use either ortho, polar tracking, or direct distance entry, but the possibility for error still exists.

The options available when dynamically moving an edge are similar to those used when moving a face. See **Figure 11-23A**. First, select the edge. Then hover over the base edge grip to display the base grip shortcut menu. Selecting the **Extend Adjacent Faces** option and moving the edge maintains the orientation of the moved edge, but its length is modified. See **Figure 11-23B**. This is because the planes and orientation of adjacent faces are maintained. Selecting the **Move Edge** option and moving the edge maintains the length and orientation of the moved edge. However, the shape and planes of adjacent faces are changed. See **Figure 11-23C**. Selecting the **Allow Triangulation** option and moving the edge maintains the length and orientation of the moved edge. But, if the move alters the planes of adjacent faces, those faces may become *nonplanar*. In other words, the face may now be located on two or more planes. If this happens, adjacent faces are divided into triangles, **Figure 11-23D**. This is visible when the object in **Figure 11-23D** is displayed in two orthographic views. See **Figure 11-24**.

Figure 11-22.
A—Select the edge subobject to be moved. B—The edge is moved. Notice how the size of the primitive used to subtract the cutout is not affected.

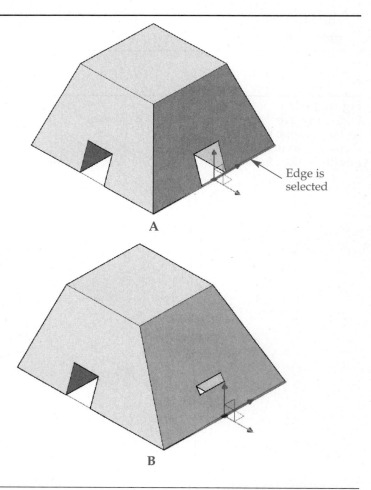

Edge is selected

A

B

The [Ctrl] key is used to access options when dynamically moving an edge. For example, after selecting the edge, pick the gizmo. Then, press and release the [Ctrl] key to cycle through the options.

Figure 11-23.
Moving an edge. A—Selecting the edge and then hovering over the base grip displays the base grip shortcut menu. B—Using the **Extend Adjacent Faces** option maintains the orientation of the moved edge, but its length is modified because the planes of adjacent faces are maintained. C—When using the **Move Edge** option, the edge maintains its length and orientation, but the shape and planes of adjacent faces are changed. D—When using the **Allow Triangulation** option, the adjacent faces may be triangulated.

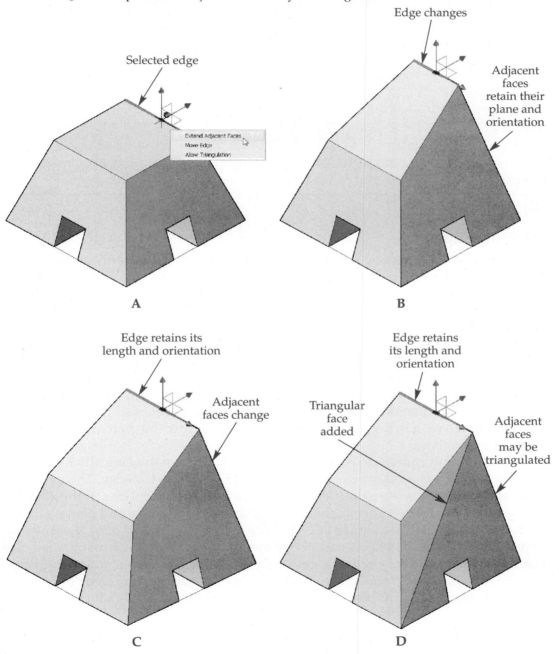

Figure 11-24.
Triangulated faces are clear in plan views. A—Front plan view. B—Side plan view.

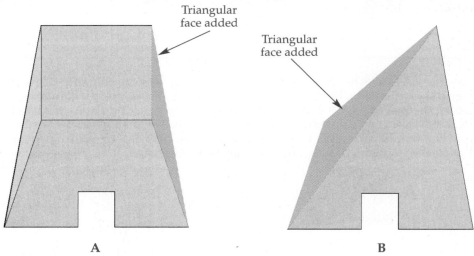

Triangular face added

Triangular face added

A

B

PROFESSIONAL TIP

You can quickly enter the **MOVE**, **ROTATE**, or **SCALE** command for a subobject by selecting the subobject, right-clicking, and picking the command from the shortcut menu.

Rotating Edges

Before you select an edge to rotate, do a little planning. Since there are a wide variety of edge rotation options, it will save time if you first decide on the location of the base point about which the edge will rotate. Next, determine the direction and angle of rotation. Based on these criteria, choose the option that will accomplish the task the quickest.

To rotate an edge, enter the **ROTATE** or **3DROTATE** command. Then, pick the edge using the [Ctrl] key. Select a base point and then enter the rotation. You can also select the edge, pick the edge grip, and cycle to the **ROTATE** mode.

Edges are best rotated using the rotate gizmo. It provides a graphic visualization of the axis of rotation. If you select a dynamic UCS while using the **3DROTATE** command, you have a variety of rotation axes to use because the gizmo can be located on a temporary plane.

There are a few options to achieve different results when dynamically rotating an edge. The [Ctrl] key is used to access these options. First, select the edge. Then, pick the gizmo and press and release the [Ctrl] key to cycle through the options. If the [Ctrl] key is not pressed, the rotated edge maintains its length, but the shape and planes of adjacent faces are changed. See **Figure 11-25A**. If the [Ctrl] key is pressed once, the length of the rotated edge is modified because the planes of adjacent faces are maintained. See **Figure 11-25B**. If the [Ctrl] key is pressed twice, the rotated edge maintains its length, but if the rotation causes faces to become nonplanar, the adjacent faces may be triangulated. See **Figure 11-25C**. Pressing the [Ctrl] key a third time resets the function.

Scaling Edges

Only linear (straight-line) edges can be scaled. Circular edges, such as the ends of cylinders, can be modified using grips or the **SOLIDEDIT** command. These tools can be used to change the diameter or establish taper angles. See Chapter 12 for a complete discussion of the **SOLIDEDIT** command.

Figure 11-25.
Rotating an edge. A—The [Ctrl] key is not pressed. B—The [Ctrl] key is pressed once. Notice the top edge of the dark face. C—The [Ctrl] key is pressed twice.

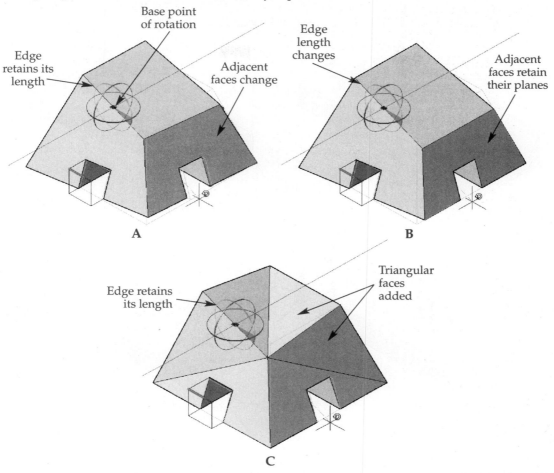

To scale a linear edge, enter the **SCALE** command. Select the edge using the [Ctrl] key. Pick a base point for the operation and enter a scale factor. You can also select the edge, pick the edge grip, and cycle to the **SCALE** mode. However, using the scale gizmo may be the best option.

The direction of the scaled edge is related to the base point you select. The base point remains stationary, while the vertices in either direction are scaled. If you enter the **SCALE** command, you are prompted for the base point. If you select the edge grip, the grip becomes the base point. The differences in opposite end and midpoint scaling of an edge are shown in **Figure 11-26**.

There are a few options to achieve different results when dynamically scaling an edge. The [Ctrl] key is used to access these options. First, select the edge. Then, pick the scale gizmo and press and release the [Ctrl] key to cycle through the options.

If the [Ctrl] key is not pressed, the edge is scaled. The shape and planes of adjacent faces are changed to match the scaled edge. See **Figure 11-27A**.

If the [Ctrl] key is pressed once, the edge is, in effect, not scaled. This is because the planes of adjacent faces are maintained.

If the [Ctrl] key is pressed twice, the edge is scaled, as are edges attached to the modified edge. However, if the scaling causes faces to become nonplanar, they may be triangulated. See **Figure 11-27B**.

Figure 11-26.
The differences in opposite end and midpoint scaling of an edge. A—The original object.
B—The edge is scaled down with a base point on the left corner. C—The edge is scaled down
to the same scale factor, but the base point is on the right corner. D—The edge is scaled down
to the same scale factor with the base point at the middle of the edge.

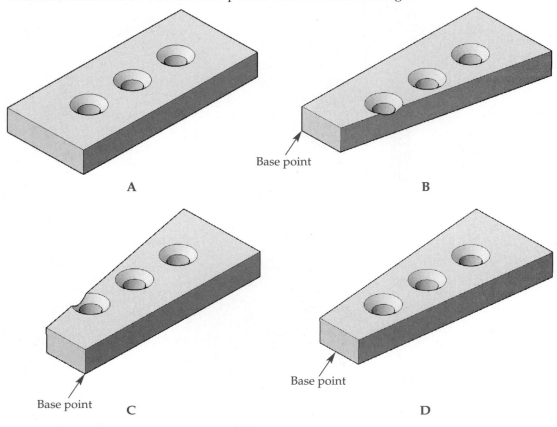

Figure 11-27.
Scaling the front edge with the base point at the middle of the edge. The original object is
shown in Figure 11-26A. A—If the [Ctrl] key is not pressed, the edge is scaled and the shape
and planes of adjacent faces are changed. B—If the [Ctrl] key is pressed twice, the edge
is scaled, as are edges attached to it. If the scaling causes faces to become nonplanar, the
adjacent faces may be triangulated.

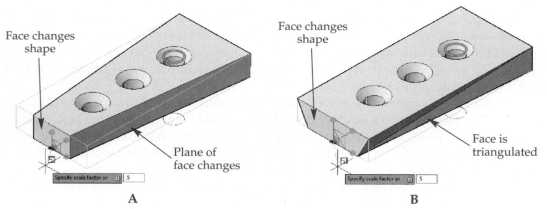

Coloring Edges

To change the color of an edge, use the [Ctrl] key to select the edge. Next, open the
Properties palette. In the **General** category, pick the drop-down list for the Color property.

Select the desired color or pick Select Color… and select a color from the **Select Color** dialog box. Edges cannot have materials assigned to them.

Deleting Edges

Edges can be deleted in certain situations. In order for an edge to be deleted, it must completely divide two faces that lie on the same plane. If this condition is met, the **ERASE** command or the [Delete] key can be used to remove the edge. The two faces become a single face.

Exercise 11-4   www.g-wlearning.com/CAD/

Complete the exercise on the companion website.

Vertex Subobject Editing

The modification of a single vertex involves moving the vertex and stretching all edges and planar faces attached to it. Vertex grips are circular and located on the vertex, as shown in **Figure 11-28**. Use the vertex subobject filter to assist in selecting vertices. A single vertex cannot be rotated or scaled, but you can select multiple vertices and perform rotating and scaling edits. When editing multiple vertices in this manner, you are, in effect, editing edges.

As with other subobject editing functions performed on a 3D solid primitive, the solid's history is removed when a vertex is modified. The solid can no longer be edited using the primitive grips; only a single base grip is displayed. Further editing of the solid must be done with the **SOLIDEDIT** command, discussed in Chapter 12, or through subobject editing.

Moving Vertices

To move a vertex, select it using the [Ctrl] key. If the move gizmo is not displayed, select it from the gizmo drop-down list in the **Selection** panel or the **Show gizmos** flyout. You can use the move gizmo, the **MOVE** command, or standard grip editing modes to move the vertex. Hovering over the base vertex grip displays a shortcut menu with two move options. See **Figure 11-29A**. The **Move Vertex** option allows the vertex to be moved without triangulating adjacent faces, but the faces may change shape. See **Figure 11-29B**. In some cases, AutoCAD may deem it necessary to triangulate faces. Using the **Allow Triangulation** option results in the triangulation of adjacent faces when the vertex is moved. See **Figure 11-29C**. Pressing the [Ctrl] key during the edit cycles through the move options.

Figure 11-28.
Vertex grips are circular and placed on the vertex.

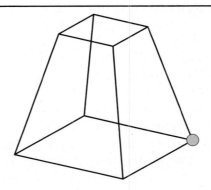

Figure 11-29.
Moving a vertex. A—The original object. B—Using the **Move Vertex** option moves the vertex and changes some of the adjacent faces. B—When using the **Allow Triangulation** option, adjacent faces are triangulated.

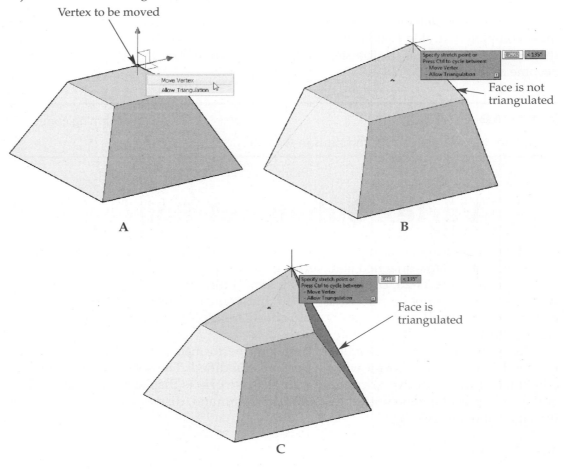

If you are dragging a vertex and faces become triangulated, you can transparently change your viewpoint to see the effect of the triangulation. Press and hold the [Shift] key. At the same time, press and hold the mouse wheel button. Now, move the mouse to change the viewpoint. This is a transparent instance of the **3DORBIT** command.

Rotating Vertices

As previously stated, a single vertex cannot be rotated or scaled, but two or more vertices can be. Since two vertices define a line, or edge, any edit is an edge modification. However, the process is slightly different from the edge modifications described earlier in this chapter.

To rotate an edge by selecting its endpoints, press the [Ctrl] key and select each vertex. See **Figure 11-30A**. You may need to use the [Ctrl]+[W] key combination to turn on selection cycling. Notice that grips appear at each selected vertex, but the edges between the vertices are not highlighted.

The **ROTATE** command can now be used to rotate the vertices (if **PICKFIRST** is set to 1). However, a more efficient method for rotating vertices is to use the **3DROTATE** command. The combination of the rotate gizmo and the UCS icon enables you to graphically view the rotation plane. Once the command is initiated, select the vertices and move the base point

Figure 11-30.
To rotate or scale vertices, multiple vertices must be selected. In effect, the edges are modified. A—Vertices are selected to be rotated. B—The rotate gizmo is placed at the base of rotation and the rotation axis is selected. C—The vertices are rotated.

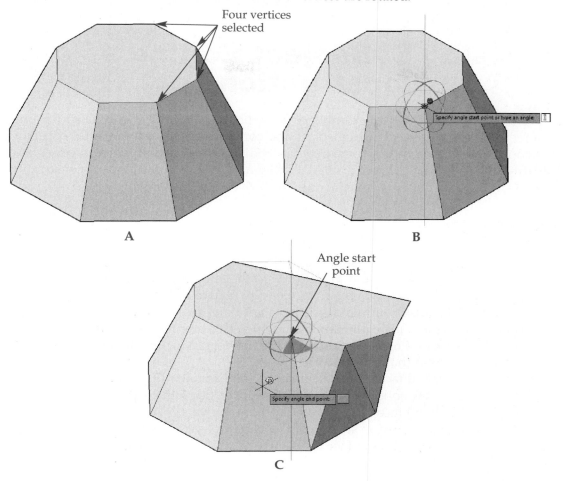

of the rotate gizmo if needed. See **Figure 11-30B**. Then, select the axis of revolution. Finally, pick the angle start point and enter the rotation. See **Figure 11-30C**.

If the [Ctrl] key is not pressed while dynamically rotating the vertices, the area of the selected vertices does not change and adjacent faces are triangulated. This is because the edges of the adjacent faces are attached to the selected vertices, so their edge length changes as the selected edge is rotated. If the [Ctrl] key is pressed once, the adjacent faces are not triangulated unless necessary, but the faces may change shape.

NOTE

If the selected edge does not dynamically rotate at the "angle end point" prompt, then the desired rotation is not possible.

Scaling Vertices

As mentioned earlier, it is not possible to scale a single vertex. However, two or more vertices can be selected for scaling. This, in effect, scales edges. The selection methods are the same as discussed for rotating vertices and the use of the [Ctrl] key while dragging produces the same effects. As the pointer is dragged, the dynamic display of scaled edges may be difficult to visualize. Therefore, it is best to use a scale factor or the **Reference** option to achieve properly scaled edges.

Using Subobject Editing as a Construction Tool

This section provides an example of how subobject editing can be used not only as a method for changing existing solids and composites, but as a powerful construction tool. Some of the procedures of subobject editing, such as editing faces, edges, and vertices, are used to construct an HVAC assembly. The entire model is constructed from a single, solid cube. This is the only primitive you will draw.

Editing Faces

1. Begin by setting the units to architectural and drawing a 24″ cube. Display the model from the southeast isometric viewpoint.
2. Select the left-hand face and move it 60″ to the left. Also, select the front face and move it out 12″. This forms the first duct. See **Figure 11-31**.
3. Using the **EXTRUDE** command, select the left-hand face and extrude it 36″ to create a new solid. This is a tee junction from which two branches will extend.
4. Select the front face of the new solid and extrude it 28″ with a taper angle of 10° to create a new solid that is a reducer.
5. Select the left-hand face of the tee junction and extrude it 20″ with a taper angle of 10° to create a new solid that is a second reducer. See **Figure 11-32**.
6. Select the left-hand face of the 20″ reducer and move it 3 17/32″ along the positive Z axis. This places the top surface of the reducer level with the trunk of the duct. Next, extrude the left-hand face of this reducer 60″ into a new solid.

Figure 11-31.
The left-hand face of the cube is moved 60″. The front face is then moved 12″.

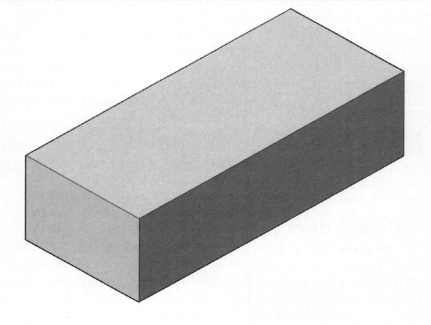

Figure 11-38.
Using the **PRESSPULL** command to expand a run of ductwork along angled walls. A—The original design. B—After expanding the first face, the adjacent face is selected using the [Ctrl] key. A second face at the corner duct is also selected and extruded. C—The resulting design. D—The front wall ducts are expanded and unioned to complete the design.

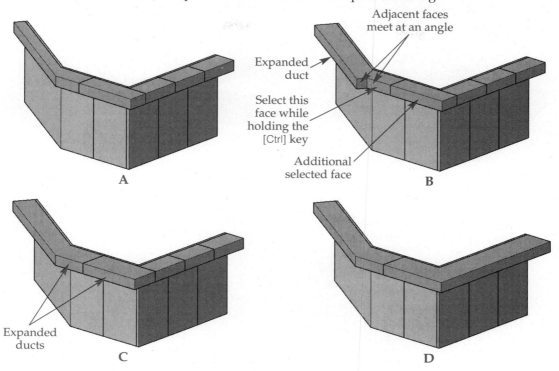

An example of using the **OFFSETEDGE** command is shown in **Figure 11-39**. Before specifying the point through which the offset object is created, you can use the **Corner** option to create the offset object with round or sharp corners. Results after using the **Round** and **Sharp** options are shown in **Figure 11-39B**. In both examples shown, the new offset object is a closed polyline.

Once the offset object is created, you can use the **EXTRUDE** or **PRESSPULL** command to create a new 3D feature. See **Figure 11-39C**. In each example shown, an extrusion is created with the **PRESSPULL** command.

Exercise 11-6

www.g-wlearning.com/CAD/

Complete the exercise on the companion website.

Exploding a Solid

A solid can be exploded. This turns the solid into surfaces and/or regions. Flat surfaces on the solid are turned into regions. Curved surfaces on the solid are turned into surfaces. To explode a solid, select the **EXPLODE** command. Then, pick the solid(s) to explode and press [Enter].

Exercise 11-7

www.g-wlearning.com/CAD/

Complete the exercise on the companion website.

Figure 11-39.
Using the **OFFSETEDGE** command. A—After entering the command, select a point on the planar face with the edges to offset. B—The offset object can have rounded or sharp corners. In each example shown, a closed polyline is created. C—The new polyline object can be used for modeling purposes. Shown are results after using the **PRESSPULL** command. Extruding in a negative direction subtracts material. Extruding in a positive direction adds material.

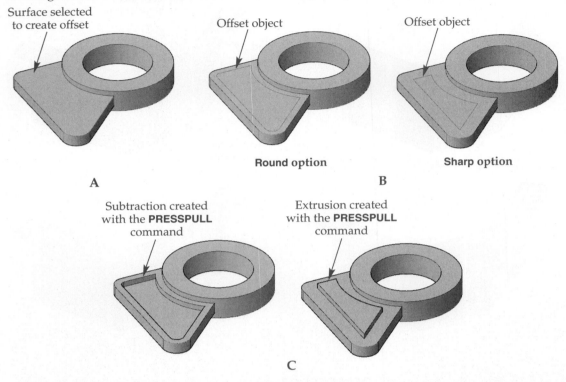

Working with Associative Arrays

A 3D array constructed with the **ARRAYRECT**, **ARRAYPOLAR**, or **ARRAYPATH** command can be created as an associative or non-associative array. Arrays are introduced in Chapter 8. An associative array acts as a single object, much like a block. If you try to use a Boolean operation with an associative array, AutoCAD will not allow the operation to perform. As with a block, an array is a collection of objects that are defined as one object. Think of a 3D associative array as a group of solids in a plastic "wrapper." This type of object cannot be used in a Boolean operation. For example, if you attempt to subtract an associative polar array of cylinders from a plate, AutoCAD will display the following message:

No solids, surfaces, or regions selected.

To use arrayed objects for a Boolean operation, create the array as a non-associative array or explode the array using the **EXPLODE** command. See **Figure 11-40**. Exploding an array creates individual objects that can then be used for other modeling purposes. **Figure 11-40C** shows the result after using the **SUBTRACT** command to subtract the individual cylinders from the base plate.

Exercise 11-8

www.g-wlearning.com/CAD/

Complete the exercise on the companion website.

Figure 11-40.
Using arrayed objects in a Boolean operation. A—A base plate with a single cylinder. B—A polar array is created from the cylinder. If the array is created as an associative array, it cannot be subtracted from the base plate in a Boolean operation. C—The result after exploding the array and subtracting the individual cylinders from the base plate. Shown is a shaded view of the model.

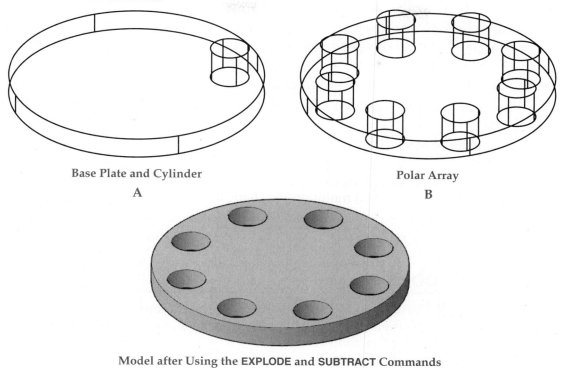

Base Plate and Cylinder
A

Polar Array
B

Model after Using the **EXPLODE** and **SUBTRACT** Commands
C

Chapter Review

Answer the following questions using the information in this chapter.

1. How do you select a subobject?
2. How do you deselect a subobject?
3. When grip editing, two types of grips appear on the object. Name the two types of grips.
4. If you have a cylinder primitive with a height of 10 units, but the height requirement has changed to 15 units, explain the procedure for adding 5 units to the cylinder height.
5. How can you change the radius of a fillet or the distances of a chamfer?
6. When moving a face on a solid primitive, how can you accurately control the axis of movement?
7. When moving a face on a solid primitive, which option maintains the planes of adjacent faces while modifying the size of the face?
8. Describe a major difference of function between the **ROTATE** and **3DROTATE** commands.
9. Which system variable enables you to use the **3DROTATE** command in a 3D view even if you select the **ROTATE** command?

10. How does the location and shape of an edge grip differ from a face grip?

11. What is the most efficient tool to use when rotating an edge and how is it displayed?

12. What is the only type of edge that can be scaled?

13. What is the only editing function that can be done when editing a single vertex?

14. How are two or more vertices selected for editing?

15. What is created when offsetting an edge of a solid with the **OFFSETEDGE** command?

16. Which option of the **OFFSETEDGE** command is used to create round corners on the resulting offset object?

17. What is the function of the **PRESSPULL** command?

18. What type of object is created when using the **PRESSPULL** command to extrude an arc?

19. What are three methods for selecting multiple objects when using the **PRESSPULL** command?

20. What key is used when selecting an object with the **PRESSPULL** command in order to extrude a face and maintain the shape and orientation of adjacent faces?

21. When a solid object is exploded, what happens to the flat surfaces of the solid?

22. When a solid object is exploded, what happens to the curved surfaces of the solid?

Drawing Problems

1. Architectural Draw the bookcase shown using the dimensions given. The final result should be a single solid object. Then, use grip and subobject editing procedures to edit the object as follows.
   A. Change the width of the bookcase to 3′.
   B. Change the height of the bookcase by eliminating the top section. The resulting height should be 3′-2″.
   C. Save the drawing as P11-1.

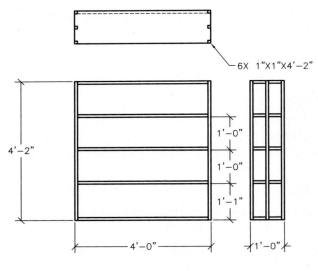

ALL WOOD THICKNESS IS 1″

2. **Architectural** Open problem P11-1. Save it as P11-2. Use primitive and subobject editing procedures to create the following edits.
   A. Change the depth of the top of the bookcase to 6-1/2".
   B. Change the depth of the bottom of the bookcase to 24".
   C. Reduce the height of the front uprights so they are flush with the top surface of the next lower shelf.
   D. Extend the front of the middle shelf to match the front of the bottom. Add two uprights at the front corners between the bottom and this shelf.
   E. Save the drawing.

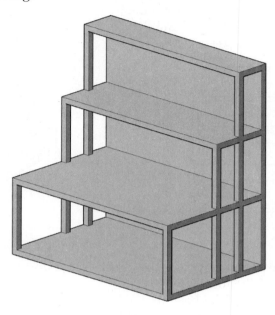

3. **Mechanical** Draw the mounting bracket shown below. Then, use primitive and subobject editing procedures to create the following edits.
   A. Change the 3.00" dimension to 3.50".
   B. Change the 2.50" dimension in the front view to 2.75".
   C. Change the location of the slot in the auxiliary view from .60" to .70" and change the length of the slot to 1.15".
   D. Change the width of each foot in the top view from 2.00" to 1.50". The overall dimension (5.00") should not change.
   E. Change the angle of the bend from 15° to 45°.
   F. Save the drawing as P11-3.

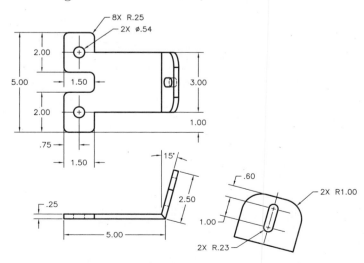

4. **General** Draw as a single composite solid the desk organizer shown in the ortho-
graphic views. Then, use primitive and subobject editing procedures to create the
following edits. The final object should look like the shaded view.
   A. Change the 3″ height to 3.25″.
   B. Change the 2″ height to 1.85″.
   C. Increase the thickness of the long compartment divider to .5″. The increase in
      thickness should be evenly applied along the centerline of the divider. Locate
      three evenly spaced, ⌀5/16″ × 1.5″ holes in this divider.
   D. Angle the top face of the rear compartments by 30°. The height of the rear of
      the organizer should be approximately 4.5″ and all corners on the bottom of
      the organizer should remain square.
   E. Save the drawing as P11-4.

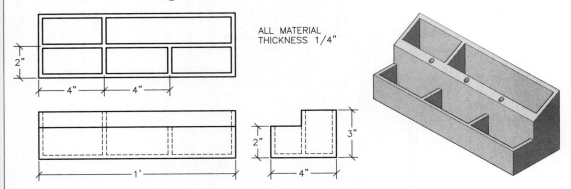

5. **General** Draw the pencil holder shown. Then, use primitive and subobject editing
procedures to create the following edits.
   A. Change the depth of the base to 4.000″. The base should be rectangular, not
      square, and the grooves should become shorter.
   B. Change the height of the top groove from .250″ to .125″.
   C. Change the diameter of two holes from ⌀.450″ to ⌀.625″.
   D. Change the diameter of the other two holes from ⌀.450″ to ⌀1.000″.
   E. Rotate the top face 15° away from the side with the grooves. The planes of the
      adjoining faces should not change. Refer to the shaded view.
   F. Save the drawing as P11-5.

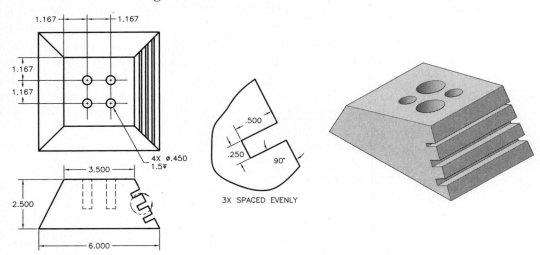

CHAPTER 12

Solid Model Editing

Learning Objectives

After completing this chapter, you will be able to:

✓ Change the shape and configuration of solid object faces.
✓ Copy and change the color of solid object faces and edges.
✓ Break apart a composite solid composed of physically separate entities.
✓ Extract edges from a 3D solid using the **XEDGES** command.
✓ Use the **SOLIDEDIT** command to construct and edit a solid model.

AutoCAD provides expanded capabilities for editing solid models. As you saw in the previous chapter, grips can be used to edit a solid model. Also, the subobjects that make up a solid, such as faces, edges, and vertices, can be edited. A single command, **SOLIDEDIT**, allows you to edit faces, edges, or the entire body of the solid.

NOTE

Mesh objects cannot be modified using the **SOLIDEDIT** command. The mesh object must be converted to a solid first. If you select a mesh for editing with the **SOLIDEDIT** command, you are given the option of converting it to a solid, as long as the display of the dialog box has not been turned off.

Overview of the SOLIDEDIT Command

The **SOLIDEDIT** command allows you to edit the faces, edges, and body of a solid. Many of the subobject editing functions discussed in Chapter 11 can also be performed with the **SOLIDEDIT** command. The options of the **SOLIDEDIT** command can be accessed in the **Solid Editing** panel on the **Home** or **Solid** tab of the ribbon or by typing SOLIDEDIT. See **Figure 12-1**. The quickest method of entering the command is by using the ribbon. For example, directly select a face editing option from the drop-down list in the **Solid Editing** panel, as shown in **Figure 12-1**.

Figure 12-1.
Accessing the **SOLIDEDIT** command options.

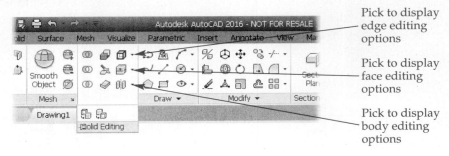

Pick to display edge editing options

Pick to display face editing options

Pick to display body editing options

Select the face editing option   Pick   Expanded help text

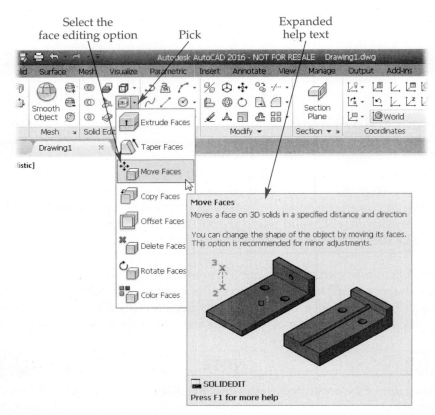

When the **SOLIDEDIT** command is entered at the keyboard, you are first asked to select the component of the solid with which you wish to work. Specify **Face**, **Edge**, or **Body**. The editing options for the selected component are then displayed and are the same as those seen in **Figure 12-1**.

The editing function is directly entered when the option is selected from the ribbon. This is why using the ribbon is the most efficient method of entering the **SOLIDEDIT** command options.

The following sections provide an overview of the solid model editing functions of the **SOLIDEDIT** command. Each option is explained and the results of each are shown. A tutorial later in the chapter illustrates how these options can be used to construct a model.

AutoCAD displays a variety of error messages when invalid solid editing operations are attempted. Rather than trying to interpret the wording of these messages, just realize that what you tried to do will not work. Actions that may cause errors include trying to rotate a face into other faces or extruding and tapering an object at too great of an angle. When an error occurs, try the operation again with different parameters or determine a different approach to solving the problem in order to maintain the design intent.

Face Editing

The basic components of a solid are its faces. Many of the **SOLIDEDIT** options are for editing faces. All eight face editing options prompt you to select faces. It is important to make sure you select the correct part of the model for editing. Remember the following three steps when using any of the face editing options.

1. First, select a face to edit. If you pick an edge, AutoCAD selects the two faces that share the edge. If this happens, use the **Remove** option to deselect the unwanted face. A more intuitive approach is to select the open space of the face as if you were touching the side of a part. AutoCAD highlights only that face.

2. Adjust the selection set at the Select faces or [Undo/Remove/ALL]: prompt. The following options are available.
   - **Undo.** Removes the previous selected face(s) from the selection set.
   - **Remove.** Allows you to select faces to remove from the selection set. This is only available when **Add** is current.
   - **All.** Adds all faces on the model to the selection set. This is only available after selecting at least one face. It can also be used to remove all faces if **Remove** is current.
   - **Add.** Allows you to add faces to the selection set. This is only available when **Remove** is current.

3. Press [Enter] to continue with face editing.

Extruding Faces

An extruded face is moved, or stretched, in a selected direction. The extrusion can be straight or have a taper. To extrude a face, select the command and pick the **Face>Extrude** option. Remember, the option is directly entered when picking the button in the ribbon. You are then prompted to select the face(s) to extrude. Nonplanar (curved) faces cannot be extruded. As you pick faces, the prompt verifies the number of faces selected. For example, when an edge is selected, the prompt reads 2 faces found. When done selecting faces, press [Enter] to continue.

Next, the height of the extrusion needs to be specified. A positive value adds material to the solid, while a negative value subtracts material from the solid. A taper can also be given.

Ribbon

Home
> Solid Editing
Solid
> Solid Editing

Extrude Faces

Type

SOLIDEDIT

SOLIDEDIT

```
Specify height of extrusion or [Path]: (enter the height)
Specify angle of taper for extrusion <0>: (enter an angle or accept the default)
Solid validation started.
Solid validation completed.
Enter a face editing option
[Extrude/Move/Rotate/Offset/Taper/Delete/Copy/coLor/mAterial/Undo/eXit] <eXit>: X↵
Solids editing automatic checking: SOLIDCHECK=1
Enter a solids editing option [Face/Edge/Body/Undo/eXit] <eXit>: X↵
```

Figure 12-2 shows an original solid object and the result of extruding the top face with a 0° taper angle and a 30° taper angle. It also shows the original solid object with two adjacent faces extruded with 15° taper angles.

In addition to extruding a face perpendicular to itself, the extruded face can follow a path. Select the **Path** option at the Specify height of extrusion or [Path]: prompt. The path of extrusion can be a line, circle, arc, ellipse, elliptical arc, polyline, or spline. The extrusion height is the exact length of the path. See **Figure 12-3**.

Exercise 12-1

www.g-wlearning.com/CAD/

Complete the exercise on the companion website.

Figure 12-2.
Extruding faces on an object. A—The original object. B—The top face is extruded with a 0° taper angle. C—The top face of the original object is extruded with a 30° taper angle. D—The top and right-hand faces of the original object are extruded with 15° taper angles.

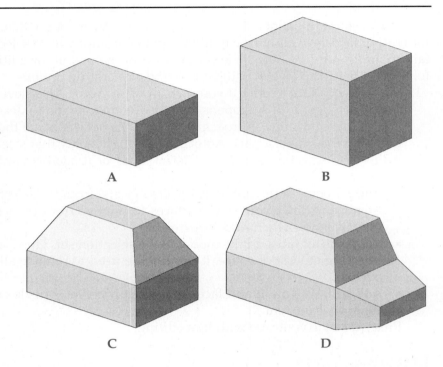

A   B

C   D

Figure 12-3.
The path of extrusion can be a line, circle, arc, ellipse, elliptical arc, polyline, or spline. Here, the paths are shown in color.

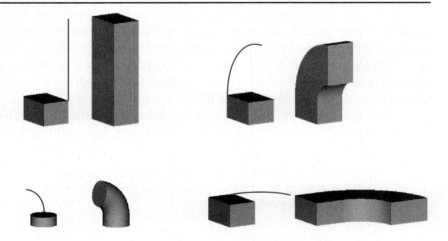

Moving Faces

The **Move** option moves a face in the specified direction and lengthens or shortens the solid object. In another application, a solid model feature (such as a hole) that has been subtracted from an object to create a composite solid can be moved with this option. Object snaps may interfere with the function of this option, so they may need to be toggled off during the operation.

To move a face, select the command and pick the **Face>Move** option. If the button is picked in the ribbon, the option is directly entered. You are then prompted to select the face(s) to move. When done selecting faces, press [Enter] to continue. Next, you are prompted to select a base point of the operation:

> Specify a base point or displacement: *(pick a base point)*
> Specify a second point of displacement: *(pick a second point or enter coordinates)*
> Solid validation started.
> Solid validation completed.
> Enter a face editing option
> [Extrude/Move/Rotate/Offset/Taper/Delete/Copy/coLor/mAterial/Undo/eXit] <eXit>: **X**↵
> Solids editing automatic checking: SOLIDCHECK=1
> Enter a solids editing option [Face/Edge/Body/Undo/eXit] <eXit>: **X**↵

When adjacent faces are perpendicular, the edited face is moved in a direction so the new position keeps the face parallel to the original. See **Figures 12-4A** and **12-4B**. Faces that are normal to the current UCS can be moved by picking a new location or entering a direct distance. If you are moving a face that is not normal to the current

Figure 12-4.
A—The hole will be moved using the **Move** option of the **SOLIDEDIT** command. B—The hole is moved. C—When the angled face is moved, a portion of it is altered to be coplanar with the vertical face. D—If the angled face is moved more, it becomes completely coplanar to the vertical face. This is a new, single face.

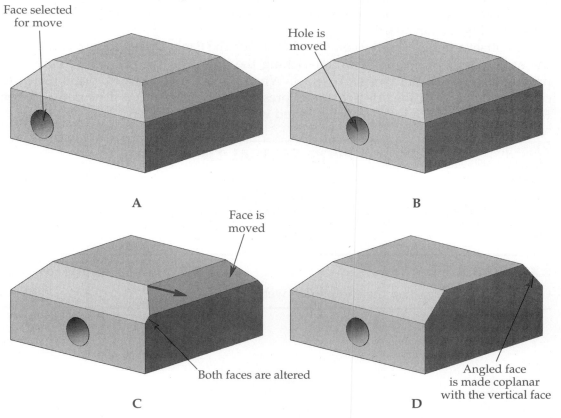

Face selected for move

Hole is moved

A

B

Face is moved

Both faces are altered

C

Angled face is made coplanar with the vertical face

D

UCS, you can enter coordinates for the second point of displacement, but it may be easier to first use the UCS icon grips or the **Face** option of the **UCS** command to align the UCS with the face to be moved.

When adjacent faces join at angles other than 90°, the moved face will be relocated as previously stated, but only if the movement is less than the dimensional offset of the two faces. For example, in **Figure 12-4B**, the top edge of the angled face is in .5″ from the vertical face. If the angled face is moved outward a distance of less than .5″, it is altered as shown in **Figure 12-4C**. A portion of the angled face becomes coplanar with the vertical face. If the angled face is moved outward a distance greater than .5″, it is altered so that it forms a single plane with the adjacent face. What has actually happened is that the angled face is moved beyond the adjacent face, while remaining parallel to its original position. Thus, in effect, it has disappeared because the adjacent, vertical face cannot be altered. See **Figure 12-4D**. In this example, the angled face was moved .75″. The new vertical face that is created can now be moved.

Exercise 12-2

www.g-wlearning.com/CAD/

Complete the exercise on the companion website.

Offsetting Faces

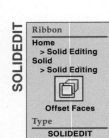

SOLIDEDIT

Ribbon
Home
> Solid Editing
Solid
> Solid Editing

Offset Faces
Type
SOLIDEDIT

The **Offset** option may seem the same as the **Extrude** option because it moves faces by a specified distance or through a specified point. Unlike the **OFFSET** command in AutoCAD, this option moves all selected faces a specified distance. It is most useful when you wish to change the size of features such as slots, holes, grooves, and notches in solid parts. A positive offset distance increases the size or volume of the solid (adds material). A negative distance decreases the size or volume of the solid (removes material). Therefore, if you wish to make the width of a slot wider, provide a negative offset distance to decrease the size of the solid. Picking points to set the offset distance and direct distance entry are always taken as a positive value, so negative values must be entered using the keyboard.

To offset a face, select the command and pick the **Face>Offset** option. Remember, the option is directly entered when picking the button in the ribbon. You are then prompted to select the face(s) to offset. When done selecting faces, press [Enter] to continue. Next, enter the offset distance, press [Enter], and then exit the command. See **Figure 12-5** for examples of features edited with the **Offset** option.

PROFESSIONAL TIP

Nonplanar (curved) faces cannot be extruded, but can be offset. Using the **Offset** option, you can, in effect, "extrude" a nonplanar face.

Exercise 12-3

www.g-wlearning.com/CAD/

Complete the exercise on the companion website.

Figure 12-5.
Offsetting faces. A—The original objects. The hole is selected to offset. The interior of the L is also selected to offset. B—A positive offset distance increases the size or volume of the solid. C—A negative offset distance decreases the size or volume of the solid.

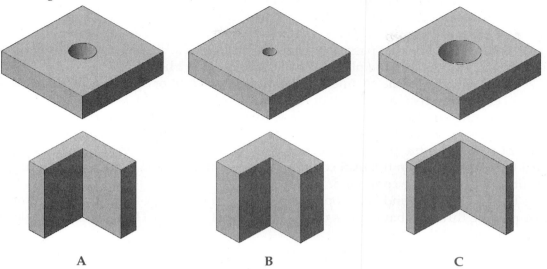

A   B   C

Deleting Faces

The **Delete** option deletes selected faces. This is a quick way to remove features such as chamfers, fillets, holes, and slots. To delete a solid face, select the command and pick the **Face>Delete** option. If the button is picked in the ribbon, the option is directly entered. You are then prompted to select the face(s) to delete. When done selecting faces, press [Enter] to continue and then exit the command. When a face is deleted, existing faces extend to fill the gap. No additional faces are created. For instance, the inclined surface of a wedge cannot be deleted, as there are no existing faces that can be extended to fill the gap. When deleting the face that is a chamfered or filleted edge, the adjacent edges are extended to fill the gap. See **Figure 12-6**.

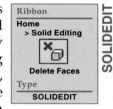

Rotating Faces

The **Rotate** option rotates a face about a selected axis. To rotate a solid face, select the command and pick the **Face>Rotate** option. Remember, the option is directly entered when picking the button in the ribbon. You are then prompted to select the face(s) to rotate. When done selecting faces, press [Enter] to continue. There are several methods by which a face can be rotated.

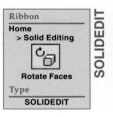

Figure 12-6.
Deleting faces. A—The original object with three rounds. B—The faces of two rounds have been deleted.

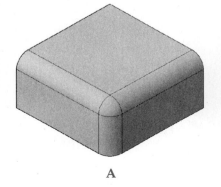

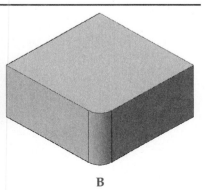

A   B

The **2points** option is the default. Pick two points to define the "hinge" about which the face will rotate. Then, provide the rotation angle and exit the command.

The **Axis by object** option allows you to use an existing object to define the axis of rotation. You can select the following objects.

- **Line.** The selected line becomes the axis of rotation.
- **Circle, arc, or ellipse.** The Z axis of the object becomes the axis of rotation. This Z axis is a line that passes through the center of the circle, arc, or ellipse and is perpendicular to the plane on which the 2D object lies.
- **Polyline or spline.** A line connecting the polyline or spline's start point and endpoint becomes the axis of rotation.

After selecting an object, enter the angle of rotation and exit the command.

When you select the **View** option, the axis of rotation is perpendicular to the current view, with the positive direction coming out of the screen. This axis is identical to the Z axis when the **UCS** command **View** option is used. Next, enter the angle of rotation and exit the command.

The **Xaxis**, **Yaxis**, and **Zaxis** options prompt you to specify a point. The X, Y, or Z axis passing through that point is used as the axis of rotation. Then, enter the angle of rotation and exit the command.

Figure 12-7 provides several examples of rotated faces. Notice how the first and second pick points determine the direction of positive and negative rotation angles.

NOTE

A positive rotation angle moves the face in a clockwise direction looking from the first pick point to the second. Conversely, a negative angle rotates the face counterclockwise. If the rotated face will intersect or otherwise interfere with other faces, an error message indicates that the operation failed or no solution was calculated. In this case, you may wish to try a negative angle if you previously entered a positive one. In addition, you can try using the opposite edge of the face as the axis of rotation by picking the appropriate points.

Figure 12-7.
When rotating faces, the first and second pick points determine the direction of positive and negative rotation angles.

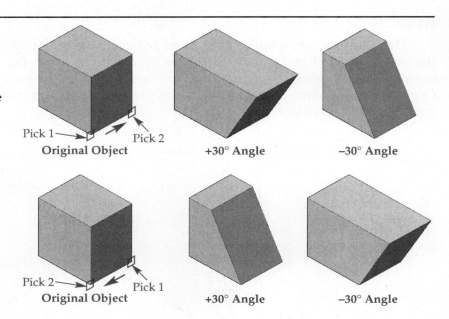

Pick 1 ⟶ Pick 2
Original Object

+30° Angle

−30° Angle

Pick 2 ⟶ Pick 1
Original Object

+30° Angle

−30° Angle

Tapering Faces

Ribbon

Home
> Solid Editing

Solid
> Solid Editing

Taper Faces

Type
SOLIDEDIT

SOLIDEDIT

The **Taper** option tapers a face at the specified angle, from the first pick point to the second. To taper a solid face, select the command and pick the **Face>Taper** option. If the button is picked in the ribbon, the option is directly entered. You are then prompted to select the face(s) to taper. When done selecting faces, press [Enter] to continue:

Specify the base point: *(pick the base point)*
Specify another point along the axis of tapering: *(pick a point along the taper axis)*
Specify the taper angle: *(enter a taper value)*

Tapers work differently depending on whether the faces being tapered describe the outer boundaries of the solid, a cavity, or a removed portion of the solid. A positive taper angle always removes material. A negative taper angle always adds material. For example, if a positive taper angle is entered for a solid cylinder, the selected object is tapered in on itself from the base point along the axis of tapering, thus removing material. A negative angle tapers the object out away from itself to increase its size along the axis of tapering, thus adding material. See **Figure 12-8**.

On the other hand, if the face (or faces) of a feature such as a hole or slot are tapered, a positive taper angle increases the size of the feature along the axis of tapering. For example, if a round hole is tapered using a positive taper angle, its diameter increases

Figure 12-8.
Tapering faces.
A—The original objects. The dark face of the box and the circumference of the cylinder are selected. B—Positive taper angle.
C—Negative taper angle.

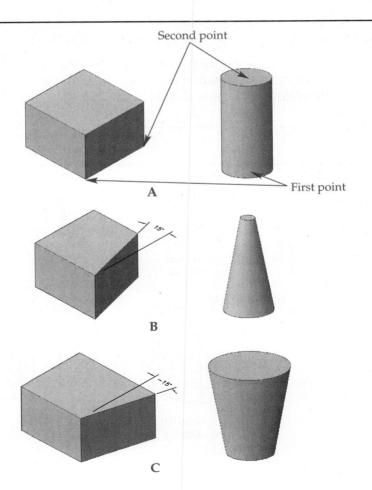

Second point

First point

A

B

C

from the base point along the axis of tapering, thus removing material from the solid. Conversely, if the same round hole is tapered using a negative taper angle, its diameter decreases from the base point along the axis of tapering, thus adding material to the solid. **Figure 12-9** shows some examples of tapering features.

Exercise 12-5

www.g-wlearning.com/CAD/

Complete the exercise on the companion website.

Copying Faces

SOLIDEDIT

Ribbon
Home
> Solid Editing

Copy Faces

Type
SOLIDEDIT

The **Copy** option copies a face to the location or coordinates given. The copied face is *not* part of the original solid model. It is actually a region, which can later be extruded, revolved, swept, etc., into a solid. This may be useful when you wish to construct a mating part in an assembly that has the same features on the mating faces or the same outline. This option is quick to use because you can pick a base point on the face, and then enter a single direct distance value for the displacement. Be sure an appropriate UCS is set if you wish to use direct distance entry.

To copy a solid face, select the command and pick the **Face>Copy** option. Remember, the option is directly entered when picking the button in the ribbon. You are then prompted to select the face(s) to copy. When done selecting faces, press [Enter] to continue. You are prompted for a base point for the copy. Pick this point and then pick a second point of displacement or press [Enter] to use the first point as a displacement. See **Figure 12-10** for examples of copied faces.

PROFESSIONAL TIP

Copied faces can also be useful for creating additional views. For example, you can copy a face to create a separate plan view with dimensions and notes. A copied face can also be enlarged to show details and to provide additional notation for design or assembly.

Figure 12-9.
If a hole or slot is tapered using a positive taper angle, its diameter or width increases from the base point along the axis of tapering, thus removing material from the solid. A negative taper angle increases the volume of the solid.

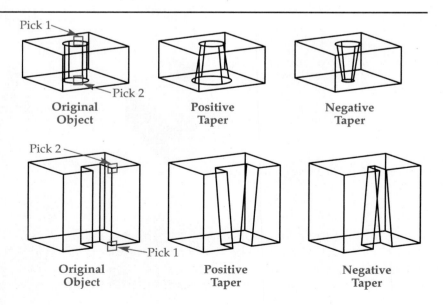

Original Object · Positive Taper · Negative Taper

Original Object · Positive Taper · Negative Taper

Figure 12-10.
A face can be quickly copied by picking a base point on the face and then entering a direct distance value for the displacement.

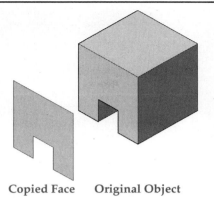
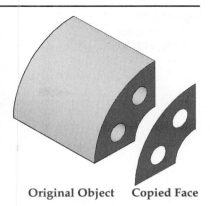

Copied Face   Original Object   Original Object   Copied Face

Coloring Faces

You can quickly change a selected face to a different color using the **Color** option. Select the command and pick the **Face>Color** option. If the button is picked in the ribbon, the option is directly entered. You are then prompted to select the face(s) to color. When done selecting faces, press [Enter] to continue. Next, select the desired color from the **Select Color** dialog box that is displayed. Remember, the color of the object (or face) determines the shaded color.

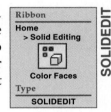

Ribbon
Home
> Solid Editing

Color Faces
Type
SOLIDEDIT

SOLIDEDIT

PROFESSIONAL TIP

Use the **Color** option to enhance features on a solid. For example, if you have a model that has holes or slots, enhance just the color of these features.

NOTE

The **Material** option of the **SOLIDEDIT** command allows you to assign a material to selected faces on a solid. Materials are discussed in Chapter 16.

Exercise 12-6   www.g-wlearning.com/CAD/

Complete the exercise on the companion website.

Edge Editing

Edges can be edited in only two ways with the **SOLIDEDIT** command. They can be copied from the solid. Also, the color of an edge can be changed.

Copying an edge is similar to copying a face. To copy a solid edge, select the command and pick the **Edge>Copy** option. Remember, the option is directly entered when picking the button in the ribbon. See **Figure 12-11.** You are then prompted to select the edge(s) to copy. When done selecting edges, press [Enter] to continue. You are prompted for a base point for the copy. Pick this point and then pick a second point of displacement or press [Enter] to use the first point as a displacement. The edge is copied as a line, arc, circle, ellipse, or spline.

Ribbon
Home
> Solid Editing

Copy Edges
Type
SOLIDEDIT

SOLIDEDIT

Figure 12-11.
Selecting edge-editing options.

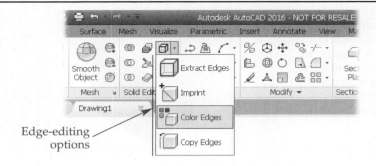

Edge-editing options

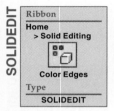

SOLIDEDIT

Ribbon

Home
> Solid Editing

Color Edges

Type

SOLIDEDIT

To color a solid edge, select the command and pick the **Edge>Color** option. You are then prompted to select the edge(s) to color. When done selecting edges, press [Enter] to continue. Next, select the desired color from the **Select Color** dialog box that is displayed and pick the **OK** button. The edges are now displayed with the new color. You may need to set a wireframe or hidden visual style current to see the change.

Extracting a Wireframe

XEDGES

Ribbon

Home
> Solid Editing
Solid
> Solid Editing

Extract Edges

Type

XEDGES

The **XEDGES** command creates copies of, or extracts, all of the edges on a selected solid. You can also extract edges from a surface, mesh, region, or subobject. Once the command is initiated, you are prompted to select objects. Select one or more solids and press [Enter]. The edges are extracted and placed on top of the existing edges. See **Figure 12-12**. The new objects are created on the current layer.

Straight edges and the curved edges where cylindrical surfaces intersect with flat or other cylindrical surfaces are the only edges extracted. Spheres and tori have no edges that can be extracted. The round bases of cylinders and cones are the only edges of those objects that will be extracted.

Figure 12-12.
Extracting edges with the **XEDGES** command. A—The original object. B—The extracted wireframe (edges).

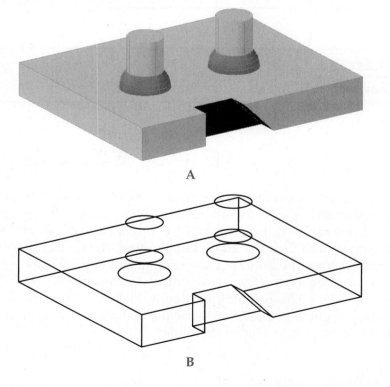

A

B

Body Editing

The body editing options of the **SOLIDEDIT** command perform editing operations on the entire body of the solid model. The body options are **Imprint**, **Separate**, **Shell**, **Clean**, and **Check**. The next sections cover these body editing options.

The **Imprint** option is a body editing function. However, since it modifies a body by adding edges, the option is located in the edges drop-down list in the **Solid Editing** panel on the **Home** tab. On the **Solid** tab, it is a separate button in the **Solid Editing** panel. In both places, the **Separate**, **Shell**, **Clean**, and **Check** options are found in the body drop-down list.

Imprint

Arcs, circles, lines, 2D and 3D polylines, ellipses, splines, regions, bodies, and 3D solids can be imprinted onto a solid, if the object intersects the solid. The imprint becomes a face on the surface based on the overlap between the two intersecting objects. Once the imprint has been made, the new face can be modified.

To imprint an object on a solid, select the **SOLIDEDIT** command and pick the **Body>Imprint** option. If IMPRINT is entered at the keyboard or the button is picked on the ribbon, the option is directly entered. Once the option is activated, you are prompted to select the solid. This is the object on which the other objects will be imprinted. Then, select the objects to be imprinted. You have the option of deleting the source objects.

The imprinted face can be modified using face editing options. **Figure 12-13** illustrates objects imprinted onto a solid model. Two of these are then extruded into the solid to create holes. The **PRESSPULL** command is used on the third object to create a cylindrical feature.

Ribbon
Home
> Solid Editing
Solid
> Solid Editing

Imprint
Type
**IMPRINT
SOLIDEDIT**

IMPRINT

Figure 12-13.
Imprinted objects form new faces that can be modified. A—A solid box with three objects on the plane of the top face. B—The objects are imprinted, then two of the new faces are extruded through the solid and used in subtraction operations. The **PRESSPULL** command is used on the third new face to create the cylindrical feature.

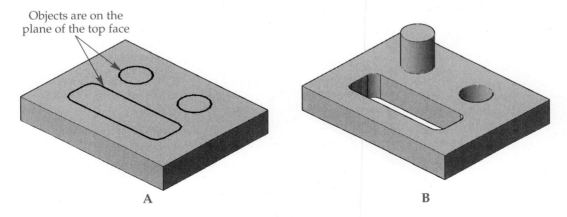

Objects are on the plane of the top face

A

B

Remember that objects are drawn on the XY plane of the current UCS unless you enter a specific Z value. Therefore, before you draw an object to be imprinted onto a solid model, be sure you have set an appropriate UCS for proper placement of the object by using the UCS icon grips or a dynamic UCS. Alternately, you can draw the object on the XY plane and then move the object onto the solid object.

Separate

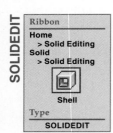

SOLIDEDIT

Ribbon
Home
> Solid Editing
Solid
> Solid Editing
Separate
Type
SOLIDEDIT

The **Separate** option separates two objects that are both a part of a single solid composite, but appear as separate physical entities. This can happen when modifying solids using the Boolean commands. The **Separate** option may be seldom used, but it has a specific purpose. If you select a solid model and an object physically separate from the model is highlighted, the two objects are parts of the same composite solid. If you wish to work with them as individual solids, they must first be separated.

To separate a solid body, select the command and pick the **Body>Separate** option. If the button is picked in the ribbon, the option is directly entered. You are then prompted to select a 3D solid. After you pick the solid, it is automatically separated. No other actions are required and you can exit the command. However, if you select a solid in which the parts are physically joined, AutoCAD indicates this by prompting The selected solid does not have multiple lumps. A "lump" is a physically separate solid entity. In order to separate a solid, the solid must be composed of multiple lumps. See **Figure 12-14.**

Shell

SOLIDEDIT

Ribbon
Home
> Solid Editing
Solid
> Solid Editing
Shell
Type
SOLIDEDIT

A *shell* is a solid that has been "hollowed out." The **Shell** option creates a shell of the selected object using a specified offset distance, or thickness. To create a shell of a solid body, select the command and pick the **Body>Shell** option. Remember, the option is directly entered when picking the button in the ribbon. You are prompted to select the solid. Only one solid can be selected.

After selecting the solid, you have the opportunity to remove faces. If you do not remove any faces, the new solid object will appear identical to the old solid object when shaded or rendered. The thickness of the shell will not be visible. If you wish to create a hollow object with an opening, select the face to be removed (the opening).

Figure 12-14.
A—After the cylinder is subtracted from the box, the remaining solid is considered one solid. B—Use the **Separate** option to turn this single solid into two solids.

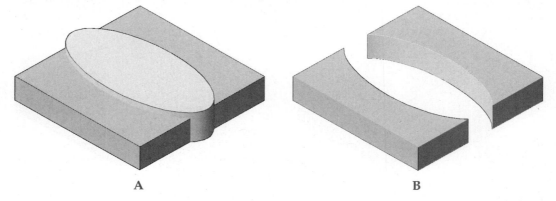

A   B

After selecting the object and specifying any faces to be removed, you are prompted to enter the shell offset distance. This is the thickness of the shell. A positive shell offset distance creates a shell on the inside of the solid body. A negative shell offset distance creates a shell on the outside of the solid body. See **Figure 12-15**. If you shell a solid that contains internal features, such as holes, grooves, and slots, a shell of the specified thickness is placed around those features. This is shown in **Figure 12-16**.

If the shell operation is not readily visible in the current view, you can rotate the view by pressing the [Shift] key and pressing the mouse wheel button at the same time to enter the transparent **3DORBIT** command. You can also rotate the view using the view cube, or you can see the results by picking the **X-Ray Effect** button in the **Visual Styles** panel on the **Visualize** tab of the ribbon.

Figure 12-15.
A—The right-front, bottom, and left-back faces (marked here by gray lines) are removed from the shell operation. B—The resulting object after the shell operation.

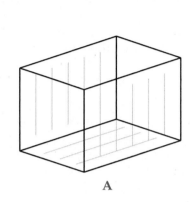
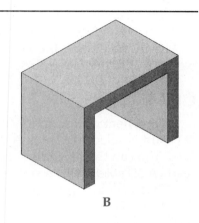

A

B

Figure 12-16.
If you shell a solid that contains internal features, such as holes, grooves, and slots, a shell of the specified thickness is also placed around those features.
A—Solid object with holes subtracted.
B—Wireframe display after shelling with a negative offset.
C—The Conceptual visual style is set current.

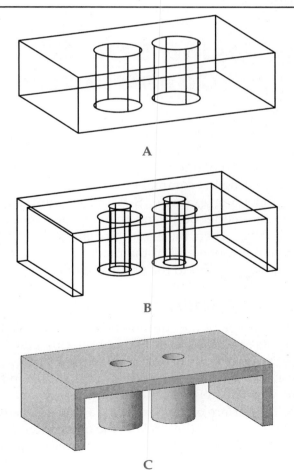

A

B

C

Exercise 12-8

www.g-wlearning.com/CAD/

Complete the exercise on the companion website.

Clean

SOLIDEDIT

Ribbon
Home
> Solid Editing
Solid
> Solid Editing

Clean

Type
SOLIDEDIT

The **Clean** option removes all unused objects and shared edges. Imprinted objects are not removed. Select the command and pick the **Body>Clean** option. If the button is picked in the ribbon, the option is directly entered. Then, pick the solid to be cleaned. No further input is required. You can then exit the command.

Check

SOLIDEDIT

Ribbon
Home
> Solid Editing
Solid
> Solid Editing

Check

Type
SOLIDEDIT

The **Check** option simply determines if the selected object is a valid 3D solid. If a true 3D solid is selected, AutoCAD displays the prompt This object is a valid ShapeManager solid. You can then exit the command. If the object selected is not a 3D solid, the prompt reads A 3D solid must be selected. You are then prompted to select a 3D solid. To access the **Check** option, select the command and pick the **Body>Check** option. Remember, the option is directly entered when picking the button in the ribbon. Then, select the object to check.

Using SOLIDEDIT as a Construction Tool

This section provides an example of how the **SOLIDEDIT** command options can be used not only to edit, but also to construct a solid model. This approach makes it easy to design and construct a model without selecting a variety of commands. It also gives you the option of undoing a single editing operation or an entire editing session without ever exiting the command.

In the following example, **SOLIDEDIT** command options are used to imprint shapes onto the model body and then extrude those shapes into the body to create countersunk holes. Then, the model size is adjusted and an angle and taper are applied to one end. Finally, one end of the model is copied to construct a mating part and a shell operation is applied to the original model.

Creating Shape Imprints on a Model

The basic shape of the solid model in this tutorial is drawn as a solid box. Then, shape imprints are added to it. Throughout this exercise, you may wish to change the UCS to assist in the construction of the part.

1. Draw a solid box using the dimensions shown in **Figure 12-17**.
2. Set the Wireframe visual style current.
3. On the top surface of the box, locate a single ∅.4 circle using the dimensions given. Then, copy or array the circle to the other three corners as shown in the figure.
4. Use the **Imprint** option to imprint the circles onto the solid box. Delete the source objects.

Figure 12-17.
The initial setup for the tutorial model.

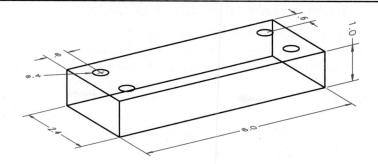

Extruding Imprints to Create Features

The imprinted 2D shapes can now be extruded to create new 3D solid features on the model. Use the **Face>Extrude** option to extrude all four imprinted circles.

1. When you select the edge of the first circle, all features on that face are highlighted, but only the circle you picked and the top face have actually been selected. If you pick inside the circle, only the circle is selected and highlighted. In either case, be sure to also pick the remaining three circles.
2. Remove the top face of the box from the selection set, if needed.
3. The depth of the extrusion is .16 units. Remember to enter –.16 for the extrusion height since the holes remove material. The angle of taper for extrusion should be 35°. Your model should look like **Figure 12-18A**.
4. Extrude the small diameter of the four tapered holes so the holes intersect the bottom of the solid body. Select the holes by picking the small diameter circles. Instead of calculating the distance from the bottom of the chamfer to the bottom surface, you can simply enter a value that is greater than this distance, such as the original thickness of the object. Again, since the goal is to remove material, use a negative value for the height of the extrusion. There is no taper angle. Your model should now look like **Figure 12-18B**.

Moving Faces to Change Model Size

The next step is to use the **Face>Move** option to decrease the length and thickness of the solid body.

1. Select either end face and the two holes nearest to it. Be sure to select the holes *and* the countersinks. Move the two holes and end face two units toward the other end, thus changing the object length to four units.
2. Select the bottom face and move it .5 units up toward the top face, thus changing the thickness to .5 units. See **Figure 12-19**.

Figure 12-18.
A—The imprinted circles are extruded with a taper angle of 35°. B—Holes are created by further extrusion with a taper angle of 0°.

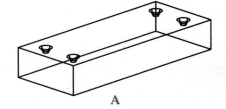

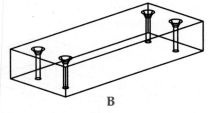

A

B

Figure 12-19.
The length of the object is shortened and the height is reduced.

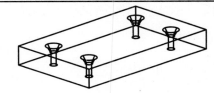

Offsetting a Feature to Change Its Size

Now, the **Face**>**Offset** option is used to increase the diameter of the four holes and to adjust a rectangular slot that will be added to the solid.

1. Using **Face**>**Offset**, select the four small hole diameters. Be sure to remove from the selection set any other faces that may be selected.
2. Enter an offset distance of –.05. This increases the hole diameter and decreases the solid volume. Exit the **SOLIDEDIT** command.
3. Select the **RECTANG** command. Set the fillet radius to .4 and draw a 2 × 1.6 rectangle centered on the top face of the solid. See **Figure 12-20A**.
4. Imprint the rectangle on the solid. Delete the source object.
5. Extrude the rectangle completely through the solid (.5 units). Remember to remove from the selection set any other faces that may be selected.
6. Offset the rectangular feature using an offset distance of .2 units. You will need to select all faces of the feature. This decreases the size of the rectangular opening and increases the solid volume. Your drawing should appear as shown in **Figure 12-20B**.

Tapering Faces

One side of the part is to be angled. The **Face**>**Taper** option is used to taper the left end of the solid.

1. Using **Face**>**Taper**, pick the face at the left end of the solid.
2. Pick point 1 in **Figure 12-21A** as the base point and point 2 as the second point along the axis of tapering.
3. Enter a value of –10 for the taper angle. This moves the upper-left end away from the solid, creating a tapered end. The result is shown in **Figure 12-21B**.

Figure 12-20.
A—The diameter of the holes is increased and a rectangle is imprinted on the top surface.
B—The rectangle is extruded to create a slot.

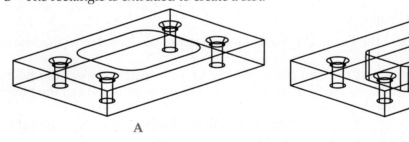

A   B

Figure 12-21.
A—The left end of the object to be tapered. Notice the pick points. B—Top view of the object showing the applied taper.

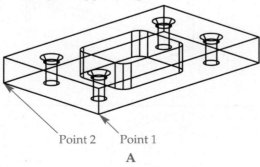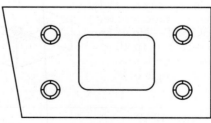

A
**Isometric View Before
Applying Taper**

B
**Top View After
Applying Taper**

Rotating Faces

Next, use the **Face>Rotate** option to rotate the tapered end of the object. The top edge of the face will be rotated away from the holes, adding volume to the solid.

1. Using **Face>Rotate**, pick the face at the left end of the solid.
2. Pick point 1 in **Figure 12-22A** as the first axis point and point 2 as the second point.
3. Enter a value of –30 for the rotation angle. This rotates the top edge of the tapered end away from the solid. See **Figure 12-22B**.

Copying Faces

A mating part will now be created. This is done by first copying the face on the tapered end of the part.

1. Using the **Face>Copy** option, pick the angled face on the left end of the solid.
2. Pick one of the corners as a base point and copy the face one unit to the left. This face can now be used to create a new solid. See **Figure 12-23A**.
3. Draw a line four units in length on the negative X axis from the lower-right corner of the copied face. Use the **EXTRUDE** command on the copied face to create a new solid. Select the **Path** option and use the line as the extrusion path. See **Figure 12-23B**. If you do not use the **Path** option, the extrusion is projected perpendicular to the face.

Figure 12-22.
The tapered end of the object is modified by rotating the face. A—Pick the points shown. B—The resulting solid.

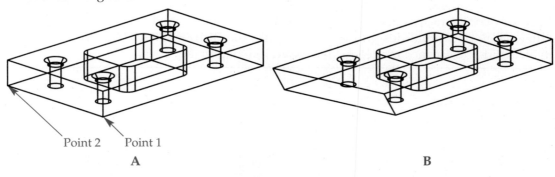

Point 2   Point 1

A

B

Figure 12-23.
Creating a mating part. A—The angled face is copied. B—The copied face is extruded into a solid.

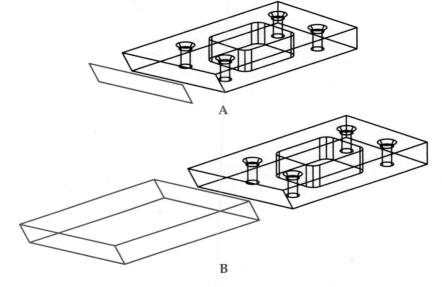

A

B

The **Face>Extrude** option of the **SOLIDEDIT** command cannot be used to turn a copied face into a solid body.

Creating a Shell

The bottom surface of the original solid will now be shelled out. Keep in mind that features such as the four holes and the rectangular slot will not be cut off by the shell. Instead, a shell will be placed around these features. This becomes clear when the operation is performed.

1. Select the **Shell** option and pick the original solid.
2. Pick the lower-left and lower-right edges of the solid. See **Figure 12-24A**. This removes the two side faces and the bottom face.
3. Enter a shell offset distance of .15 units. The shell is created and the model should appear similar to **Figure 12-24A**.
4. Use the view cube or **3DORBIT** command to view the solid from the bottom. Also, set the Conceptual visual style current. Your model should look like the one shown in **Figure 12-24B**.

The **SURFEXTRACTCURVE** command allows you to extract curves such as lines, splines, polylines, or arcs from an existing surface, 3D solid, or face of a 3D solid. This enables the designer to create wireframe geometry that displays the contours of a surface. Once these new contours are created, the designer can use the geometry for 3D model design edits or for conceptual design purposes. Using the **SURFEXTRACTCURVE** command on 3D solid and surface models is discussed in Chapter 10.

Figure 12-24.
A—The shelled object. B—The viewpoint is changed and the Conceptual visual style is set current.

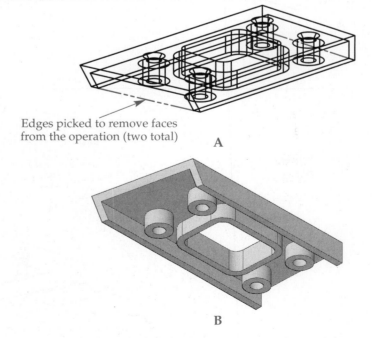

Edges picked to remove faces
from the operation (two total)

A

B

Chapter Review

Answer the following questions using the information in this chapter.

1. What three components of a solid model can be edited using the **SOLIDEDIT** command?
2. When using the **SOLIDEDIT** command, how many faces are highlighted if you pick an edge?
3. How do you deselect a face that is part of the selection set when using the **SOLIDEDIT** command?
4. How can you select a single face?
5. Name the objects that can be used as the path of extrusion when extruding a face.
6. What is one of the most useful aspects of the **Offset Faces** option?
7. How do positive and negative offset distance values affect the volume of the solid?
8. How is a single object, such as a cylinder, affected by entering a positive taper angle when using the **Taper Faces** option?
9. When a shape is imprinted onto a solid body, which component of the solid does the imprinted object become and how can it be used?
10. What is the purpose of the **XEDGES** command?
11. How does the **Shell** option affect a solid that contains internal features such as holes, grooves, and slots?
12. How can you determine if an object is a valid 3D solid?
13. Describe two ways to change the view of your model while you are inside of a command.
14. How can you extrude a face in a straight line, but not perpendicular to the face?

Drawing Problems

1. General Complete the tutorial presented in this chapter. Then, perform the following additional edits to the original solid.
   A. Lengthen the right end of the solid by .5 units.
   B. Taper the right end of the solid with the same taper angle used on the left end, but taper it in the opposite direction.
   C. Fillet the two long, top edges of the solid using a fillet radius of .2 units.
   D. Rotate the face at the right end of the solid with the same rotation angle used on the left end, but rotate it in the opposite direction.
   E. Save the drawing as P12-1.

Problems 2–6. Follow the instructions to complete each problem. Do not add dimensions to the models. For objects that have dimensions with tolerances, use the base dimension from which the upper and lower limits are calculated to create the model. Save each drawing as P12-(problem number).

2. **Mechanical** Construct the solid part shown using as many **SOLIDEDIT** options as possible. After completing the object, make the following modifications.
   A. Lengthen the 1.250″ diameter feature by .250″.
   B. Change the .750″ diameter hole to .625″ diameter.
   C. Change the thickness of the .250″ thick flange to .375″ (toward the bottom).
   D. Extrude the end of the 1.250″ diameter feature .250″ with a 15° taper inward.
   E. Save the drawing as P12-2.

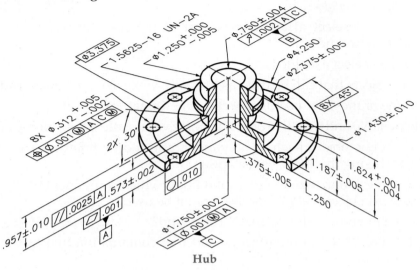

Hub

3. **Mechanical** Construct the solid part shown. Then, perform the following edits on the solid using the **SOLIDEDIT** command.
   A. Change the diameter of the hole to 35.6/35.4.
   B. Add a 5° taper to each inner side of each tooth (the bottom of each tooth should be wider while the top remains the same).
   C. Change the width of the 4.8/4.0 key to 5.8/5.0.
   D. Save the drawing as P12-3.

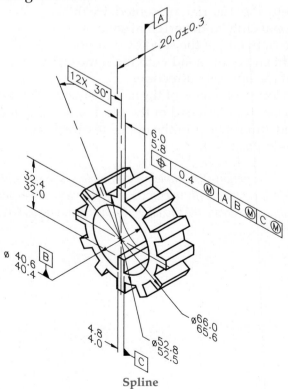

Spline

Figure 13-10.
A—Dimensions placed in the plane of the top surface. B—Dimensions placed using two UCSs that are perpendicular to the top face.

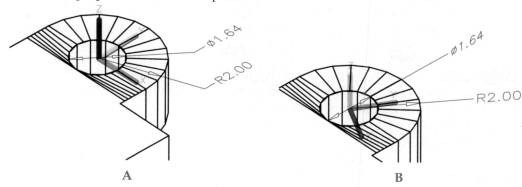

A

B

Placing Leaders and Radial Dimensions in 3D

Although standards such as ASME Y14.5 should be followed when possible, the nature of 3D drawing and the requirements of the project may determine how dimensions and leaders are placed. Remember, the most important aspect of dimensioning a 3D drawing is its presentation. Is it easy to read and interpret?

Leaders and radial dimensions can be placed on or perpendicular to the plane of the feature. **Figure 13-10A** shows the placement of leaders and dimensions on the plane of the top surface. **Figure 13-10B** illustrates the placement of leaders and dimensions on two planes that are perpendicular to the top surface of the object. Remember that text, dimensions, and leaders are always created on the XY plane of the current UCS. Therefore, to create the layout in **Figure 13-10B**, you must use more than one UCS.

Exercise 13-2

www.g-wlearning.com/CAD/

Complete the exercise on the companion website.

Chapter Review

Answer the following questions using the information in this chapter.

1. How can you create 3D text with thickness?
2. If text is placed using the wrong UCS, how can it be edited to appear on the correct one?
3. How can text be placed horizontally based on your viewpoint if the object is displayed in 3D?
4. Name three items that should be a part of a 3D dimensioning template drawing.
5. What is *information grouping*?

Drawing Problems

For the following problems, create solid models as instructed. If appropriate, apply geometric constraints to the base 2D drawing. If you are using constraints and create profiles that can be converted into regions for solid modeling, apply geometric constraints and save the drawing with a different name than the final solid model.

1. **Mechanical** This is a two-view orthographic drawing of a window valance mounting bracket. Create it as a solid model. Use solid primitives and Boolean commands as needed. Use the dimensions given. Similar holes have the same offset dimensions. Create new UCSs as needed. Display an appropriate pictorial view of the drawing. Then, add dimensions. Finally, add the material note so it is plan to the 3D view. Plot the drawing to scale on a C-size sheet of paper. Save the drawing as P13-1.

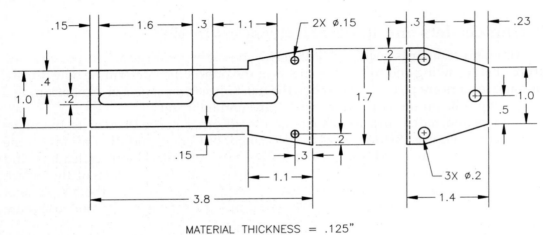

MATERIAL THICKNESS = .125"

2. **Mechanical** Open P5-3. If you have not completed this problem, construct it using the directions for the problem in Chapter 5. Display the 3D view in a single viewport. Then, add dimensions to the 3D view, as shown in Chapter 5. Plot the drawing on a C-size sheet of paper. Save the drawing as P13-2.

3. **Architectural** Create the end table as a solid model using solid primitives and Boolean commands as needed. The result should be a single object. As a test of your object-editing skills, try drawing the entire model by starting with only a single rectangle. You can copy, resize, extrude, and move objects as you create them from the single rectangle. Use the dimensions given and the following information to construct the model.
   A. Table height is 24″.
   B. Top of bottom shelf is 5″ off the floor.
   C. Table legs must be located no less than 1/2″ from the tabletop edge.
   D. Shelf must be no closer than .75″ from the outside of table legs.
   E. Dimension the table as shown.
   F. Save the drawing as P13-3.

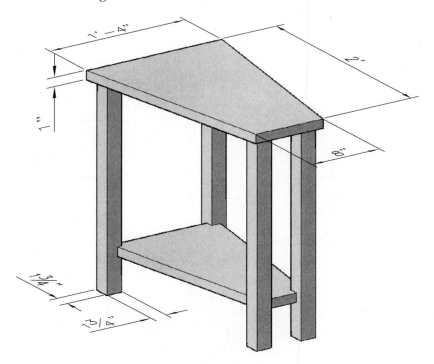

4. **Architectural** Shown are the profiles of a roof gutter (for the collection of rain-water) and a gutter downspout. Draw the profiles in 3D using the dimensions shown. Use the following additional information to construct a 3D model like the one shown in the shaded view.
   A. Offset the gutter profile to create a material thickness of .025″. Be sure to close the ends to create a closed polyline so a 3D solid is created when it is extruded.
   B. Extrude the gutter profile 12″ to create a one-foot section.
   C. Relocate the downspout profile on the underside of the gutter.
   D. Construct an extrusion path for the downspout. Refer to the shaded view, but use your own design.
   E. Extrude the downspout profile along the path.
   F. Dimension the end of the gutter profile in 3D.
   G. Save the drawing as P13-4.

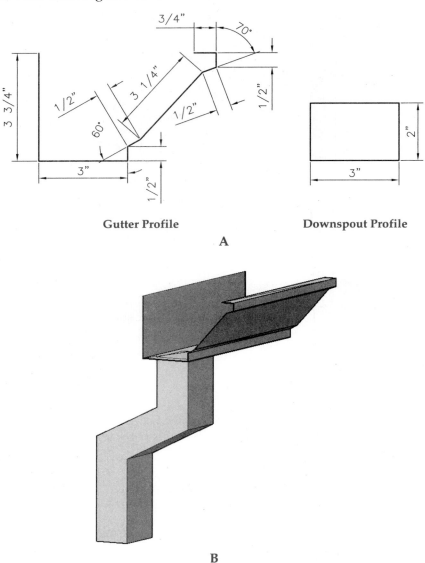

Gutter Profile   Downspout Profile

A

B

5. **Mechanical** Open P5-4. If you have not completed this problem, construct it using the directions for the problem in Chapter 5. Display the 3D view in a single viewport. Then, add dimensions to the 3D view. Plot the drawing on a C-size sheet of paper. Save the drawing as P13-5.

Problems 6 and 7 are mechanical parts. Create a solid model of each part. Dimension each model. Place the title of each model so it is plan to the pictorial view. Plot the finished drawings on B-size paper. Save each drawing as **P13**-*(problem number).*

6. Mechanical

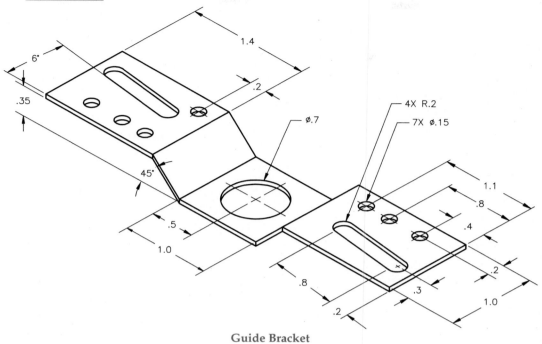

Guide Bracket

7. Mechanical

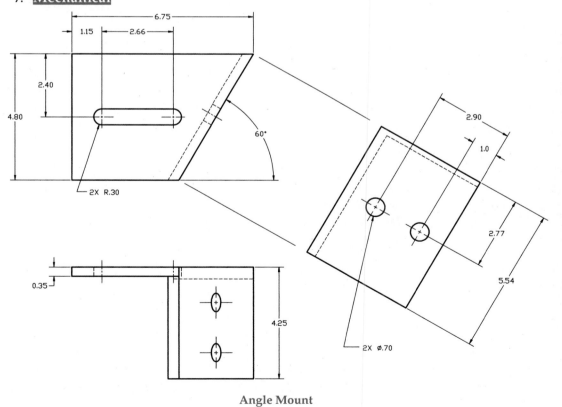

Angle Mount

Three-dimensional models are not typically dimensioned, but there are cases where dimensions may be applied to assist workers on the shop floor. Whenever a 3D model is dimensioned, follow accepted drafting practices when possible, but you may need to violate certain rules in order to clearly describe the part.

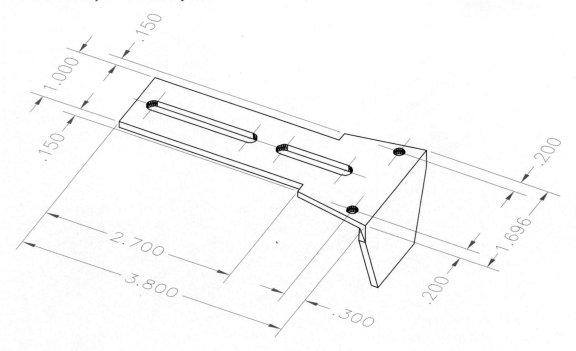

CHAPTER 14

Model Documentation, Analysis, and Point Clouds

Learning Objectives

After completing this chapter, you will be able to:

✓ Create model documentation using drawing views.
✓ Create section views.
✓ Create detail views.
✓ Create auxiliary views.
✓ Construct a 3D section plane through a 3D model.
✓ Adjust the size and location of section planes.
✓ Create a dynamic section of a 3D model.
✓ Construct 2D and 3D section blocks.
✓ Analyze solid and surface models.
✓ Exchange solid model file data with other programs.
✓ Work with point clouds.

This chapter discusses the tools available in AutoCAD for creating drawing views and showing section views of objects. This chapter also discusses AutoCAD tools for conducting model analysis and working with point clouds.

Introduction to Model Documentation

There are multiple techniques available to create drawing views from 3D models in AutoCAD. Using these techniques, orthographic, isometric, section, and auxiliary views can be created in order to produce multiview drawings. The techniques available include the following.

- Use the **VIEWBASE**, **VIEWPROJ**, **VIEWSECTION**, and **VIEWDETAIL** commands to create multiview drawings that are derived from and associated to a 3D solid or surface model. The drawing views are placed in a paper space layout. This is the most efficient method to create multiview drawings in AutoCAD.
- Use the **SECTIONPLANE** command to show the internal features of a 3D solid model. This command can create 2D and 3D section views on an object.

- Use the **SOLVIEW** command to create a multiview drawing from a solid model. This command allows you to create a layout containing orthographic, section, and auxiliary views. The **SOLDRAW** command can then be used to complete profile and section views. The **SOLVIEW** and **SOLDRAW** commands are not discussed in this chapter. The **VIEWBASE**, **VIEWPROJ**, **VIEWSECTION**, and **VIEWDETAIL** commands are typically more useful and provide greater flexibility in editing.

Drawing Views

You can create 2D drawing views from AutoCAD 3D solid or surface models using the **VIEWBASE** command. This command allows you to quickly create a multiview drawing layout. Drawing views created with the **VIEWBASE** command are created in paper space (layout space). After creating a layout of views, you can dimension them using AutoCAD dimensioning commands.

NOTE

The **VIEWBASE** command can be used to create drawing views from AutoCAD 3D models or from Autodesk Inventor® part (IPT), assembly (IAM), or presentation (IPN) files. This chapter covers creating drawing views from AutoCAD 3D models only.

Drawing views created with the **VIEWBASE** command are *associative*. This means that they are linked to the model from which they are created and update to reflect changes to the model geometry. This capability allows you to keep drawing views up-to-date when design changes are required.

When using the **VIEWBASE** command, you create a *base view* of the model and then have the option to create additional views that are projected from the base view without exiting the command. You can create both orthographic and isometric views. A base view created with the **VIEWBASE** command is defined as a *parent view*. Views projected from the base view inherit the properties of the base view, such as the drawing scale and display properties, and are placed in orthographic alignment with the base view. If the base view is moved, any projected views are moved with it to maintain the parent-child relationship.

You can also create drawing views from models imported with the **IMPORT** command. The **IMPORT** command is discussed later in this chapter. Drawing views can also be created from an assembly or subassembly to create an assembly drawing. This chapter discusses creation of drawing views from part models.

Drawing View and Layout Setup

As previously mentioned, drawing views created with the **VIEWBASE** command are created in paper space. To activate paper space, pick a layout tab. Then, before creating views, set up the page layout as required. If a default viewport appears in the layout, *delete the viewport*. Then, right-click on the active layout tab and select **Page Setup Manager...** to access the layout settings. Select the desired paper size, drawing orientation, and other settings. If you have a template drawing already created, you can create a new layout based on the template by right-clicking on a layout tab and selecting **From Template....**

By default, drawing views created with the **VIEWBASE** command are generated using third-angle projection. This is the ANSI standard. The default projection used for drawing views can be set using the **VIEWSTD** command. This command

and other drawing view commands are accessed from the **Layout** tab of the ribbon. See **Figure 14-1**. Pick the dialog box launcher button at the lower-right corner of the **Styles and Standards** panel to initiate the **VIEWSTD** command. This opens the **Drafting Standard** dialog box, **Figure 14-2**.

The options in the **Projection type** area determine the type of projection used for new drawing views. By default, the **Third angle** button is selected. If the **First angle** button is selected, views are created using first-angle projection. Select this option to create views in accordance with ISO standards. The options in the **Thread style** area determine the way threaded ends are shown in section views and the type of circular thread representation used for views showing threaded features. The thread representation options are used when creating views from Autodesk Inventor models or imported models containing threaded features.

The options in the **Shading/Preview** area of the **Drafting Standard** dialog box determine the type of preview that appears when placing a drawing view. When **Shaded** is selected in the **Preview type** drop-down list, a shaded view appears. This is the default option. This option provides a preview image of the view orientation before the view is placed. When the **Bounding box** option is selected, a preview box representing the extents of the view appears. However, no preview image is displayed. The **Shaded view quality** option is set to 100 dpi by default. A higher setting will provide better resolution quality when a shaded preview is displayed. However, for all practical purposes, a setting of 150 dpi or higher does not produce a dramatic difference in resolution quality.

Figure 14-1.
The AutoCAD drawing view commands are accessed from the **Layout** tab of the ribbon.

Commands for creating drawing views

Pick to open the **Drafting Standard** dialog box

Figure 14-2.
The **Drafting Standard** dialog box.

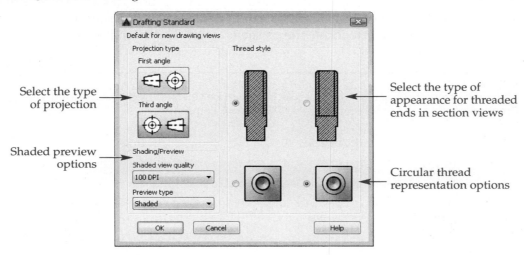

Select the type of projection

Shaded preview options

Select the type of appearance for threaded ends in section views

Circular thread representation options

Creating Drawing Views

After setting up a layout and making the appropriate drawing view settings, you are ready to place the base view. As previously discussed, drawing views can be created from a 3D solid or surface model. Drawing views created in the layout are generated from the model you have created in model space. If no model exists in the drawing and you initiate the **VIEWBASE** command, the **Select File** dialog box appears. From this dialog box, you can select a model created in Autodesk Inventor.

When you select the **VIEWBASE** command from the ribbon, a preview representing a scaled orthographic view of the model appears. This is the *base view* of the model. You can then specify the location of the base view or select an option. To select an option, use dynamic input or make a selection from the **Drawing View Creation** contextual ribbon tab. See **Figure 14-3**.

The **Select** option switches you to model space and allows you to add objects to or remove objects from the base view. If necessary, use the appropriate option to select solids or surfaces to add or remove. Use the **Layout** option to return to the layout after making changes. The **Orientation** option can be used to select a different view from the default base view. The **Orientation** options correspond to AutoCAD's six orthographic and four isometric preset views. These views are based on the WCS. By default, the front view is used as the base view by AutoCAD. Depending on the construction of your model, you may want to use a different view, such as the top view, to serve as the base view. In **Figure 14-4**, the top view of the model is selected as the base view. This view will establish the front view on the orthographic drawing. In most cases, the front orthographic view on the drawing describes the most critical contour of the model. You will typically use the front or top view of the 3D model for the front orthographic view on the drawing.

The **Type** option of the **VIEWBASE** command is available before picking the initial location of the base view. This option is used to specify whether projected views are placed after placing the base view. By default, this option is set to **Base and Projected**. With this option, you can place additional, projected views while the **VIEWBASE** command is still active. Projected views that you place are *projected* from the base view and oriented in the proper orthographic alignment. As previously discussed, projected views have a parent-child relationship with the base view. The base view is the *parent* view and any projected views are *child* views. If the base view is moved, for example, the projected view is moved with it. Projected views are not created with the **VIEWBASE** command when the **Type** option is set to **Base only**. If you place a base view in this manner and later decide to create projected views from the base view, you can use the **VIEWPROJ** command. The **VIEWPROJ** command allows you to place projected views from any selected view.

After you pick the location for the base view, you can press [Enter] to place projected views (if the **Type** option is set to **Base and Projected**). Drag the cursor away from the base view and then pick to locate the projected view. The direction in which you move

VIEWBASE

Ribbon
Layout
> Create View

Base View from
Model Space
Type
VIEWBASE

Figure 14-3.
The view orientation, scale, and other settings can be selected from the **Drawing View Creation** contextual tab when using the **VIEWBASE** command.

Select the view
orientation

Pick to select
the scale

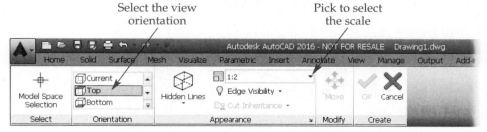

Figure 14-4.
Placing a base view with the **VIEWBASE** command. A—The original model drawn in model space.
B—The preview of the base view in paper space. The Top view option is selected for the base
view because it is the most descriptive view of the model. C—The resulting base view.

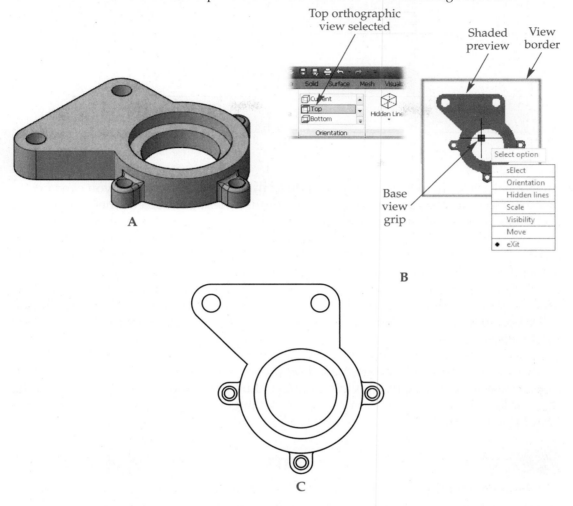

A

B

C

the cursor determines the orientation of the projected view. You can continue placing
projected views, or you can press [Enter] to end the command. To end the command
without placing projected views, press [Enter] twice.

The display style of the base view can be changed by using the **Hidden Lines**
option. The four display style options available are **Visible Lines**, **Visible and Hidden
Lines**, **Shaded with Visible Lines**, and **Shaded with Visible and Hidden Lines**. See
Figure 14-5. These options can be selected using dynamic input or the drop-down list
in the **Appearance** panel in the **Drawing View Creation** contextual ribbon tab.

The **Scale** option is used to set the scale of the base view. You can select a scale
from the drop-down list in the **Appearance** panel in the **Drawing View Creation** contextual ribbon tab. You can also specify the scale by typing a value.

The **Move** option allows you to move the base view after picking the initial location on screen. When you select the **Move** option, the view is reattached to the cursor so
that you can move it to another location. After you pick a new location, the **VIEWBASE**
options are again made available.

The **Visibility** option allows you to control the display of drawing geometry in
the view. The **Interference edges** option controls whether both object and hidden lines are
displayed for interference edges. By default, this option is set to **No**. The **Tangent edges**
option controls whether tangential edges are displayed to show the intersection of

Figure 14-5.
Display style options available for drawing views. A—**Visible Lines**. B—**Visible and Hidden Lines**. C—**Shaded with Visible Lines**. D—**Shaded with Visible and Hidden Lines**.

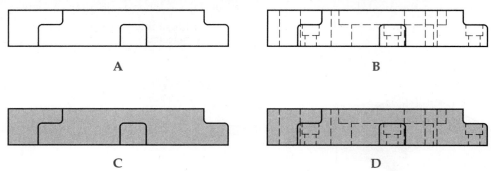

surfaces. By default, this option is set to **No**. If this option is set to **Yes**, you can specify whether tangent edges are shortened to distinguish them from object lines that overlap. The **Bend extents** option is only available when working with a model that includes a view with sheet metal bends. The **Thread features** option controls thread displays on models with screw thread features. The **Presentation trails** option is used to control the display of trails in views created from presentation files.

Additional drawing view options can be accessed in the **Appearance** panel of the **Drawing View Creation** contextual ribbon tab. Selecting the dialog box launcher button at the bottom of the panel opens the **View Options** dialog box. The options in the **View Justification** drop-down list determine the justification of the view. The *justification* refers to how the view is "anchored." When changes are made to the model, such as a change in size, the view updates based on the justification setting. If the justification is set to **Fixed**, geometry in the view unaffected by the edit does not change from the original location. If the justification is set to **Centered**, the geometry updates about the center point of the view.

NOTE

The **Representation** option of the **VIEWBASE** command is used to create a view based on a model representation and is only available when working with models created in Autodesk Inventor.

Working with Drawing Views

Creating a base view creates an AutoCAD *drawing view* object. A drawing view has a view border, a base grip, and a parameter grip for changing the scale. The scale and certain other properties of the view can be edited, as discussed later in this chapter. However, the content of the drawing view *cannot* be edited.

When a base view is created, new layers are created by AutoCAD for the drawing view geometry. The layers are created based on support for the type of geometry represented. Object lines (visible lines) in the view are placed on a newly created layer named MD_Visible. Hidden lines in the view are placed on a newly created layer named MD_Hidden, and so on. The drawing view object is placed on the current layer or on the 0 layer. Additional layers may be created by AutoCAD, depending on the type of model, display style used, and edges displayed in the view. However, the layers are only created to organize the drawing view geometry. The layer properties can be modified to change the appearance of the drawing view geometry, but the geometry cannot be otherwise modified.

In **Figure 14-6**, a multiview drawing layout is created by placing three projected views after placing the base view. The projected views include two orthographic views and an isometric view. Notice that wireframe styles are used for the orthographic views and a shaded style is used for the isometric view. Drawing views can be edited after placing them to change properties as needed. In addition, as previously discussed, drawing views update when changes are made to the model. This is discussed in more detail in the next section.

PROFESSIONAL TIP

You can create more than one base view in a layout. Additional base views can be used to create assembly or subassembly drawings.

Updating Drawing Views

Drawing views maintain an associative relationship with the model from which they are created. However, it is important to note that this associative relationship is controlled by the model, not the drawing view. When making design changes, you make changes to the model geometry, *not* the drawing view.

During the design process, it is often necessary to make modifications. If you have created drawing views from a model, and then make modifications to the model, the derived base view and any views projected from the base view are updated automatically by default. This behavior is controlled by the **VIEWUPDATEAUTO** system variable. When this system variable is set to 1 (on), drawing views automatically update when you open a drawing file or activate a layout with drawing views. The

---

Figure 14-6.
A multiview drawing layout consisting of front, top, right-side, and isometric views. The top view of the model is used as the front view (base view) in the drawing. Notice the different display style options that are used.

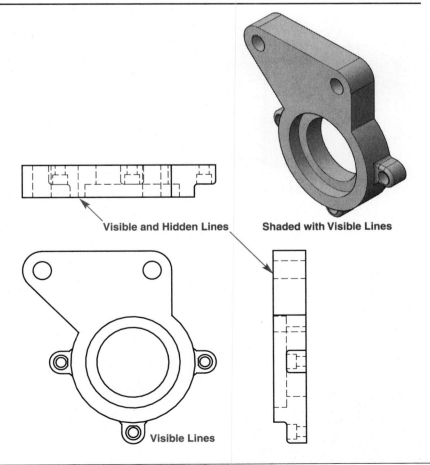

Visible and Hidden Lines

Shaded with Visible Lines

Visible Lines

VIEWUPDATEAUTO system variable setting is indicated by the **Auto Update** button located on the **Update** panel of the **Layout** tab on the ribbon. Refer to **Figure 14-1**. With the default setting of 1, the button has a blue background.

Drawing views can be updated manually if needed. For example, if the **VIEWUPDATEAUTO** system variable is set to 0 (off) and changes are made to the model, drawing views are not updated by AutoCAD automatically. If you make changes to a model in model space and then switch to a layout with drawing views of the model, red markers appear to indicate that the views are *out-of-date*. See **Figure 14-7**. A balloon notification also appears on the status bar. You can pick the link in the balloon notification or use one of AutoCAD's view update commands to update the views.

In **Figure 14-7A**, the six arrayed holes have been modified by changing the diameter from ∅1.8 to ∅.9. In **Figure 14-7B**, the drawing views appear with markers to indicate they are out-of-date. To update the views, select **Update all Views** from the **Update View** drop-down list in the **Update** panel on the **Layout** tab of the ribbon. After updating the views, the holes are updated to the modified diameter. See **Figure 14-7C**. In this example, all of the views are updated at once. This is the preferred method for updating views. However, you can also update views individually by selecting **Update View** from the **Update View** drop-down list in the **Update** panel on the **Layout** tab of the ribbon. Selecting **Update View** is the same as using the **VIEWUPDATE** command. When using this command, you are prompted to select each view to update.

Update All Views

Ribbon
Layout
> Update
Update all Views

VIEWUPDATE

Ribbon
Layout
> Update
Update View
Type
VIEWUPDATE

Figure 14-7.
Updating drawing views after editing the 3D model. A—The holes in the flanged coupler have been edited from ∅1.8 to ∅.9. B—After switching to the layout, markers appear to indicate that the drawing views are out-of-date. C—The views are updated to reflect the changes in the model.

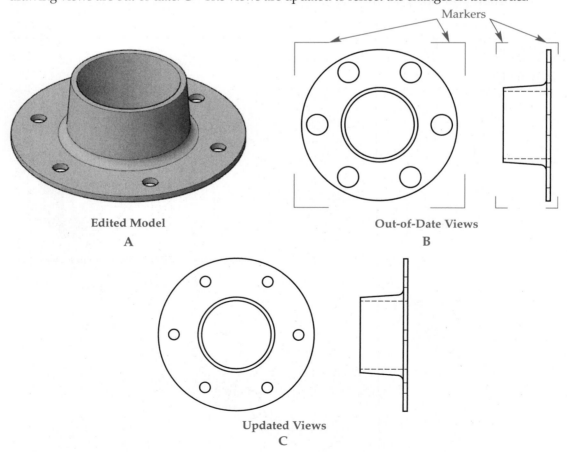

Markers

Edited Model
A

Out-of-Date Views
B

Updated Views
C

Drawing views are different from viewports. You cannot pick inside of the view and edit or modify the drawing geometry by any standard AutoCAD editing technique. Drawing views can only be updated by changing the 3D model geometry.

In order to modify solid primitive subobjects in a composite solid model, the solid history must be recorded.

Editing Drawing Views

Drawing views can be edited after being created. Like other AutoCAD objects, drawing views can be moved or rotated. In addition, certain properties of drawing views, such as the display style, edge visibility, and scale, can be modified.

The **VIEWEDIT** command can be used to edit the properties of a drawing view. You can quickly initiate this command by double-clicking on a view. The **Drawing View Editor** contextual ribbon tab appears, **Figure 14-8**. The editing options are similar to the options available when you create a base view.

As previously discussed, projected views inherit the properties of the base view when created. If you select a projected view with the **VIEWEDIT** command, you can change the display style, edge visibility, and scale.

A quick way to change the scale of a view is to use the scale parameter grip. This grip appears when you single-click on a view. Pick on the grip to access a different scale. To move a drawing view, single-click on the view and then pick on the base grip to move the view directly. When moving a parent view, any child views will move accordingly to maintain alignment. When moving a child view, you can move the view, but it cannot be moved out of alignment with the parent view. This applies to orthogonal views. If you move an isometric view, it is not aligned to other views and can be moved freely around the layout.

Picking on a drawing view grip and right-clicking displays a shortcut menu. The **Stretch** option can be used to move the drawing view. The **Rotate** option can be used to rotate the drawing view. You can rotate the view dynamically using the cursor or you can specify a rotation angle. If a drawing view is rotated, any parent-child relationships that exist between the view and other drawing views are broken.

You can break the alignment between parent and child views when moving a child view. To do this, select the drawing view grip and press [Shift] once. The view is free to move to a different location. To reestablish the alignment, press [Shift] again. The alignment is restored and the child view cannot move out of alignment with the parent view.

Figure 14-8.
The **Drawing View Editor** contextual ribbon tab.

View property options

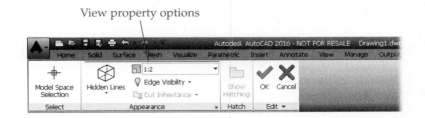

Section Views

A *section view* shows the internal features of an object along a section line (cutting plane). A section view is projected from an existing view, such as an orthographic top view. The existing view serves as a parent view. To create a section view, you pick points on the parent view to define the section line (cutting-plane line). You can also select an object, such as a line or polyline, to define the section line. Section views are created using the **VIEWSECTION** command. This command can be used to create full, half, offset, or aligned sections from an AutoCAD 3D model or an Autodesk Inventor file.

Section views created with the **VIEWSECTION** command are created in the same paper space layout as other drawing views. Section views are *associative*. A section view is linked to the parent view that creates the section view. As with other types of drawing views, section views are updated automatically when model changes are made if the **VIEWUPDATEAUTO** system variable is set to 1.

By default, a section identifier is placed with the section line and a section view label is placed with the section view when you create the view. See **Figure 14-9**. The

Figure 14-9.
A section view created with the **VIEWSECTION** command. Shown is a full section created from the top view of the model. A—The model drawing. B—After creating the top view as a base view, a section view is created by drawing the section line through the center of the object. Note the placement of the section identifier and section view label.

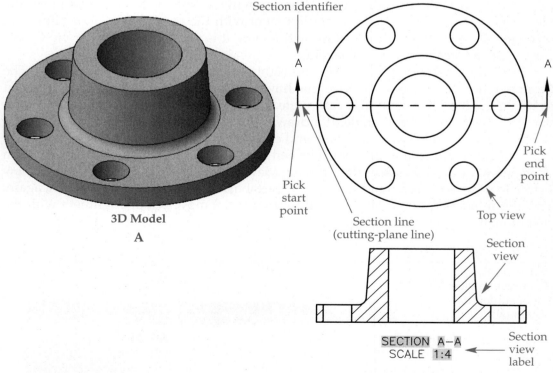

To create a half section, select the **VIEWSECTION** command, select the parent view, and access the **Half** option. The option is accessed directly from the ribbon by selecting the **Half** option from the **Section** drop-down list in the **Create View** panel of the **Layout** tab. Three points are required to define the section line. Use object snaps and object snap tracking to assist in specifying each point. If you pick an incorrect point, use the **Undo** option. After drawing the final segment of the section line, pick to locate the view. You can then select an option or press [Enter] to exit the command. You can adjust options using the **Section View Creation** contextual ribbon tab, dynamic input, or the command line. The options are the same as those available when creating a full section.

An example of using the **Depth** option to adjust a half section is shown in **Figure 14-15**. In **Figure 14-15A**, the half section is created with the section line drawn through the middle of the object. In **Figure 14-15B**, the **Depth** option is selected to move the depth line to the back end of the hole feature. In this case, adjusting the depth of the section line helps clarify the interior detail of the part.

Ribbon

Layout
> Create View

Section View Half

Type
VIEWSECTION

VIEWSECTION

Exercise 14-3

www.g-wlearning.com/CAD/

Complete the exercise on the companion website.

Figure 14-15.
Using the **Depth** option to adjust the visibility of objects in a section view. A—A top view and half section view of a model. The **Depth** option is set to **Full**. B—The half section view is adjusted by moving the depth line in the top view. Notice that the back leg no longer appears. The depth line is shown for reference only and does not appear after exiting the command.

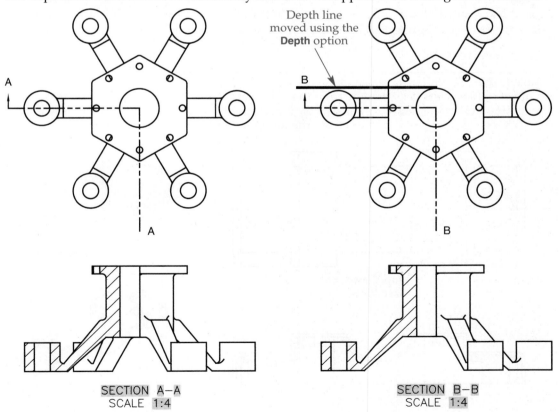

Depth line moved using the **Depth** option

SECTION A—A
SCALE 1:4

SECTION B—B
SCALE 1:4

Half Section View
A

Half Section View after Adjusting the Depth
B

Offset Section

VIEWSECTION

Ribbon

Layout
> Create View

Section View
Offset

Type

VIEWSECTION

An offset section *shifts* (offsets) the section line to pass through certain features of a part or assembly for better clarification of detail. Typically, the section line consists of several segments drawn through features such as holes and bosses. See **Figure 14-16**.

To create an offset section, select the **VIEWSECTION** command, select the parent view, and access the **Offset** option. The option is accessed directly from the ribbon by selecting the **Offset** option from the **Section** drop-down list in the **Create View** panel of the **Layout** tab. Then, pick the points to define the section line. Select as many points as needed to define the section. Use object snaps and object snap tracking as needed. If you pick an incorrect point, use the **Undo** option. After drawing the final segment of the section line, select the **Done** option. Then, pick to locate the view. You can then select an option or press [Enter] to exit the command. The options are the same as those available when creating a full section.

Exercise 14-4

www.g-wlearning.com/CAD/

Complete the exercise on the companion website.

Aligned Section

An aligned section is made by passing two nonparallel cutting planes through an object. The resulting section view shows features that are oriented at an angle rotated into the same cutting plane. See **Figure 14-17**.

The purpose of an aligned section is to show the true size and shape of a feature. In **Figure 14-17A**, an aligned section is created to show the true size and shape of the right arm. Notice that the right arm is rotated into the center cutting plane to project the view. For this view, the **Projection** option is set to **Normal**. This is conventional practice. In **Figure 14-17B**, the right arm is projected from its true position and

---

Figure 14-16.
An offset section view is created by drawing an offset section line through several features. The section line in this drawing consists of five segments. Shown is a layout including a top view, an offset section view created from the top view, and an isometric section view created from the offset section.

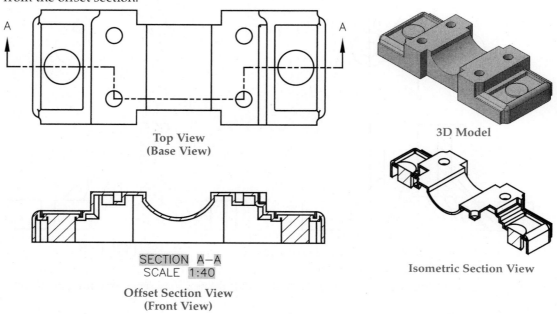

Top View
(Base View)

3D Model

SECTION A–A
SCALE 1:40

Offset Section View
(Front View)

Isometric Section View

Figure 14-17.
An aligned section view is created by drawing an angled section line through nonparallel features. A—The **Normal** projection method is used to project the right arm in its true shape and size. Notice that the right arm rotates into the center cutting plane to make the projection. This is conventional practice. B—Using the **Orthogonal** option creates a true projection of the right arm and results in foreshortening in the section view.

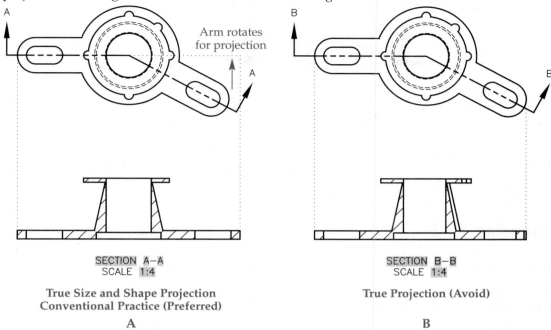

SECTION A–A
SCALE 1:4

True Size and Shape Projection
Conventional Practice (Preferred)

A

SECTION B–B
SCALE 1:4

True Projection (Avoid)

B

appears foreshortened in the front view. For this view, the **Projection** option is set to **Orthogonal**. In this instance, this is *not* the preferred method to create the view.

To create an aligned section, select the **VIEWSECTION** command, select the parent view, and access the **Aligned** option. The option is accessed directly from the ribbon by selecting the **Aligned** option from the **Section** drop-down list in the **Create View** panel of the **Layout** tab. Then, pick the points to define the section line. Use object snaps and object snap tracking as needed. After drawing the final segment of the section line, select the **Done** option. Then, pick to locate the view. You can then select an option or press [Enter] to exit the command. The options are the same as those available when creating a full section.

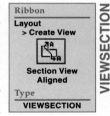

Ribbon
Layout
> Create View
Section View
Aligned
Type
VIEWSECTION

VIEWSECTION

In **Figure 14-17**, notice that the angled segment of the section line extends beyond the lower right arm. To orient the end of the section line in this manner, it may be necessary to edit the section line using grips to stretch the vertex. Another option is to draw the section line first and use the **Object** option of the **VIEWSECTION** command to create the section. This is discussed next.

Creating a Section View from an Object

You can select an object in the paper space layout to use as the section line when creating a section view. This is a useful method when it is difficult to locate points using the **VIEWSECTION** command. To use an object as the section line, select the **VIEWSECTION** command, select the parent view, and access the **Object** option. The option is accessed directly from the ribbon by selecting the **From Object** option from the **Section** drop-down list in the **Create View** panel of the **Layout** tab. Then, select the object and press [Enter]. Pick a point to locate the view. The object you select determines the type of section created. In **Figure 14-18**, a polyline is drawn in the desired location prior to accessing the **Object** option of the **VIEWSECTION** command.

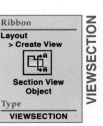

Ribbon
Layout
> Create View
Section View
Object
Type
VIEWSECTION

VIEWSECTION

Figure 14-18.
A section view can be created from an object using the **Object** option of the **VIEWSECTION** command. The polyline shown here provides an alternate way to create the aligned section for the object shown in Figure 14-17.

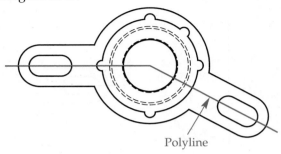

Polyline

The **Object** option provides an alternate way to create the section view and may be easier than picking points. When using the **Object** option, the selected object is automatically deleted after creating the section view.

Exercise 14-5

www.g-wlearning.com/CAD/

Complete the exercise on the companion website.

Constraining Section Lines

When picking points to define a section line, you have the option to apply constraints to control the location of the line. If **Infer Constraints** is activated, constraints are automatically applied when you pick points. The **Infer Constraints** button on the status bar controls whether constraints are inferred when drawing new objects. The **Infer Constraints** button does not appear on the status bar by default. Select the **Infer Constraints** option in the status bar **Customization** flyout to add the **Infer Constraints** button to the status bar. Inferring constraints provides a way to constrain the geometry of the section line to the geometry in the drawing view. This is similar to applying constraints automatically when drawing objects in model space.

Constraints can also be applied to the section line after creating the section view. The **VIEWSYMBOLSKETCH** command is used to constrain a section line using geometric and dimensional constraints. After accessing this command, select the section line. The **Parametric** tab is displayed and symbol sketch mode is activated. In this mode, objects other than the drawing view geometry and section line are faded. A polyline replaces the section line. You can use the appropriate tools to constrain the section line to the drawing view geometry. You can also add construction geometry if needed to apply constraints. Geometry that is added is only displayed in symbol sketch mode and does not display when you exit the command. When finished adding constraints, pick the **Finish Symbol Sketch** button in the **Exit** panel of the **Parametric** tab. When prompted, specify whether to save or discard changes made in symbol sketch mode.

VIEWSYMBOLSKETCH

Ribbon
Layout
> Modify View

View Sketch

Type
VIEWSYMBOL-
SKETCH

PROFESSIONAL TIP

Constrain the section line if future design modifications change the drawing view geometry size or location. For example, apply a fix constraint to constrain the section line to the center of the sectioned object.

Excluding Components from Sectioning

Certain features in section views, such as fasteners, are not shown sectioned. For example, components such as screws, pins, and thin-walled objects in an assembly are shown without section lines. This practice conforms with drafting standards. When creating a section view from a parent view that includes items such as fasteners and shafts, you can use the **VIEWCOMPONENT** command to control how sectioning is applied.

To exclude items from sectioning, access the **VIEWCOMPONENT** command. Select the components in the parent view and then select the **None** option. If you select the **Section** option, the components are included in sectioning. Selecting the **Slice** option creates a thin section at the section line and removes the visibility of objects behind the section line.

Ribbon

Layout
> Modify View

Edit Components

Type
VIEWCOMPONENT

VIEWCOMPONENT

Editing Section Views

The **VIEWEDIT** command, introduced earlier in this chapter, can be used to edit the properties of a section view. You can quickly initiate this command by double-clicking on a section view. This displays the **Section View Editor** contextual ribbon tab. Many of the same options used when creating a section view are available in this tab. Additional options may be available depending on the type of view selected.

The **Cut Inheritance** option is available when a view created from a section view inherits the section cut. An example of this is an isometric view projected from a section view. By default, the isometric view shows the section cut with the sectioned portion hatched. To remove the section cut from the isometric view, expand the **Cut Inheritance** drop-down list in the **Section View Editor** contextual ribbon tab and uncheck the **Section cut** option. This resets the view to an isometric view without sectioning.

The **Defer Updates** option, available in the **Edit** panel of the **Section View Editor** contextual ribbon tab, allows you to make edits without previewing the changes dynamically. The effects of changes you make are not displayed until you exit the command by picking the **OK** button in the **Edit** panel. By default, the **Defer Updates** option is inactive. For very large drawings, activating this option may save time by preventing dynamic previews from occurring. For very small drawings, deferring updates provides only a marginal benefit.

As previously discussed, you can edit a section view by using grips to edit the location of the section line. If a section line is constrained, press the [Shift] key to relax constraints and break the alignment with the constrained position.

NOTE

The hatch pattern applied to the section view can be edited in the same manner as a hatch pattern applied in a 2D drawing. Double-click on the hatch to display the **Drawing View Hatch Editor** contextual ribbon tab. You can also hover over the center grip of the hatch pattern to access grip editing options.

Exercise 14-6

www.g-wlearning.com/CAD/

Complete the exercise on the companion website.

Detail Views

A *detail view* shows a selected portion of a view to clarify model details. A detail view is projected from a parent view and is typically shown at a larger scale. As with other types of projected views, the detail view is linked to the parent view.

Detail views are created using the **VIEWDETAIL** command. A detail view is created by drawing a circular or rectangular boundary to define the extents of the view. You can create detail views from an AutoCAD 3D model or an Autodesk Inventor file.

Detail views are associative. As with other types of drawing views, detail views are updated automatically when model changes are made if the **VIEWUPDATEAUTO** system variable is set to 1.

Examples of detail views are shown in **Figure 14-19**. Creating a detail view places a *detail boundary* representing the "cutout" area in the parent view. The detail boundary includes a detail identifier. The detail view is placed in another location on the drawing and includes a view label. The detail identification is automatically incremented when you place additional detail views. The appearance of elements in the detail identifier and detail view label is controlled by the detail view style. A *detail view style* defines settings such as the text style and height, symbol size, and detail boundary appearance.

The **VIEWDETAILSTYLE** command is used to create and modify detail view styles. This command accesses the **Detail View Style Manager** dialog box, **Figure 14-20**. Picking the **New...** button allows you to create a new detail view style. Picking the **Modify...** button opens the **Modify Detail View Style** dialog box for the selected style. The tabs

VIEWDETAILSTYLE

Ribbon

Layout
> Styles and
Standards

Detail View Style

Type
VIEWDETAILSTYLE

Figure 14-19.
A layout of drawing views of a chair including the top, front, and side views and two detail views. The detail views have a larger scale and show details of the seat and leg. Each detail view is drawn with a circular boundary.

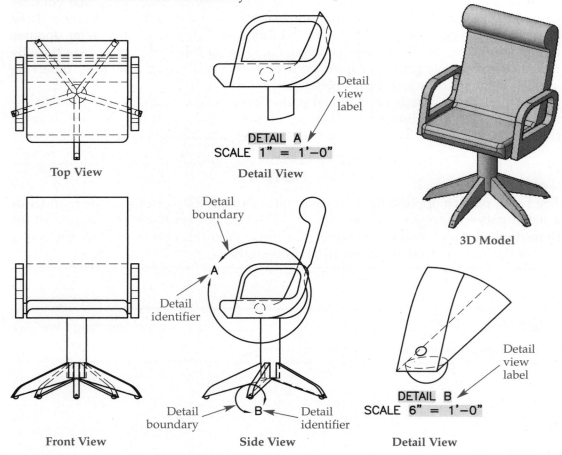

Top View

Detail View

Detail view label

DETAIL A
SCALE 1" = 1'-0"

3D Model

Front View

Detail boundary

Detail identifier

Side View

Detail boundary

Detail identifier

Detail View

Detail view label

DETAIL B
SCALE 6" = 1'-0"

in the **New Detail View Style** dialog box or the **Modify Detail View Style** dialog box are used to make settings for the detail identifier, detail boundary, and detail view label. These tabs are shown in **Figure 14-20**. As with other types of styles, develop standards for detail views in accordance with company or industry standards.

Creating a Detail View

Creating a detail view is similar to creating a section view. To create a detail view, select the **VIEWDETAIL** command and then select the parent view. The default method for creating the view is to create a circular detail boundary. Refer to **Figure 14-19**. This is the preferred display for most detail views. You can change the boundary type to rectangular using the **Boundary** option. With the **Rectangular** option, a rectangular detail boundary is drawn and the detail view has a rectangular outline. See **Figure 14-21**. Selecting one of the options from the **Detail** drop-down list in the **Create View** panel of the **Layout** tab on the ribbon begins the command and sets the appropriate boundary type. If you are creating a circular detail boundary, select the parent view and then pick a point to specify the center of the view. At the next prompt, drag the cursor or enter a value to set the size of the boundary. Then, pick a point to locate the view. A rectangular detail boundary is created in the same manner. After locating the view,

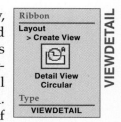

Ribbon
Layout
> Create View

Detail View
Circular

Type
VIEWDETAIL

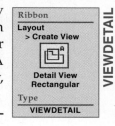

Ribbon
Layout
> Create View

Detail View
Rectangular

Type
VIEWDETAIL

Figure 14-20.
The **Detail View Style Manager** dialog box is used to create or modify detail view styles.
A—Pick the **New...** button to create a new detail view style or pick the **Modify...** button to modify an existing style. B—The **Identifier** tab. C—The **Detail Boundary** tab. D—The **View Label** tab.

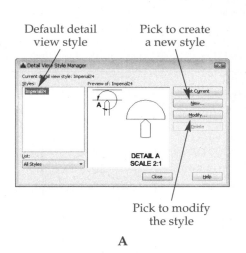

A

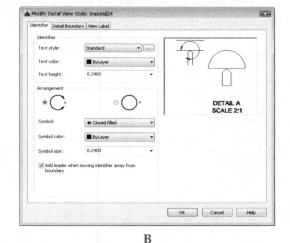

B

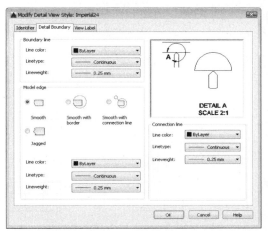

C

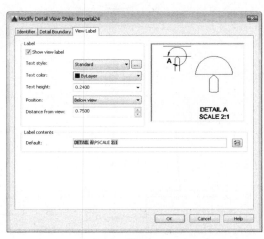

D

Chapter 14 Model Documentation, Analysis, and Point Clouds **363**

you can select an option or press [Enter] to exit the command. You can adjust options using the **Detail View Creation** contextual ribbon tab. You can also use dynamic input or the command line.

The **Hidden Lines**, **Scale**, **Visibility**, and **Move** options are the same as those previously discussed for section views. The **Model Edge** option is used to adjust the edges of the detail view and set border display and leader options. See **Figure 14-22**. The **Smooth** option creates a smooth edge for the view. This is the default option. The **Smooth with Border** option creates a smooth edge for the view and draws a circular or rectangular border, depending on the type of boundary specified. The **Smooth with Connection Line** option creates a smooth edge, draws a circular or rectangular border, and attaches a leader from the detail symbol in the parent view to the detail view. The **Jagged** option creates the view with a jagged edge. With this option, no border is displayed and the view does not have a leader attached. The **Annotation** option allows you to adjust the view identifier and specify whether a view label is shown.

Once created, detail views can be edited by editing the detail boundary or detail identifier. To edit the detail boundary, select the boundary and hover over one of the four boundary grips to display a shortcut menu. The options allow you to stretch the boundary and change the boundary type to circular or rectangular. Hovering over the detail identifier grip allows you to move the identifier or reset the identifier to the initial position.

You can also edit a detail view by using the **VIEWEDIT** command or by selecting the view to display grips. A detail view has a base grip providing access to the standard grip editing options and a parameter grip for changing the scale. The detail view label is an mtext object. It can be moved by selecting the label and then selecting the base grip.

Figure 14-21.
A layout of drawing views including a detail view drawn using a rectangular detail boundary.

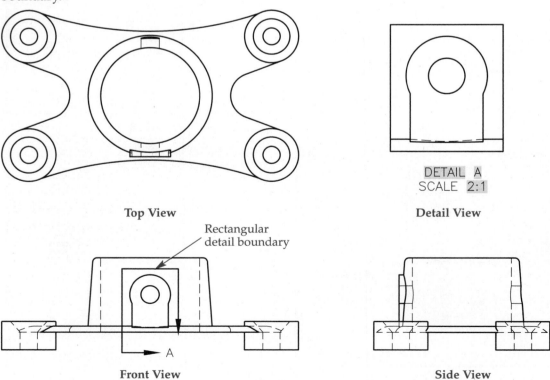

Figure 14-22.
The **Model Edge** option of the **VIEWDETAIL** command is used to adjust the model edges in the detail view. Included are settings for creating a border and a connection leader. A—**Smooth** option. B—**Smooth with Border** option. C—**Jagged** option. D—**Smooth with Connection Line** option.

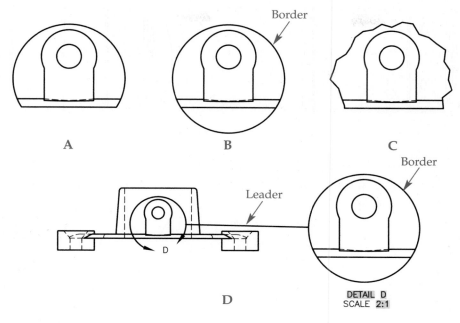

DETAIL D
SCALE 2:1

NOTE

The **Properties** palette can also be used to make modifications to the detail boundary and identifier. Deleting the detail boundary deletes the detail view.

Exercise 14-7   www.g-wlearning.com/CAD/

Complete the exercise on the companion website.

Auxiliary Views

An undocumented feature in AutoCAD is the ability to create an auxiliary view by using a section view. Often, a multiview drawing contains inclined surfaces that do not describe the true size or shape of features in a regular orthographic view. To establish an auxiliary view, you can draw a full section line using the **Full** option of the **VIEWSECTION** command. When specifying the section line, pick two points on the inclined surface. If needed, draw a parallel construction line across the inclined surface and use it to create the section line. The auxiliary view plane is oriented parallel to the inclined edge of the surface and the view is created perpendicular to the surface. To remove the display of the section line and view label from the drawing, freeze the MD_Annotation layer. See **Figure 14-23**. Notice that since the section line does not intersect the object, no hatch pattern is created.

Exercise 14-8   www.g-wlearning.com/CAD/

Complete the exercise on the companion website.

Figure 14-23.
An auxiliary view can be created to show the true shape of an inclined surface by drawing a full section view. A—A full section view is generated from the front view of the object by drawing a section line on the inclined surface. B—The completed drawing after freezing the MD_Annotation layer.

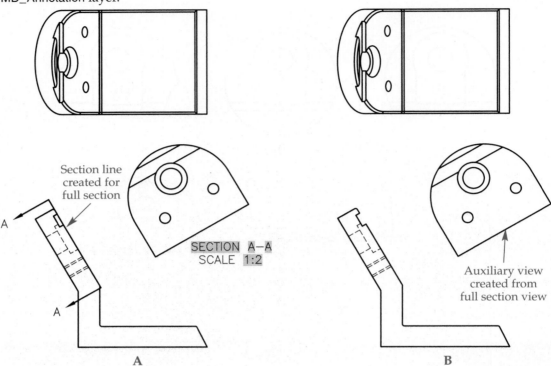

A

B

Dimensioning Drawing Views

After placing drawing views on a paper space layout, you can dimension each view as needed in the layout. See **Figure 14-24**. When dimensioning drawing views, make sure that the **DIMASSOC** system variable is set to 2 so that associative dimensions are created. Associative dimensions are associated to the dimensions of the model and update when the physical model changes. However, depending on edits to the model geometry, some dimensions can become disassociated.

Associative dimensions added to a drawing view are attached to the underlying object geometry. As the size or location of an object changes, the associative dimensions automatically adjust to their new size or location. Associative dimensions can become disassociated when a 3D model is modified or updated and the dimensions describing the underlying geometry lose their attached location. The **Annotation Monitor** is used to monitor and notify you of any changes in the associativity of dimensions placed in drawing views. This feature can be turned on by picking the **Annotation Monitor** button on the status bar. Picking this button displays the **Annotation Monitor** icon to the right of the button. See **Figure 14-25A**. The **Annotation Monitor** is turned off by default, but it is automatically activated when a model is edited and dimensions are updated. The status of the **Annotation Monitor** is controlled by the **ANNOMONITOR** system variable.

The **Annotation Monitor** icon provides feedback regarding the state of associative dimensions. If any associative dimensions become disassociated, the icon turns red and a balloon notification appears. See **Figure 14-25B**. In addition, yellow alert icons appear in the layout next to the disassociated dimensions. You can click the balloon notification link to delete the dimensions all at once, or you can pick on individual alert icons in the layout to update the dimensions.

Figure 14-24.
A layout of dimensioned drawing views including a top view and a half section view used for the front view.

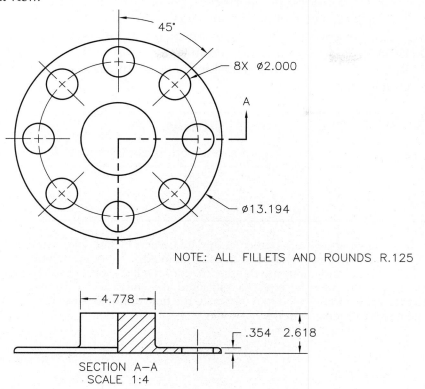

45°

8X ⌀2.000

A

⌀13.194

NOTE: ALL FILLETS AND ROUNDS R.125

4.778

.354 2.618

SECTION A—A
SCALE 1:4

Figure 14-25.
The **Annotation Monitor** is used to monitor the status of associative dimensions placed in drawing views. A—Picking the **Annotation Monitor** button on the status bar turns the feature on and displays the **Annotation Monitor** icon as shown. B—The **Annotation Monitor** icon turns red and a balloon notification appears to indicate dimensions have become disassociated as a result of a model edit.

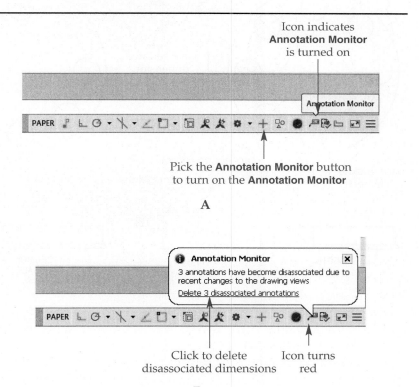

Icon indicates **Annotation Monitor** is turned on

Annotation Monitor

PAPER

Pick the **Annotation Monitor** button to turn on the **Annotation Monitor**

A

ⓘ Annotation Monitor ✕
3 annotations have become disassociated due to recent changes to the drawing views
Delete 3 disassociated annotations

PAPER

Click to delete disassociated dimensions

Icon turns red

B

In **Figure 14-26A**, the model has been edited by deleting the center cylinder in **Figure 14-24** and changing the thickness of the base. The drawing views have been automatically updated and yellow alert icons appear to indicate the dimensions that have become disassociated. Picking on an alert icon displays a shortcut menu with options to reassociate or delete the dimension. Selecting the **Reassociate** option initiates the **DIMREASSOCIATE** command and allows you to pick points to redefine the dimension. In **Figure 14-26B**, two dimensions have been reassociated and one dimension has been deleted.

PROFESSIONAL TIP

Dimensioning in a paper space layout is referred to as *transspatial dimensioning*.

Figure 14-26.
By default, drawing views are automatically updated when the model is edited. However, dimensions may become disassociated as a result of an edit. A—After deleting the cylinder in the center of the part and changing the thickness of the base, yellow alert icons appear to indicate dimensions that have become disassociated. B—The cylinder dimension is deleted and the **Reassociate** option is used to reassociate the hole diameter dimension in the top view and the height dimension in the front view.

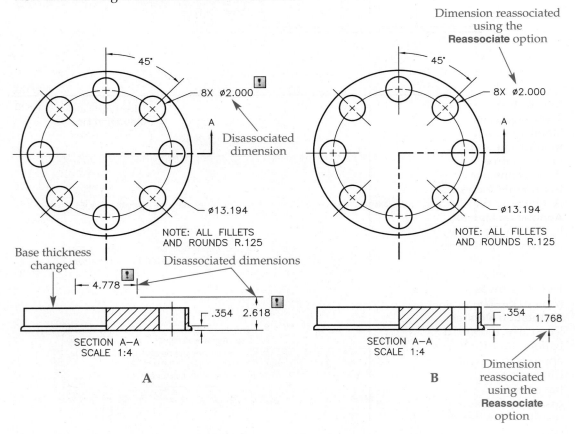

Using the EXPORTLAYOUT Command

You can use the **EXPORTLAYOUT** command to export a layout containing drawing views into a new drawing (DWG) file. However, this technique will break the associativity between the 3D model and the drawing views. The advantage to using this technique is that the drawing geometry can be edited in the same way as any other AutoCAD 2D geometry in model space. The disadvantage is that the 2D geometry has lost any associativity back to the 3D model. When using the **EXPORTLAYOUT** command, you save the exported layout as a DWG file. The drawing views become blocks in the new drawing file.

NOTE

The **FLATSHOT** command can also be used to create a multiview orthographic drawing from a 3D model. However, the resulting drawing views do not have associative properties. The **FLATSHOT** command creates a flat projection of the 3D objects in the drawing from the current viewpoint. The view is created in model space. The view that is created is composed of 2D geometry and is projected onto the XY plane of the current UCS.

Creating Section Planes

The **SECTIONPLANE** command offers a powerful visualization and display tool. It allows you to construct a section plane, known as an AutoCAD *section object*, that can then be used as a cutting plane to cut through a 3D model. Section planes can be used to cut through solids, surfaces, meshes, regions, and point clouds. Working with point clouds is discussed later in this chapter.

Once the section object is drawn, it can be moved to any location, jogs can be added to it, and it can be rendered "live" so that internal features and sectioned material are dynamically visible as the cutting plane is moved. The section object stores object property information. A variety of settings allow you to customize the appearance of section features. Additionally, you can generate 2D sections/elevations or 3D sections that can be inserted into the drawing as a block.

Once the **SECTIONPLANE** command is initiated, you are prompted to select a face or the first point on the section object. The **Draw section** option allows you to define a section object by picking multiple points. The **Orthographic** option allows you to locate a section object by specifying an orthographic plane. The **Type** option is used to set the initial display state for the section object when it is created. These options and additional sectioning options are discussed in the following sections.

Ribbon

Home
> Section
Solid
> Section

Section Plane

Type
SECTIONPLANE
SPLANE

SECTIONPLANE

Picking a Face to Construct a Section Plane

The simplest way to create a section plane is to pick a flat face on the 3D object. Access the **SECTIONPLANE** command, move the pointer to the face you wish to select, and then pick it. A transparent section object is placed on the selected face and the model is cut at the plane. See **Figure 14-27**. The section plane can now be moved to create a section anywhere along the 3D model.

The automatic cut that appears on the model when creating a section object is the default display and is called *live sectioning*. Live sectioning is discussed later in this chapter. To turn live sectioning on or off, select the section object and pick the **Live Section** button in the **Section Plane** contextual ribbon tab. See **Figure 14-28**. You can

Figure 14-27.
Creating a section object on a face. A—The object before the face is selected. B—The section object is created.

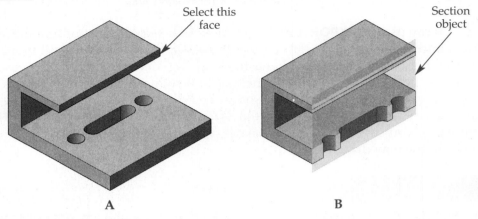

Select this face

Section object

A

B

Figure 14-28.
The **Section Plane** contextual ribbon tab contains display settings and editing options for section objects.

Section object display and editing options

Pick to create a 2D or 3D section

Pick to turn live sectioning on or off

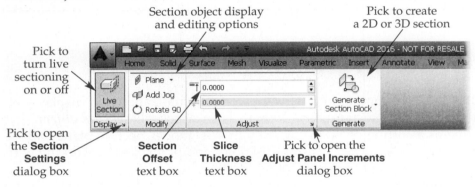

Pick to open the **Section Settings** dialog box

Section Offset text box

Slice Thickness text box

Pick to open the **Adjust Panel Increments** dialog box

also turn live sectioning on or off by selecting the section object, right-clicking, and selecting **Activate live sectioning** from the shortcut menu. The **Section Plane** contextual ribbon tab and the shortcut menu contain options for editing section objects and generating section views, as discussed later in this chapter.

Picking Two Points to Construct a Section Plane

A second method for defining a section plane is to pick two points through which the section object passes. The section object is perpendicular to the XY plane of the current UCS. When the command is initiated, pick the first point, which cannot be on a face. See P1 in **Figure 14-29**. It may be best to turn off the dynamic UCS function or you could end up picking a face as the first point instead of a point. After picking the first point, move the pointer and notice that the section plane rotates about the first point. Next, pick the second point (P2) to define a line that cuts through the model. After the second point is picked, the section object is created. The section plane extends just beyond the edges of the model.

Picking Multiple Points to Construct a Section Plane

The previous method accepts only two points to construct a single section plane. Using the **Draw section** option, you can specify multiple points in order to create section plane *jogs*. A section object drawn in this manner can represent an offset or aligned section plane. As discussed earlier in this chapter, offset and aligned sections are used for objects containing features lying in different planes.

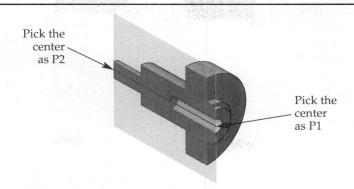

Figure 14-29.
Creating a section object by selecting two points.

Pick the center as P2

Pick the center as P1

Once the command is initiated, select the **Draw section** option. Pick the start point, using object snaps if necessary. See **Figure 14-30**. Continue picking points as needed. After picking the last point to define the section plane, press [Enter]. You are then prompted to specify a point in the direction of the section view. This point is on the opposite side of the section object from the viewer. Pick a point on the model using object snaps if necessary. The section plane is created.

Exercise 14-9

www.g-wlearning.com/CAD/

Complete the exercise on the companion website.

Creating Orthographic Section Planes

The **Orthographic** option enables you to quickly place a section plane through the front, back, top, bottom, left, or right side of the object. See **Figure 14-31**. The origin is the center point of all objects in the model. Once the command is initiated, select the **Orthographic** option. Then, specify which orthographic plane you want to use as the section plane. The section object is created and all objects in the drawing are affected by it. You can move the section plane by selecting a grip and dragging, as discussed later in this chapter.

Figure 14-30.
Using the **Draw section** option of the **SECTIONPLANE** command to create a section object with multiple segments.

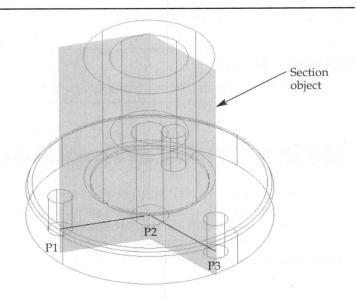

Section object

P1

P2

P3

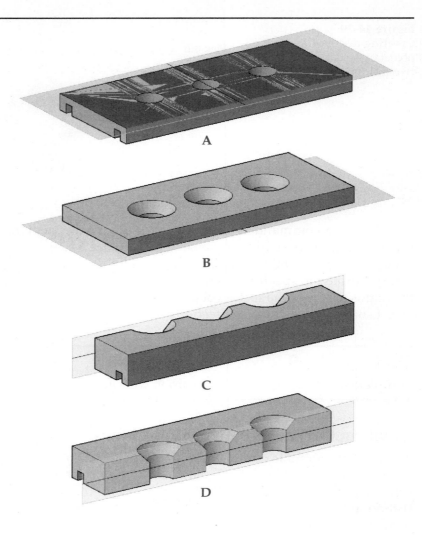

Figure 14-31.
Examples of orthographic **SECTIONPLANE** options. A—Top. B—Bottom. C—Left. D—Right.

You may encounter a situation in which there is more than one solid on the screen and you want to use the **Orthographic** option to create a section object based on just one object. In this case, create a new layer, move objects you do not want to section to this layer, and then freeze the layer. The section object will be created based on the object that is visible. However, if the section plane passes through the objects on the frozen layers, those objects will be sectioned when the layers are thawed. A section plane affects all visible objects.

Exercise 14-10

www.g-wlearning.com/CAD/

Complete the exercise on the companion website.

Using and Editing Section Planes

A wide range of display options and settings are available for section objects. In addition, section objects can be edited to achieve the desired model display. Sectioning options can be accessed from the **Section Plane** contextual ribbon tab, the right-click shortcut menu, and the **Properties** palette. Display and editing functions applying to section objects are discussed in the following sections.

Section Object States

There are four possible states for the section object created by the **SECTIONPLANE** command—section plane, section slice, section boundary, and section volume. See **Figure 14-32**. The section object can be changed from one state to another. As previously discussed, the **Type** option can be used to set the section object state when using the **SECTIONPLANE** command. Depending on which state is active, the section object produces different results on the 3D model. To change the section object state, select the section object and then select an option from the **Section Type** drop-down list in the **Modify** panel in the **Section Plane** contextual ribbon tab. You can also pick the triangular menu grip on the section object to access the options, as discussed later in this chapter.

When the section object is created with the **Type** option set to **Plane**, the object is in the *section plane state*. This is the default setting. A transparent plane is displayed on each segment of the section object. The transparent plane is called the *section plane indicator*. A line called the *section line* connects the pick points (or the edges of the section object). The section plane extends infinitely in the section object's Z direction and along the direction of the object segment (unless connected to other segments). A section object with the section plane state active is shown in **Figure 14-32B**.

The *section slice state* creates a thin cut (slice) through the 3D model. See **Figure 14-32C**. A 2D box enclosing the sectioned object represents the thickness of the slice. To set the section slice state when creating the section object, set the **Type** option to **Slice**. To change the thickness of the slice, select the section object and change the value in the **Slice Thickness** text box in the **Adjust** panel in the **Section Plane** contextual ribbon tab.

When the section object is in the *section boundary state*, a transparent plane is displayed on each segment of the section object. A 2D box extends to the XY-plane boundaries of the section object. See **Figure 14-32D**. The sectioned object fits inside of this footprint. The section plane extends infinitely in the section object's Z direction. To set the section boundary state when creating the section object, set the **Type** option to **Boundary**.

Figure 14-32.
Section object states. A—The original object. B—Section plane. C—Section slice. D—Section boundary. E—Section volume.

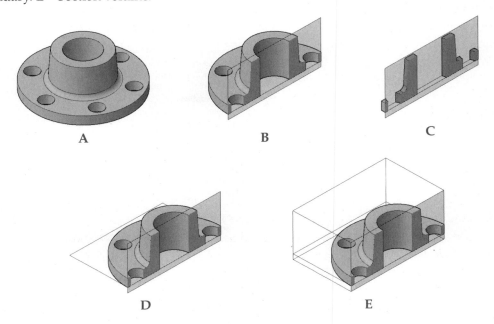

When the section object is in the *section volume state*, a transparent plane is displayed on each segment of the section object. In addition, a 3D box extends to the XYZ boundaries of the section object. The sectioned object fits inside of this box. See **Figure 14-32E**. To set the section volume state when creating the section object, set the **Type** option to **Volume**.

NOTE

A section object cannot contain jogs when using the slice section state. If the **Type** option is set to **Slice** when you create the section object, the **Draw section** option is unavailable. Select a face, pick two points, or use the **Orthographic** option to complete the command. If a section object containing jog segments is changed to the slice section state, AutoCAD displays an alert that the jogs will be removed. Use caution if you decide to proceed, as the results may be different from what you expect.

Section Object Properties

Once the section object is created, the properties of the section object can be changed. The **Section Object** category in the **Properties** palette contains properties specific to the section object. See **Figure 14-33**. These properties are described in the next sections.

Name

The default name of the first section object is Section Plane(1). Subsequent section planes are sequentially numbered, such as Section Plane(2), Section Plane(3), and so on. It may be beneficial to rename section objects so the name is representative of the section. For example, Front Half Section is much more descriptive than Section Plane(1). To rename a section object, select the Name property. Then, type a new name in the text box.

Figure 14-33.
The properties of a section object can be changed in the **Properties** palette.

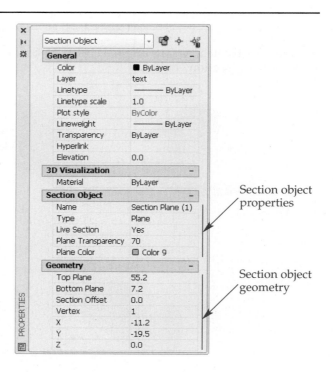

Section object properties

Section object geometry

Type

As discussed earlier, the section object is in one of four states. The four states are section plane, section slice, section boundary, and section volume. To change the state of the section object, select the Type property. Then, pick the state in the drop-down list.

Live Section

Live sectioning is a tool that allows you to dynamically view the internal features of a model as the section object is moved. This tool is discussed later in the chapter. To turn live sectioning on or off, select the Live Section property. Then, pick either Yes (on) or No (off) in the drop-down list. This is the same as turning live sectioning on or off using the ribbon or the shortcut menu.

Plane Transparency

The Plane Transparency property determines the opacity of the plane for the section object. The property value can range from 1 to 100. The lower the value, the more opaque the section plane object. See **Figure 14-34**.

Plane Color

The plane of the section object can be set to any color available in the **Select Color** dialog box. To change the color, select the Plane Color property and then select a color from the drop-down list. To select a color in the **Select Color** dialog box, pick the Select Color... entry in the drop-down list. This property only affects the plane of the section object, not the lines defining the boundary, volume, or section line. The color of these lines is controlled by the Color property in the **General** category.

Section Object Grips

When the translucent planes of a section object or the lines representing the section object state are picked, grips are displayed. The specific grips displayed are related to the current section object state. The types of grips are:

- Base grip.
- Menu grip.
- Direction grip.
- Second grip.
- Arrow grips.
- Segment end grips.

Base Grip

The *base grip* appears at the first point picked when defining the section object. See **Figure 14-35**. It is the grip about which the section object can be rotated and scaled. The section object can also be moved using this grip.

Figure 14-34.
A—The Plane Transparency property of the section plane object is set to 1 (or 1% transparent). B—The Plane Transparency property of the section plane object is set to 85 (or 85% transparent).

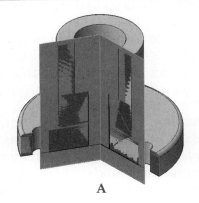
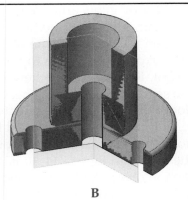

A   B

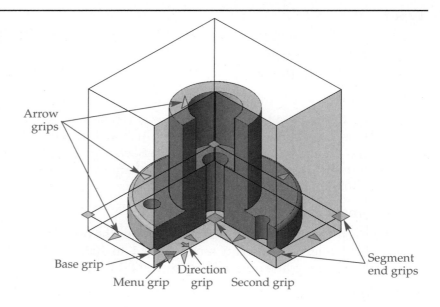

Figure 14-35.
The types of grips displayed on a section object.

Arrow grips

Base grip

Menu grip

Direction grip

Second grip

Segment end grips

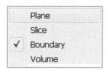

Figure 14-36.
The section state menu is used to change section object states.

	Plane
	Slice
✓	Boundary
	Volume

Menu Grip

The *menu grip* is always next to the base grip. Refer to **Figure 14-35**. Picking this grip displays the section state menu. See **Figure 14-36**. To switch the section object between states, pick the grip and then select the state from the menu.

Direction Grip

The *direction grip* indicates the direction in which the section will be viewed. Refer to **Figure 14-35**. Pick the grip to flip the view 180°. The direction grip also shows the direction of the live section. Live sectioning is discussed later in this chapter.

Second Grip

The *second grip* appears at the second point picked when defining the section object. Refer to **Figure 14-35**. The section object can be rotated and stretched about the base grip using the second grip.

Arrow Grips

Arrow grips are located on all of the lines that represent the section plane, slice, boundary, and volume. Refer to **Figure 14-35**. These grips are used to lengthen or shorten the section plane object segments or adjust the height of the section volume. The arrow grips at the top and bottom of the boundary box are used to change the height. Regardless of where the pointer is moved, the section object only extends in the segment's current plane. Changing the length of one segment of the section plane does not affect other segments.

Grip editing can be used to modify the boundary of a section object when it does not extend past the model boundary. In **Figure 14-37A**, the section object created for the model in **Figure 14-30** is displayed in the section boundary state. Notice that the section object does not extend beyond the boundary of the model. This is because the hole centers were selected to create the section object using the **Draw section** option.

Figure 14-37.
A—The section object created for the model in Figure 14-30 displayed in the section boundary state. Notice that the section plane does not extend past the model boundary. B—The arrow grip is being used to extend the left side of the section plane past the boundary of the solid model. C—The edited section object. The right side can be corrected in the same manner.

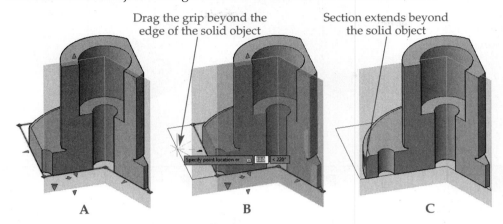

Drag the grip beyond the edge of the solid object

Section extends beyond the solid object

A   B   C

The boundary can be quickly corrected using the arrow grips. Notice in **Figure 14-37B** that the arrow grip is being used to extend the left side of the section plane past the boundary of the solid model. This allows any subsequent section views to display the entire object rather than just a portion of it. The right side of the section plane can be extended in the same manner using the opposite arrow grip.

The arrow grips located on the line segments of the section plane move the position of the section plane. As a segment of the section plane is moved, it maintains its angular relationship and connection to any adjacent section plane segment.

Segment End Grips

The *segment end grips* are located at the end of each line segment defining the section object state. Refer to **Figure 14-35**. The number of displayed segment end grips depends on whether the section object is in the section plane, section slice, section boundary, or section volume state. These grips provide access to the standard grip editing options of stretch, move, copy, rotate, scale, and mirror. If the rotate option is used, the section plane is rotated about the selected segment end grip. Moving a segment end grip can change the angle between section plane segments.

Adjusting Section Objects

Additional editing functions for section objects are available in the **Section Plane** contextual ribbon tab. Refer to **Figure 14-28**. If you need to move a section object to show a different cut in the model, you can offset the section object by a specified distance. Select the section object and enter a positive or negative offset value in the **Section Offset** text box in the **Adjust** panel. The section object is moved perpendicular to the section plane in a positive or negative direction relative to the WCS. See **Figure 14-38**. Picking the up or down arrow next to the **Section Offset** text box offsets the section plane by an incremental value in a positive or negative direction. The incremental value can be adjusted by picking the dialog box launcher button at the bottom of the **Adjust** panel to access the **Adjust Panel Increments** dialog box.

A section object can also be rotated to show a different cut in the model. To rotate a section object in 90° increments, select the **Rotate 90** option in the **Modify** panel. The section object rotates 90° about the section line, section boundary, or section volume, depending on the current section state. This provides an alternative method to using grips to rotate the section object.

Figure 14-38.
Offsetting a section object. A—The original section object. B—The section object is offset in a positive direction.

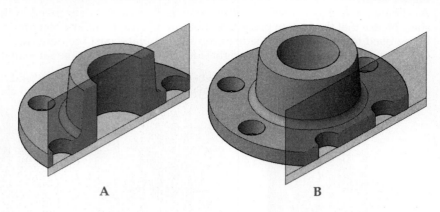

A   B

PROFESSIONAL TIP

Standard editing commands can be accessed by selecting the section object and right-clicking to display the shortcut menu. Select **Move**, **Scale**, or **Rotate** from the shortcut menu to adjust the section object location as needed.

Adding Jogs to a Section

You can quickly add a jog, or offset, to an existing section object. First, select the section object. Then, select the **Add Jog** option in the **Modify** panel in the **Section Plane** contextual ribbon tab. You can also right-click to display the shortcut menu and select **Add jog to section**. You are then prompted:

> Specify a point on the section line to add jog:

Select a point directly on the section line. If any object snap is active, the **Nearest** object snap is temporarily turned on to ensure you pick the line. Once you pick, the jog is automatically added perpendicular to the line segment. See **Figure 14-39A**.

It is not critical that you pick the exact location on the line where you want the jog to occur. Remember, grips allow you to easily adjust the section plane location. Notice in **Figure 14-39B** that the second jog barely cuts through the first hole. The intention is to run the section plane through the middle of the hole. To fix this, drag the arrow grip so the section plane segment is in the desired location, **Figure 14-39C**.

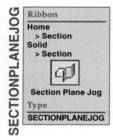

Ribbon

Home
> Section
Solid
> Section

Section Plane Jog

Type
SECTIONPLANEJOG

Exercise 14-11

www.g-wlearning.com/CAD/

Complete the exercise on the companion website.

Live Sectioning

Live sectioning is a tool that allows you to view the internal features of models that are cut by the section plane of the section object. The view is dynamically updated as the section object is moved. This tool is used to visualize internal features and for establishing section locations from which 2D and 3D section views can be created.

Live sectioning is automatically turned on when a section object is created. Live sectioning can be turned on and off for individual section objects, but only one section object can be "live" at any given time.

To turn live sectioning on or off, select the section object. Then, select the **Live Section** option in the **Display** panel in the **Section Plane** contextual ribbon tab. You can also enter the **LIVESECTION** command and select the section object to toggle the

Ribbon

Home
> Section
Solid
> Section

Live Section

Type
LIVESECTION

Figure 14-39.
Adding a jog to a section object. A—Pick a point on the section line to add a jog. B—The jog is added, but it is not in the proper location. C—Using the arrow grip, the jog is moved to the proper location.

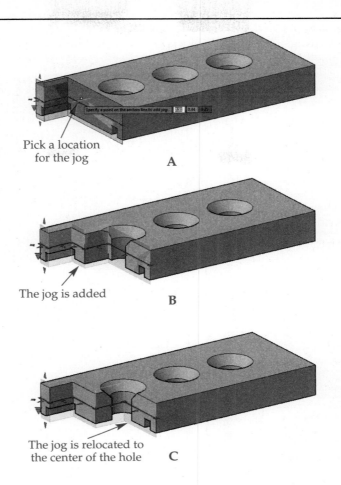

Pick a location for the jog

A

The jog is added

B

The jog is relocated to the center of the hole

C

on/off setting. When live sectioning is turned on, the material behind the viewing direction of the section plane is removed. The cross section of the 3D object is shown in gray (by default) and the internal shape of the 3D object is visible.

A wide variety of options allow you to change the appearance of not only the live sectioning display, but also of 2D and 3D section blocks that can be created from the sectioned display. These settings are found in the **Section Settings** dialog box. See **Figure 14-40**. To open this dialog box, select the section object and pick the dialog box launcher button in the lower-right corner of the **Display** panel in the **Section Plane** contextual ribbon tab. You can also pick the dialog box launcher button in the lower-right corner of the **Section** panel in the **Home** or **Solid** tab of the ribbon.

To change the settings for live sectioning, pick the **Live Section settings** radio button at the top of the **Section Settings** dialog box. The categories displayed in the dialog box contain properties related to live sectioning. Settings for 2D and 3D sections and elevations are discussed later in this chapter.

The three categories for live sectioning are **Intersection Boundary**, **Intersection Fill**, and **Cut-away Geometry**. To display a brief description of any property, hover the cursor over the property in the **Section Settings** dialog box. The description is displayed in a tooltip. A check box at the bottom of the **Section Settings** dialog box allows you to apply the properties to all section objects or to just the selected section object.

PROFESSIONAL TIP

Live sectioning can be quickly turned on or off by double-clicking on the section plane object.

Figure 14-40.
Section settings. A—For a 2D block. B—For a 3D block. C—For live sectioning.

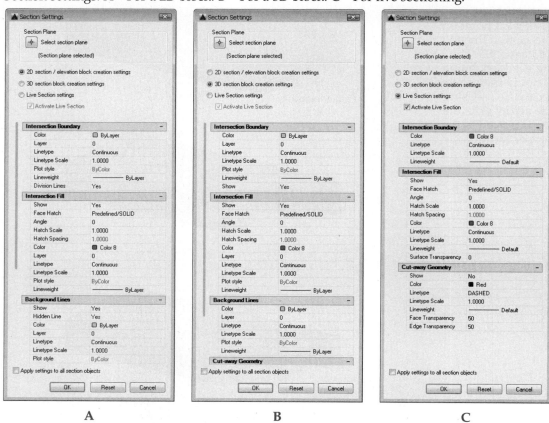

A   B   C

Intersection Boundary

The intersection boundary is where the model is intersected by the section object. It is represented by line segments. You can set the color, linetype, linetype scale, and line weight of the intersection boundary lines.

Intersection Fill

The intersection fill is the material visible on the model surface where the section object cuts. It is displayed as a solid fill, by default. Any hatch pattern available in AutoCAD can be used as the intersection fill. See **Figure 14-41A**. The angle, hatch scale, hatch spacing, and color can be set. In addition, the linetype, linetype scale, and line weight can be changed. The fill pattern can even be set to be transparent.

Cutaway Geometry

The cutaway geometry is the part of the model removed by the live sectioning. By default, this geometry is not displayed. Changing the Show property to Yes displays the geometry. See **Figure 14-41B**. You can set the color, linetype, linetype scale, and line weight of the lines representing the cutaway geometry. In addition, the Face Transparency and Edge Transparency properties allow you to create a see-through effect, as seen in **Figure 14-42**. Each of these two properties is set to 50 by default.

PROFESSIONAL TIP

You can also display the cutaway geometry without using the **Section Settings** dialog box. Select the section object, right-click, and pick **Show cut-away geometry** in the shortcut menu. Live sectioning must be on.

Figure 14-41.
Live sectioning settings. A—The intersection fill can be displayed as a hatch pattern in any specified color. B—The cutaway geometry removed by the live sectioning is displayed.

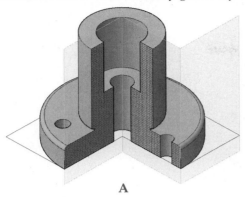
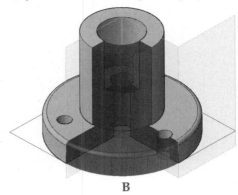

A

B

Figure 14-42.
The cutaway geometry is displayed with 100% transparent faces and solid black lines.

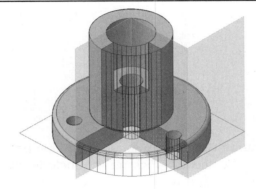

Exercise 14-12

www.g-wlearning.com/CAD/

Complete the exercise on the companion website.

Generating 2D and 3D Sections and Elevations

The **SECTIONPLANETOBLOCK** command provides a fast and efficient method of creating sections. The sections can be either 2D or 3D. Not only can the sections be displayed on the current drawing, they can also be exported as a file that can then be used in any other drawing or document for display, technical drawing, or manufacturing purposes.

Creating Sections

To create a section, enter the **SECTIONPLANETOBLOCK** command. You can also select the section object and then select the **Generate Section Block** option in the **Generate** panel in the **Section Plane** contextual ribbon tab. The **Generate Section/Elevation** dialog box is displayed. See **Figure 14-43**. In this dialog box, you can specify whether the section will be 2D or 3D, select what is included in the section, and specify a destination for the section. To expand the dialog box, pick the **Show details** button, which looks like a down arrow.

To create a 2D section, pick the **2D Section/Elevation** radio button in the **2D/3D** area of the dialog box. A 2D section is projected onto the section plane, but is placed flat on the XY plane of the current UCS. To create a 3D section, pick the **3D Section** radio button. A 3D section is placed so its surfaces are parallel to the corresponding cut surfaces on the 3D object.

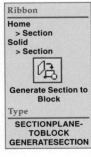

Ribbon

Home
> Section

Solid
> Section

Generate Section to Block

Type

SECTIONPLANE-TOBLOCK
GENERATESECTION

SECTIONPLANETOBLOCK

Figure 14-43.
The expanded
**Generate Section/
Elevation** dialog box.

Pick 2D or 3D

Select objects
to include

Select where
the section will
be placed

Pick to expand
or collapse the
dialog box

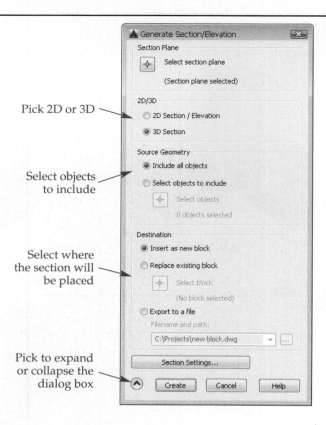

In the **Source Geometry** area of the dialog box, you can specify which geometry is included in the section. Picking the **Include all objects** radio button includes all 3D solids, surfaces, and regions in the section. To limit the section to certain objects, pick the **Select objects to include** radio button. Then, pick the **Select objects** button, select the objects on-screen, and press [Enter]. The number of selected objects is then displayed in the dialog box.

The **Destination** area of the dialog box is where you specify how the section will be placed. To place the section into the current drawing, pick the **Insert as new block** radio button. To update an existing section block, pick the **Replace existing block** radio button. Then, pick the **Select block** button, select the block on-screen, and press [Enter]. You will need to do this if the section object is changed. To save the section to a file for use in other drawings, pick the **Export to a file** radio button. Then, enter a path and file name in the text box.

Once all settings have been made, pick the **Create** button. The section is attached to the cursor and can be placed like a regular block. See **Figure 14-44.** Additionally, the options available are the same as if a regular block is being inserted. Once the block is inserted, it can be moved, rotated, and scaled as needed.

Section Settings

Picking the **Section Settings...** button at the bottom of the **Generate Section/ Elevation** dialog box opens the **Section Settings** dialog box discussed earlier. Using this dialog box, you can adjust all of the properties associated with the type of section being created. Depending on whether the **2D Section/Elevation** or **3D Section** radio button is selected in the **Generate Section/Elevation** dialog box, the appropriate categories and properties are displayed in the **Section Settings** dialog box. Refer to **Figure 14-40.**

The categories discussed earlier in relation to the **Live Section settings** radio button are available. Also, two additional categories are displayed for 2D and 3D sections:

- **Background Lines.** Available for 2D and 3D sections.
- **Curve Tangency Lines.** Only available for 2D sections.

Figure 14-44.
Inserting a 2D
section block.

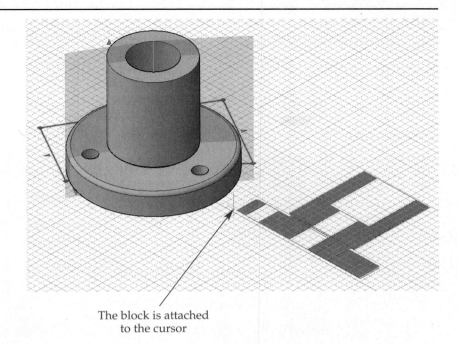

The block is attached
to the cursor

These categories are discussed below. Examples of 2D and 3D sections inserted as blocks in the drawing are shown in **Figure 14-45**. Notice how properties can be set to show cutaway geometry in a different color and to change the section pattern, color, and linetype scale.

Background Lines. The properties in the **Background Lines** category provide control over the appearance of all lines that are not on the section plane. You can choose to have visible background lines, hidden background lines, or both displayed. They can be emphasized with color, linetype, or line weight. The layer, linetype scale, and plot style can also be changed. These settings are applied to both visible and hidden background lines.

Curve Tangency Lines. The properties in the **Curve Tangency Lines** category apply to lines of tangency behind the section plane. For example, the object shown in **Figure 14-45** has a round on the top of the base. This results in a line of tangency behind the section plane where the round meets the vertical edge. You can have these lines displayed or suppressed. In general, lines of tangency are not shown in a section view. If you choose to display these lines, you can set the color, layer, linetype, linetype scale, plot style, and line weight of the lines.

NOTE

When a 3D section is created, you must turn off live sectioning to see the complete sectioned object in the block. With live sectioning on, only the cut surfaces appear in the block.

Updating the Section View

Once the section view is created, it is not automatically updated if the section object is changed. To update the section view, select the section object (not the block) and select the **Generate Section Block** option from the **Generate** panel in the **Section Plane** contextual ribbon tab. Then, in the **Destination** area of the **Generate Section/Elevation** dialog box, pick the **Replace existing block** radio button. If necessary, pick the **Select block** button and select the section block in the drawing. If you want to change the appearance of the section view, pick the **Section Settings...** button and adjust the properties as needed. Finally, pick the **Create** button in the **Generate Section/Elevation** dialog box to update the section block.

Figure 14-45.
A—The section object is created. B—A 2D section block is inserted into the drawing and the view is made plan to the block. C—A 3D section block is inserted into the drawing. Notice how the hatch pattern is displayed. D—The 3D section block is updated and now the cutaway geometry is displayed.

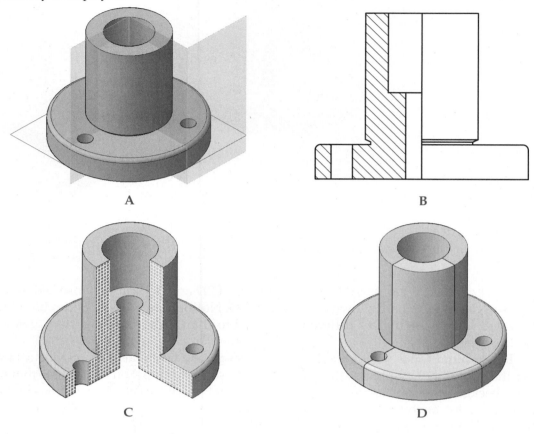

A   B

C   D

Exercise 14-13

www.g-wlearning.com/CAD/

Complete the exercise on the companion website.

Model Analysis

AutoCAD provides a number of tools for analyzing solid and surface models. Model analysis is conducted to evaluate various data and determine whether design changes are needed prior to manufacturing. Model analysis tools available in AutoCAD are discussed in the following sections.

Solid Model Analysis

MASSPROP

Type
MASSPROP

The **MASSPROP** command allows you to analyze a solid model for its physical properties. The data obtained from **MASSPROP** can be retained for reference by saving the data to a file. The default file name is the drawing name. The file is an ASCII text file with a .mpr (mass properties) extension. The analysis can be used for third-party applications to produce finite element analysis, material lists, or other testing studies.

Once the command is initiated, you are prompted to select objects. Pick the objects for which you want the mass properties displayed and press [Enter]. AutoCAD analyzes the model and displays the results in the **AutoCAD Text Window**. See **Figure 14-46**. The following properties are listed.

Figure 14-50.
Draft analysis. A—The original solid model. B—The resulting analysis based on angle values of 4.0 and −4.0. The top of the model is within the draft parameters. Areas along the side and bottom are shaded blue to indicate the draft is at the lowest angle specified.

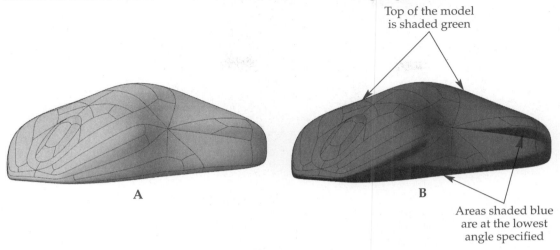

Top of the model is shaded green

Areas shaded blue are at the lowest angle specified

A

B

Solid Model File Exchange

AutoCAD drawing files can be converted to files that can be used for testing and analysis. Use the **ACISOUT** or **EXPORT** command to create a file with a .sat extension. These files can be imported into AutoCAD with the **ACISIN** or **IMPORT** command.

Solids can also be exported for use with stereolithography software. These files have a .stl extension. Use the **3DPRINT** command to create STL files.

Importing and Exporting Solid Model Files

A solid model is frequently used with analysis and testing software or in the manufacture of a part. The **ACISOUT** and **EXPORT** commands allow you to create a type of file that can be used for these purposes. Once the **ACISOUT** command is initiated, you are prompted to select objects. After selecting objects and pressing [Enter], a standard save dialog box is displayed. See **Figure 14-51**. When using the **EXPORT** command, the standard save dialog box appears first. After entering a file name and selecting a file type (SAT), you are then prompted to select objects.

An SAT file can be imported into AutoCAD and automatically converted into a drawing file using the **ACISIN** or **IMPORT** command. Once either command is initiated, a standard open dialog box appears. Change the file type to SAT, locate the file, and pick the **Open** button.

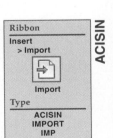

Stereolithography Files

Stereolithography (SLA) is an additive manufacturing process that creates various plastic 3D model prototypes using a computer-generated solid model, a laser, and a vat of liquid polymer. This technology is also referred to as *rapid prototyping* or *3D printing*. Some additive manufacturing processes, such as fused-deposition modeling (FDM), add material in "layers" from a filament that is extruded through a heated nozzle. Other additive manufacturing processes are selective laser sintering (SLS), 3D printing (3DP), multi-jet modeling (MJM), and electron beam melting (EBM). Using one of these processes, a prototype 3D model can be designed and formed in a short amount of time without using standard subtractive manufacturing processes.

Figure 14-51.
Exporting an ACIS file.

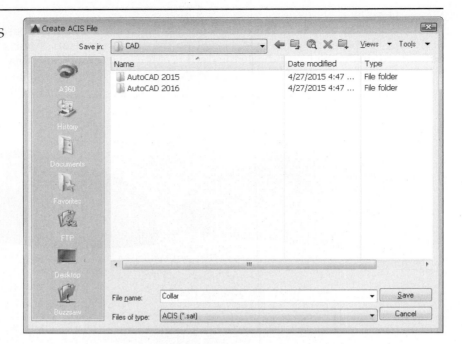

Most CAD software today can create a stereolithograph file (STL file). AutoCAD can export a drawing file to the STL format, but *cannot* import STL files.

Using the 3DPRINT Command to Create STL Files

The **3DPRINT** command is used to create an STL file. Solids and watertight meshes can be selected for use with the command. A *watertight mesh* is completely closed and contains no openings. Watertight meshes are converted to 3D solids when using the **3DPRINT** command.

When the **3DPRINT** command is initiated, a dialog box appears with two options: **Learn about preparing a 3D model for printing** and **Continue**. Select the **Continue** option. Then, select the solids or watertight meshes and press [Enter]. The **Send to 3D Print Service** dialog box is displayed, **Figure 14-52**. This dialog box allows you to change the scale or select other objects. When done, pick the **OK** button to display a standard save dialog box. Type the file name in the **File name:** text box and pick the **Save** button. Next, a web page is displayed (if you are connected to the Internet). This page offers options for 3D printer service providers. Unless you are sending the file to a service provider, close this window. The STL file is created and ready to be sent to your additive manufacturing machine to create the model.

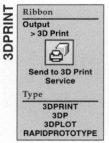

3DPRINT

Ribbon
Output
> 3D Print

Send to 3D Print Service

Type

3DPRINT
3DP
3DPLOT
RAPIDPROTOTYPE

Figure 14-52.
This dialog box is displayed when creating an STL file with the **3DPRINT** command.

Pick to select other objects   Change the scale if needed   Use to navigate the preview

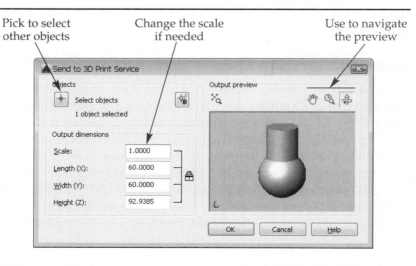

When using the **3DPRINT** command, the **FACETRES** system variable is automatically set to 10 for the creation of an STL file. The system variable is reset after the command is completed.

Exercise 14-14

www.g-wlearning.com/CAD/

Complete the exercise on the companion website.

Point Clouds and Reality Capture

A *point cloud* is a digital representation of an existing object that is constructed of large numbers of points in 3D space. A point cloud provides accurate 3D data that can be used for design and modeling purposes in a variety of applications, including building construction, surveying, manufacturing, accident and crime scene reconstruction, gaming, and computer animation.

AutoCAD provides tools for working with point clouds to help in 3D visualization, design, and model creation. You can use point clouds to:

- Visualize as-built conditions.
- Create new vector-based drawings using the point cloud data as a reference.
- Apply color mappings and transparency for improved visualization.
- Crop regions to show what is needed or not needed for different projects.
- Generate new geometry using extraction commands.

Working with point clouds in AutoCAD begins with converting raw data from *scan files* into a supported point cloud file format. The resulting point cloud file can then be attached to a drawing file for reference purposes. Working with a point cloud is much like working with an external reference file.

The process of importing and converting point cloud data is referred to as *reality capture*. The following sections provide an introduction to point clouds and the tools available for working with point clouds in AutoCAD.

Laser Scanning

A *laser scanner* is a device used to capture data points from an existing structure or product and create *scan files*. Laser scanning is used by engineers, designers, and animators. The scanning project is planned so that an entire building, facility, mechanical part, assembly, or geographic feature can be scanned with a single scan or multiple overlapping scans. The scanner can be mounted on a tripod or desktop, or held by hand. The scanner rapidly records millions of data points to create the scan file. Multiple scan files may be created, depending on the size of the project. Each data point in a scan file has XYZ values, but the resulting data is different from vector data, which has from-to XYZ values.

Scan files are produced in different file formats. Some of the more common scan file formats, such as Faro (.fls and .fws), Leica (.ptg, .pts, and .ptx), and Lidar (.las), are associated with different laser scanner manufacturers. In order to be used in AutoCAD, raw scan files must first be converted to a point cloud file. Point cloud files are created using the Autodesk ReCap program.

Autodesk ReCap

Autodesk ReCap is a reality capture software program that indexes and registers point clouds. It is a separate application program that is installed when you install AutoCAD. Autodesk ReCap provides the ability to process extremely large files containing raw scanned data. It contains tools for enhancing and modifying point cloud data for use in design work. Files created in Autodesk ReCap can be used in AutoCAD, Autodesk Inventor, Autodesk Revit, and other Autodesk software, and can be shared online with anyone using Autodesk A360.

Autodesk ReCap converts scan files to point cloud files through a process called *indexing*. Indexed files are saved as reality capture scan (RCS) files. In a typical work-flow, RCS files are associated with a reality capture project (RCP) file. An RCP file is used to organize information from multiple scan files in a point cloud project. An RCP file references the individual RCS files, but does not contain the files.

The process of working with scan files in Autodesk ReCap consists of the following steps.

1. Create a new project (RCP file).
2. Import scan files.
3. Register the scan files, if necessary.
4. Index the scan files.
5. Clean up the files by removing points, creating scan regions, and changing the display colors.

Using Autodesk ReCap

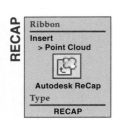

Ribbon

Insert
> Point Cloud

Autodesk ReCap

Type

RECAP

Autodesk ReCap provides an interface for working with scan files and point cloud data. To open Autodesk ReCap, select the **Autodesk ReCap** button in the **Point Cloud** panel on the **Insert** ribbon tab. The home screen appears, **Figure 14-53**. The initial tools that are available allow you to start a new project, open an existing project, set system preferences, and access the help system. Any recent projects appear as thumbnails along the bottom of the home screen.

To start a new project, select **New project** at the top of the screen. On the next screen, you are prompted to enter a name for the project and assign a folder location for files associated with the project. Enter the new project name and select the folder where the files will be saved. Then, select **Proceed**. The **Import files** screen appears and prompts you to select the scan files to import. You can select individual files, select a folder containing files, or drag the files to the screen. If you choose to select individual files or a folder containing files, use the **Import Point Clouds** dialog box to select the files. You can select from a variety of file formats. Select one or more files and pick the **Open** button. After selecting files, an icon representing each file appears near the top of the **Import files** screen. See **Figure 14-54**. Each icon has a circular progress bar indicating the import status. You can continue working while the files are importing. If you have selected multiple files, you can choose to register the scan files. Registration is discussed later in this chapter. You can also adjust the scan settings used for the point cloud. The scan settings are discussed next.

Figure 14-53.

The Autodesk ReCap home screen. To start a new project, select **New project**.

Pick to start a new project

Pick to open an existing project

Pick to set system preferences

Pick to access the help system

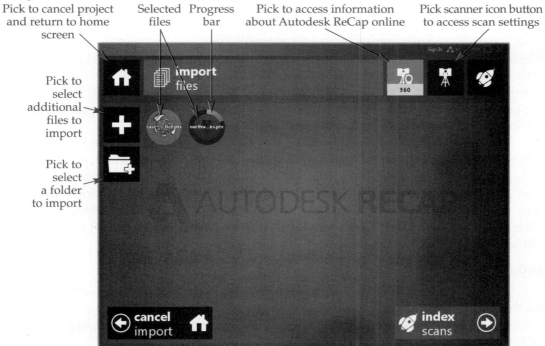

Figure 14-54.

The **Import files** screen. Scan settings can be accessed by selecting the scanner icon button at the top of the screen.

Pick to cancel project and return to home screen

Selected files

Progress bar

Pick to access information about Autodesk ReCap online

Pick scanner icon button to access scan settings

Pick to select additional files to import

Pick to select a folder to import

Scan Settings

The point cloud scan settings allow you to control the filtering and size of the point cloud, the resolution, and the coordinate system. To access the scan settings, pick the scanner icon button at the top of the **Import files** screen. Refer to **Figure 14-54**. Then, select the **Filtering** button to access the filtering and clipping options on the **Scan settings** screen. See **Figure 14-55A**. Set the type of filtering by selecting one of the **Filter scans** options. The **Minimal** option preserves most points in the point cloud. The **Standard** option removes weak points. The **Aggressive** option removes points that are not associated with an object or surface. The clipping options define the clipping range and clipping intensity. The clipping range sliders set the minimum and maximum distances from the scanner at which points are preserved. The clipping intensity sliders set the minimum and maximum intensity values at which points are preserved. Points that are too weak or too intense are not included.

Advanced scan settings can be accessed by selecting the **Advanced** button. See **Figure 14-55B**. Picking the **Decimation** button activates the decimation grid setting, which allows you to adjust the resolution, or density, of the point cloud. The value defines the minimum cubic volume of a single point. Drag the slider or enter a value in the text box to adjust the setting. A smaller value produces more points and a higher resolution, resulting in a larger file size. A higher value produces fewer points and a lower resolution, resulting in a smaller file size.

To align the point cloud with a coordinate system, use the **Coordinate system** text boxes. If a coordinate system has been assigned, it is listed in the **Current** text box. Use the **Target** text box to convert to a different coordinate system, if needed. The **Up axis** buttons are used to specify the axis orientation of the model. Pick the **X**, **Y**, or **Z** button to specify which axis is used to set the "up" direction of the model. The Z axis is the default "up" axis. Picking the **Flip axis** button flips the "up" axis orientation. Picking the **3D preview** button displays a preview of the specified orientation.

Once you are done adjusting settings, pick **Import files** to return to the **Import files** screen. You can then register the scan files or proceed to indexing the scan files, as discussed next.

Scan Registration

Registration is the process of aligning scan files so that different files in the project use the same coordinate system. This provides a way to accurately link together scans that are generated from different scanner positions. When scan files are importing, you can choose to skip registration, register the scan files manually, or have the program register the scan files automatically. If you choose to register the scan files manually, registration mode is activated and you are prompted to select a scan file to serve as the base file. Then, you select three matching pairs of points between the base file and another scan file. After selecting the points, a preview of the alignment appears and you are prompted to approve the registration. The process is repeated to register the remaining scan files.

If you do not want to register the scan files, you can skip registration. This allows you to start indexing files without performing registration. If you decide to have the program register the scan files automatically, registration is completed once the scan files are imported. During automatic registration, the scan files are processed so that they align properly when indexed together. Once registration is completed, a report is generated and you have the opportunity to preview the alignment. If the results are sufficient, you can proceed to indexing the scans.

Indexing Scans

Select **Index scans** to begin file indexing. An icon representing each file appears on the **Index scans** screen and the indexing progress is reported at the bottom of the screen. Once the files are indexed, select **Launch project** to begin working with the point cloud.

Figure 14-55.
Point cloud scan settings can be adjusted on the **Scan settings** screen. A—Select the **Filtering** button to set filtering and clipping options. B—Select the **Advanced** button to access settings for the decimation grid, coordinate system, and axis orientation.

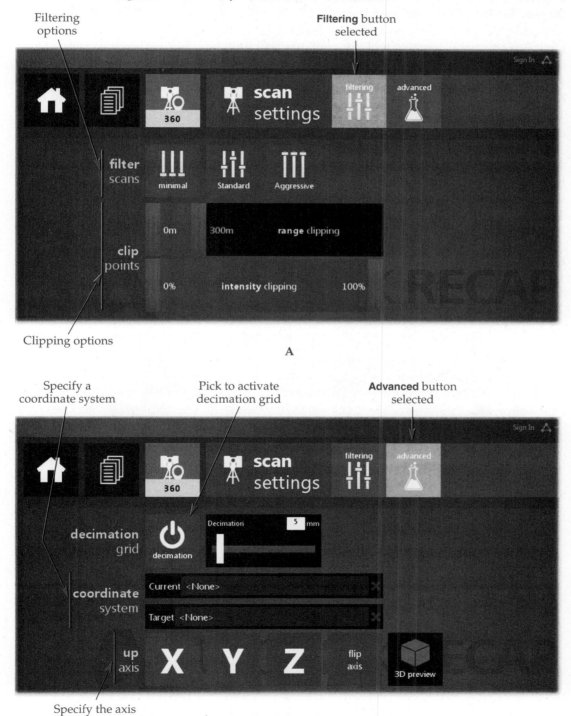

Project Screen

The **Project** screen appears when you launch or open a project. It allows you to make edits and changes to the display. See **Figure 14-56**. For example, you can control the point display, view the point cloud in different colors and lighting, create scan regions, and use standard navigation tools to adjust the view. After making changes, you can then save the project.

The left side of the **Project** screen includes four tile menus and the **Feedback** button below the tile menus. To expand a tile menu, hover over the tile menu button. The expanded menus are shown in **Figure 14-57**. Many of the buttons in the tile menus display a flyout when you hover over the button. The **Home** tile menu contains tools for managing the project. The **Display Settings** tile menu contains tools for changing the color display, applying lighting, changing the point size and intensity, and toggling the display of navigation tools. The **Limit Box** tile menu contains tools for configuring a limit box. A *limit box* is a cropped area defining the width, length, and depth of a three-dimensional portion of the point cloud. A limit box is typically used to create a scan region, as discussed later in this section. The **Navigation** tile menu contains the standard navigation tools, such as zoom, pan, and orbit. The view cube, which is displayed by default in the upper-right corner of the **Project** screen, can also be used to adjust the view.

The **Project** screen contains additional tools at the bottom of the screen. The context-sensitive menu in the middle displays tools appropriate for the current mode, view, or selection. The **Selection Tool** flyout contains tools for making a window, fence, or plane selection in the point cloud. You can use these tools to delete points or create a clipping boundary. The **Annotation Tool** flyout contains tools for measuring a linear distance, measuring an

Figure 14-56.
The **Project** screen appears when you launch a project or open an existing project. Shown is a point cloud of the interior of a college building. (*Pittsburg State University School of Construction*)

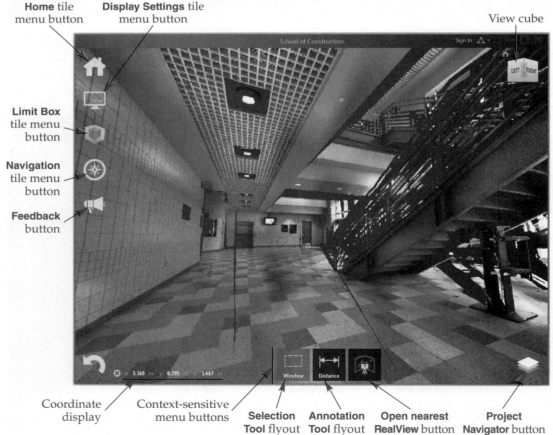

Home tile menu button

Display Settings tile menu button

View cube

Limit Box tile menu button

Navigation tile menu button

Feedback button

Coordinate display

Context-sensitive menu buttons

Selection Tool flyout

Annotation Tool flyout

Open nearest RealView button

Project Navigator button

Figure 14-57.
Tile menus that appear on the **Project** screen. Hovering over a tile menu button displays the menu. Hovering over a button in the menu displays a flyout. A—**Home** tile menu. B—**Display Settings** tile menu. C— **Limit Box** tile menu. D—**Navigation** tile menu.

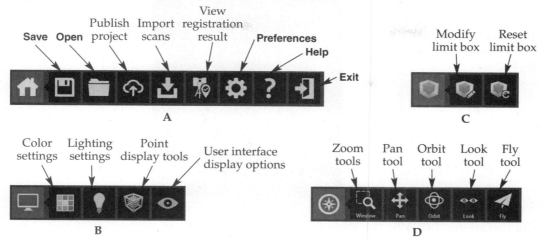

angular value, and inserting a note at a selected point. Picking the **Open nearest RealView** button displays a panoramic view of the point cloud. See **Figure 14-58**.

The **Project Navigator** is displayed at the lower-right corner of the **Project** screen when you hover over the **Project Navigator** button. See **Figure 14-59**. Items in the **Project Navigator** are organized into three categories: **View States**, **Scan Regions**, and

Figure 14-58.
Displaying a panoramic view of the point cloud shown in Figure 14-56. Notice how the view is rotated to show more of the supporting structure of the stairs. (*Pittsburg State University School of Construction*)

Figure 14-59.
The **Project Navigator**, displayed at the lower-right corner of the **Project** screen, provides tools for creating view states and scan regions.

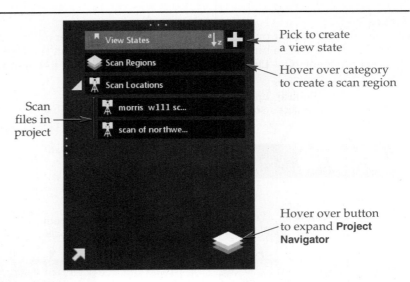

Pick to create a view state

Hover over category to create a scan region

Scan files in project

Hover over button to expand **Project Navigator**

Scan Locations. Initially, the **View States** and **Scan Regions** categories do not contain any items. The **Scan Locations** category lists all scan files used in the project. The **View States** category is used to create and display view states. A *view state* is similar to a named view in AutoCAD and allows you save the current view displayed on the **Project** screen. To create a view state, pick the plus icon to the right of the **View States** category. Enter a name for the new view state and press [Enter]. You can then pick on the name of the view state to restore it after making display changes. The **Scan Regions** category is used to create and display scan regions. A *scan region* is a portion of the point cloud representing a three-dimensional area. Before creating a scan region, define a limit box to represent the extents. Then, pick the plus icon to the right of the **Scan Regions** category. Enter a name for the new scan region and press [Enter]. When you hover over the name of the scan region, the corresponding portion of the point cloud is highlighted.

Once you are done making changes, pick the **Save** button in the **Home** menu to save the project. The point cloud is now ready to be inserted in AutoCAD.

Inserting a Point Cloud File

POINTCLOUDATTACH

Ribbon
Insert
> Point Cloud

Attach

Type
POINTCLOUD-ATTACH
PCATTACH

Inserting a point cloud file into an AutoCAD drawing file allows you to view the point cloud model and use it for reference or design purposes. Access the **POINTCLOUDATTACH** command to attach a point cloud file. In the file selection dialog box that appears, navigate to the point cloud file and select **Open**. You can attach an RCS or RCP file. After selecting a file, the **Attach Point Cloud** dialog box appears with a preview of the point cloud model. Use this dialog box to specify insertion options. The options are similar to those available when attaching an external reference. Specify the path type, insertion point, scale, rotation, and geometric location, if available from the point cloud file. Check the **Lock point cloud** check box if you want to lock the position of the point cloud and prevent it from being moved or rotated. Check the **Zoom to point cloud** check box to automatically zoom to the point cloud. This is the default option. Selecting the **Show Details** button displays additional information about the point cloud, including the number of points.

The maximum number of points displayed for point clouds is controlled by the **Maximum point cloud points per drawing** value in the **3D Modeling** tab of the **Options** dialog box. The maximum value is 25 million. Adjust the initial setting if needed before accessing the **POINTCLOUDATTACH** command. Set the maximum value to 25 million to get the largest density possible for the point cloud.

Point Cloud Display Options

After the point cloud is attached, you can adjust the viewpoint and display as needed. To access point cloud display options, select the point cloud to access the **Point Cloud** contextual ribbon tab. See **Figure 14-60**. The **Display**, **Visualization**, **Cropping**, and **Options** panels provide display options and tools. The options in the **Extract** and **Section** panels are used to extract 2D drawing geometry and create section objects, as discussed later in this chapter.

The tools in the **Display** panel allow you to adjust the point size and density. Use the slider bars to adjust the values, if needed. The navigation tools in the **Display** panel can be used to manipulate the viewpoint or change to a parallel or perspective projection.

The options in the **Visualization** panel allow you to adjust the display color, transparency, and lighting. The options in the **Stylization** drop-down menu are used to set the color style. The **Scan Colors** option uses the original scan colors. The **Object Color**

Figure 14-60.
The **Point Cloud** contextual ribbon tab. (*Model courtesy of Laser Design and GKS Services*)

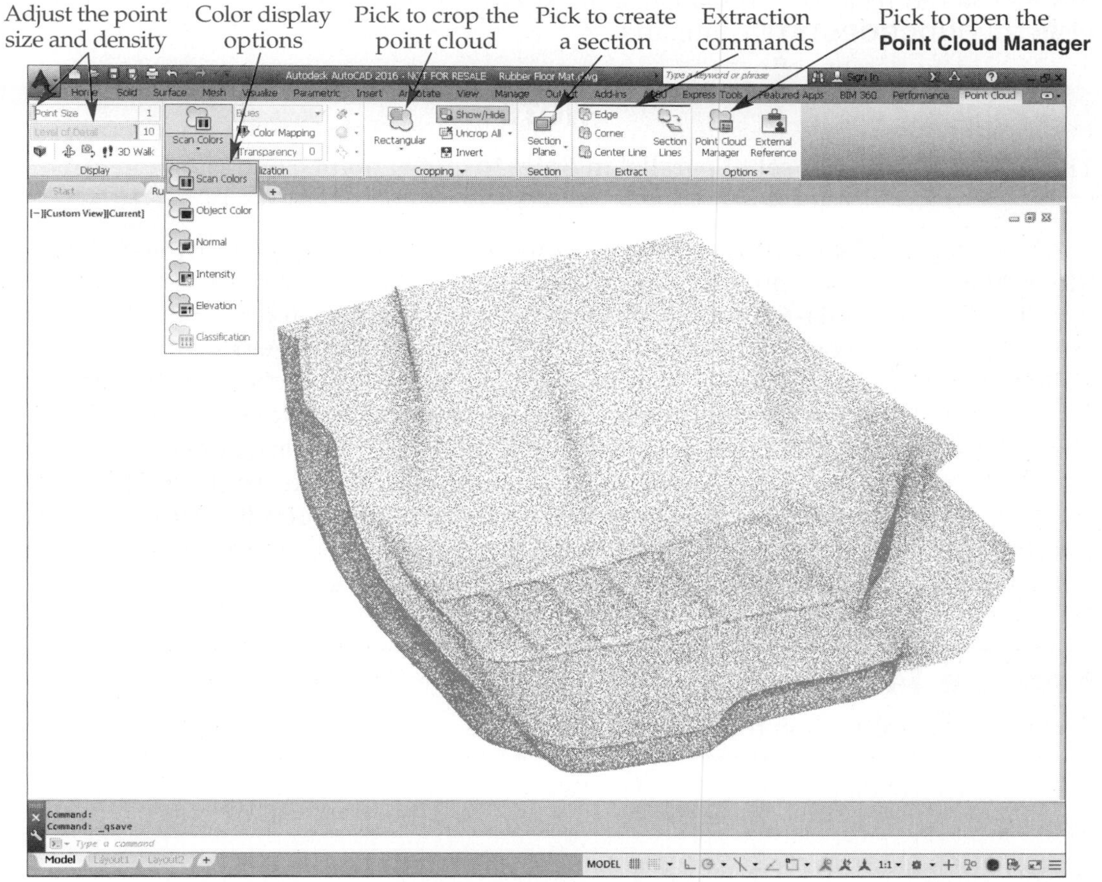

Adjust the point size and density   Color display options   Pick to crop the point cloud   Pick to create a section   Extraction commands   Pick to open the **Point Cloud Manager**

option uses the color assigned to the point cloud object. The **Normal** option assigns colors based on the normal direction of each point. The colors used are based on values assigned to the X, Y, and Z axes. The **Intensity** option assigns colors based on point intensity values. The **Elevation** option assigns colors based on the Z coordinate values of points. The **Classification** option assigns colors based on point classifications. This option is available if classifications are defined in the scan file. Picking the **Color Mapping** button in the **Visualization** panel opens the **Point Cloud Color Map** dialog box, which can be used to customize the **Intensity**, **Elevation**, and **Classification** color styles. Color schemes available for the **Intensity**, **Elevation**, and **Classification** color styles can be selected from the **Color Schemes** drop-down list.

The **Transparency** option is used to apply transparency to the point cloud. Use the slider or the text box to adjust the value. The value can range from 0 to 90. As with other transparency settings in AutoCAD, the higher the transparency value, the greater the transparency.

The **Lighting** drop-down list can be used to adjust the lighting applied to the scene. The default option is **No Lighting**. If you select the **Single Sided** or **Double Sided** option, you can set the shading and light source using the **Shading** and **Light Source** drop-down lists. Lighting is discussed in detail in Chapter 17.

The options in the **Cropping** panel allow you to crop the point cloud using a rectangular, polygonal, or circular boundary. Cropping enables you to show relevant areas without displaying the entire point cloud. See **Figure 14-61**. When defining the cropping boundary, you can specify whether to keep the points inside or outside the crop. Once the crop is defined, you can invert, hide, or remove the crop. You can also save the cropped view as a crop state. To save a crop state, expand the **Cropping** panel and pick the **New Crop State** button to activate the **POINTCLOUDCROPSTATE** command. When prompted, enter a name for the crop state. The **Crop State** drop-down list in the expanded **Cropping** panel lists all crop states for the point cloud and can be used to restore a previously created crop state. To delete a crop state, use the **POINTCLOUDCROPSTATE** command.

POINTCLOUDCROPSTATE

Ribbon

Point Cloud
> Cropping

New Crop State

Type

**POINTCLOUD-
CROPSTATE**

NOTE

The projection must be set to parallel in order to crop a point cloud.

POINTCLOUDMANAGER

Ribbon

Point Cloud
> Options

**Point Cloud
Manager**

Type

**POINTCLOUD-
MANAGER**

Selecting the **Point Cloud Manager** button in the **Options** panel displays the **Point Cloud Manager**, which provides information about all point clouds attached to the drawing. See **Figure 14-62**. Any scans and scan regions associated with the point cloud are listed. You can control the display of each item by picking the **On/Off** button to the right of the item. To isolate the display of a scan region, right-click on the name and select **Isolate** from the shortcut menu. This displays the scan region and hides other items. The **On** and **Off** options can also be selected from the shortcut menu.

By default, the point cloud is framed by a bounding box. You can control the display of the bounding box by selecting an option from the **Show Bounding Box** drop-down list, which is located in the expanded **Options** panel. Selecting the **External Reference** button in the **Options** panel displays the **External References** palette.

Exercise 14-15

www.g-wlearning.com/CAD/

Complete the exercise on the companion website.

Figure 14-61.
Cropping a point cloud. A—A portion of the interior of a building has been cropped to show a wall corner. B—Further cropping is used to isolate a portion of the end wall. (*Pittsburg State University School of Construction*)

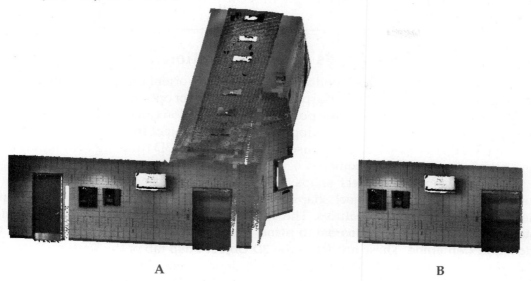

A   B

Working with Point Clouds

As you have seen, point clouds provide a powerful visualization tool for use in AutoCAD modeling applications. However, point clouds can also be used as a reference in creating new drawing geometry. For example, you can create new 2D geometry based on the existing features in a point cloud, such as the walls and floors in a building. If you are using the 2D geometry to create profile shapes, you can then construct 3D models by extruding or applying other 3D modeling techniques. Several methods for creating new drawing geometry based on point clouds are available in AutoCAD. You can use an existing surface in a point cloud as a drawing surface to construct 2D or 3D objects. In addition, you can extract 2D geometry from point clouds by using extraction commands. Another technique is to create a section object that cuts through a point cloud and then extract 2D geometry from the resulting section boundary. These methods are discussed in the following sections.

Figure 14-62.
The **Point Cloud Manager** with the default tree view mode active.

Scan regions defined in point cloud

Pick to display list view

Pick a button to toggle on/off status

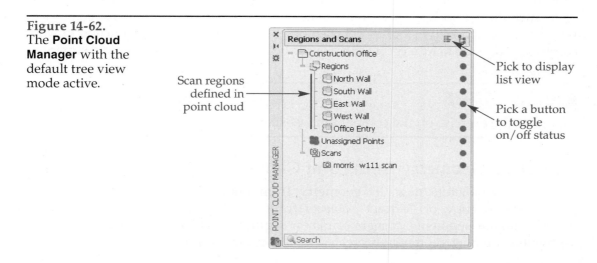

NOTE

Hardware acceleration must be enabled in order to work with point clouds. For optimal results, make sure your computer's graphics card is updated with the most current drivers installed.

Point Cloud Drawing Aids and Construction Tools

As with other types of 3D model surfaces, you can orient the UCS to a planar face in a point cloud for construction purposes. Use the **Object** option of the **UCS** command to orient the UCS to a surface in the point cloud. This option can be quickly accessed by picking on the UCS icon, right-clicking, and selecting **Object** from the shortcut menu. Once the UCS is properly oriented, use standard 2D and 3D drawing commands and object snaps to create new geometry.

AutoCAD provides object snaps for accurately selecting existing points in a point cloud. Point cloud object snaps assist in creating new geometry and taking measurements of existing features. The available point cloud object snaps are **Node**, **Intersection**, **Edge**, **Corner**, **Nearest to plane**, **Perpendicular to plane**, **Perpendicular to edge**, and **Centerline**. These are set in the **3D Object Snap** tab of the **Drafting Settings** dialog box, introduced in Chapter 1. The point cloud object snaps can also be activated using the **3D Object Snap** flyout on the status bar. The **3D Object Snap** flyout does not appear on the status bar by default. Select the **3D Object Snap** option in the status bar **Customization** flyout to add the **3D Object Snap** flyout to the status bar.

NOTE

In order to use some of the point cloud object snaps available in AutoCAD, the point cloud must contain *segmentation data*. Segmentation data is obtained by identifying groups of neighboring points that make up the same planar or non-planar surface. Segmentation data is processed when scan files are indexed in Autodesk ReCap. Accurate processing of segmentation data is dependent on the type of laser scanner used and the work that is scanned. To determine whether the point cloud contains segmentation data, select the point cloud, open the **Properties** palette, and check the status of the Segmentation property in the **Misc** category.

Another way to orient the UCS to a planar surface on a point cloud is to use a dynamic UCS. To activate the dynamic UCS function, press the [F6] key or the [Ctrl]+[D] key combination. Dynamic UCS behavior when detecting a point cloud surface is similar to that when working with solid models. Hover the cursor over the appropriate surface to align the XY plane with the surface and then draw the new object. When the command is completed, the UCS reverts to its previous location. In order for the dynamic UCS function to work correctly, make sure to turn off 3D object snaps before hovering over the desired surface.

Exercise 14-16   www.g-wlearning.com/CAD/
Complete the exercise on the companion website.

Extracting Geometry from Point Clouds

You can generate new 2D geometry from point clouds by using extraction commands to copy or "extract" objects from existing surfaces. You can extract edges of planar segments, corners of planar segments, and centerlines of cylindrical segments. The new geometry is created on the current layer and can be modified in the same way as other 2D objects. The **PCEXTRACTEDGE, PCEXTRACTCORNER,** and **PCEXTRACTCENTERLINE** commands are used to extract geometry from point clouds. These commands are accessed in the **Extract** panel in the **Point Cloud** contextual ribbon tab. Refer to **Figure 14-60.**

The **PCEXTRACTEDGE** command is used to extract an edge where two planar surfaces intersect. Access the command and hover the cursor over the first planar surface. When the surface becomes highlighted, select the surface. Then, hover the cursor over a second planar surface that is adjacent to the first and select the surface. The edge at the intersection of the two surfaces is extracted as a line.

The **PCEXTRACTCORNER** command is used to extract a point where three planar surfaces intersect. Access the command and hover the cursor over the first planar surface. When the surface becomes highlighted, select the surface. Then, hover the cursor over a second planar surface that is adjacent to the first and select the surface. In the same fashion, select a third planar surface that is adjacent to the first two surfaces. A point object is generated at the intersection of the three surfaces.

The **PCEXTRACTCENTERLINE** command is used to extract a line along the center axis of a cylindrical object. Access the command and hover the cursor over a cylindrical surface. When the surface becomes highlighted, select the surface. A line coinciding with the center axis of the surface is extracted.

Point Cloud Sections

Section objects, discussed earlier in this chapter, can be applied to point clouds. Using the **SECTIONPLANE** command, a point cloud can be sectioned by picking a face, picking points, or creating an orthographic section plane. The resulting section can be modified by adding jogs, offsetting or rotating the section plane, or using grips to edit the section plane location. As when sectioning other types of objects, live sectioning is turned on when the section plane is created. The **Section** panel in the **Point Cloud** contextual ribbon tab provides access to the orthographic section plane options. Picking the **2 Point** option activates the **SECTIONPLANE** command.

Once the section plane is created, you can use it to extract the lines forming the section of the point cloud. This is an alternative to using other point cloud extraction commands. To extract section lines, make sure that live sectioning is turned on for the section object. Then, select the point cloud and access the **PCEXTRACTSECTION** command. The **Extract Section Lines from Point Cloud** dialog box appears, **Figure 14-63.** This dialog box is used to specify settings for the extracted geometry. Selecting the **Entire cross section** option in the **Extract** area extracts geometry from the area in which the section plane cuts the point cloud. Selecting the **Perimeter only** option extracts geometry along the section plane perimeter. The options in the **Output geometry** area are used to set the layer and color assigned to the geometry and specify whether lines or polylines are extracted. If you are extracting polylines, you can set the resulting polyline width. The **Preview result** check box located at the bottom of the dialog box is checked by default. This allows you to preview the extraction results and adjust settings before completing the command.

The slider in the **Maximum points to process** area is used to specify the number of points processed. A higher value increases the accuracy of the extracted geometry, but also increases processing time. At higher settings, the extraction may take a very long time to process, depending on your computer. A time estimate appears below the slider and updates as you move the slider.

The **Extraction tolerances** area contains tolerance settings that specify how the geometry is generated. The **Minimum line length:** option sets the minimum length of extracted objects. Enter a value or pick the button next to the text box to set the length by picking two points in the drawing. The **Connect lines tolerance:** option sets the distance below which neighboring line and arc segments are joined. Enter a value or

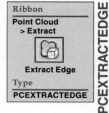

PCEXTRACTEDGE

Ribbon
Point Cloud
> Extract

Extract Edge

Type
PCEXTRACTEDGE

PCEXTRACTCORNER

Ribbon
Point Cloud
> Extract

Extract Corner

Type
PCEXTRACT-
CORNER

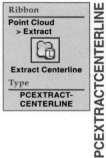

PCEXTRACTCENTERLINE

Ribbon
Point Cloud
> Extract

Extract Centerline

Type
PCEXTRACT-
CENTERLINE

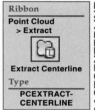

PCEXTRACTSECTION

Ribbon
Point Cloud
> Extract

Extract Section
Lines

Type
PCEXTRACT-
SECTION

Figure 14-63.
The **Extract Section Lines from Point Cloud** dialog box.

Specify the type of extraction to perform

Layer and color settings

Specify the object type to extract

Tolerance options

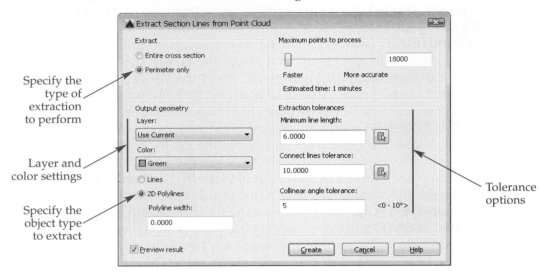

pick the button next to the text box to set the distance by picking two points in the drawing. The **Collinear angle tolerance:** option controls the number of object segments generated based on the angle between neighboring segments. The higher the setting, the fewer segments generated. The setting can range from 0 to 10.

After making the appropriate settings, pick **Create** to close the dialog box and begin processing. Once processing is completed, a preview of the extracted geometry is displayed and you are prompted to accept the results. Select the **Accept** option to complete the operation. Selecting the **Settings** option redisplays the **Extract Section Lines from Point Cloud** dialog box, where you can change the settings to produce different results. Selecting the **Undo** option deletes the extracted geometry and allows you to continue the command. You can use the **Settings** option to adjust the extraction settings or exit the command.

In **Figure 14-64A**, the point cloud has been sectioned using an orthographic section plane. The results after using the **PCEXTRACTSECTION** command are shown in **Figure 14-64B**. In this example, polylines are extracted along the entire cross section. The resulting polylines can be edited by using grips or standard editing commands, such as **PEDIT** and **JOIN**.

Figure 14-64.
Extracting section lines from a point cloud. A—A point cloud sectioned with an orthographic section object using the **Top** option. B—The drawing after extracting polylines from the point cloud using the **Entire cross section** option. The point cloud is shown with transparency applied for illustration purposes only. *(Model courtesy of Laser Design and GKS Services)*

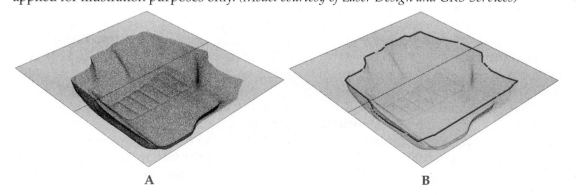

A   B

Exercise 14-17

www.g-wlearning.com/CAD/

Complete the exercise on the companion website.

Chapter Review

Answer the following questions using the information in this chapter.

1. Which option of the **VIEWBASE** command is used to change the default view when placing the base view?

2. Explain how to change the default projection used by AutoCAD when placing drawing views.

3. Explain two ways to place a projected view.

4. Describe two ways to locate the section line when creating a section view with the **VIEWSECTION** command.

5. When creating a section view with the **VIEWSECTION** command, what controls the appearance of elements in the section identifier and section view label?

6. What command is used to edit a section view?

7. What are the two types of detail boundaries that can be used when creating a detail view?

8. What option of the **VIEWDETAIL** command allows you to create a smooth edge for the detail view and include a border?

9. To what value should the **DIMASSOC** system variable be set in order to dimension drawing views with associative dimensions?

10. What does the **SECTIONPLANE** command create?

11. How is the **Face** option of the **SECTIONPLANE** command used?

12. Which option of the **SECTIONPLANE** command is used to create sections with jogs?

13. Which section object grips are used to accomplish the following tasks?
    A. Change the section object state.
    B. Lengthen or shorten the section object segment.
    C. Rotate the section view 180°.

14. How is live sectioning turned on or off?

15. Which category in the **Section Settings** dialog box provides control over the material that is removed by the section object?

16. What are the two types of section view blocks that can be created from a section object?

17. What is the function of the **MASSPROP** command?

18. What is the extension of the ASCII file that can be created by **MASSPROP**?

19. What command is used to graphically analyze surface continuity in a model?

20. Briefly explain how to set values defining the acceptable curvature range when using the **ANALYSISCURVATURE** command.

21. Which commands export and import solid models?

22. Which type of file has an .stl extension?

23. What is the purpose of the **3DPRINT** command?

24. What is a *point cloud*?

25. Name at least three applications for which a point cloud can provide accurate data for design and modeling purposes.

26. In Autodesk ReCap, what is the name of the process in which scan files are converted to point cloud files?

27. What is *registration*?

28. What is a *scan region*?

29. What command is used to attach a point cloud in AutoCAD?

30. What command is used to extract a point object from a point cloud where three planar surfaces intersect?

Drawing Problems

1. **Mechanical** In this problem, you will create a multiview layout of the base bracket by placing drawing views with the **VIEWBASE** command. Set up a page layout using the ANSI B (17.00 × 11.00 Inches) paper size. Select the appropriate scale for the orthographic views and the isometric view. Edit the views as needed to look like the layout shown.
   A. Construct the model shown using the dimensions provided.
   B. Create the front, top, right-side, and isometric views shown.
   C. The isometric view should be a shaded view.
   D. Dimension the drawing in the layout using the dimensions given. Make sure to use associative dimensions.
   E. Save the drawing as P14-1.

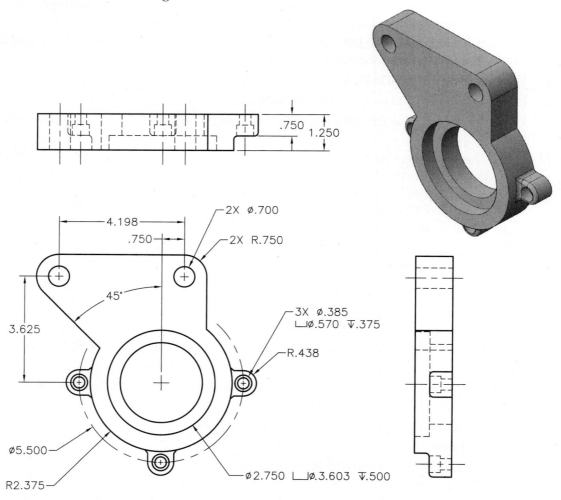

NOTE: ALL FILLETS AND ROUNDS R.125 UNLESS NOTED

2. **Mechanical** In this problem, you will create a multiview layout of the flanged coupler by placing drawing views with the **VIEWBASE** command. Set up a page layout using the ANSI B (17.00 × 11.00 Inches) paper size. Select the appropriate scale for the orthographic views and the isometric view. Edit the views as needed to look like the layout shown.
   A. Construct the model shown using the dimensions provided.
   B. Create the front, right-side, and isometric views shown.
   C. The isometric view should be a shaded view.
   D. Dimension the drawing in the layout using the dimensions given. Make sure to use associative dimensions.
   E. Save the drawing as P14-2.

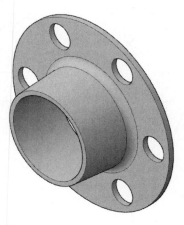

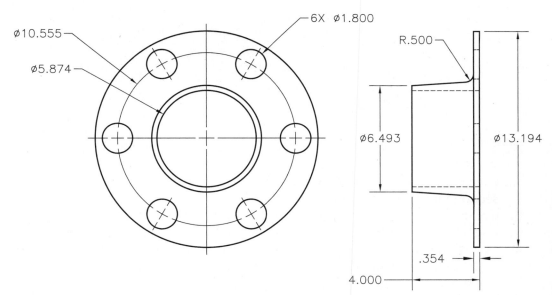

3. **General** Open one of your solid model problems from a previous chapter and do the following.
   A. Create a multiview layout of the model. Use three orthographic views and one isometric view.
   B. Create a page layout and use an appropriate scale for the views.
   C. Use the **VIEWBASE** command to create the views.
   D. The isometric view should be shaded.
   E. Dimension the orthographic views.
   F. Save the drawing as P14-3.

4. **Mechanical** In this problem, you will complete a drawing you created in Exercise 14-3 by adding dimensions to the views.
   A. Open drawing EX14-3.
   B. Referring to the layout shown, dimension the top and front views. Make sure to use associative dimensions.
   C. Add the note and the centerlines shown.
   D. Save the drawing as P14-4.

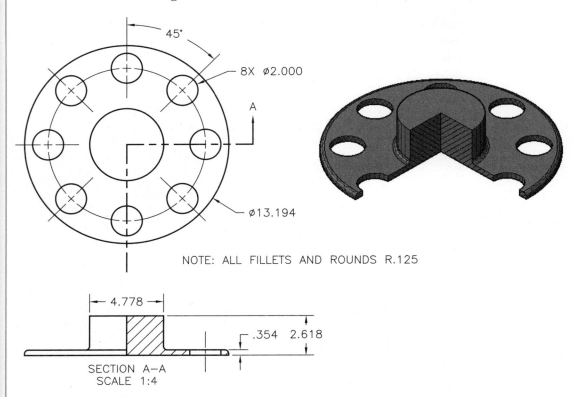

SECTION A—A
SCALE 1:4

NOTE: ALL FILLETS AND ROUNDS R.125

5. **General** Open one of your solid model problems from a previous chapter and do the following.
   A. Create a multiview layout of the model. Create three views, including one section view, and an isometric section view.
   B. Create a page layout and use an appropriate scale for the views.
   C. Use the **VIEWBASE** and **VIEWSECTION** commands to create the views.
   D. The isometric view should be an isometric section view projected from the section view and should be shown shaded.
   E. Dimension the views.
   F. Save the drawing as P14-5.

6. General Open one of your solid model problems from a previous chapter and do the following.
   A. Use the **Face** option of the **SECTIONPLANE** command to create a section object.
   B. Alter the section so that the section plane object cuts through features of the model.
   C. Change the section settings to display an ANSI hatch pattern.
   D. Save the drawing as P14-6.

7. General Open one of your solid model problems from a previous chapter and do the following.
   A. Construct a section through the model using the **Draw section** option of the **SECTIONPLANE** command. Cut through as many features as possible.
   B. Display cutaway geometry with a 50% transparency.
   C. Display section lines using an appropriate hatch pattern.
   D. Generate a 3D section block that displays the cutaway geometry in a color of your choice.
   E. Create a layout with a viewport for the 3D block displayed at half the size of the original model.
   F. Save the drawing as P14-7.

8. General Choose five solid model problems from previous chapters and copy them to a new folder. Then, do the following.
   A. Open the first drawing. Export it as an SAT file.
   B. Do the same for the remaining four files.
   C. Compare the sizes of the SAT files with the DWG files. Compare the combined sizes of both types of files.
   D. Begin a new drawing and import one of the SAT files.

Drawing Problems - Chapter 14

9. **Mechanical** Draw the object shown as a solid model. Only half of the object is shown; draw the complete object. Do not dimension the object. Then, do the following.
   A. Construct a section plane that creates a full section, as shown.
   B. Display the intersection fill as an ANSI hatch pattern.
   C. Activate live sectioning and view the cutaway geometry with a high level of transparency.
   D. Save the drawing as P14-9.

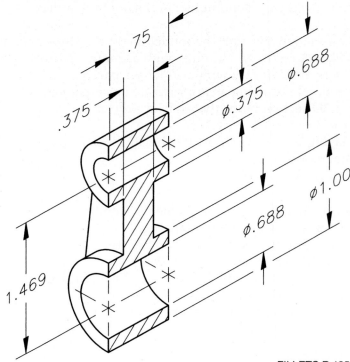

FILLETS R.125

10. **Mechanical** Draw the object shown as a solid model. Do not dimension the object. Then, do the following.
    A. Construct a section object that creates a full section along the centerline of the hole.
    B. Generate a 2D section and display it on the drawing at half the size of the original. Specify section settings as desired.
    C. Generate a 3D section and display it on the drawing at half the size of the original. Do not display cutaway geometry. Specify section settings as desired.
    D. Activate live sectioning. Do not display the cutaway geometry.
    E. On the original solid model, display the intersection fill as an ANSI hatch pattern.
    F. Save the drawing as P14-10.

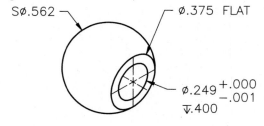

11. **Mechanical** Draw the object shown as a solid model. Only half of the object is shown; draw the complete object. Do not dimension the object. Then, do the following.
   A. Construct a section plane that creates a half section.
   B. Display the intersection fill as an ANSI hatch pattern.
   C. Activate live sectioning and view the cutaway geometry with a low level of transparency.
   D. Generate a 2D section and display it on the drawing at full size. Specify section settings as desired.
   E. Generate a 3D section and display it on the drawing at half the size of the original. Do not display cutaway geometry. Specify section settings as desired.
   F. Save the drawing as P14-11.

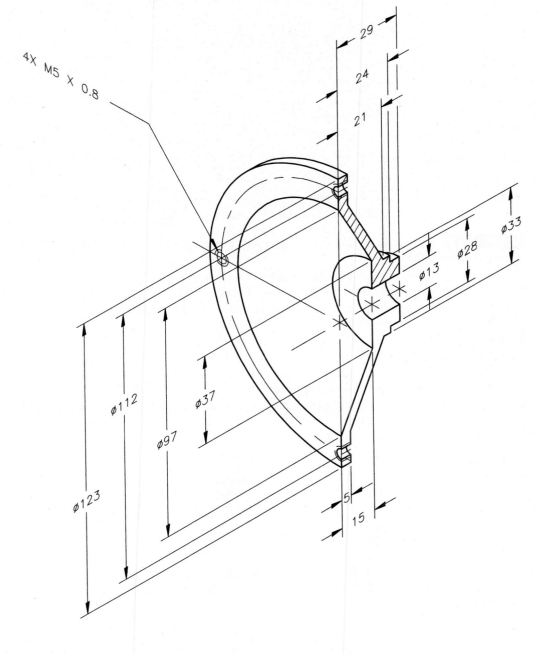

12. **Mechanical** Draw the object shown as a solid model. Use your own dimensions. Then, do the following.
   A. Construct a section plane that creates an offset section. The section should pass through the center of two holes in the base and through the large central hole.
   B. Display the intersection fill as an ANSI hatch pattern.
   C. Activate live sectioning and view the cutaway geometry with a low level of transparency in the color red.
   D. Generate a 3D section of the sectioned solid model and save it as a block.
   E. Create a two-view orthographic layout. Use the orthographic presets to create a top view and a front view. Use an appropriate scale to plot on a B-size sheet.
   F. Create a third floating viewport and insert the 3D section block scaled to half the size of the drawing.
   G. Save the drawing as P14-12.

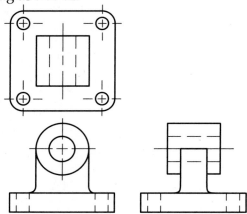

13. **Mechanical** Draw the object shown as a solid model. Do not dimension the object. Then, do the following.
   A. Construct a section plane that creates a half section.
   B. Display the intersection fill as an ANSI hatch pattern.
   C. Activate live sectioning and view the cutaway geometry with a low level of transparency in the color red.
   D. Alter the section plane to create the section shown.
   E. Generate a 3D section and save it as a block.
   F. Create a two-view orthographic layout. Use the orthographic presets to create a top view and a front view. Use an appropriate scale to plot on an A-size sheet.
   G. Create a third floating viewport and insert the 3D section block scaled to half the size of the drawing.
   H. Save the drawing as P14-13.

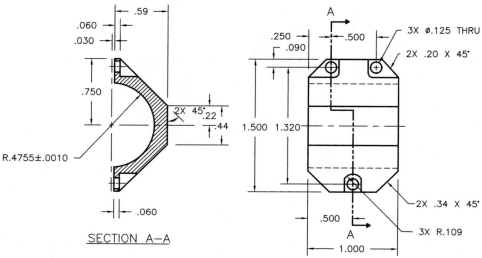

SECTION A—A

15

Visual Style Settings and Basic Rendering

Learning Objectives

After completing this chapter, you will be able to:

✓ Control the display of solid models.
✓ Describe the **Visual Styles Manager** palette.
✓ Change the settings for visual styles.
✓ Create custom visual styles.
✓ Export visual styles to a tool palette.
✓ Render a scene using sunlight.
✓ Save a rendered image from the **Render** window.

There are many options available in AutoCAD for controlling the display of models. These settings help with your visualization while you are working on a model and may also impact printed output. In Chapter 1, you were introduced to the default visual styles. In this chapter, you will learn about all visual style settings and how to redefine the visual style. You will also learn how to create your own visual style. Finally, this chapter provides an introduction to lights and rendering.

Controlling Solid Model Display

AutoCAD solid models can be displayed in wireframe form, with hidden lines removed, or in shaded form by using a shaded visual style. The 2D Wireframe visual style is the default display when a drawing is started based on the acad.dwt template. Wireframe displays are the quickest to display. When a drawing is started based on the acad3D.dwt or acadiso3D.dwt template, the default display is the Realistic visual style, which is a shaded display. The following sections introduce display options available in wireframe, hidden, and shaded displays.

Isolines

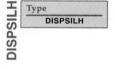

The appearance of a solid model in a wireframe display is controlled by the **ISOLINES** system variable. *Isolines* represent the edges and curved surfaces of a solid model. This setting does *not* affect the final rendered model. However, if the Show property in the **Edge Settings** category in the **Visual Styles Manager** palette is set to Isolines, isolines are displayed when the visual style is set current. The default **ISOLINES** value is four. It can have a value from zero to 2047. All solid objects in the drawing are affected by changes to the **ISOLINES** value, as are all visual styles set to display isolines. **Figure 15-1** illustrates the difference between **ISOLINES** settings of four and 12.

The setting of the **ISOLINES** system variable can be changed in the **Visual Styles Manager** palette. You can also type **ISOLINES** and enter a new value, or you can change the **Contour lines per surface** setting in the **Display resolution** area of the **Display** tab in the **Options** dialog box. See **Figure 15-2**. The settings in the **Display** tab in the **Options** dialog box can also be used to control the **DISPSILH** and **FACETRES** system variables, discussed next.

Creating a Display Silhouette

In the 2D Wireframe visual style, a model can appear smooth with only a silhouette displayed, similar to the Hidden visual style. This is controlled by the **DISPSILH** (display silhouette) system variable. The **DISPSILH** system variable has two values, 0 (off) and 1 (on). **Figure 15-3** shows solids with **DISPSILH** set to 0 and 1 after setting the 2D Wireframe visual style current and then using **HIDE**.

The setting can be changed by typing **DISPSILH** and entering a new value. You can also set the variable using the **Draw true silhouettes for solids and surfaces** check box in the **Display performance** area of the **Display** tab in the **Options** dialog box. Refer to **Figure 15-2**. A preferred technique is to set **ISOLINES** to 0 and **DISPSILH** to 1 to

Figure 15-1.
Isolines define curved surfaces.
A—**ISOLINES** = 4.
B—**ISOLINES** = 12.

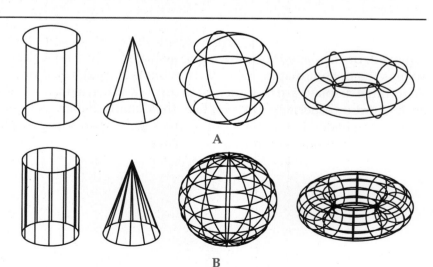

Figure 15-2.
The **ISOLINES**, **FACETRES**, and **DISPSILH** values can be set in the **Options** dialog box.

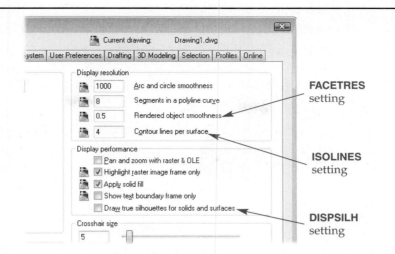

FACETRES setting

ISOLINES setting

DISPSILH setting

create a true display of the silhouette edge of a 3D model. For the 2D Wireframe visual style, the **DISPSILH** variable can also be set in the **Visual Styles Manager** palette. The variable is controlled by the Draw true silhouettes property in the **2D Wireframe options** category. Setting this property to Yes turns on silhouettes.

Controlling Surface Smoothness

The smoothness of curved surfaces in hidden, shaded, and rendered displays is controlled by the **FACETRES** system variable. This variable determines the number of polygon faces applied to the model. The value can range from .01 to 10.0 and the default value is .5. This system variable can be changed by typing **FACETRES** or by changing the **Rendered object smoothness** setting in the **Display resolution** area of the **Display** tab in the **Options** dialog box. Refer to **Figure 15-2**. For the 2D Wireframe visual style, the variable can be set in the **Visual Styles Manager** palette. The Solid smoothness property in the **Display resolution** category controls the variable. **Figure 15-4** shows the effect of two different **FACETRES** settings.

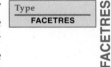

FACETRES

Figure 15-3.
A—The **HIDE** command used when **DISPSILH** is set to 0. Objects are displayed faceted.
B—The **HIDE** command used when **DISPSILH** is set to 1. Facets are eliminated and only the silhouette is displayed.

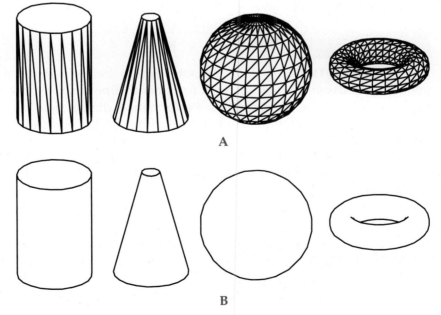

A

B

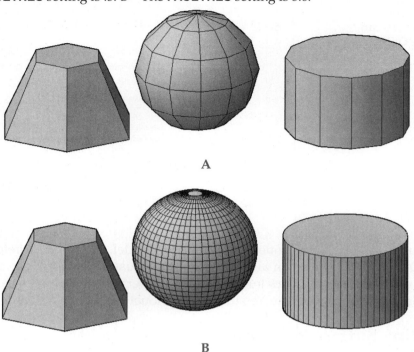

A

B

Controlling Line Display

The smoothness of 2D lines displayed in the 2D Wireframe visual style is controlled by the **LINESMOOTHING** system variable. By default, this system variable is set to 1 (on). At this setting, antialiasing is turned on. *Aliasing* refers to the jagged or "stair-stepped" appearance of pixels defining diagonal or curved edges at a lower resolution. *Antialiasing* is a method used to lessen this effect by shading adjacent pixels. Turning on the **LINESMOOTHING** system variable will improve the appearance of 2D object lines in wireframe displays.

NOTE

The **LINESMOOTHING** system variable cannot be turned off unless high quality geometry is disabled. The settings for smooth line display and high quality geometry are located in the **Graphics Performance** dialog box, accessed by using the **GRAPHICSCONFIG** command or right-clicking on the **Hardware Acceleration** button on the status bar and selecting **Graphics Performance…**. The **Smooth line display** check box turns **LINESMOOTHING** on and off and is only available when the **High quality geometry (for capable devices)** check box is unchecked. Both check boxes are checked by default.

Exercise 15-1

www.g-wlearning.com/CAD/

Complete the exercise on the companion website.

Overview of the Visual Styles Manager

The **Visual Styles Manager** palette provides access to all of the visual style settings. This palette is a floating window similar to the **Properties** palette. See **Figure 15-5**. Changes made in the **Visual Styles Manager** redefine the visual style in the current drawing.

At the top of the **Visual Styles Manager** are image tiles for the defined visual styles. See **Figure 15-6**. The default visual styles are 2D Wireframe, Conceptual, Hidden, Realistic, Shaded, Shaded with Edges, Shades of Gray, Sketchy, Wireframe, and X-Ray. User-defined visual styles also appear as image tiles. The image on the tile is a preview of the visual style settings. Selecting an image tile provides access to the properties of the visual style in the palette below. The name of the currently selected visual style appears below the image tiles and the corresponding image tile is surrounded by a yellow border.

To set a different visual style current using the **Visual Styles Manager**, double-click on the image tile. You can also select the image tile and pick the **Apply Selected Visual Style to Current Viewport** button immediately below the image tiles. An icon consisting of a small image of the button is displayed in the image tile of the visual style that is current in the active viewport, as shown in **Figure 15-6**. A drawing icon appears in the image tile if the visual style is current in a viewport that is not active. The AutoCAD icon appears in the image tiles of the default visual styles.

You can also set a different visual style current by selecting the style from the **Visual Style Controls** flyout in the viewport controls located in the upper-left corner of the drawing window. See **Figure 15-7**. Selecting **Visual Styles Manager...** in the **Visual Style Controls** flyout opens the **Visual Styles Manager**.

Ribbon
Visualize
> Visual Styles
> Visual Style
Manager...
Type
VISUALSTYLES
VSM

VISUALSTYLES

Figure 15-5.
The **Visual Styles Manager** palette.

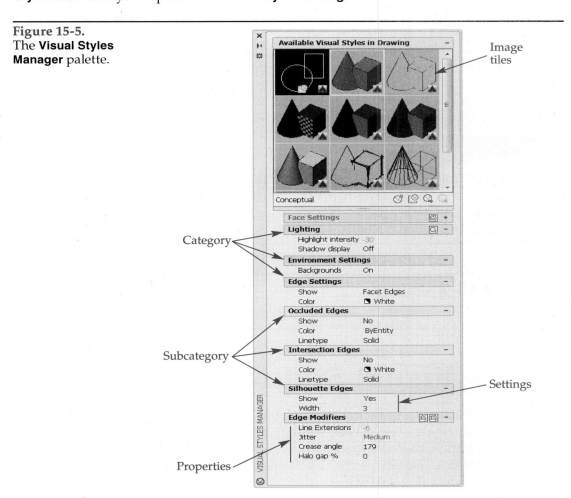

Figure 15-6.
The image tiles correspond to the visual styles. The image on the tile is a preview of the visual style's settings.

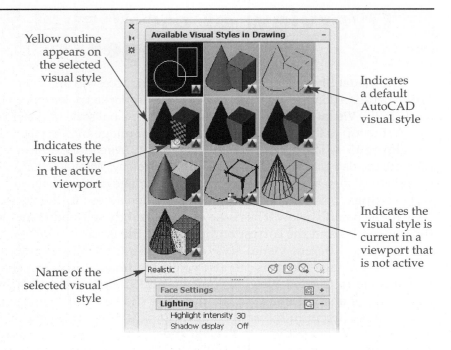

Yellow outline appears on the selected visual style

Indicates a default AutoCAD visual style

Indicates the visual style in the active viewport

Indicates the visual style is current in a viewport that is not active

Name of the selected visual style

Figure 15-7.
A visual style can be set current by selecting it from the **Visual Style Controls** flyout in the viewport controls.

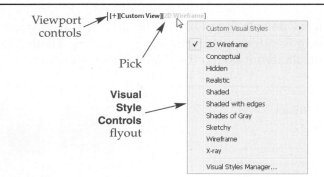

Viewport controls

Pick

Visual Style Controls flyout

NOTE

The command line offers an additional way to set a visual style current. When you type the letters in a visual style name on the command line, visual styles that match the letters you type appear in the suggestion list. This method can only be used on the command line and is not available with dynamic input.

Exercise 15-2

www.g-wlearning.com/CAD/

Complete the exercise on the companion website.

Visual Style Settings

The **Visual Styles** panel on the **Visualize** tab of the ribbon provides several settings for altering the visual style. These settings are also available in the **Visual Styles Manager**. In addition, there are settings in the **Visual Styles Manager** that are not available on the ribbon. The next sections discuss settings available in the **Visual Styles Manager** for the default visual styles.

Changing any setting in the **Visual Styles Manager** *redefines* the visual style in the current drawing. Changes made using the ribbon are temporary and are not kept when a different visual style is set current. This is important to remember.

2D Wireframe

When the 2D Wireframe visual style is set current, lines and curves are used to show the edges of 3D objects. Assigned linetypes and lineweights are displayed. All edges are visible as if the object is constructed of pieces of wire soldered together at the intersections (thus, the name *wireframe*). Either the 2D or 3D wireframe UCS icon is displayed. OLE objects will display normally as long as they are parallel to the viewing plane. If they are not parallel, only the OLE frame is visible. In addition, the drawing window display changes to the 2D model space context and parallel projection. For the 2D Wireframe visual style, the **Visual Styles Manager** displays the following categories. See **Figure 15-8**.

- **2D Wireframe Options**
- **2D Hide—Occluded Lines**
- **2D Hide—Intersection Edges**
- **2D Hide—Miscellaneous**
- **Display Resolution**

2D Wireframe Options

The Contour lines property controls the **ISOLINES** system variable. The setting is 4 by default. Isolines are suppressed when the **HIDE** command is used with the 2D Wireframe visual style set current. The Draw true silhouettes property controls the **DISPSILH** system variable. It is set to No by default, which is equivalent to a **DISPSILH** setting of 0.

2D Hide—Occluded Lines

The Color property in this category controls the **OBSCUREDCOLOR** system variable. This property determines the color of *occluded* lines (obscured lines hidden from the view) when the **HIDE** command is used. The default setting is ByEntity. This means that, when displayed, obscured lines are shown in the same color as the object.

The Linetype property controls the **OBSCUREDLTYPE** system variable. This property determines whether obscured lines are displayed and in which linetype they are displayed. The default setting is Off, which means that obscured lines are not displayed when the **HIDE** command is used. Selecting a linetype allows you to have obscured lines displayed instead of removed from the view. The available linetypes are: Solid,

Figure 15-8.
The categories and properties available for the 2D Wireframe visual style.

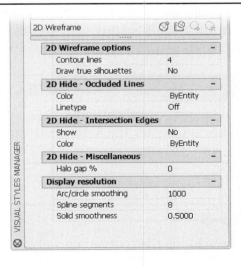

Dashed, Dotted, Short Dash, Medium Dash, Long Dash, Double Short Dash, Double Medium Dash, Double Long Dash, Medium Long Dash, and Sparse Dot. When a linetype is selected from the drop-down list, obscured lines are displayed in that linetype after the **HIDE** command is used.

NOTE

The linetypes available in the Linetype property drop-down list are not the same as the linetypes loaded into the **Linetype Manager** dialog box. They are independent of the zoom level, which means the dash size will stay the same when zooming in and out.

2D Hide—Intersection Edges

This category is used to toggle the display of polylines at the intersection of 3D surfaces and set the color of the lines. The Show property controls the **INTERSECTIONDISPLAY** system variable. This property determines whether polylines are displayed at the intersection of non-unioned 3D surfaces. The default setting is No, which means that polylines are not displayed when the **HIDE** command is used.

The Color property in this category controls the **INTERSECTIONCOLOR** system variable. This property determines the color of the polylines displayed at intersection edges. By default, the setting is ByEntity. This means that, when displayed, the polylines at intersection edges are shown in the same color as the object.

2D Hide—Miscellaneous

The Halo gap % property controls the **HALOGAP** system variable. This property determines the gap that is displayed where one object partially obscures another (between the foreground edge and where the background edge starts to show). The default setting is 0 and the value can range from 0 to 100. The value refers to a percentage of one unit. The gap is only displayed when the **HIDE** command is used. It is not affected by the zoom level.

Display Resolution

The Arc/circle smoothing property controls the zoom percentage set by the **VIEWRES** command. This determines the resolution of circles and arcs when hardware acceleration is disabled. The value can range from 1 to 20,000. The higher the value, the higher the resolution of circles and arcs.

The Spline segments property controls the **SPLINESEGS** system variable. This property determines the number of line segments in a spline-fit polyline. The value can range from −32,768 to 32,767.

The Solid smoothness property controls the **FACETRES** system variable. As previously discussed, the default setting is .5 and the value can range from .01 to 10.0. A higher **FACETRES** value will create a smoother finish on 3D printed models.

NOTE

Polygon faces will not be visible if the Draw true silhouettes property is set to Yes. However, a higher setting for the Solid smoothness property will make curved edges smoother.

Exercise 15-3

www.g-wlearning.com/CAD/

Complete the exercise on the companion website.

Conceptual

When the Conceptual visual style is set current, objects are smoothed and shaded. The shading is a transition from cool to warm colors. The transitional colors help highlight details. The previous projection is retained and that context is set current.

NOTE

The categories and properties for the shaded visual styles (those other than 2D Wireframe) are the same. These settings are discussed in the Common Visual Style Settings section later in this chapter.

Hidden

The Hidden visual style removes obscured lines from your view and makes 3D objects appear solid. The previous projection is retained and that context is set current. The benefit of using Hidden is that you get sufficient 3D display, but it does not push the graphics system too hard. Objects are not shaded or colored. This is very useful when working on complex drawings and/or using a slow computer.

Realistic

As with the Conceptual visual style, the objects have smoothing and shading applied to them when the Realistic visual style is set current. In addition, if materials are applied to the objects, the materials are displayed. The previous projection is retained and that context is set current. This visual style is good for adjusting textures and patterns on materials and for a final look at the scene before rendering.

Shaded

This style is very similar to Realistic, except the lighting is smooth instead of smoothest and textures are turned off.

Shaded with Edges

This style is the same as Shaded, except isolines are displayed with the default number of 4. Intersection edges are turned on and displayed as solid and white. Silhouette edges are also displayed with a width setting of 3.

Shades of Gray

This style is similar to the other shaded styles, except object colors are displayed as monochrome. Materials and textures are turned off. Facet edges are turned on, as are silhouette edges. This style is good for adjusting lights and shadows. It is much easier to see what is happening with lights when all objects are the same color.

Sketchy

The Sketchy visual style has most of the same settings as the Hidden visual style. However, objects look like they are hand-sketched. Line extensions and jitter are turned on, as are silhouette edges.

Wireframe

The Wireframe visual style is almost identical in appearance to the 2D Wireframe visual style. The main differences are that the 3D mode UCS icon is displayed and OLE objects display normally regardless of the viewing angle. The previous projection is retained and that context is set current. When working in 3D, a wireframe view is sometimes necessary to select objects normally hidden from your view.

X-Ray

This visual style is very similar to the Shaded with Edges visual style, except the Opacity property is set to 50%. This makes objects appear somewhat transparent.

PROFESSIONAL TIP

Depending on the visual style chosen, drawing window colors for individual elements such as the background, grid lines, crosshairs, etc., may need to be adjusted to make it easier to see and work with the objects in the drawing. Access the **Options** dialog box and select the **Colors...** button in the **Display** tab to change these settings in the **Drawing Window Colors** dialog box. Colors used with 2D model space, 3D parallel projection, and 3D perspective projection are controlled separately.

Common Visual Style Settings

With the exception of 2D Wireframe, all of the visual styles share similar categories and settings in the **Visual Styles Manager**. The visual styles have **Face Settings**, **Lighting**, **Environment Settings**, and **Edge Settings** categories. See **Figure 15-9**. These categories and the properties available in them are discussed in the next sections.

Face Settings

The Face style property controls the **VSFACESTYLE** system variable. The options have slightly different names, but are the same as the **No Face Style**, **Realistic Face Style**, and **Warm-Cool Face Style** options available by selecting the buttons in the **Visual Styles** panel on the **Visualize** tab of the ribbon. Remember, though, selecting a button on the ribbon does not redefine the visual style in the current drawing. Rather, it is a temporary change to the viewport display.

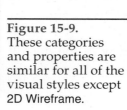

VSFACESTYLE

Ribbon

Visualize
> Visual Styles

No Face Style

Realistic Face
Style

Warm-Cool Face
Style

Type

VSFACESTYLE

Figure 15-9.
These categories and properties are similar for all of the visual styles except 2D Wireframe.

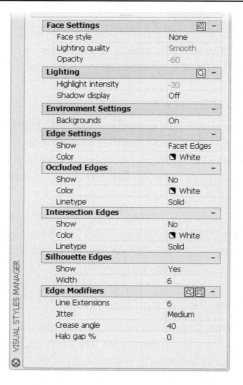

The Lighting quality property controls the **VSLIGHTINGQUALITY** system variable. This property determines whether curved surfaces are displayed smooth or as a series of flat faces. Lighting quality is unavailable if the Face style property is set to None.

The Color property controls the **VSFACECOLORMODE** system variable. This property determines how color is applied to the faces of an object. The effects can be temporarily applied by picking a button in the face colors flyout in the **Visual Styles** panel on the **Visualize** tab of the ribbon. The choices are:

- Normal. The object color is applied to faces.
- Monochrome. One color is applied to all faces. This also displays and enables the Monochrome color property.
- Tint. A combination of the object color and a specified color is applied to faces. This also displays and enables the Tint color property.
- Desaturate. The object color is applied to faces, but the saturation of the color is reduced by 30%.

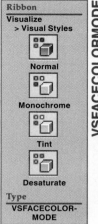

The Monochrome color and Tint color properties control the **VSMONOCOLOR** system variable. This system variable determines the color that is applied when the face Color property is set to Monochrome or Tint.

The Opacity property controls the **VSFACEOPACITY** system variable. This property determines how transparent or opaque faces are in the viewport. The value can range from –100 to 100. When the setting is 1, the faces are completely transparent. When the setting is 100, the faces are completely opaque. Settings below 0 set the value, but turn off the effect. To quickly turn the effect on or off, pick the **Opacity** button on the **Face Settings** category title bar. See **Figure 15-10**. This changes the value from negative to positive, or vice versa. This property cannot be changed if the Face style property is set to None.

Opacity may also be controlled by picking the **X-Ray Effect** button in the **Visual Styles** panel on the **Visualize** tab of the ribbon. The adjacent **Opacity** slider sets the value. However, remember, picking a button or making settings on the ribbon is only temporary. The change is not saved to the visual style.

The Material display property controls the **VSMATERIALMODE** system variable. When set to Off, objects display in their assigned color. When the setting is changed to Materials, the objects display the color of the material, but not the textures. When the setting is changed to Materials and textures, full materials are displayed.

CAUTION

Displaying materials and textures on 3D objects in a complex drawing will slow system performance. Set the Material display property to Materials and textures only when it is absolutely necessary.

Figure 15-10.
Picking the **Opacity** button on the **Face Settings** category title bar turns the effect of opacity on or off.

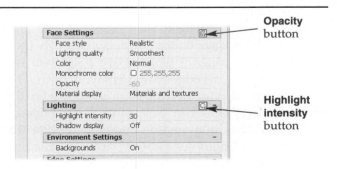

Lighting

The Highlight intensity property controls the **VSFACEHIGHLIGHT** system variable. This property determines the size of the highlight on faces to which no material is assigned. A small highlight on an object makes it look smooth and hard. A large highlight on an object makes it look rough or soft. The initial value is 30 or –30; the value can range from –100 to 100. The higher the setting is above 0, the larger the highlight. Settings below 0 set the value, but turn off the effect. To quickly turn the effect on or off, pick the **Highlight intensity** button on the **Lighting** category title bar. Refer to **Figure 15-10**. This changes the value from negative to positive or vice versa. This property cannot be changed if the Face style property is set to None.

The Shadow display property controls the **VSSHADOWS** system variable. This property controls if and how shadows are cast when the visual style is set current. The options have slightly different names, but are the same as those available by selecting a button in the shadows flyout in the **Lights** panel on the **Visualize** tab of the ribbon. Remember, however, a setting made using the ribbon is temporary and does not redefine the visual style.

If the property is set to Ground shadow, objects cast shadows on the ground, but not onto other objects. The "ground" is the XY plane of the WCS. The Mapped Object shadows setting, which corresponds to the **Full Shadows** button on the ribbon, only works if lights have been placed in the scene and hardware acceleration is enabled.

VSSHADOWS

Ribbon
Visualize
> Lights

Shadows Off

Ground Shadows

Full Shadows

Type
VSSHADOWS

Environment Settings

The Backgrounds property controls the **VSBACKGROUNDS** system variable. This property determines whether the preselected background is displayed in the viewport. Backgrounds can only be assigned to a view when a named view is created or edited. After the view is created, restore the view to display the background.

Edge Settings

Properties in the **Edge Settings** category determine the appearance of model edges. This category includes the **Occluded Edges**, **Intersection Edges**, **Silhouette Edges**, and **Edge Modifiers** subcategories. Refer to **Figure 15-5**.

The Show property controls the **VSEDGES** system variable. This property determines how edges on solid objects are represented when the visual style is set current. The options are the same as those available by selecting a button in the edge flyout in the **Visual Styles** panel on the **Visualize** tab of the ribbon. The settings of Isolines and Facet Edges determine how edges and curved surfaces are displayed on a 3D model. Setting this property to None turns off isolines and facets and displays no edges. If the Face style property is set to None, the Show property cannot be set to None.

The Color property in the **Edge Settings** category controls the **VSEDGECOLOR** system variable. This property determines the color of all edges on objects in the drawing. It is disabled when the **Show** property is set to None.

The Number of lines and Always on top properties are displayed when the Show property is set to Isolines. The Number of lines property controls the **ISOLINES** system variable. The Always on top property controls the **VSISOONTOP** system variable. This property determines if isolines are displayed when the model is shaded or hidden. When set to Yes, all edges are displayed, even if they are physically hidden from view.

Additional edge settings are available in the **Edge Settings** category when the Show property is set to Isolines or Facet Edges. These settings are available in the **Occluded Edges**, **Intersection Edges**, and **Edge Modifiers** subcategories. The settings in the **Silhouette Edges** subcategory are available regardless of the Show property setting. The settings in the subcategories are discussed as follows.

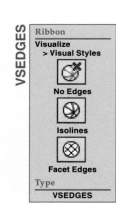

VSEDGES

Ribbon
Visualize
> Visual Styles

No Edges

Isolines

Facet Edges

Type
VSEDGES

Figure 15-11.
A—Occluded lines are not shown. B—The Show property is set to Yes and occluded lines are shown.

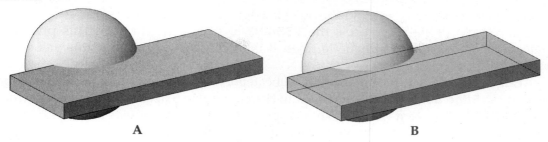

A   B

Occluded Edges Subcategory. This subcategory is not available if the Show property in the **Edge Settings** category is set to None. The Show property in this subcategory controls the **VSOCCLUDEDEDGES** system variable. This property determines whether or not occluded edges are displayed in a hidden or shaded view. See **Figure 15-11**.

The Color property in this subcategory controls the **VSOCCLUDEDCOLOR** system variable. The Linetype property controls the **VSOCCLUDEDLTYPE** system variable. These properties are similar to the properties for occluded lines discussed earlier in this chapter in the 2D Hide—Occluded Lines section.

Intersection Edges Subcategory. This subcategory is not available if the Show property in the **Edge Settings** category is set to None. The Show property in this subcategory controls the **VSINTERSECTIONEDGES** system variable. This property determines whether lines are displayed where one 3D object intersects another 3D object. See **Figure 15-12**.

The Color property in this subcategory controls the **VSINTERSECTIONCOLOR** system variable. The Linetype property controls the **VSINTERSECTIONLTYPE** system variable. These properties are similar to the properties for intersection lines discussed earlier in this chapter in the 2D Hide—Intersection Edges section.

PROFESSIONAL TIP

Setting the intersection edges Color property to a color that contrasts with the objects in your model is a good way to quickly check for interference between 3D objects.

Figure 15-12.
A—A line does not appear where these two objects intersect. B—The Show property is set to Yes and a line appears at the intersection.

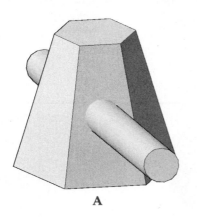
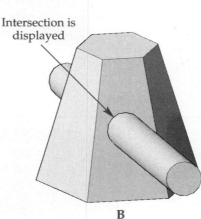

Intersection is
displayed

A   B

Silhouette Edges Subcategory. This subcategory is available for each setting for the Show property in the **Edge Settings** category. The Show property in this subcategory controls the **VSSILHEDGES** system variable. It determines whether or not silhouette edges are displayed around the outside edges of all objects.

The Width property controls the **VSSILHWIDTH** system variable. This property determines the width of silhouette lines. It is measured in pixels and the value can range from 1 to 25.

Edge Modifiers Subcategory. This subcategory is not displayed if the Show property in the **Edge Settings** category is set to None. The Line Extensions property controls the **VSEDGELEX** system variable. This property can be used to create a hand-sketched appearance by extending the ends of edges. See **Figure 15-13A**. In order to make changes to this property, the **Line Extensions edges** button must be on in the **Edge Modifiers** subcategory title bar. The value for the Line Extensions property can range from –100 to 100, which is the number of pixels. The higher the setting, the longer the extension. A negative value sets the extension length, but turns off the property. Picking the button makes the value positive and applies the effect (or makes the value negative and turns off the effect).

The Jitter property controls the **VSEDGEJITTER** system variable. Jitter makes edges of objects look as if they were sketched with a pencil. See **Figure 15-13B**. In order to make changes to this property, the **Jitter edges** button must be on in the **Edge Modifiers** subcategory title bar. There are four settings from which to select: Off, Low, Medium, and High. The number of sketched lines increases at each higher setting.

When the Show property in the **Edge Settings** category is set to Facet Edges, the Crease angle and Halo gap % properties are displayed in the **Edge Modifiers** subcategory. The Crease angle property controls the **VSEDGESMOOTH** system variable. This property determines how facet edges are displayed based on the angle between adjacent facets. See **Figure 15-14**. The value can range from 0 to 180. This is the number of degrees between facets at which a line is displayed. The Halo gap % property is similar to the setting discussed earlier in the 2D Hide—Miscellaneous section; however, this property controls the **VSHALOGAP** system variable.

Figure 15-13.
A—Line extensions have been turned on for this visual style. B—Jitter has been turned on for this visual style.

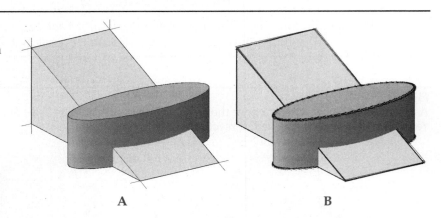

A   B

Figure 15-14.
A—The current visual style is set to Conceptual and the Crease angle property is set to 179. B—The Crease angle property is set to 10. Notice the edges between facets.

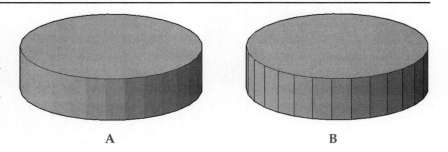

A   B

Creating Your Own Visual Style

As you saw in the previous sections, you can customize the default AutoCAD visual styles. However, you may also want to create a number of different visual styles to quickly change the display of the scene. Custom visual styles are easy to create.

To create a custom visual style, open the **Visual Styles Manager**. Then, pick the **Create New Visual Style** button below the image tiles. You can also right-click in the image tile area and select **Create New Visual Style...** from the shortcut menu. In the **Create New Visual Style** dialog box that appears, type a name for the new style and give it a description. See **Figure 15-15**. Then, pick the **OK** button to create the new visual style.

An image tile is created for the new visual style. The name and description of the visual style appear as help text when the cursor is over the image tile. Select the image tile to display the default properties for the new visual style. Then, change the settings as needed to meet your requirements.

Custom visual styles are only saved in the current drawing. They are not automatically available in other drawings. To use the new visual styles in any drawing, they must be exported to a tool palette. This is discussed in the next section.

PROFESSIONAL TIP

To return one of AutoCAD's visual styles to its default settings, right-click on the image tile in the **Visual Styles Manager** and select **Reset to default** from the shortcut menu.

Exercise 15-4   www.g-wlearning.com/CAD/

Complete the exercise on the companion website.

Steps for Exporting Visual Styles to a Tool Palette

To have custom visual styles available in other drawings, export them to a tool palette. Use the following procedure.

1. Create and customize a visual style as described in the previous section.
2. Open the **Tool Palettes** window.
3. Right-click on the **Tool Palettes** title bar and pick **New Palette** from the shortcut menu.
4. Type the name of the new palette, such as My Visual Styles, in the text box that appears. See **Figure 15-16A**.
5. The new palette is added and active. You are ready to export your custom visual styles into it.
6. Select the image tile of the visual style in the **Visual Styles Manager**. Remember, a yellow border appears around the selected image tile.
7. Pick the **Export the Selected Visual Style to the Tool Palette** button below the image tiles in the **Visual Styles Manager**. You can also right-click on the image tile and select **Export to Active Tool Palette** from the shortcut menu.

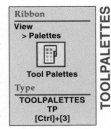

```
Ribbon
View
  > Palettes

  Tool Palettes
Type
TOOLPALETTES
TP
[Ctrl]+[3]
```
TOOLPALETTES

Figure 15-15.
Creating a new visual style.

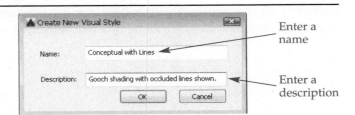

Enter a name

Enter a description

Chapter 15 Visual Style Settings and Basic Rendering **427**

Figure 15-16.
A—Creating a new tool palette on which to place visual style tools. B—A visual style has been copied to the tool palette as a tool.

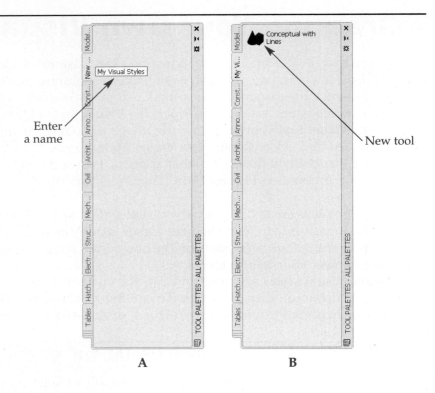

Enter a name

My Visual Styles

Conceptual with Lines

New tool

A　　　　　　　　B

A new tool now appears in the palette with the same image, name, and description as the visual style in the **Visual Styles Manager**. See **Figure 15-16B**. Selecting the tool applies the visual style to the current viewport. You can also right-click on the tool to display a shortcut menu. Using this menu, you can apply the visual style to the current viewport, apply it to all viewports, or add the visual style to the current drawing. The shortcut menu also allows you to rename the tool, access the properties of the visual style, and delete the visual style from the palette.

PROFESSIONAL TIP

A visual style can be added as a tool on a tool palette by dragging its image tile from the **Visual Styles Manager** and dropping it onto the tool palette.

Exercise 15-5

www.g-wlearning.com/CAD/

Complete the exercise on the companion website.

Deleting Visual Styles from the Visual Styles Manager

Custom visual styles can be deleted from the **Visual Styles Manager**. Pick the image tile of the visual style you want to delete. Then, pick the **Delete the Selected Visual Style** button below the image tiles. You can also right-click on the image tile and select **Delete** from the shortcut menu. You are *not* warned about the deletion. The default AutoCAD visual styles cannot be deleted, nor can a visual style that is currently in use.

Plotting Visual Styles

A visual style not only affects the on-screen display, it also affects plots. To plot objects with a specific visual style, use the following guidelines.

Plotting a Visual Style from Model Space

Open the **Plot** dialog box and expand it by picking the **More Options** (>) button. Then, select the desired display from the **Shade plot** drop-down list in the **Shaded viewport options** area. If the desired visual style is current in the viewport, you can also select As displayed from the drop-down list. Finally, plot the drawing.

Plotting a Visual Style from Layout (Paper) Space

When plotting from layout (paper) space, the shade plot properties of the viewports govern how the viewport is plotted. The viewports can be set to plot visual styles in three different ways.

In the first method, select the viewport in layout space and right-click to display the shortcut menu. Pick **Shade plot** to display the cascading menu. Then, select the appropriate visual style.

In the second method, use the **Properties** palette to set the Shade plot property of the viewport. To do this, select the viewport in layout space and open the **Properties** palette. Pick the Shade plot property in the **Misc** category and change the setting to the desired option. The Shade plot property is also available in the **Quick Properties** palette.

In the third method, use the **Visual styles** suboption of the **Shadeplot** option of the **MVIEW** command. When prompted to select objects, pick the border of the viewport. Do not pick the objects in the viewport.

NOTE

The visual style of the viewport may also be selected when you use the **VPORTS** command to create a viewport configuration in the **Viewports** dialog box. Select the viewport in the **Preview** area of the dialog box. Then, pick the visual style desired from the **Visual Style:** drop-down list at the bottom of the dialog box. The **VPORTS** command can be used in model space or layout (paper) space.

Introduction to Rendering

Visual styles provide a way to plot your 3D scene to paper or a file, but control over the appearance is limited to the visual style settings. In Chapter 1, you were briefly introduced to the **RENDER** command. The **RENDER** command allows you to create photorealistic images with complete control over the scene. In this chapter, you will be introduced to AutoCAD's rendering and lighting tools. Materials, lights, and more advanced rendering features are discussed in later chapters.

When you render a scene, you are making a realistic image of your design that can be printed, displayed on a web page, or used in a presentation. To create an attractive rendering, you have to figure out what view you want to display, where the lights should be placed, what types of materials need to be applied to the 3D objects, and the kind of output that is needed. This section shows you how to create a quick rendering of your scene.

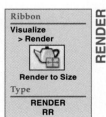

Ribbon
Visualize
> Render

Render to Size
Type
RENDER
RR

Introduction to Lights

Lights provide the illumination to a scene and are essential for rendering. There are three types of lighting in AutoCAD—default lighting, sunlight, and user-created lighting. AutoCAD automatically creates one or two default light sources in every scene. These lights ensure that all surfaces on the model are illuminated and visible. The types of lighting are discussed in more detail in Chapter 17.

A scene can be rendered with the default lights, but the results are usually not adequate to produce a photorealistic image. Normally, user-created lighting or sunlight is added to the scene and adjusted to obtain the desired results. When a light is added to a scene, the default lights must be turned off. The first time you add a light, you receive a warning to this effect (unless the warning has been disabled).

In this section, you will learn how to add sunlight to the scene. Chapter 17 provides detailed information on lighting. Sunlight is produced by an automated distant light. Sunlight can be turned on by picking the **Sun Status** button in the **Sun & Location** panel on the **Visualize** tab of the ribbon. The button background is blue when sunlight is on. See **Figure 15-17**.

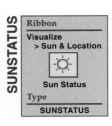

If the **Default Lighting** button is on in the **Lights** panel when the **Sun Status** button is turned on, a warning dialog box is displayed. This dialog box gives you choices to either turn off default lighting or keep it on. You cannot see the effects of sunlight with default lighting turned on, so it is recommended to turn it off.

Another warning dialog box is displayed informing you that sunlight requires a different exposure setting from other lights. Different types of lighting require different exposure settings and AutoCAD is recommending you to adjust the exposure settings before switching to sunlight. Exposure settings are discussed in Chapter 18. For now, choose to keep exposure settings the way they are.

Once sunlight is turned on, you can make adjustments to visualize different effects of the sunlight settings on the objects in your scene. Shadows will display in the viewport only if the current visual style is set to display shadows or if shadows are turned on by selecting one of the options in the shadows flyout in the **Lights** panel in the **Visualize** tab of the ribbon. Full shadows will be shown in the rendering.

The **Date** and **Time** sliders in the **Sun & Location** panel on the **Visualize** tab of the ribbon are active when sunlight is turned on. You can drag the sliders to adjust the date and time. The current date and time are displayed on the right-hand end of the sliders. As you drag the sliders, the lighting and shadows in the scene change to reflect the settings.

Figure 15-17.
The **Lights** and **Sun & Location** panels on the **Visualize** tab of the ribbon. A—Turn off default lighting to see the effects of sunlight in the scene. B—Use the **Date** and **Time** sliders in the **Sun & Location** panel to adjust the date and time.

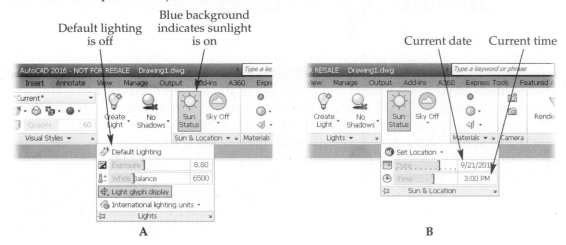

A

B

Turning on **Sky Background and Illumination** adds more realism to the scene by filling in the shadow areas with indirect light. See **Figure 15-18**. Picking the flyout under the **Sky Off** button provides access to this setting. The projection must be set to perspective to access the sky settings.

NOTE

If the scene is washed-out with light after turning on sunlight, you can adjust the exposure by using the **Light Exposure** slider in the **Lights** panel in the **Visualize** tab of the ribbon. A higher exposure setting will darken the scene.

Rendering the Scene

The **Render** panel on the **Visualize** tab of the ribbon is shown in **Figure 15-19**. If you pick the **Render to Size** button, the **Render** window appears and AutoCAD immediately starts rendering the viewport. The rendering is processed over a number of levels, with each level improving the quality of the image. A progress bar at the bottom of the window indicates which level is being processed and its percentage of completion. The **Render** window is explained more in the next section.

The **Render to Size** flyout contains options for rendering the scene at different resolutions. The default size is 800 × 600 pixels. Larger resolutions will result in longer rendering times.

Figure 15-18.
A scene rendered with sunlight. In addition, sky background and illumination has been turned on. Notice that objects in the scene cast and receive shadows.

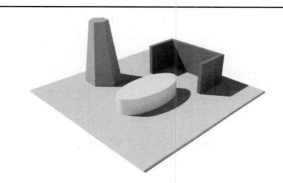

Figure 15-19.
The **Render** panel on the **Visualize** tab of the ribbon.

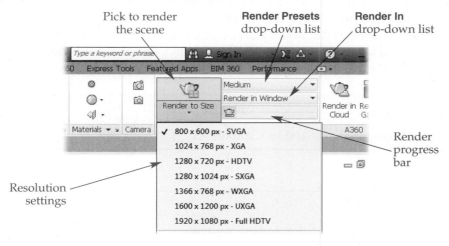

Also in the **Render** panel, you will find the **Render Presets** drop-down list. This is located in the upper-right corner of the panel. This drop-down list gives you a selection of rendering presets based on image quality. The options are:

- Low
- Medium
- High
- Coffee-Break Quality
- Lunch Quality
- Overnight Quality

The Low, Medium, and High presets apply 1, 5, and 10 levels, respectively, of processing during the rendering. The more levels, the longer the rendering takes to complete. The other three presets produce much better quality and are classified by the amount of time the rendering takes to complete. The Coffee-Break Quality preset takes 10 minutes, the Lunch Quality preset takes one hour, and the Overnight Quality preset takes 12 hours, producing the highest-quality rendering.

The options in the **Render In** drop-down list allow you to render the scene to the **Render** window (discussed in the next section) or directly to the viewport. When rendering to the viewport, you can render the entire viewport or a portion of it. If you render with the **Render in Viewport** option, the rendered scene appears in the viewport. The **Render in Region** option prompts you to pick two points of a crop window defining an area of the scene to render. Panning or zooming in the viewport after rendering will cause the rendering to disappear, but this temporary rendering may be enough to identify problem areas and make changes before the final rendering.

PROFESSIONAL TIP

Rendering a complex drawing may take a very long time and you do not want to repeat it because of some small error. It is important to make sure that everything in the scene is perfect before the final rendering. By rendering a cropped region and using lower-quality renderings, you can verify the appearance of any questionable areas without performing a full rendering.

Introduction to the Render Window

The **Render** window has two areas. See **Figure 15-20**. The image pane is where the rendering appears. Immediately below it, the history and statistics pane shows a list of all of the images rendered during the drawing session, with the most recent at the top.

You can zoom into the image in the image pane for detailed inspection. Use the mouse scroll wheel or the zoom buttons at the top of the window. Pick the save button at the top of the window to save the image to an image file. If there is more than one rendered image shown in the history and statistics pane, select the appropriate image before saving. The history and statistics pane shows the path and file name for saved images.

NOTE

Advanced rendering is discussed in Chapter 18.

Exercise 15-6

Complete the exercise on the companion website.

Figure 15-20.
The **Render** window.

Image pane

History and statistics pane

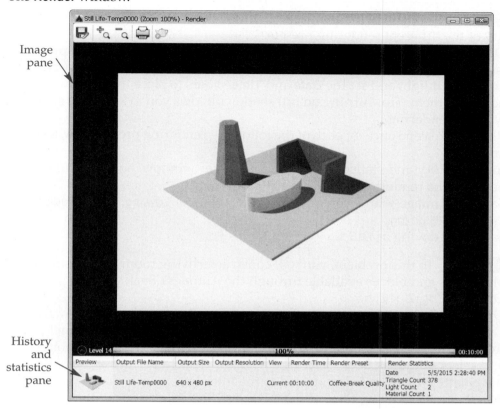

Chapter Review

Answer the following questions using the information in this chapter.

1. Which AutoCAD system variable is used to control the number of isolines displayed on a solid model?

2. What is the **Visual Styles Manager**?

3. Name the 10 default AutoCAD visual styles that can be edited in the **Visual Styles Manager**.

4. Describe the difference between setting the Lighting quality property to Smooth and Faceted.

5. What does the Desaturate setting of the Color property do?

6. What has to be added to a scene before full shadows are displayed?

7. If you want to make your scene look hand sketched, but the Line Extensions and Jitter properties are not available, what other setting(s) do you have to change?

8. How do you set a visual style to display silhouette edges?

9. List the four settings for the Jitter property.

10. What is an *intersection edge*?

11. How do you make your own visual styles available in other drawings?

12. Which visual styles cannot be deleted?

13. How can you turn on sunlight?

14. Explain the function of the **Render in Region** option in the **Render** panel on the ribbon.

15. Name the two main areas of the **Render** window.

16. How can you save a rendered image in the **Render** window?

Drawing Problems

1. **General** This problem demonstrates the differences in rendering time and image quality of the different rendering presets.
   A. Open any 3D drawing from a previous chapter and display an appropriate isometric view. Change the projection to perspective, if it is not already current.
   B. Turn on sunlight and set the **Date** and **Time** sliders to place the shadows where you want them. Tip: Turning on full shadows allows you to locate the shadows without rendering.
   C. Render the scene once for each of the following rendering presets: Low, Medium, and High.
   D. In the history and statistics pane of the **Render** window, note the differences between the rendering times for each rendering.
   E. Save each image with a corresponding name: P15-1-Low.jpg, P15-1-Medium.jpg, and P15-1-High.jpg.
   F. Save the drawing as P15-1.

2. **Architectural** In this problem, you will construct a living room scene using some simple shapes and blocks available through the Autodesk Seek website.
   A. Draw a 12′ × 12′ × 1″ box.
   B. Draw two boxes to represent walls, 12′ × 4″ × 8′. Position them as shown.
   C. Open the **Content Explorer** by selecting **Explore** from the **Content** panel on the **Add-ins** tab on the ribbon. Pick on the drop-down list next to the home icon and select **Autodesk Seek**. Select the Autodesk Seek hyperlink at the bottom of the palette (you must be connected to the Internet).
   D. Search for 3D drawing files for the following objects, download the files, open them, and then drag and drop them into the scene: sofa, table, end table, lamp, book shelf, and chair. The blocks do not have to be exactly the same as shown here and may need to be scaled up or down. Position them as shown.
   E. To create a more realistic scene, change the projection to perspective and move your viewing angle to a more natural height.
   F. Apply each of the default visual styles to the viewport and plot each one. Use the **As displayed** option in the **Plot** dialog box. Note the differences in each one.
   G. Save the drawing as P15-2.

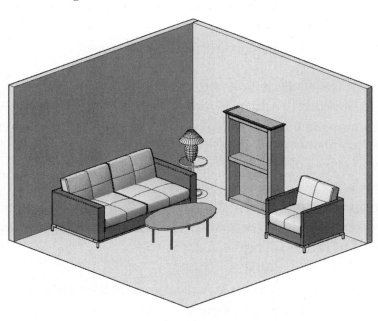

3. **Architectural** In this problem, you will create a new visual style to display the scene as if it is hand sketched.
   A. Open drawing P15-2.
   B. Create a new visual style named Hand Sketched with a description of Displays objects as sketched.
   C. Change the Line Extensions and Jitter settings to make the scene look as shown.
   D. Change any other settings you like.
   E. Plot the scene. Select the new visual style in the **Shade plot** drop-down list.
   F. Export the new visual style to a tool palette so that it can be used in other drawings.
   G. Save the drawing as P15-3.

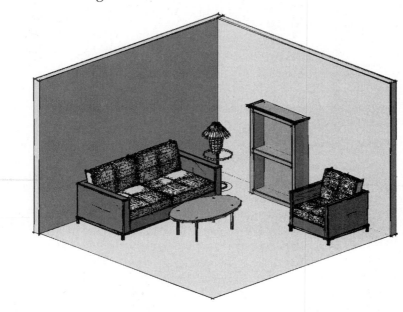

4. **Mechanical** In this problem, you will create a realistic image of the car fender that you created in Chapter 7.
A. Open drawing P7-4.
B. Freeze any layers needed so that only the fender is displayed. Display the fender in the color you want it to be.
C. Draw a planar surface to represent the ground.
D. Set the Realistic visual style current. Then, turn on the highlight intensity and full shadows. Also, set the Show property in the **Edge Settings** category to None.
E. Turn on sunlight and sky background and illumination. Adjust the **Date** and **Time** sliders to make the shadows look as shown.
F. Render the scene and save it as a JPEG image. Name the file P15-4.jpg.
G. Save the drawing as P15-4.

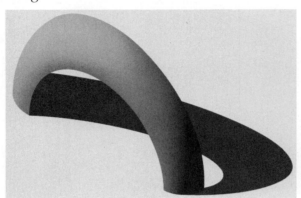

CHAPTER 16

Materials in AutoCAD

Learning Objectives

After completing this chapter, you will be able to:

✓ Attach materials to the objects in a drawing.
✓ Change the properties of existing materials.
✓ Create new materials.

A *material* is simply an image stretched over an object to make it appear as though the object is made out of wood, marble, glass, brick, or various other materials. AutoCAD provides an assortment of materials that can be used in your drawings to create a realistic scene. The materials are grouped into categories to make them easier to find.

Materials are easy to attach. They can be dragged and dropped onto the objects, attached to all selected objects, and even attached based on the object's layer. Once the material is attached, you can adjust how the material is *mapped* to the object. If the current visual style is set to display materials in the viewport, you can immediately see the effects on the object. The properties of a material can also be changed to make it look shinier, softer, smoother, rougher, and so on. When you finally render the scene, you will see the full effect of the materials.

The *materials library* is the location where all materials are stored. When you install AutoCAD, the Autodesk Material Library is installed. The images in the Base Resolution Image Library are low resolution (512 × 512) and are used with AutoCAD materials. The Medium Resolution Image Library contains medium-resolution images (1024 × 1024) that are good for close-up work or large-scale model rendering. The Medium Resolution Image Library is an additional software option that you must install after installing AutoCAD. When you attempt to render, you may be asked if you want to download the Medium Resolution Image Library. Just follow the prompts to accomplish this task.

Materials Browser

Ribbon

Visualize
> Materials

[icon]

Materials Browser

Type

MATBROWSER-
OPEN
RMAT
MAT

MATBROWSEROPEN

In AutoCAD, the **Materials Browser** palette is used to manage the materials library. In this chapter, the **Materials Browser** palette is referred to as the *materials browser*. The materials browser provides access to all materials that are available in the Autodesk libraries and from other sources. The **Materials** panel in the **Visualize** tab of the ribbon contains buttons for accessing the materials browser and the materials editor. See **Figure 16-1**.

There are two main areas in the materials browser, **Figure 16-2**. The **Document Materials** area contains materials that have been selected for use in the current drawing. The **Library** area shows all available libraries from which materials may be selected. These areas are discussed in more detail in the next sections.

Figure 16-1.
The **Materials Browser** button is located on the **Materials** panel in the **Visualize** tab of the ribbon. The materials editor is displayed by picking the dialog box launcher on the panel.

Figure 16-2.
The materials browser contains all of the available materials.

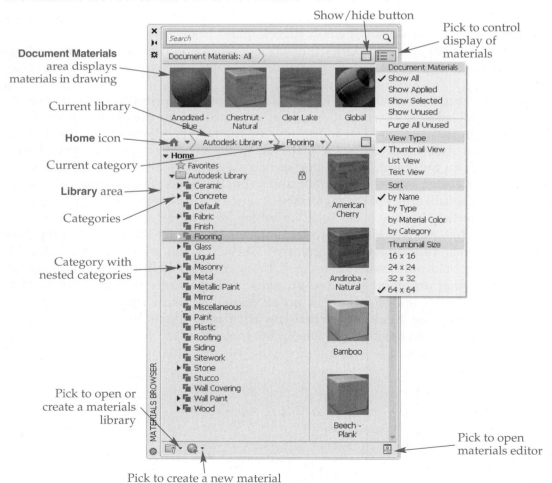

Document Materials Area

Materials are added to the **Document Materials** area from the **Library** area. Hovering over a material swatch in the **Library** area displays two options. See **Figure 16-3**. Picking **Adds material to document** adds the material to the **Document Materials** area. Picking **Adds material to document and displays in editor** adds the material to the **Document Materials** area and opens it in the materials editor (discussed later). After you add a material to the document, a swatch for the material appears in the **Document Materials** area.

Picking the button on the right end of the **Document Materials** title bar opens a drop-down menu. The options in this menu allow you to control which materials are displayed in the **Document Materials** area. You can show all materials, only the materials that are applied to objects in the drawing, selected materials, or unused materials. Materials may be sorted by name, type, color, or category. They can be displayed in thumbnail, list, or text view, and the size of thumbnails can be changed. Refer to **Figure 16-2**. Picking the show/hide button on the **Document Materials** title bar hides the display of the **Library** area. Pick the button again to expand the **Library** area.

Right-clicking on a material in the **Document Materials** area displays a shortcut menu. There are several options available in the shortcut menu:

- **Assign to Selection.** This is active if one or more objects are selected in the scene. The material will be applied to the selected object(s).
- **Select Objects Applied To.** All objects in the drawing that have the material applied to them are selected in the drawing window.
- **Edit.** This displays the materials editor for editing the selected material. The materials editor is discussed later in this chapter.
- **Duplicate.** A copy of the material is added to the drawing. A new material is added with the same name as the original material, but with a sequential number added to the name.
- **Rename.** This option is used to change the name of the material. It is a good idea to rename all of your materials to meaningful names.
- **Delete.** This option removes the material from the drawing. If it is currently being used in the drawing, a warning alerts you to this. Continuing with the deletion removes the material from the objects in the drawing as well as from the **Document Materials** area. It is not, however, removed from the library.
- **Add to.** This option displays a cascading menu with two choices. The material may be added to the Favorites library in the **Library** area. This is excellent for organizing materials that you frequently use and want to make available for quick access in other drawings. You can also add the material to the active tool palette.
- **Purge All Unused.** This option removes from the drawing all materials not currently assigned to an object.

Figure 16-3.
Hovering over a material in the **Library** area displays two buttons that allow you to add the material to the document. Picking the button on the right adds the material to the document and displays it in the materials editor.

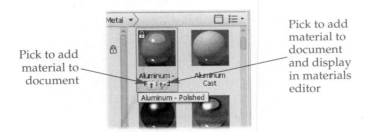

Pick to add material to document

Pick to add material to document and display in materials editor

Library Area

The **Library** area of the materials browser contains the open library files. By default, two libraries exist when AutoCAD is installed: Favorites and Autodesk Library. The name of the current library is displayed on the **Library** area title bar. You can select a different library from the drop-down list on the left end of the title bar. Picking the **Home** icon next to the drop-down list deselects the current library and displays the available libraries in the right-hand column of the **Library** area.

The Autodesk Library is composed of the materials installed when AutoCAD was installed and cannot be edited. Favorites can be used to collect and organize commonly used materials, as well as user defined materials. Favorites is empty by default, but you may add your own categories and materials to it. Custom materials are only available in the drawing in which they were created unless saved to Favorites or another user defined materials library. Custom materials that you create in a drawing but do not save to a materials library are called *embedded materials*.

To expand a library in the tree view, pick the triangle next to the library name. Libraries can have categories within them. If the category contains nested categories, a triangle appears next to the name. Refer to **Figure 16-2**. Pick the triangle to expand the item. Categories in the Autodesk Library are organized by material type. Categories in Favorites and user defined libraries must be created by the user. As you make selections in the tree view, the **Library** area title bar updates to display the name of the current item.

Materials are displayed in the right-hand column and change to reflect whatever category is selected. Picking the button on the right end of the **Library** area title bar opens a drop-down menu similar to the drop-down menu in the **Document Materials** area. The options in the **Library** section allow you to select which library is displayed. The view type, sort, and thumbnail size options are the same as those previously discussed for materials displayed in the **Document Materials** area. Picking the show/hide button on the **Library** area title bar hides the tree view and changes the display to a single window displaying materials.

The libraries that you assemble can be saved externally and opened in other drawings. The button with the folder icon at the lower-left corner of the materials browser provides options for managing materials libraries and categories:

- **Open Existing Library.** Allows you to select a library file (*.adsklib) and display it in the materials browser.
- **Create New Library.** Gives you the option of saving a library you assembled in the materials browser. Library files are saved with the .adsklib file extension.
- **Remove Library.** Used to delete libraries from the materials browser.
- **Create Category.** Allows you to add categories to your library.
- **Delete Category.** Used to remove categories.
- **Rename.** Allows you to rename libraries and categories.

The **Creates a new material in the document** button at the bottom of the materials browser provides a useful way to create your own materials based on the standard Autodesk material types. Selecting **Concrete**, **Metal**, **Plastic**, etc. from the drop-down list creates a material with the default settings for that material type. The materials editor is opened for you to modify the new material to your liking. Picking the **Opens/Closes material editor** button at the lower-right corner of the materials browser also opens the materials editor. Refer to **Figure 16-2**.

NOTE

You cannot rename or remove the default Autodesk libraries. Categories in the Autodesk Library cannot be renamed or removed, and you cannot add categories. Categories can be added to the Favorites library.

Applying and Removing Materials

To attach a material to an object in the drawing, you can drag the material from the materials browser and drop it onto an object. To apply a material only to a face on an object, hold the [Ctrl] key and pick the face. To apply a different material to an object, simply drag the new material to the object and pick.

You can use the **MATERIALATTACH** command to assign materials to the layers in your drawing. Once a material is assigned to a layer, any object on that layer is displayed in the material, as long as the object's material property is set to ByLayer. When objects are created in AutoCAD, the default "material" assigned to them is ByLayer. If your objects are organized on layers, this is the easiest way to attach materials. You can override the layer material by applying a material to individual objects.

Figure 16-4 shows the **Material Attachment Options** dialog box displayed by the **MATERIALATTACH** command. The list on the left side of the dialog box shows the materials loaded into the drawing. The right side of the dialog box shows the layers in the drawing and the material attached to each layer. When no material is attached to a layer, the material is listed as Global. The Global material is a "blank" material in every drawing. To attach a material to a layer, drag the material from the list on the left and drop it onto the layer name on the right. To remove a material from a layer, pick the **X** button next to the material name on the right side of the dialog box.

The **Remove Materials** button on the **Materials** panel in the **Visualize** tab of the ribbon allows you to quickly set an object's material back to ByLayer. When the button is picked, a paintbrush selection cursor is displayed. If you select an object that has a material specifically attached to it, the material is removed. Selecting an object already set to ByLayer has no effect.

Ribbon
Visualize
> Materials

Attach By Layer
Type
MATERIALATTACH

MATERIALATTACH

Ribbon
Visualize
> Materials

Remove Materials

Remove Materials

Figure 16-4.
Attaching materials to layers.

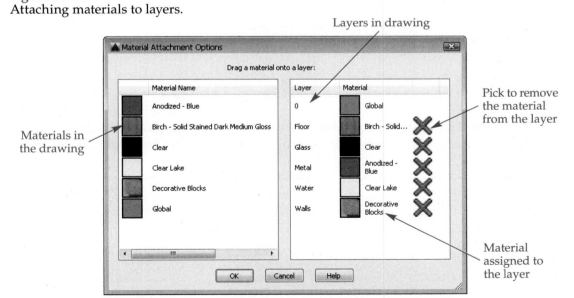

A material can also be removed using the **Properties** palette. To remove a material from an object or subobject, simply change its Material property in the **3D Visualization** category to Global. If a material has not been assigned to the object's layer, the property can also be set to ByLayer. See **Figure 16-5**.

PROFESSIONAL TIP

A material can also be applied to an object by selecting the object in the drawing window, right-clicking on the material swatch in the materials browser, and selecting **Assign to Selection**. A material can be loaded into the drawing without attaching it to an object by picking the material and dragging and dropping it into a blank area of the drawing. This makes the material available in the drawing.

Exercise 16-2

www.g-wlearning.com/CAD/

Complete the exercise on the companion website.

Material Display Options

As you learned in the previous chapter, visual styles control how materials are displayed in the viewport. The Material display property of a visual style can be set to display materials and textures, materials only, or neither materials nor textures. The **Materials** panel on the **Visualize** tab of the ribbon has three buttons in a flyout that correspond to, but override, this property setting:

- **Materials/Textures Off.** Objects are displayed in their assigned colors.
- **Materials On/Textures Off.** Objects are displayed in the basic color of the material, but no other material details are displayed.
- **Materials/Textures On.** Objects are displayed with the effects of all material properties visible.

VSMATERIALMODE

Ribbon
Visualize
> Materials

Materials/Textures
Off

Materials On/
Textures Off

Materials/Textures
On

Type
VSMATERIALMODE

Exercise 16-3

www.g-wlearning.com/CAD/

Complete the exercise on the companion website.

Materials Editor

The *materials editor* allows you to create new materials and edit existing materials to your liking. The next sections discuss creating and editing materials.

Figure 16-5.
Removing a material from an object.
A—The material is assigned. B—The material is removed.

Material assigned

Material removed

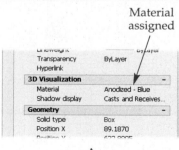

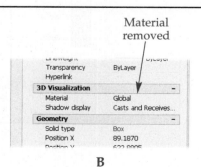

A   B

The **Materials Editor** palette is displayed by picking the dialog box launcher button at the lower-right corner of the **Materials** panel in the **Visualize** tab of the ribbon. See **Figure 16-6**. You can also type the **MATEDITOROPEN** command, double-click on a material swatch in the materials browser, or pick the **Opens/Closes material editor** button in the lower-right corner of the browser. In addition, you can hover over a material swatch in the **Document Materials** area to display a small pencil icon. Picking this icon opens the materials editor.

A preview of the material appears at the top of the materials editor. Picking the drop-down menu button at the lower-right corner of the material swatch displays three menus. The options in these menus allow you to change the geometry, type of lighting, and quality used for the preview.

The **Scene** menu provides 13 shapes from which to select: **Sphere, Cube, Cylinder, Canvas, Plane, Object, Vase, Draped Fabric, Glass Curtain Wall, Walls, Pool of Liquid, Utility**, and **Torus**. The **Environment** menu provides four basic lighting options that allow you to preview a material in a setting similar to where you may be using it. The **Grid Light** option simulates overhead neutral-color lighting that you might find in an office environment. The **Plaza** option simulates an outdoor lighting environment with a bluish tint and strong shadows. The **Snowfield** option produces soft, even light that appears to be coming from everywhere. The **Warm Light** option produces a yellowish cast similar to incandescent indoor lighting.

The options in the **Render Settings** menu set the quality of the preview image. The **Quick** option provides a good quality preview, but the environment settings are poorly shown. The **Draft** option displays the environment nicely and will take a few seconds to update (depending on material complexity and computer speed). The **Production** option provides the highest quality, but the resulting display may take a few minutes to update. The updates occur whenever you change a material property.

Figure 16-6.
The materials editor. A—A preview of the material appears at the top of the palette. Picking the drop-down menu button at the lower-left corner of the palette displays the drop-down menu shown in B. B—This drop-down menu provides options for creating a new material.

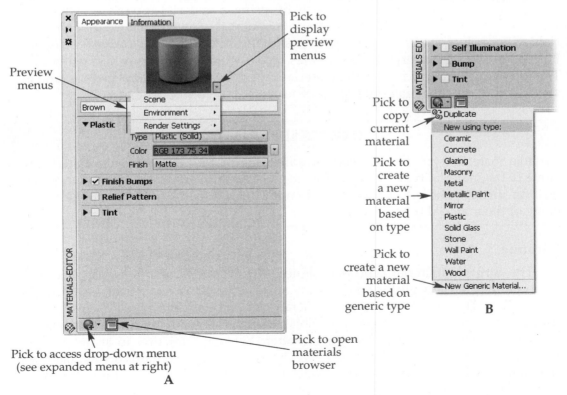

The name of the material can be changed by editing it in the text box located below the material preview. You can also rename the material using the **Name** text box in the **Information** tab.

Materials created or edited in the materials editor are then added to the **Document Materials** area of the materials browser. The materials browser can be opened from the materials editor by picking the **Opens/Closes material browser** button at the bottom of the **Appearance** tab. Refer to **Figure 16-6A**.

Creating and Modifying Materials

There are three different approaches to creating new materials. You can duplicate an existing material. You can start with an existing material type as a template. Finally, you can start with a generic material, which is like starting from scratch. To access these options, use the drop-down menu at the lower-left corner of the materials editor.

New Material from an Existing Material

By far the easiest way to create your own material is to start with an existing material, create a duplicate, and make any needed modifications. Look through the materials library in the materials browser to find a material that is close to what you want and add the material to the **Document Materials** area. Hover over the material swatch in the **Document Materials** area and select the pencil icon to launch the materials editor.

The material selected in the **Document Materials** area of the materials browser is displayed in the materials editor. Pick **Duplicate** from the drop-down menu at the lower-left corner of the materials editor. Refer to **Figure 16-6B**. This creates a duplicate material with the same name and a sequential number. Change the name and the properties to your liking (discussed later) and close the materials editor. The new material is in the **Document Materials** area of the materials browser ready to use.

New Material Using an Existing Material Type

Another way to create a material is by using a material type as a template for your new material. Using the drop-down menu at the lower-left corner of the materials editor, select one of the material types under the **New using type:** option. Refer to **Figure 16-6B**. The material type provides certain default settings as a starting point. The following sections discuss the material types, their possible applications, and any special properties the material type may have.

PROFESSIONAL TIP

Often, a material type can be used for a material that is completely unrelated to the name of the material type. For example, the concrete material type may actually serve well for dirt or sand. Be creative and do not limit yourself to what is implied by the name of the material type.

Ceramic

The *ceramic* material type is designed for ceramic floors. However, this material type may serve well for other glossy surfaces, such as countertops, bathtubs, or dinnerware. The Type property can be Ceramic or Porcelain. The Finish property can be High Gloss/Glazed, Satin, or Matte. The **Finish Bumps** category contains a Type property that can be Wavy or Custom. The **Relief Pattern** category contains an Image property. The image determines the relief pattern. The **Tint** category contains a Tint Color property that allows you to assign a tint color defined by its hue and saturation value mixed with white. This property is also available in the other material types.

Concrete

The *concrete* material type works well for concrete floors, walls, and sidewalks. This material type can be used for anything constructed of concrete, such as an in-ground swimming pool. The Sealant property can be None, Epoxy, or Acrylic. The **Finish Bumps** category contains a Type property that can be Broom Straight, Broom Curved, Smooth, Polished, or Stamped/Custom. The **Weathering** category contains a Type property that can be Automatic or Custom-Image (based on a selected image).

Glazing

The glass in windows is called glazing. Therefore, the *glazing* material type is designed mostly for use on windows or thin glass objects. The solid glass material type is designed for thicker glass. The Color property can be Clear, Green, Gray, Blue, Blue-green, Bronze, or Custom. The Reflectance value determines how reflective the material is. The Sheets of Glass property simulates the effect of multiple panes of glass.

Masonry

Masonry includes brick walls, cobblestone, tile flooring, and so on. The *masonry* material type is designed for use on these objects. The Type property can be CMU (concrete masonry unit) or Masonry. The Finish property can be Glossy, Matte, or Unfinished. The **Relief Pattern** category contains an Image property. The image determines the relief pattern.

Metal

The *metal* material type works well for mechanical parts and other objects made from different types of metals. This material type primarily is for raw, or unfinished, metal. The Type property can be Aluminum, Anodized Aluminum, Chrome, Copper, Brass, Bronze, Stainless Steel, or Zinc. The Finish property can be Polished, Semi-polished, Satin, or Brushed. The **Relief Pattern** category contains a Type property that can be Knurl, Diamond Plate, Checker Plate, or Custom-Image (based on a selected image). The **Cutouts** category contains a Type property that can be Staggered Circles, Straight Circles, Squares, Grecian, Cloverleaf, Hexagon, or Custom. Depending on which type of metal is selected, additional properties may be available for editing.

Metallic Paint

The *metallic paint* material type is for objects made of metal, but with a finish applied. This includes objects like car parts, lawn furniture, and kitchen appliances. The **Flecks** category contains Color and Size properties. The **Pearl** category includes a Type property that can be Chromatic or Second Color. The **Top Coat** category includes a Type property that can be Car Paint, Chrome, Matte, or Custom. The **Top Coat** category also includes a Finish property that can be Smooth or Orange Peel.

Mirror

The *mirror* material type is designed for very reflective objects, such as mirrors. It can also be used for water, glass, or any object that should have a high reflectivity. The Color property determines the color of the mirror. As discussed earlier, the **Tint** category is also available.

Plastic

The *plastic* material type is designed for use on plastic objects. The plastic can be opaque, translucent, glossy, or textured. The Type property can be Plastic (Solid), Plastic (Transparent), or Vinyl. The Finish property can be Polished, Glossy, or Matte. The **Finish Bumps** category contains an Image property, as does the **Relief Pattern** category. The two images do not have to be the same.

Solid Glass

The *solid glass* material type is intended for thick glass objects or a volume of liquid, such as a glass of water. The Color property can be Clear, Green, Gray, Blue, Blue-green, Bronze, or Custom. The Reflectance property determines the degree of reflectivity of the material. The Refraction property can be Air, Water, Alcohol, Quartz, Glass, Diamond, or Custom. The Roughness property determines the polish on the material. The **Relief Pattern** category contains a Type property that can be Rippled, Wavy, or Custom.

Stone

The *stone* material type works well for stone walls, stone walkways, and marble countertops. The **Stone** category contains an Image property, which is the image applied to the material. This can be a selected image or a specified texture. The **Stone** category also contains a Finish property that can be Polished, Glossy, Matte, or Unfinished. The **Finish Bumps** category contains a Type property that can be Polished Granite, Stone Wall, Glossy Marble, or Custom. The **Relief Pattern** category contains an Image property. This can be a selected image or a specified texture.

Wall Paint

The *wall paint* material type works well for interior or exterior painted walls and other objects. The Finish property can be Flat/Matte, Eggshell, Platinum, Pearl, Semi-gloss, or Gloss. The Application property can be Roller, Brush, or Spray.

Water

The *water* material type is designed for any liquid. Pools, reflecting ponds, rivers, lakes, and oceans are examples of where this material type may be used. The Type property can be Swimming Pool, Generic Reflecting Pool, Generic Stream/River, Generic Pond/Lake, or Generic Sea/Ocean. The Color property (when available) can be Tropical, Algae/Green, Murky/Brown, Generic Reflecting Pool, Generic Stream/River, Generic Pond/Lake, Generic Sea/Ocean, or Custom. The Wave Height property sets the amplitude of ripples in the liquid.

Wood

The *wood* material type works well for various finishes used for flooring, furniture, wood trim, and so on. The **Wood** category contains an Image property, which is the image applied to the material. This can be a selected image or a specified texture. When the **Stain** check box is checked, the color of stain can be set. The **Wood** category also contains a Finish property, which can be Unfinished, Glossy Varnish, Semi-gloss Varnish, or Satin Varnish. The Used For property can be Flooring or Furniture. The **Relief Pattern** category contains a Type property that can be Based on Wood Grain or Custom (which is based on a selected image).

Generic

The generic material type is the "blank canvas" material. All properties are available to create any material needed. This material type is used to create a material from scratch, as discussed in the next section.

PROFESSIONAL TIP

AutoCAD provides fantastic-looking materials to dress up the scene and make it look real. However, after you get comfortable with creating materials, start a library of your own materials. If your project is presented to a customer along with projects from competitors, and your competitors are using standard AutoCAD materials, your project will stand out from the crowd.

Exercise 16-4

Complete the exercise on the companion website.

New Material from Scratch

Creating a material from scratch gives you complete control over the material. Although there are many options to consider, creating a material from scratch may be the only way to get the exact material you need to complete your scene. To start creating a material from scratch, select the generic material type by selecting **New Generic Material...** from the drop-down menu at the lower-left corner of the materials editor. Refer to **Figure 16-6B**. The **New Generic Material...** option is also available from the materials browser. The properties and settings are discussed in the next sections.

NOTE

New materials are placed in the **Document Materials** area and are only available in the current drawing unless added to the Favorites library or a user defined library.

Generic Category

The *diffuse color* is the color of the object in lighted areas, or the perceived color of the material. See **Figure 16-7**. It is the predominant color you see when you look at the object. The Color property sets the diffuse color of the material. In AutoCAD, the ambient and specular colors are determined by the diffuse color.

The *ambient color* is the color of the object where light does not directly provide illumination. It can be thought of as the color of an object in shadows. In nature, shadows cast by an object typically contain some of the ambient color.

A full-color image is provided on the companion website.   www.g-wlearning.com/CAD/

Figure 16-7.
The three colors of a material are illustrated here. In AutoCAD, the Color property sets the diffuse color. The ambient and specular colors are based on the diffuse color.

Diffuse color
(Color property)

Specular color

Ambient color

The *specular color* is the color of the highlight (the shiny spot). It is typically white or a light color. The amount of specular color shown is determined by the glossiness and reflectivity of the material and the intensity of lighting in the scene.

You have two options for controlling the Color property of the material. By default, the material color is dark gray (RGB 80 80 80). Pick the current color setting in the edit box to display the **Select Color** dialog box and choose the color you want. See **Figure 16-8**. Picking the button to the right of the color edit box displays a drop-down list. If you select **Color by Object** in the drop-down list, the base color of the object is used as the diffuse color.

The Image property allows you to apply an image or texture to control the appearance of the material. Picking in the image preview area displays a standard open dialog box. When you select an image, it is applied to the material and is displayed in the material preview area, **Figure 16-9**. Also, the texture editor is displayed with the

Figure 16-8.
Setting a color for a material.

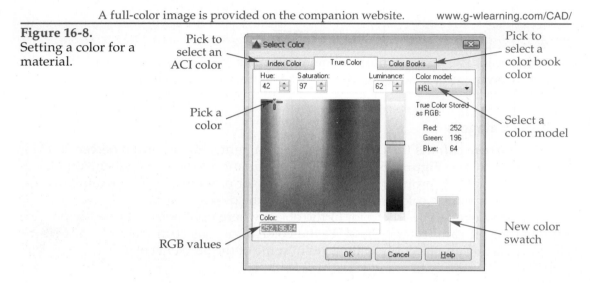

Figure 16-9.
The property settings in the **Generic** category of a material created from scratch.

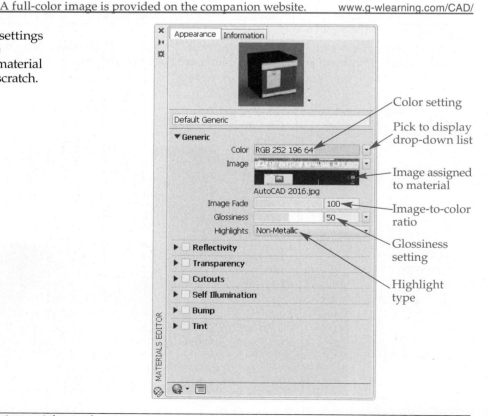

image loaded and ready for editing, if necessary. The button to the right of the image preview area is used to apply a texture instead of an image. Textures and texture editing are discussed later in this chapter.

When an image, such as a bitmap image or digital photograph, is applied, the material color is replaced with the image. **Figure 16-10** shows an object with a computer screen image applied to the material attached to it. The Image Fade property controls the ratio between the image and the object color. When set to 100, the image completely replaces the object color.

The Glossiness property controls how shiny the material appears, **Figure 16-11**. It is a measure of the surface roughness of a material. A setting of 100 specifies a very shiny material, such as a smooth surface. A setting of 0 specifies a matte (dull) material. The button to the right of the setting allows you to add a texture or image to the Glossiness property. The pattern is applied to the glossiness effect. **Reflectivity** must be turned on for the glossiness effect to be seen. Refer to the next section.

The Highlights property determines how the shiny areas are created. Choices for this property are Metallic or Non-metallic. Highlights are brighter when Metallic is selected.

A full-color image is provided on the companion website.   www.g-wlearning.com/CAD/

Figure 16-10.
A material with an image applied instead of a color to simulate a tablet screen. A—The screen object's material does not have an image applied. B—A bitmap image has been applied to the image component of the screen object's material. C—The finished tablet.

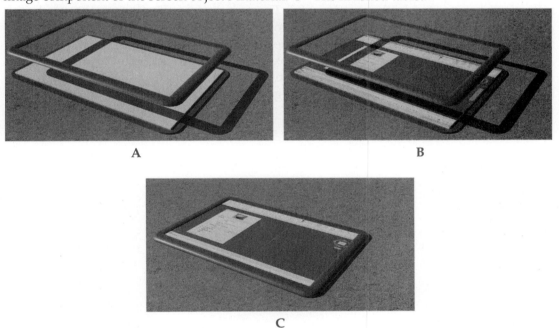

A full-color image is provided on the companion website.   www.g-wlearning.com/CAD/

Figure 16-11.
Three different glossiness settings are illustrated here.

| Glossiness = 25 | Glossiness = 50 | Glossiness = 75 |
| A | B | C |

Highlights can be seen everywhere. Look around you right now at edges and inclined surfaces. The diffuse color of the surface typically has little to do with the color of the highlight. The color of the light source usually determines the predominant highlight color. Most highlights are white or near white because most light sources are white or nearly white. However, highlights in the interior of a home may have a yellow cast to them because incandescent light bulbs generally cast yellow light. Compact fluorescent lights cast a slightly different color, although many are designed to cast the same color as an incandescent bulb. Outside with a clear sky and bright sun, highlights may have a slight blue cast. Keep these points in mind when creating your own materials. These small details are what make a scene realistic.

Reflectivity Category

Reflectivity is a measure of how much light is bounced off the surface. There are two basic property settings for reflectivity, Direct and Oblique. See **Figure 16-12**. The Direct property controls how much light is reflected back for surfaces that are more or less facing the camera. The Oblique property controls how much light is reflected back when the surface is at an angle to the camera.

Each property has a slider/text box that controls the amount of reflectivity. No reflections are created with a setting of 0. The maximum reflections are created with a setting of 100. Object color, lighting, surroundings, and other factors also determine just how reflective a material appears in the scene.

The buttons to the right of the text boxes allow you to add a texture or image to control the reflection. The white areas of the texture or image have a reflection. The black areas do not have a reflection. The degree of reflectivity varies for gray areas and the grayscale values of colors.

Figure 16-12.
Setting the reflectivity for a material.

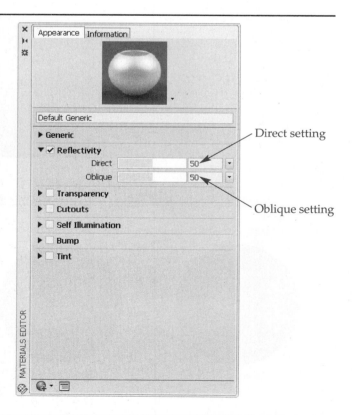

Transparency Category

Transparency is a measure of how much light the material allows to pass through it. Glass, water, crystal, and some plastics, along with other materials, are nearly completely transparent. **Figure 16-13** shows an example of using a transparent material to show the internal workings of a mechanical assembly. The **Transparency** category contains a number of properties that combine to create any transparent or semitransparent material, **Figure 16-14**.

The amount of light that passes through the material is controlled by the Amount property. When this property is set to 0, the material is opaque. A setting of 100 creates a completely transparent material.

A full-color image is provided on the companion website.   www.g-wlearning.com/CAD/

Figure 16-13.
The material used for the housing on this mechanism has a transparency setting of about 50.

Figure 16-14.
Setting transparency for a material.

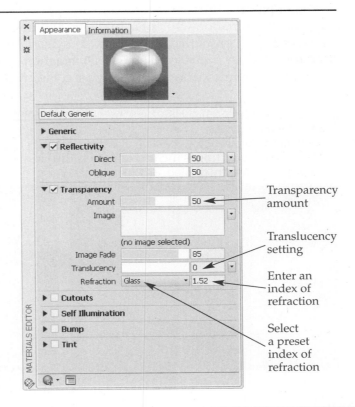

Transparency amount

Translucency setting

Enter an index of refraction

Select a preset index of refraction

The Image property is used to add a transparency map to the material. White areas in the image or texture are transparent. Black areas are opaque. All other colors produce varying degrees of transparency based on their grayscale values. The Amount property is applied to the transparency map. Maps are discussed later in this chapter.

The Image Fade property determines how much of an impact the transparency map has on the transparency of the material. A setting of 100 means the transparency is completely see-through based on the transparency map. As the setting is decreased, a higher percentage of transparency is determined by the Amount property.

Translucency is a quality of transparent and semitransparent materials that causes light to be diffused (scattered) as it passes through the material. See **Figure 16-15**. This makes any object with the material applied to it appear as if it is being illuminated from within, or glowing. The thicker the material, the more pronounced the effect. When the Translucency property setting is 0, light appears to travel through the material, lighting the opposite side. A setting of 100 creates a material similar in appearance to frosted glass.

Refraction, sometimes specified as the *index of refraction (IOR)*, is a measure of how much light is bent (refracted) as it passes through transparent or semitransparent materials. Refraction is what causes objects to appear distorted when viewed through a bottle or glass of water. See **Figure 16-16**. The Refraction property sets the IOR. The higher the value, the more that light is bent as it passes through the material. The IOR of water is 1.3333. You can enter a value in the text box or select a preset IOR by picking the name that is displayed to the left of the text box.

Cutouts Category

The Cutouts property allows you to select an image or texture to use for a pattern of cutouts (holes), **Figure 16-17**. Black areas in the image will appear to be see-through, as if there is no object in those areas. White areas in the image have the normal material colors. This is similar to using a transparency map, but without the other transparency settings. **Figure 16-18** shows an example of a cutout map applied to a material.

Self Illumination Category

Self illumination is an effect of a material producing illumination. See **Figure 16-19**. For example, the surface of a neon tube glows. However, in AutoCAD, a material with self illumination will not actually add illumination to a scene. This effect can be simulated with properly placed light sources. *Luminance* is defined as the value of light reflected off a surface. The **Self Illumination** category contains several properties related to self illumination and luminance, **Figure 16-20**.

A full-color image is provided on the companion website. www.g-wlearning.com/CAD/

Figure 16-15.
The effect of translucency. A—The glass material has a translucency setting of zero. B—When the translucency setting is increased, light is diffused within the material. Notice how the glass appears slightly frosted.

A B

Figure 16-16.
The effect of refraction. A—The transparent material on the sphere has a refraction setting of zero. B—When the refraction setting is increased, the cylinder behind the sphere is distorted as light is refracted by the material.

A   B

Figure 16-17.
Adding a cutout map to a material.

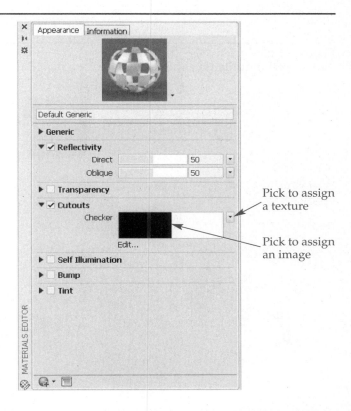

Figure 16-18.
The effect of applying an image to the Cutouts property. A—This black and white image will be used as the cutout map. B—The material on the plane is completely opaque. C—When the cutout map is applied to the material, the dark areas of the map produce transparent areas on the object.

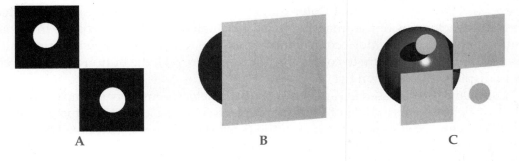

A   B   C

Figure 16-19.
The effect of self illumination/luminance. A—The globe of this light bulb does not have any self illumination. B—Self illumination is applied to the globe material.

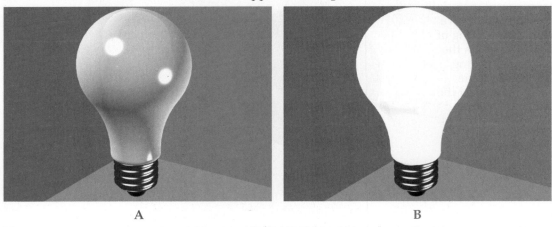

A   B

Figure 16-20.
Setting self illumination and luminance for a material.

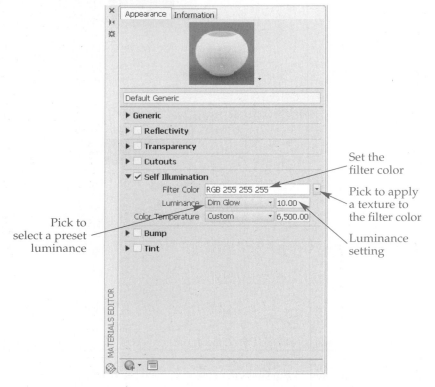

The Filter Color property controls the color of the self illumination effect. Pick in the edit box to open the **Select Color** dialog box. An image may be selected instead of a color. Black areas of the image are not illuminated. White areas of the image are illuminated. Grayscale values control how much the rest of the image illuminates the material. To apply a texture to control the illumination, pick the button next to the edit box to display a drop-down list.

Luminance is expressed in candelas per square meter (cd/m^2). For example, 1 cd/m^2 is the equivalent of one candela of light radiating from a surface area that is one square meter. You can enter a specific value in the Luminance property text box. Or, you can pick the name displayed next to the text box to display a drop-down list with choices for typical materials. Some of these choices include Dim Glow, LED Panel, and Cell Phone Screen.

The Color Temperature property determines the warmth or coolness value of the color. The value is expressed as degrees Kelvin. Candles and incandescent bulbs are warm. Fluorescent lights and TV screens are cool. The drop-down list gives you choices for typical objects with different color temperatures, but you can enter a value directly in the text box.

Bump Category

The **Bump** category contains settings for making some areas of the material appear raised and other areas depressed, **Figure 16-21**. The image or texture used for this effect is called a *bump map*. The black, white, and grayscale values of the map are used to determine raised and depressed areas. Dark areas of the map appear raised and light areas appear depressed.

For example, to show the texture of a brick wall, you could physically model the grooves into the wall. This would take a lot of time to model and would immensely increase the rendering time because of the increased complexity of the geometry. Using the properties in the **Bump** category is an easier and more efficient way to accomplish the same task. **Figure 16-22** shows a bump map used to represent an embossed stamp on a metal case.

To apply an image as a bump map, pick the Image property swatch to display a standard open dialog box. To apply a texture as a bump map, pick the button next to the edit box to display a drop-down list.

The Amount property determines the relative height of the bump pattern. A setting of 0 results in a flat material, or no bumps. A setting of 1000 creates the maximum difference between low and high areas of the pattern.

Figure 16-21.
Making bump settings for a material.

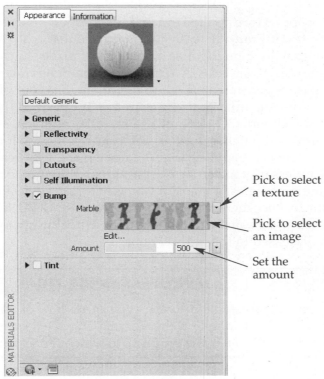

Pick to select a texture

Pick to select an image

Set the amount

Figure 16-22.
The effect of a bump map. A—This image will be used as the bump map. B—When applied to the material, the bump map simulates an embossed stamp on the metal case.

A

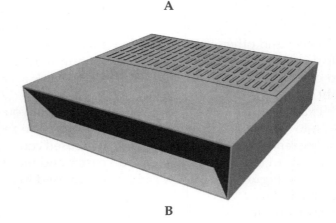

B

Maps

As discussed in previous sections, images and textures may be applied to materials to enhance the color, glossiness, reflectivity, transparency, translucency, cutouts, self illumination, and bumpiness. The applied image or texture is called a *map*. A material that has a map applied to at least one of its properties is called a *mapped material*.

An image or texture applied to a material property may be 2D or 3D. A *2D texture*, also called a *texture map*, is really just an image that is stretched over the surfaces of the object. It may be thought of as a fixed set of pixels, kind of like a mosaic pattern, applied to the object's surfaces. A *3D texture*, sometimes referred to as a *procedural map*, is mathematically generated based on the colors that you select. This texture extends through the object, similar to textures in the "real world." For example, if you attach AutoCAD's wood map (3D texture) to a material property and assign the material to an object with a cutout, the grain in the cutout will match the grain on the exterior. An image (2D texture) applied to this same object will "reapply" itself to the cutout surface, not necessarily matching the pattern on the adjacent surfaces.

There are nine types of maps in AutoCAD that can be applied to material properties. The image, checker, gradient, and tiles maps are 2D texture maps. The marble, noise, speckle, waves, and wood maps are 3D texture maps. Each map has unique settings, as discussed in the next sections.

PROFESSIONAL TIP

The term *map* may be applied to the type of texture or the property to which the texture is applied. For example, a checker map may be used as a bump map. *Checker* is the type of map and *bump* is the property to which the texture is applied.

Figure 16-27.
There are 11 different gradient types available for use in a gradient map. The four shown here are applied to the same object with the same lighting and mapping coordinates. A—Linear asymmetrical. B—Box. C—Diagonal. D—Light normal.

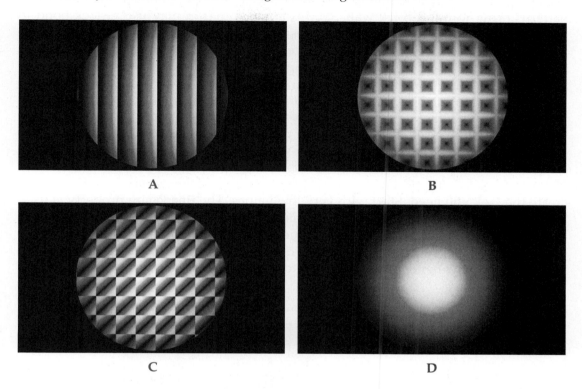

The Interpolation property controls the transition of colors from one node to the next. Transitions are applied to the nodes from left to right, regardless of the node number. The options are:

- **Ease in.** Shifts the transition closer to the node on the right.
- **Ease in out.** Shifts the transition toward the node, but it remains more or less centered on the node.
- **Ease out.** Shifts the transition closer to the node on the left.
- **Linear.** This is the default. The transition is constant from one node to the next.
- **Solid.** No transition between nodes. There is an abrupt change at each node.

The Position property is simply the position of the selected node in the ramp. The node on the left is at the 0 position and the node on the right is at the 1.000 position. Nodes in between will be at varying values between 0 and 1.000. You can also change this value by dragging the nodes left or right.

Picking the **Invert Gradient** button reverses all color values, inverting the gradient pattern. In effect, the ramp is flip-flopped from left to right.

Noise may be added to the gradient map to create an uneven appearance. The properties in the **Noise** and **Noise Threshold** categories are similar to those for the noise map. The noise map is described later in this chapter.

The properties in the **Transforms, Position, Scale,** and **Repeat** categories are the same as described for an image map. Refer to the Image section for details on these properties.

Tiles

A *tiles map* is a pattern of rectangular, colored blocks surrounded by colored grout lines. This may be the most versatile map in the whole collection. Tiles are used to simulate tile floors, ceiling grids, hardwood floors, and many different types of brick walls. When you specify Tiles as the map for a property, the texture editor displays the properties for the map, **Figure 16-28**.

You first need to define the pattern for the map. The **Pattern** category contains properties for defining the pattern. For the Type property, select one of the seven predefined tile patterns or Custom to create your own. The names of the predefined patterns bring to mind brick walls. For example, a mason may use a stack bond to build a brick wall. However, remember these are only *patterns*. You can also use a brick pattern to create tile floors and acoustic ceiling panels. Four of the tile patterns are shown in **Figure 16-29**. The Tile Count property sets how many tiles are in each row and column before the pattern repeats.

As the name implies, the properties in the **Tile Appearance** category determine what the tiles look like. You can choose any color you wish or apply a texture or image to the tiles. Ceramic tile floors look more realistic if each tile is slightly different in color. The Color Variance property can be used to alter the color of random tiles to create a more realistic appearance. The Fade Variance property is used to fade the color of random tiles. You will have to experiment with the color variance and fading to create the look you need. Start with very low values. The Randomize property is used to alter the random color variation in the tiles. This variation is automatically applied, but entering a different random seed changes the pattern. If your scene has more than one object with this material applied to it, duplicate the material and change the seed number of the new material, then apply it to the other object.

Figure 16-28.
Adjusting the properties of a tiles map.

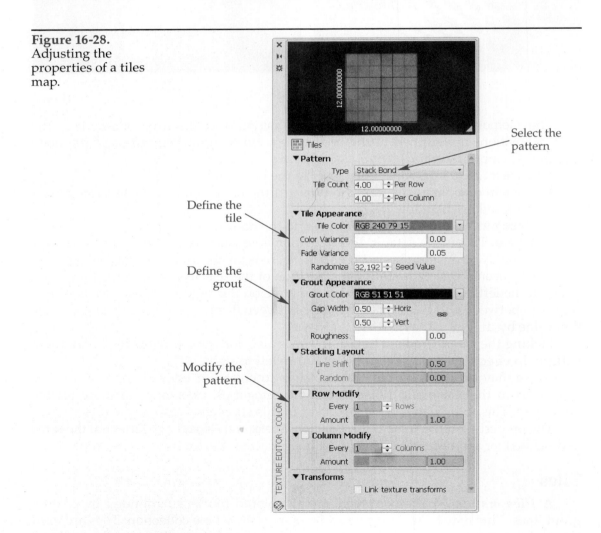

Figure 16-29.
A tiles map can have a custom pattern or predefined pattern. Four of the predefined patterns are shown here.
A—Running bond.
B—English bond.
C—Stack bond.
D—Fine running bond.

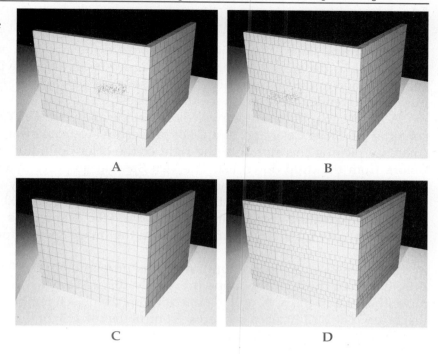

A B C D

The properties in the **Grout Appearance** category control what the grout looks like. The grout is the line between the tiles. The Grout Color property is set to dark gray by default, but any color, image, or texture may be used. The Gap Width property determines how wide the grout lines are in relation to the tiles. There are horizontal and vertical settings. In some cases, such as for a hardwood floor material, you will have to set the scale differently on the horizontal and vertical axes to make the gap thicker in one direction.

The properties in the **Stacking Layout** category are only available for a custom pattern. The Line shift property changes the location of the vertical grout lines in every other row to create an alternate pattern of tiles. The default value is 0.50 and the range is from 0.00 to 100.00. The Random property randomly moves the same lines. This works nicely for hardwood floor materials. The default value is 0.00 and the range is 0.00 to 100.00.

The properties in the **Row Modify** and **Column Modify** categories are available with all tile pattern types, but may be disabled by default. To enable the settings, pick the check box by the category name. The settings in these areas allow you to change the number of grout lines in the horizontal and vertical directions to create your own pattern. The two Every properties determine which rows and columns will be changed. When set to 0, no changes take place in the row or column. When set to 1, every row or column will be changed. When set to 2, every other row or column will be changed, and so on. The value must be a whole number. The Amount property controls the size of the tiles in the row or column. A setting of 1 means that the tiles remain their original size. A setting of 0.50 makes the tiles one-half of their original size, a setting of 2 makes the tiles twice their original size, and so on. A setting of 0.00 completely turns off the row or column and the underlying material color shows through.

The properties in the **Transforms**, **Position**, **Scale**, and **Repeat** categories are the same as described for an image map. Refer to the Image section for details on these properties.

Marble

A *marble map* is a 3D map based on the colors and values you set. It is used to simulate natural stone. When you specify Marble as the map for a property, the texture editor displays the properties for the map, **Figure 16-30.** The viewport may not reflect changes made in the texture editor, even if set to display materials and textures. You may have to render the scene to see the changes.

A marble map is based on two colors—stone and vein. The **Appearance** category contains the Stone Color and Vein Color properties. You can swap the vein and stone colors by picking the button to the right of the color definition and selecting **Swap Colors** from the drop-down list. The Vein Spacing property determines the relative distance between each vein in the marble. The Vein Width property determines the relative width of each vein. Each of these settings can range from 0.00 to 100.00.

The **Link texture transforms** check box in the **Transforms** category works as described earlier for an image map. Refer to the Image section for details on texture transforms.

Since this is a 3D (procedural) map, the properties in the **Position** category are different from those of the maps previously discussed. The three Offset properties move the map in the X, Y, and Z directions on the object. Simply enter a value in the text boxes. The XYZ Rotation properties control the rotation of the map around the X, Y, or Z axis. You can move the sliders or enter an angle in the text boxes.

NOTE

The mathematical calculations that create a 3D (procedural) map are based on the world coordinate system. If you move or rotate the object, a different result is produced.

Figure 16-30.
Adjusting the properties of a marble map.

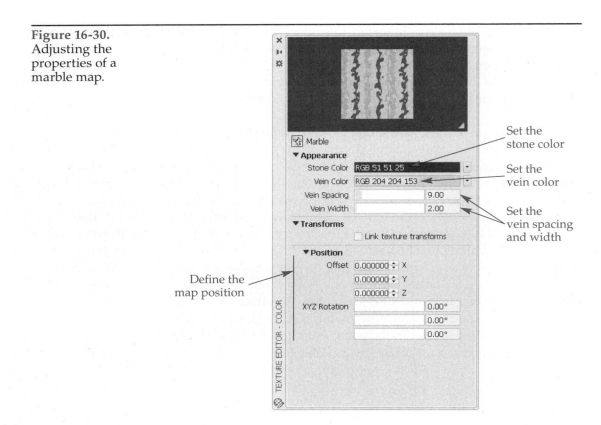

Set the stone color

Set the vein color

Set the vein spacing and width

Define the map position

Noise

A *noise map* is a 3D map based on a random pattern of two colors used to create an uneven appearance on the material. It is most often used to simulate materials such as concrete, soil, asphalt, grass, and so on. When you specify Noise as the map for a property, the texture editor displays the properties for the map, **Figure 16-31**.

The properties in the **Appearance** area control how the noise looks. First, you need to select the type of noise. The options for the Noise Type property are:

- **Regular.** This is "plain" noise and useful for most applications.
- **Fractal.** This creates the noise pattern using a fractal algorithm. When this is selected, the Levels property in the **Noise Threshold** category is enabled.
- **Turbulence.** This is similar to fractal noise, except that it creates fault lines.

The Size property controls the size scale of the noise. The larger the value, the larger the size of the noise. The default value is 1.00 and the value can range from 0.00 to 1 billion. The Color 1 and Color 2 properties control the color of the pattern of noise. You can assign a color, image, or texture to the property. To swap the color definitions, pick the button next to the properties and select **Swap Colors** from the drop-down list.

The properties in the **Noise Threshold** category are used to fine-tune the noise effect. The properties in this category are:

- **Low.** The closer this setting is to 1.00, the more dominant color 1 is. The default setting is 0.00 and it can range from 0.00 to 1.00.
- **High.** The closer this setting is to 0.00, the more dominant color 2 is. The default setting is 1.00 and it can range from 0.00 to 1.00.
- **Levels.** Sets the energy amount for the fractal and turbulence types. Lower values make the fractal noise appear blurry and the turbulence lines more defined. The default setting is 3.00 and it can range from 1.00 upward.
- **Phase.** Randomly changes the noise pattern with each value. This allows you to have materials with the same noise map settings look slightly different. You should have different patterns on different materials. This adds a level of realism to your scene.

Figure 16-31.
Adjusting the properties of a noise map.

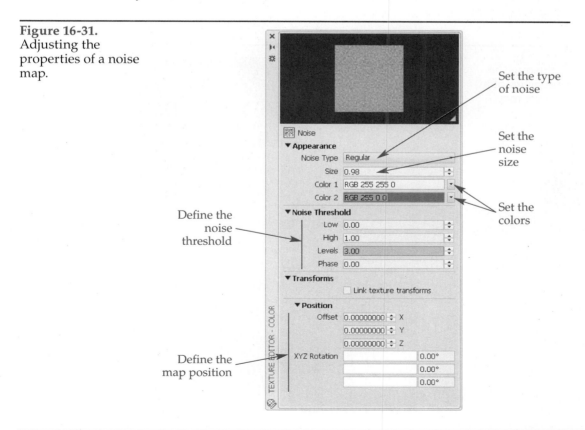

Set the type of noise

Set the noise size

Set the colors

Define the noise threshold

Define the map position

The properties in the **Transforms** and **Position** categories are the same as described for a marble map. Refer to the Marble section for details on these properties.

Speckle

A *speckle map* is a 3D map based on a random pattern of dots created from two colors. This map is very useful for textured walls, sand, granite, and so on. When you specify Speckle as the map for a property, the texture editor displays the properties for the map, **Figure 16-32**.

The settings for a speckle map are very simple. In the **Appearance** category, pick colors for the Color 1 and Color 2 properties. You cannot use maps, only colors. To swap the color definitions, pick the button next to the properties and select **Swap Colors** from the drop-down list. The Size property controls the size of the speckles.

The properties in the **Transforms** and **Position** categories are the same as described for a marble map. Refer to the Marble section for details on these properties.

Waves

A *waves map* is a 3D map in a pattern of concentric circles. Imagine dropping two or three stones into a pool of water and watching the ripples intersect with each other. A number of wave centers are randomly generated and a pattern created by the overlapping waves is the result. As the name implies, the waves map is usually used to simulate water. When you specify Waves as the map for a property, the texture editor displays the properties for the map, **Figure 16-33**.

In the **Appearance** category, pick colors for the Color 1 and Color 2 properties. You cannot use maps, only colors. To swap the color definitions, pick the button next to the properties and select **Swap Colors** from the drop-down list. The Distribution property can be set to 2D or 3D. This setting determines how the wave centers are distributed on the object. Selecting 3D means that the wave centers are randomly distributed over the surface of an imaginary sphere. This distribution affects all sides of an object. On the other hand, selecting 2D means that the wave centers are distributed on the XY plane. This is much better for nearly flat surfaces, such as the surface of a pond or lake.

Figure 16-32.
Adjusting the properties of a speckle map.

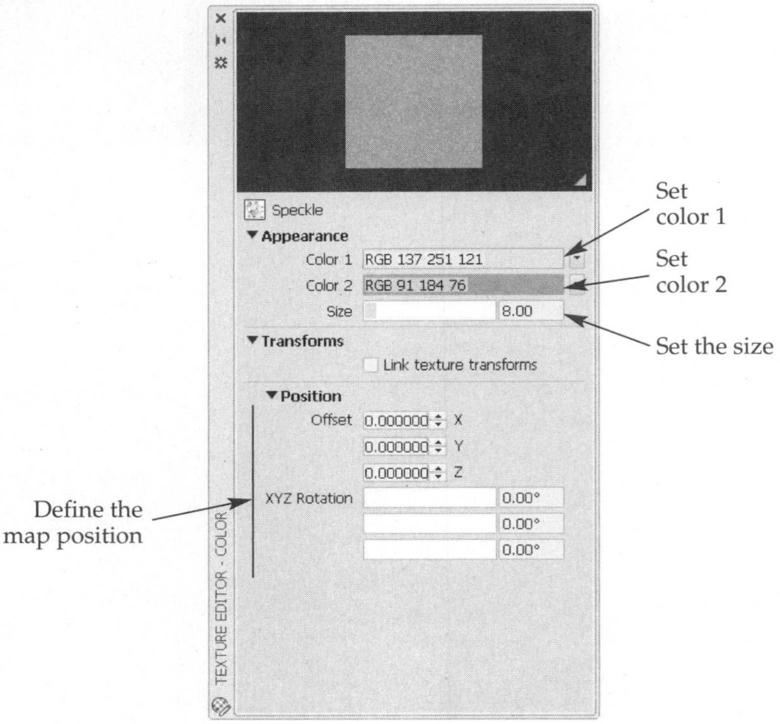

Figure 16-33.
Adjusting the properties of a waves map.

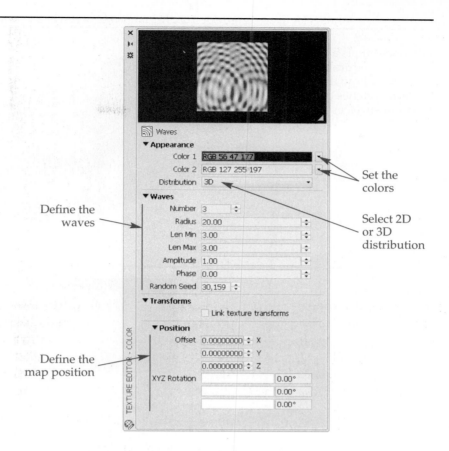

Set the colors

Select 2D or 3D distribution

Define the waves

Define the map position

The properties in the **Waves** category define the pattern of waves. The Number property sets the number of wave centers that are generating the waves. The Radius property sets the radius of the circle or sphere from which the waves originate. The Len Min and Len Max properties define the minimum and maximum interval for each wave. The Amplitude property can be thought of as the "power" of the wave. The default value is 1.00, but the value can range from 0.00 to 10000.00. A value less than 1.00 makes color 1 more dominant. For a value greater than 1.00, color 2 is more dominant. The Phase property is used to shift the pattern and the Random Seed property is used to redistribute the wave centers.

The properties in the **Transforms** and **Position** categories are the same as described for a marble map. Refer to the Marble section for details on these properties.

PROFESSIONAL TIP

Remember, you can change the material swatch geometry in the materials editor, such as to a cube, sphere, or cylinder. Some maps, like a waves map, are easier to understand when displayed on a cube.

Wood

A *wood map* is a 3D map that generates a wood grain based on the colors and values you select. See **Figure 16-34**. When you specify Wood as the map for a property, the texture editor displays the properties for the map, **Figure 16-35**.

A wood map is based on two colors. The Color 1 and Color 2 properties in the **Appearance** category are used to specify these colors, usually one dark and one light color. To swap the color definitions, pick the button next to the properties and select

---

Figure 16-34.
A wood map is a 3D, or procedural, map. Note how the pattern matches on adjacent surfaces.

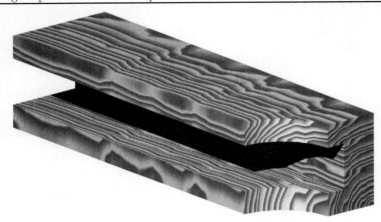

Figure 16-35.
Adjusting the properties of a wood map.

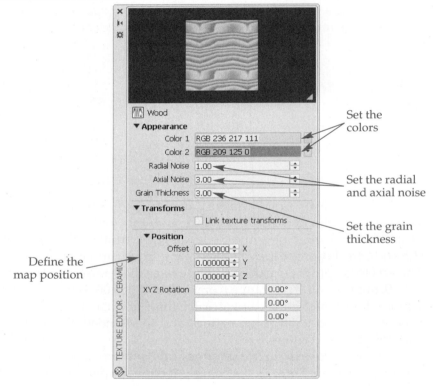

Set the colors

Set the radial and axial noise

Set the grain thickness

Define the map position

Swap Colors from the drop-down list. The Radial Noise property determines the waviness of the wood's rings. The rings are found by cutting a tree crosswise. The Axial Noise property determines the waviness of the length of the tree trunk. The Grain Thickness property determines the relative width of the grain.

The properties in the **Transforms** and **Position** categories are the same as described for a marble map. Refer to the Marble section for details on these properties.

Exercise 16-5

www.g-wlearning.com/CAD/

Complete the exercise on the companion website.

Adjusting Material Maps

Simply applying a map to a material property rarely results in a realistic scene when the scene is rendered. The maps usually need to be adjusted to produce the desired results. Maps can be adjusted at the material level or the object level. A combination of these two adjustments is usually required to produce a photorealistic rendering.

Material-Level Adjustments

Material-level adjustments involve changing the properties of the map in the material definition. Map properties are discussed in previous sections. Sometimes, these adjustments may be enough to get the materials looking the way you want them.

Other adjustments become necessary when the same material is applied to more than one object in the same scene. If you make changes at the material level, they affect all objects with that same material. If you need different objects to have different settings, then you will have to make object-level adjustments.

Object-Level Adjustments

Material mapping refers to specifying how a mapped material is applied to an object. When a mapped material is attached to an object, a default set of mapping coordinates, or simply *default mapping*, is used to apply the map to the object.

AutoCAD allows you to adjust mapping at the object level for 2D-mapped (texture-mapped) materials. The **MATERIALMAP** command applies a grip tool, or *gizmo*, based on one of four mapping types: planar, box, spherical, or cylindrical. See **Figure 16-36**. The colored edge represents the start and end of the map. For best results, select the mapping type based on the general shape of the object to which mapping is applied. Do not be afraid to experiment with other mapping types. Any mapping type can be used on any object, regardless of the object's shape. However, only one mapping type can be applied to an object at any given time.

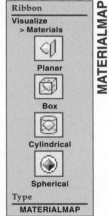

Ribbon

Visualize
> Materials

Planar

Box

Cylindrical

Spherical

Type
MATERIALMAP

MATERIALMAP

After one of the mapping types is selected, you are prompted to select the faces or objects. You can select multiple objects or faces. After making a selection, the gizmo is placed on the selection set. The command remains active for you to adjust the mapping or enter an option.

Drag the grips on the gizmo to stretch or scale the material. The effects of editing a color map are dynamically displayed if the current visual style is set to display materials and textures. Otherwise, exit the command and render the scene to see the effect of the edit. To readjust the mapping, select the same mapping type and pick the object again. The gizmo is displayed in the same location as before.

The **Move** and **Rotate** options of the command toggle between the move and rotate gizmos. Using the gizmos, you can move and rotate the map on the object. The **Reset**

Figure 16-36.
These are the four material map gizmos. From left to right: planar, box, spherical, and cylindrical. The colored edge represents the start and end of the map.

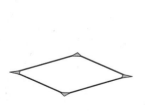 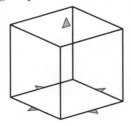 

option of the command restores the default mapping to the object. The **Switch mapping mode** option allows you to change between the four types of mapping.

If the command is typed, there is an additional option. The **Copy mapping to** option provides a quick and easy way to apply the changes made to the current object to other objects in the scene. Enter this option, select the face or object to copy from, and then select the faces or objects to copy to. This option is also available if the **Switch mapping mode** option is entered.

For example, look at **Figure 16-37A**. The grain on the stair risers is running vertically when it should run horizontally. First, apply a planar map gizmo to the bottom riser. Next, rotate the mapping 90°, **Figure 16-37B**. Finally, use the **Copy mapping to** option to copy the mapping to the other risers, **Figure 16-37C**.

NOTE

Mapping coordinates can be applied to 3D-mapped (procedural-mapped) materials, but adjusting the coordinates has no effect. This is because the 3D map is generated from mathematical calculations based on the world coordinate system. A 3D-mapped material must be adjusted at the material level.

Figure 16-37.
Correcting material mapping. A—The grain on the risers runs vertically instead of horizontally. B—Rotating the map with the rotate gizmo. C—The corrected rendering. *(Model courtesy of Arcways, Inc., Neenah, WI)*

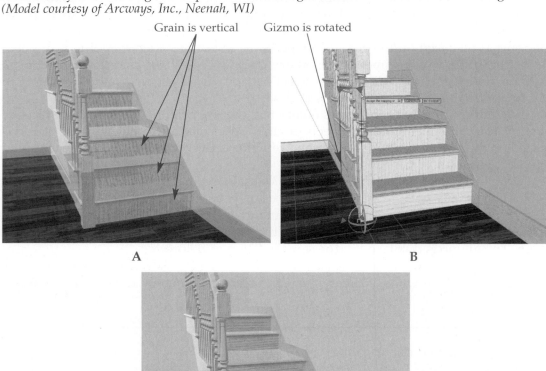

If you are using a reflectivity, self illumination, or bump map and need to adjust it at the object level, you cannot see the effects of the mapping change in the viewport. Apply the same map as a color map. Also, set the visual style to display materials and textures. Then, adjust the object mapping as needed. The edits are dynamically displayed in the viewport. When the image is in the correct location, remove the color map from the material.

Exercise 16-6

www.g-wlearning.com/CAD/

Complete the exercise on the companion website.

Chapter Review

Answer the following questions using the information in this chapter.

1. Define *material*.
2. Define *materials library*.
3. What is the Favorites section of the **Library** area of the materials browser used for?
4. Describe how to attach a material using the materials browser.
5. How can materials be attached to layers?
6. By default, which material is attached to newly created objects?
7. Which material is used as the base material for creating new materials?
8. Name the 13 shapes that can be used to display the material in the preview in the materials editor.
9. How do you know if a material in the materials browser is being used in the drawing?
10. How can the name of an existing material be changed?
11. Name the 14 basic material types.
12. When creating a material to look like plastic, what is the benefit of using the plastic material type instead of starting from scratch using the generic material?
13. In the **Reflectivity** category of the generic material, there are Direct and Oblique properties. What are these used for?
14. An image mapped to a material property will normally repeat itself to cover the entire object. What is this called and how do you turn it off?
15. When a cutout map is applied to a material, what are the different effects produced by black and white areas in the image?
16. How much illumination does a self illuminated material add to a scene?
17. How is a marble material created?
18. Explain how black and white areas of a map applied to the Transparency property affect the transparency of a material.
19. Explain what the nodes in the ramp of a gradient map are for.
20. Name the four types of mapping available for adjusting texture maps at the object level.

Drawing Problems

1. **General** In this problem, you will create a scene with basic 3D objects, attach materials to the objects, and adjust the settings of the materials.
   A. Start a new drawing and set the units to architectural.
   B. Draw a 15′ × 15′ planar surface to represent the floor.
   C. Draw two boxes to represent two walls. Make the boxes 15′ × 4″ × 9′. Position them to form a 90° corner. Alternately, draw a polysolid of the same dimensions.
   D. Draw a R2′ × 5′H cone in the center of the room.
   E. Open the materials browser and locate a material similar to the one shown on the floor. Attach this material to the floor.
   F. Locate an appropriate material for the wall and attach it.
   G. Locate an appropriate material for the cone and attach it.
   H. Turn on the sun and adjust the time to create good shadows. Refer to Chapter 15 for an introduction to sun settings.
   I. Render the scene. Save the rendering as P16-1.jpg. Save the drawing as P16-1.

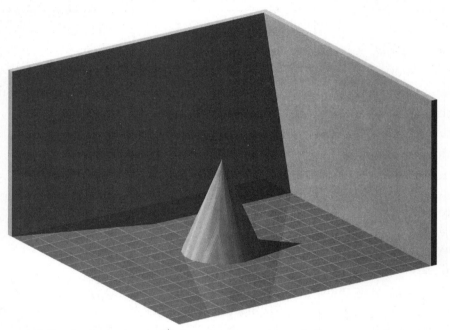

A full-color image is provided on the companion website.   www.g-wlearning.com/CAD/

2. **Architectural** In this problem, you will attach materials to the objects in an existing drawing and render the scene.
   A. Open drawing P15-2 from Chapter 15. Save the drawing as P16-2.
   B. Attach the materials of your choice to the objects in the scene. Do not be restricted by the names of the materials. For example, a concrete material may be suitable for items such as foliage or even carpet with a simple color change. Be creative.
   C. If the items in the scene were inserted using the Autodesk Seek website, they may be blocks with nested layers. Instead of exploding the blocks, use the **MATERIALATTACH** command and attach materials to the layers on which the nested objects reside.
   D. Turn on the sun and adjust the time to create good shadows. Refer to Chapter 15 for an introduction to sun settings.
   E. Render the scene. Save the rendering as P16-2.jpg. Save the drawing.

A full-color image is provided on the companion website.   www.g-wlearning.com/CAD/

3. General In this problem, you will create custom wood and marble materials.
   A. Start a new drawing and save it as P16-3.
   B. Draw two 5 × 5 × 5 boxes and position them near each other. Using other primitives, cut notches and holes in the boxes. The boxes will be used to test the custom materials.
   C. In the materials editor, create two new materials. Name one Wood-*your initials* and the other Marble-*your initials*.
   D. Attach the wood material to one of the boxes and the marble material to the other box.
   E. Render the scene and make note of the wood grain and marble veins.
   F. Use the texture editor to change the properties of the materials.
   G. Render the scene again and make note of the changes. Using the **Render in Region** button on the **Render** panel in the **Visualize** tab of the ribbon can save time when testing material changes.
   H. When you are satisfied with the materials, save the rendering as P16-3.
   I. Save the drawing.

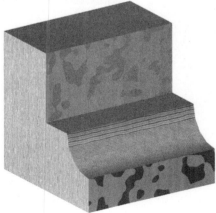

A full-color image is provided on the companion website.   www.g-wlearning.com/CAD/

4. **General** In this problem, you will create a bitmap and use it as a transparency map and a bump map.
   A. Draw a rectangle with an array of smaller rectangles inside of it as shown. Sizes are not important and the pattern can be varied if you like.
   B. Display a plan view of the rectangles. Then, copy all of the objects to the Windows Clipboard by pressing [Ctrl]+[C] and selecting the objects.
   C. Launch Windows Paint. Then, paste the objects into the blank file. Notice how the AutoCAD background outside of the large rectangle is also included.
   D. Use the select tool (rectangle) in Paint to draw a window around the large rectangle created in AutoCAD and the smaller rectangles within it. Copy this to the Windows Clipboard by pressing [Ctrl]+[C].
   E. Start a new Paint file without saving the current one and paste the image from the Clipboard into the new blank file. Now, the unwanted AutoCAD background is no longer displayed. If needed, change the small rectangles to black and the lattice to white using the tools in Paint. The colors should be the reverse of what is shown below. Then, save the image file as P16-4.bmp and close Paint.
   F. In AutoCAD, draw a solid box of any size.
   G. Using the materials editor, create a new material.
   H. Assign the P16-4.bmp image file you just created as a transparency map. Adjust the map so that it is scaled to fit to the object.
   I. Render the scene and note the effect.
   J. Turn off the transparency property by unchecking the check box in the category name.
   K. Assign the P16-4.bmp image file as a bump map. Adjust the map so that it is scaled to fit to the object.
   L. Render the scene and note the effect.
   M. Save the drawing as P16-4.

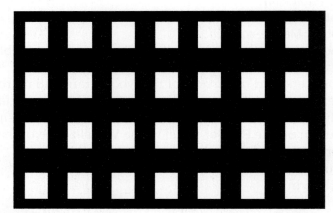

Lighting

Learning Objectives

After completing this chapter, you will be able to:

✓ Describe the types of lighting in AutoCAD.
✓ List the user-created lights available in AutoCAD.
✓ Change the properties of lights.
✓ Add a background to your scene and control its appearance.

In the movie industry, it has often been said that "Lighting is everything." That statement also rings true when creating realistic scenes in AutoCAD. If lights are used incorrectly, the scene will be washed-out with light or too dark to see anything. In Chapter 15, you were introduced to lighting. You learned how to adjust lighting by turning off the default lights and adding sunlight. In this chapter, you will learn all about the lights available in AutoCAD. You will learn lighting tips and tricks to help make the scene look its best.

Types of Lights

Ambient light is like natural light just before sunrise. It is the same intensity everywhere. All faces of the object receive the same amount of ambient light. Ambient light cannot create highlights, nor can it be concentrated in one area. AutoCAD does not have an ambient light setting. Instead, it relies on indirect illumination, which is discussed in Chapter 18.

A *point light* is like a lightbulb. Light rays from a point light shine out in all directions. A point light can create highlights. The intensity of a point light falls off, or weakens, over distance. Other programs, such as Autodesk 3ds Max®, may call these lights *omni lights*. A *target point light* is the same as a standard point light except that a target is specified. The illumination of the target point light is directed toward the target.

A *distant light* is a directed light source with parallel light rays. This acts much like the sun. Rays from a distant light strike all objects in your model on the same side and with the same intensity. The direction and intensity of a distant light can be changed.

A *spotlight* is like a distant light, but it projects in a cone shape. Its light rays are not parallel. A spotlight is placed closer to the object than a distant light. Spotlights have a hotspot and a falloff. The light from a standard spotlight is directed toward a target. A *free spotlight* is the same as a standard spotlight, but without a target.

A *weblight* is a directed light that represents real-world distribution of light. The illumination is based on photometric data that can be entered for each light. The light from a standard weblight is directed toward a target. A *free weblight* is the same as a standard weblight, but without a target point.

Properties of Lights

Several factors affect how a light illuminates an object. These include the angle of incidence, reflectivity of the object's surface, and the distance that the light is from the object. In addition, the ability to cast shadows is a property of light.

Angle of Incidence

AutoCAD renders the faces of a model based on the angle at which light rays strike the faces. This angle is called the *angle of incidence*. See **Figure 17-1**. A face that is perpendicular to light rays receives the most light. As the angle of incidence decreases, the amount of light striking the face also decreases.

Reflectivity

The angle at which light rays are reflected off a surface is called the *angle of reflection*. The angle of reflection is always equal to the angle of incidence. Refer to **Figure 17-1**.

The "brightness" of light reflected from an object is actually the number of light rays that reach your eyes. A surface that reflects a bright light, such as a mirror, is reflecting most of the light rays that strike it. The amount of reflection you see is called the *highlight*. The highlight is determined by the angle from the viewpoint relative to the angle of incidence. Refer to **Figure 17-1**.

The surface quality of the object affects how light is reflected. A smooth surface has a high specular factor. The *specular factor* indicates the number of light rays that have the same angle of reflection. Surfaces that are not smooth have a low specular factor. These surfaces are called *matte*. Matte surfaces *diffuse*, or "spread out," the light as it strikes the surface. This means that few of the light rays have the same angle of reflection. **Figure 17-2** illustrates the difference between matte and high specular finishes. Surfaces can also vary in *roughness*. Roughness is a measure of the polish on a surface. This also affects how diffused the reflected light is.

Figure 17-1.
The amount of reflection, or highlight, you see depends on the angle from which you view the object.

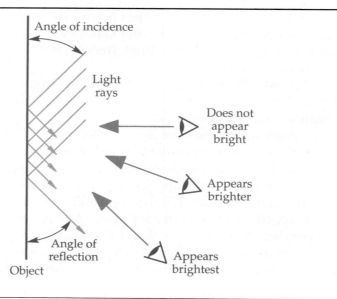

Figure 17-2.
Matte surfaces produce diffuse light. This is also referred to as having a low specular factor. Shiny surfaces evenly reflect light and have a high specular factor.

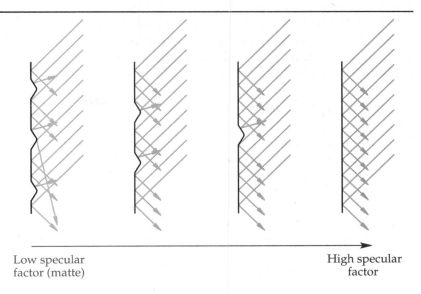

Low specular factor (matte)

High specular factor

When rendering metal and other shiny materials, it helps to have a bright object nearby to create some highlights on the object. Adding simple vertical planes to the left and right of the object (out of the camera's view) and applying a self-illuminating material to them will create some nice highlights. Photographers use this trick in the studio.

Hotspot and Falloff

A spotlight produces a cone of light. The *hotspot* is the central portion of the cone, where the light is brightest. See **Figure 17-3**. The *falloff* is the outer portion of the cone, where the light begins to blend to shadow. The hotspot and falloff of a spotlight are not affected by the distance the light is from an object. Spotlights are the only lights with hotspot and falloff properties.

A full-color image is provided on the companion website.   www.g-wlearning.com/CAD/

Figure 17-3.
The hotspot of a spotlight is the area that receives the most light. The smaller cone is the hotspot. The falloff receives light, but less than the hotspot. The larger cone is the falloff.

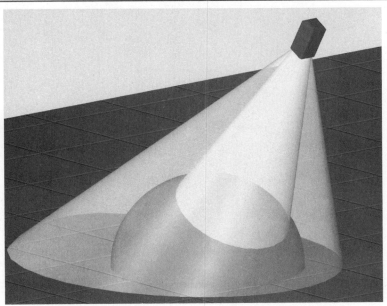

Attenuation

The farther an object is from a point light or spotlight, the less light that reaches the object. See **Figure 17-4**. The intensity of light decreases over distance. This decrease is called *attenuation*. All lights in AutoCAD, except distant lights, have attenuation. Often, attenuation is called *falloff* or *decay*. However, do not confuse this with the falloff of a spotlight, which is the outer edge of the cone of illumination. All lights that attenuate in AutoCAD are calculated using inverse square attenuation. This is a "real-world" calculation in which the illumination of an object decreases in inverse proportion to the square of the distance. For example, if an object is two units from the light, it receives $(1/2)^2$, or 1/4, of the full light. If the object is four units away, it receives $(1/4)^2$, or 1/16, of the full light.

NOTE

Different types of attenuation other than inverse square attenuation are available when creating point lights and spotlights, as discussed later in this chapter. However, these options are only available to maintain backward compatibility with older releases of AutoCAD and have no effect on the scene.

PROFESSIONAL TIP

The intensity of the sun's rays does not diminish from one point on Earth to another. They are weakened by the angle at which they strike Earth. Therefore, since distant lights are similar to the sun, attenuation is not a factor with distant lights.

A full-color image is provided on the companion website.   www.g-wlearning.com/CAD/

Figure 17-4.
Attenuation is the intensity of light decreasing over distance.

Less illumination

More illumination

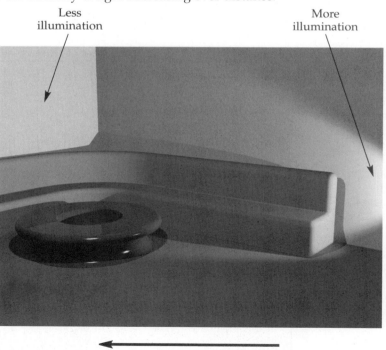

Attenuation

AutoCAD Lights

AutoCAD has three types of lighting: default lighting, sunlight with or without sky illumination, and user-created lighting. *Default lighting* is the lighting automatically available in the scene. It is composed of one or two default light sources that appear to be originating from above and to the left of the viewer. As the viewpoint is changed, the light sources follow to maintain the same illumination of the scene. There is no control over default lighting and it must be shut off whenever one of the other types of lighting is used.

As you saw in Chapter 15, *sunlight* may be added to any scene. AutoCAD uses a distant light to simulate the parallel rays of the sun. The date and time of day can be adjusted to create different sunlight illumination. *Sky illumination* may also be added with sunlight to simulate light bouncing off objects in the scene and particles in the atmosphere. This helps create a more-natural feel.

User-created lighting results when you add AutoCAD light objects to the drawing. There are four types of user-created lights: distant light, weblight, point light, and spotlight. See **Figure 17-5**. A distant light is a directed light source with parallel light rays. A weblight is a directional point light containing light intensity (photometric) data. A point light is like a lightbulb with light rays shining out in all directions. A spotlight is like a distant light, but it projects light in a cone shape instead of having parallel light rays.

When created, point lights, weblights, and spotlights are represented by *light glyphs*, or icons, in the drawing. To suppress the display of light glyphs, pick the **Light Glyph Display** button in the expanded **Lights** panel on the **Visualize** tab of the ribbon. The button is blue when light glyphs are displayed. The default lights, sun, and distant lights are not represented by glyphs.

In this section, you will learn how to add lights. You will also learn how to adjust the various properties of sunlight and AutoCAD light objects. The tools for working with lights can be accessed using the command line, tool palettes, and the **Lights** and **Sun & Location** panels on the **Visualize** tab of the ribbon. See **Figure 17-6**.

So that you will never work with a completely dark scene, default lighting is applied in the viewport and to the rendering if no other lights are added. In order for your lights to be applied, you must switch between default lighting and user lighting. To do this, pick the **Default Lighting** button in the expanded **Lights** panel on the **Visualize** tab of the ribbon. This button toggles the lighting between default lighting and whatever lights are available in the scene. When default lighting is on, the button is blue. When off, the button is not highlighted. If you elected for AutoCAD to do so, the default lighting will be automatically shut off when sunlight is turned on or a user-created light is added to the scene.

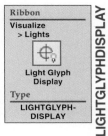

Ribbon

Visualize
> Lights

Light Glyph Display

Type

LIGHTGLYPH-
DISPLAY

LIGHTGLYPHDISPLAY

A full-color image is provided on the companion website. www.g-wlearning.com/CAD/

Figure 17-5.
AutoCAD has four types of user-created lights: distant, point, weblight, and spotlight. A weblight is really a targeted point light. It projects in all directions, but may be predominant in one direction.

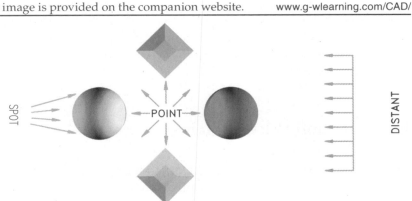

Figure 17-6.
The tools for adding and controlling lights. A—Tools in the **Lights** panel of the **Visualize** ribbon tab. B—Tools in the **Sun & Location** panel of the **Visualize** ribbon tab.

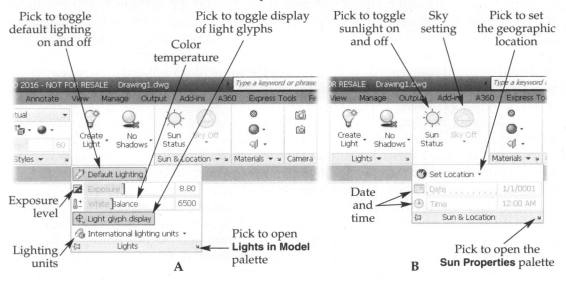

Lighting Units

There are three settings for *lighting units* available in AutoCAD: standard (generic), American (foot-candles), and International (lux). Generic lighting is the only type of lighting that was used in AutoCAD prior to AutoCAD 2008. This lighting provided very nice results, but the settings were not based on any real measurements. American or International lighting is called photometric lighting. *Photometric lighting* is physically correct and attenuates at the square of the distance from the source. For more accuracy, photometric data files can be imported from lighting manufacturers.

The **LIGHTINGUNITS** system variable sets what type of lighting is used. The setting can be changed using the **International lighting units** drop-down list in the expanded **Lights** panel on the **Visualize** tab of the ribbon. Refer to **Figure 17-6**. A setting of 1 or 2 results in photometric lighting. Entering a value of 1 specifies American lighting units. A setting of 2 specifies International lighting units. This is the default setting. The only difference between a setting of 1 and 2 is that American units are displayed as foot-candles (fc) and International units are displayed as lux (lx). A setting of 0 specifies that standard (generic) lighting is used.

NOTE

All lights in AutoCAD are calculated using photometric lighting. The standard lighting units setting is only available for backward compatibility with older releases of AutoCAD and has no effect on the scene. If you set the **LIGHTINGUNITS** system variable to 0, AutoCAD will display a message informing you that non-photometric lights are not supported by the current rendering engine.

Exposure and White Balance

The brightness and color temperature of lighting in the scene can be adjusted using the sliders in the expanded **Lights** panel on the **Visualize** tab of the ribbon. Refer to **Figure 17-6**. These are global controls that provide a quick way to adjust the lighting. The **Light Exposure** slider is used to adjust the brightness. Lower values make the

are available. The **Intensity** option provides additional options for setting the power of the light source. The **Flux** option can be used to specify the power in lumens (lm). The **Illuminance** option can be used to specify the illuminance in lux (lx) or foot-candles (fc), depending on the current lighting units. An additional option, **Distance**, is available under **Illuminance**. This defaults to one unit and is used to calculate illuminance.

NOTE

The **Shadow** and **Attenuation** options are only available for backward compatibility with older releases of AutoCAD. Shadow casting cannot be turned off and shadow-mapped shadows are created by AutoCAD using the Soft (area) type, as previously discussed. Inverse square attenuation is used regardless of the attenuation type selected. Setting the attenuation start and end limits has no effect on the scene.

Target Point Lights

The target point light is like a regular point light except that the command starts by asking for a *source* location and a *target* location. The rest of the options are the same. You can type TARGETPOINT or pick the **Targetpoint** option in the **LIGHT** command to create a target light.

PROFESSIONAL TIP

A point light may be used as an incandescent lightbulb, such as the lightbulb in a table lamp. Most of these lightbulbs cast a yellow light. In these cases, you may want to change the color of the light to a light yellow. Compact fluorescent lightbulbs may cast white, light blue, or light yellow light. LED-based lights cast a color based on the color of the LED. Other colors can be used to give the impression of heat or colored lights.

Exercise 17-3

www.g-wlearning.com/CAD/

Complete the exercise on the companion website.

Spotlights

Spotlights are user-created lights that have light rays projecting in a cone shape in one direction. See **Figure 17-14**. When a spotlight is created, the location from where the light is originating must be specified, along with the direction in which the light rays travel. A light glyph represents spotlights in the drawing. See **Figure 17-15**. Spotlights have attenuation, which means that the distance of the objects in the scene to the spotlight affects the intensity of the illumination.

The **SPOTLIGHT** or **LIGHT** command is used to create a spotlight. You are first prompted to specify the location of the light. This is from where the light rays will originate. Next, you are prompted for the target location. This is simply the location where the light is aimed.

Once the location and target are established, several options for the light are available. The **Name**, **Intensity Factor**, **Status**, **Photometry**, and **Filter Color** options work the same as the corresponding options for a point light. However, a spotlight also has hotspot and falloff settings.

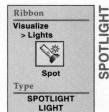

Ribbon
Visualize
> Lights
Spot
Type
SPOTLIGHT
LIGHT

SPOTLIGHT

Figure 17-14.
Three spotlights are used to simulate recessed ceiling lights. Notice how the light from each spotlight projects in a cone.

Figure 17-15.
This is the light glyph for a spotlight.

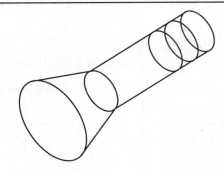

The *hotspot* is the inner cone of illumination for a spotlight. Refer to **Figure 17-16**. This is measured in degrees. To set the hotspot, enter the **Hotspot** option and then specify the number of degrees for the hotspot.

The *falloff,* not to be confused with attenuation, is the outer cone of illumination for a spotlight. Like the hotspot, it is measured in degrees. The falloff value must be greater than or equal to the hotspot value. It cannot be less than the hotspot value. In practice, the falloff value is often much greater than the hotspot value. To set the falloff, enter the **Falloff** option and then specify the number of degrees for the falloff.

Figure 17-16.
Hotspot and falloff for a spotlight are angular measurements.

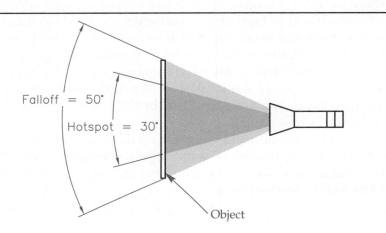

Falloff = 50°

Hotspot = 30°

Object

Once the light is created and you select it in the viewport, grips are displayed. If you hover the cursor over a grip, a tooltip is displayed indicating what the grip will modify. You can use grips to change the location of the spotlight and its target, the hotspot, and the falloff. If you hover over a falloff or hotspot grip, the current angle is displayed in the wireframe cone (if dynamic input is on).

Free Spotlight

A free spotlight is like a standard spotlight except that you do not specify a target, only the light location. The rest of the options are the same. You can type FREESPOT or pick the **Freespot** option in the **LIGHT** command to create a free spotlight. When created, a free spotlight points down the Z axis (from positive to negative) of the current UCS. A free spotlight may be easier to control than a standard spotlight because you do not have to worry about the target point. If you want to change the angle or position of the light, use the **3DMOVE** and **ROTATE3D** commands.

Exercise 17-4

www.g-wlearning.com/CAD/

Complete the exercise on the companion website.

Weblight

A photometric weblight is really just a targeted point light. The difference is that a weblight provides a more precise representation of the light. Real-world lights appear to evenly illuminate from their source, but, in reality, the shape of the light, the material used in its manufacture, and other factors make all lights distribute their energy in different ways. These data are provided by light manufacturers in the form of light distribution data. Light distribution data can be loaded into the **Photometric Web** subcategory of the **General** category in the **Properties** palette when the light is selected. Select the Web file property, then pick the button to the right and select an IES file. IES stands for Illuminating Engineering Society. The AutoCAD help documentation has additional information on IES files.

Think of the web of a weblight as a spherical cage surrounding the light source. If the light is evenly distributed from its source, the cage is a true sphere. In actuality, a light may emit more light energy in the X direction than in the Z direction. In this case, the cage bulges out further in the X direction. The position of this bulge may be important to the illumination of the scene and you may need to rotate the web to apply more or less light in one direction or another.

WEBLIGHT

Ribbon

Visualize > Lights

Weblight

Type

**WEBLIGHT
LIGHT**

The **WEBLIGHT** or **LIGHT** command is used to create a weblight. You are prompted for source and target locations. The **Name**, **Intensity Factor**, **Status**, **Photometry**, **Shadow**, and **Filter Color** options work the same as the corresponding options for the previously discussed lights. However, weblights have an additional **Web** option. When this option is activated, these options are presented:

- **File.** Allows you to select an IES file.
- **X.** Rotates the web around the X axis.
- **Y.** Rotates the web around the Y axis.
- **Z.** Rotates the web around the Z axis.
- **Exit.** Exits the **Web** option.

Point and spotlights can be converted to weblights, and vice versa, using the **Properties** palette. Simply select an existing light and open the **Properties** palette. In the **General** category, the Type property determines whether the light is a point light, spotlight, or weblight. Select the type in the drop-down list. Using the **Properties** palette with lights is discussed in detail later in this chapter.

Free Weblight

FREEWEB

Type

**FREEWEB
LIGHT**

A free weblight is the same as a standard weblight except there is no target. Only the source location is specified when placing the light. To change the location and direction of the light, use the **ROTATE3D** and **3DMOVE** commands.

Photometric Lights Tool Palette Group

Photometric lights may be easily added to the drawing using the tool palettes in the **Photometric Lights** tool palette group. This palette group contains four palettes: **Fluorescent**, **High Intensity Discharge**, **Incandescent**, and **Low Pressure Sodium**. Lights created with these tools have preset properties for **Intensity Factor** and **Filter Color**. The long fluorescent lights are weblights. The high intensity discharge, low-pressure sodium, and regular incandescent lights are point lights. The incandescent halogen lights are free spotlights. The recessed incandescent lights are a special weblight designed for recessed light fixtures.

Lights in Model and Properties Palettes

The **Lights in Model** palette is extremely useful for controlling the lights in your scene, **Figure 17-17**. Using the **Lights in Model** palette in conjunction with the **Properties** palette, you can manage and edit all of the lights in a scene.

Figure 17-17.
All user-created lights are listed in the **Lights in Model** palette.

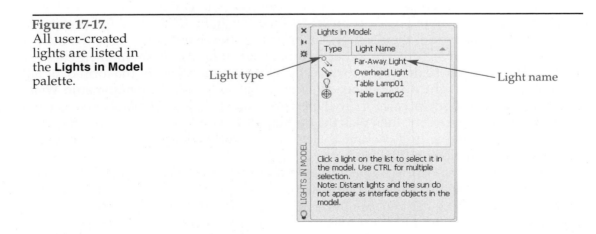

Lights in Model Palette

The **LIGHTLIST** command displays the **Lights in Model** palette. This palette can also be displayed by picking the dialog box launcher button at the lower-right corner of the **Lights** panel on the **Visualize** tab of the ribbon. All user-created lights in the scene are displayed in the list. To modify the properties of a light, either double-click on the light name or right-click on it and select **Properties** from the shortcut menu. This opens the **Properties** palette. See **Figure 17-18**. If the **Properties** palette is already open, you can simply select a light in the **Lights in Model** palette. You can select more than one light by pressing the [Ctrl] key and selecting the names in the **Lights in Model** palette, which allows you to change all of their settings at the same time. This is an excellent way to make the lighting in your scene uniform or to control a series of lights with a single edit.

A light can be deleted from the scene using the **Lights in Model** palette. To do so, simply right-click on the name of the light and select **Delete Light** from the shortcut menu. The light is removed from the drawing. Using the **UNDO** command restores the light.

If a single light is selected, the right-click menu also contains an option to control the display of the glyph for the selected light. The options are **Auto, On,** or **Off.**

Type
LIGHTLIST

LIGHTLIST

Figure 17-18.
The **Properties** palette with a weblight selected.

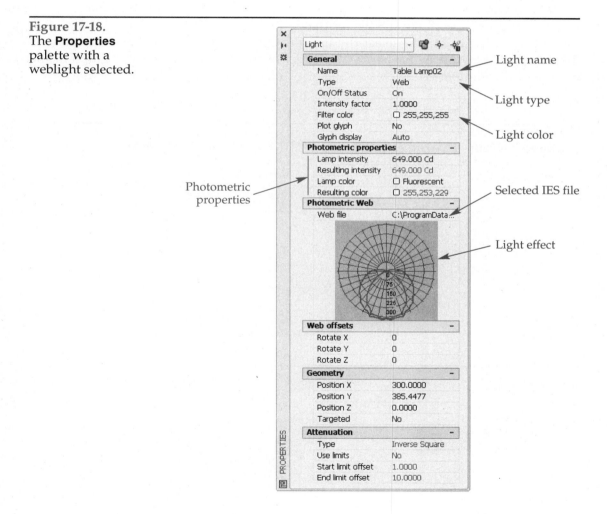

Properties Palette

The **Photometric properties** subcategory of the **General** category in the **Properties** palette has special settings for photometric lights. The Lamp intensity property determines the brightness of the light. The value may be expressed in candelas (cd), lumens (lm), or illuminance (lux) values. When you select the Lamp intensity property, a button is displayed to the right of the value. Picking this button opens the **Lamp Intensity** dialog box, **Figure 17-19**. In this dialog box, you can change the illumination units and set the intensity (Lamp intensity property). You can also set an intensity scale factor. This is multiplied by the Lamp intensity property to obtain the actual illumination supplied by the light. The read-only Resulting intensity property in the **Properties** palette displays the result.

The Lamp color property in the **Photometric properties** subcategory controls the color of the light. If you select the property, a button is displayed to the right of the value. Picking this button opens the **Lamp Color** dialog box. See **Figure 17-20**. This dialog box gives you the option to control the color of the light by either standard spectra colors or Kelvin colors. The color selected in the **Filter color:** drop-down list is applied to the color of the light. The **Resulting color:** swatch displays the color cast by the light once the filter color is applied. If the filter color is white (255,255,255), then the light color is the color cast by the light.

If a weblight is selected, the **Photometric Web** subcategory in the **General** category is where you can specify an IES file for the light. Select the Web file property, then pick the button to the right and select the IES file. Once the file is selected, its location is displayed in the Web file property. The effect of the data is shown in a graph at the bottom of the **Photometric Web** subcategory. Refer to **Figure 17-18**.

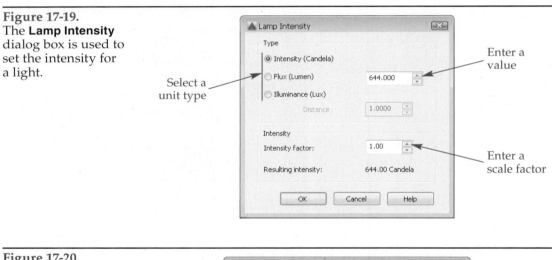

Figure 17-19.
The **Lamp Intensity** dialog box is used to set the intensity for a light.

Select a unit type

Enter a value

Enter a scale factor

Figure 17-20.
The **Lamp Color** dialog box is used to set the color for the light and a filter color, if needed.

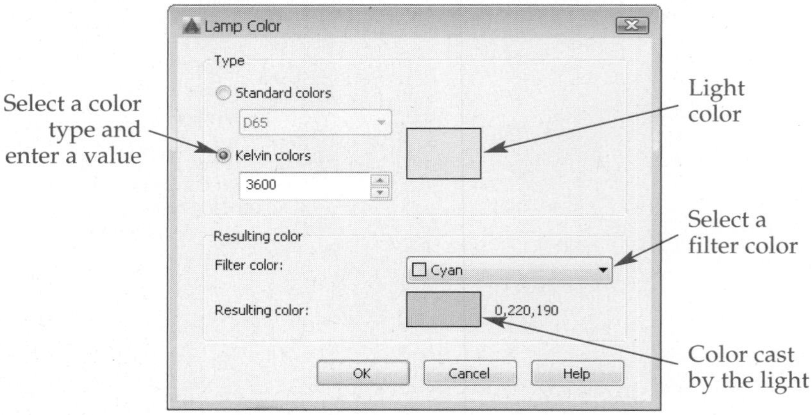

Select a color type and enter a value

Light color

Select a filter color

Color cast by the light

The **Web offsets** subcategory allows you to rotate the web around the X, Y, and Z axes, as discussed earlier in this chapter in the Weblight section.

In the **Geometry** category, you can change the X, Y, and Z coordinates of the light. You can also change the X, Y, and Z coordinates of the light's target. If the light is not targeted, the Target X, Target Y, and Target Z properties are not displayed. To change the light from targeted to free, and vice versa, select Yes or No in the Targeted property drop-down list.

The properties in the **Attenuation** category are the same as those discussed earlier for lights that have attenuation properties. The Type property is set to Inverse Square and is read-only. The attenuation limits properties are read-only and are only for backward compatibility with older releases of AutoCAD.

PROFESSIONAL TIP

It is important to give your lights names that make them easy to identify in a list. If you accept the default names for lights, they will be called Pointlight1, Spotlight5, Distantlight7, Weblight2, etc., making them difficult to identify. Use the **Name** option when creating the light or, after the light is created, the **Properties** palette to change the name of the light.

Determining Proper Light Intensity

Placing lights usually requires adjustments to produce the results you are looking for. In addition, you will also typically spend some time determining the proper light intensity and other settings. As a general guideline, the object nearest to a point light or spotlight should receive the full illumination, or full intensity, of the light. Full intensity of any light that has an attenuation property is a value of one. Remember, as discussed earlier in this chapter, AutoCAD calculates attenuation using the inverse square method. You can calculate the light intensity based on this method to establish an approximate starting point.

For example, suppose you have drawn an object and placed a point light and a spotlight. The point light is 55 units from the object. The spotlight is 43 units from the object. Use the following calculations to determine the intensity settings for the lights based on inverse square attenuation.

Since the point light is 55 units from the object, the object receives $(1/55)^2$, or 1/3025 ($55^2 = 3025$), of the light. Therefore, set the intensity of the point light to 3025 (3025/3025 = 1). Since the spotlight is 43 units from the object, the object receives $(1/43)^2$, or 1/1849 ($43^2 = 1849$), of the spotlight's illumination. Therefore, set the intensity of the spotlight to 1849 (1849/1849 = 1).

However, keep in mind that these settings are merely a starting point. You will likely spend some time adjusting lighting to produce the desired results. In some cases, it may take longer to light the scene than it did to model it.

PROFESSIONAL TIP

If there are no lights in the scene or all lights are turned off, default lighting is activated so that you can still see what you are doing. If you render a scene and the image appears black, lights in the scene may have their intensity set too low or they may be positioned too far away from the objects in the scene.

Shadows

Shadows are critical to the realism of a rendered 3D model. A model without shadows appears obviously fake. On the other hand, a model with realistic materials and shadows may be hard to recognize as computer generated. In AutoCAD, all lights, including default lights, cast shadows. You may not see shadows in the viewport depending on the current visual style settings, but shadows are always calculated in the rendered image. As previously discussed, the **Shadow** option that appears in command prompts when creating lights is only available for backward compatibility with older releases of AutoCAD. All lights display realistic shadows with soft edges.

Adding a Background

A *background* is the backdrop for your 3D model. The background can be a solid color, a gradient of colors, an image file, the sun and sky, or the current AutoCAD drawing background color. By default, the background is the drawing background.

To change the background for your drawing, you must first create a named view with the **VIEW** command. In the **View Manager** dialog box, pick the **New...** button to display the **New View/Shot Properties** dialog box. See **Figure 17-21**. The view name, category, and type are specified at the top of the dialog box. Near the bottom of the **View Properties** tab is the **Background** area. The drop-down list in this area is used to specify the type of background. The options are Default, Solid, Gradient, Image, and Sun & Sky. The Default setting uses the current AutoCAD viewport color.

VIEW

Ribbon
Visualize
> Views

View Manager

Type
VIEW
V

NOTE

An image based lighting environment background can also be used to set the background of a scene. Image based lighting and other advanced rendering functions are discussed in Chapter 18.

Figure 17-21.
The **View Properties** tab of the **New View/ Shot Properties** dialog box. A named view must be created before you can use a background in your scene.

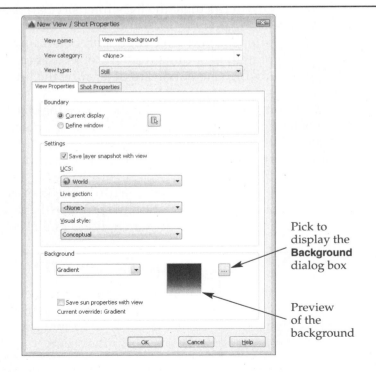

Pick to display the **Background** dialog box

Preview of the background

Before you create the named view, establish the viewpoint from which you want to see the final rendering. Set the perspective projection current, if desired. These settings are saved with the view. The **Views** drop-down list in the **Views** panel on the **Visualize** tab of the ribbon makes it very easy to recall the view. You can also make a named view current by selecting it from the **Custom Model Views** menu in the **View Controls** flyout in the viewport controls.

Solid Backgrounds

If you select Solid in the drop-down list, the **Background** dialog box is displayed. The **Type:** drop-down list in this dialog box is automatically set to Solid and the default color is displayed in the **Preview** area. See **Figure 17-22**. In the **Solid options** area of the dialog box, pick the horizontal **Color:** bar to open the **Select Color** dialog box. Then, select the background color that you desire. When the **Select Color** dialog box is closed, the color you picked is displayed in the **Preview** area of the **Background** dialog box. Pick the **OK** button to accept your changes and close the **Background** dialog box. Next, save the view, set the new view current, and pick the **OK** button to close the **View Manager** dialog box.

NOTE

Make sure to do a test rendering with the background color you selected. The result may look quite different from your expectations.

Gradient Backgrounds

A gradient background can be composed of two or three colors. If you select Gradient in the drop-down list in the **New View/Shot Properties** dialog box, the **Background** dialog box is displayed with Gradient selected and the default gradient colors displayed in the **Preview** area. See **Figure 17-23**. The **Top color:**, **Middle color:**, and **Bottom color:** swatches are displayed on the right-hand side of the **Gradient options** area. Selecting a swatch opens the **Select Color** dialog box for changing the color. To create a two-color gradient composed of the top and bottom colors, uncheck the **Three color** check box. The **Rotation:** text box provides the option of rotating the gradient. Pick the **OK** button

Figure 17-22.
Creating a solid background.

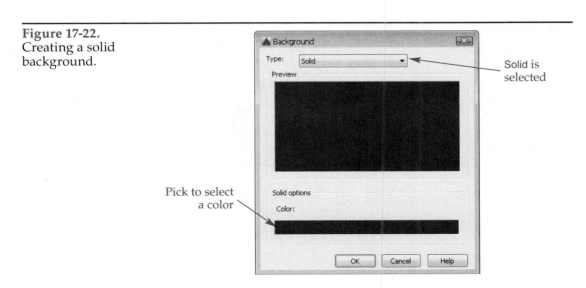

Solid is selected

Pick to select a color

Figure 17-23.
Creating a gradient
background.

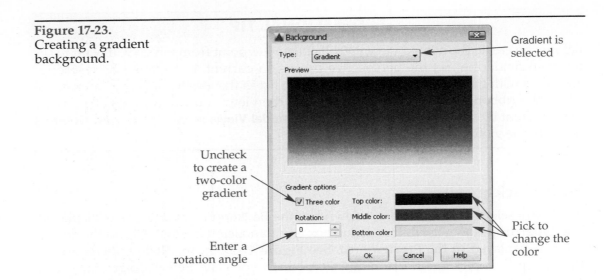

Uncheck
to create a
two-color
gradient

Enter a
rotation angle

Gradient is
selected

Pick to
change the
color

to close the **Background** dialog box. Next, save the view, set the view current, and pick the **OK** button to close the **View Manager** dialog box.

Convincing, clear blue skies can be simulated using the **Gradient** option. Initially, set the **Top**, **Middle**, and **Bottom** color values the same. Then, change the lightness (luminance) in the **True Color** tab in the **Select Color** dialog box. Preview the background and make adjustments as needed.

Using an Image as a Background

An image can be used as a background. This technique can be used to produce realistic or imaginative settings for your models. If you select Image in the drop-down list in the **New View/Shot Properties** dialog box, the **Background** dialog box is displayed with Image selected and a blank image displayed in the **Preview** area. To locate the image file, pick the **Browse...** button to display a standard open dialog box. These image file types may be used for the background: TGA, BMP, PNG, JFIF (JPEG), TIFF, GIF, and PCX. Locate the image file and select **Open**. See **Figure 17-24**.

Once the image file is selected, it must be adjusted. The **Preview** area of the **Background** dialog box shows the image with a preview of a drawing sheet. This drawing sheet indicates how the image is going to be positioned in the view. Pick the **Adjust Image...** button to open the **Adjust Background Image** dialog box. See **Figure 17-25**.

Figure 17-24.
Setting an image as
the background.

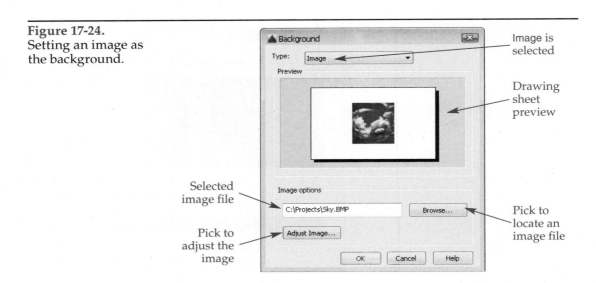

Selected
image file

Pick to
adjust the
image

Image is
selected

Drawing
sheet
preview

Pick to
locate an
image file

Figure 17-25.
Adjusting the
background image.

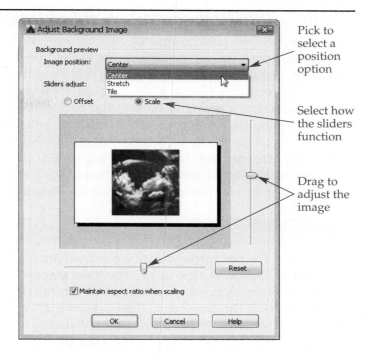

In the **Image position:** drop-down list, pick how the image is applied to the viewport. The Center option centers the image in the view without changing its aspect ratio or scale. The Stretch option centers the image and stretches or shrinks it to fill the entire view. This is one way to plot an image file from AutoCAD. The Tile option keeps the image at its original size and shape, but moves it to the upper-left corner and duplicates it, if needed, to fill the view.

After the image is positioned, use the sliders to adjust it further. The sliders are disabled if Stretch is selected in the **Image position:** drop-down list. The slider function is based on which radio button is picked above the image:

- **Offset**. The sliders move the image in the X or Y direction.
- **Scale**. The sliders scale the image in the X or Y direction. This may distort the image if it is scaled too much in one direction. To prevent distortion, check the **Maintain aspect ratio when scaling** check box at the bottom of the dialog box.

The **Reset** button is located at the bottom-right corner of the preview pane. Picking this button returns the scale and offset settings to their original values.

Once the image is adjusted, pick the **OK** button to close the **Adjust Background Image** dialog box. Then, pick the **OK** button to close the **Background** dialog box. Next, save the view, set the view current, and pick the **OK** button to close the **View Manager** dialog box.

Sun and Sky

If you select Sun & Sky in the drop-down list in the **New View/Shot Properties** dialog box, the **Adjust Sun & Sky Background** dialog box is displayed. AutoCAD uses the settings in this dialog box to simulate the sun in the sky. This dialog box has a preview tile at the top and the **General**, **Sky Properties**, **Horizon**, **Advanced**, **Sun disk appearance**, **Sun Angle Calculator**, **Rendered Shadow Details**, and **Geographic Location** categories. The settings in these categories were discussed earlier in the Sun Properties Palette section in this chapter.

Once the sky is set, pick the **OK** button to close the **Adjust Sun & Sky Background** dialog box. Then, save the view, set the view current, and pick the **OK** button to close the **View Manager** dialog box.

Changing the Background of an Existing View

To change the background of an existing named view, open the **View Manager** dialog box. Select the view name in the **Views** tree on the left-hand side of the dialog box. Then, in the **General** category in the middle of the dialog box, select the Background override property. See **Figure 17-26**. Next, pick the drop-down list for the property and make a selection. Picking None sets the background to the AutoCAD default background. Setting the property to Solid, Gradient, or Image opens the **Background** dialog box, where you can make settings for that type. Selecting Sun & Sky opens the **Adjust Sun & Sky Background** dialog box. Picking Edit opens the **Background** dialog box. Once the background type has been changed or the existing background edited, pick the **OK** button to save the view and close the **View Manager** dialog box.

NOTE

After exiting the **Background** dialog box, you are returned to the **View Manager** dialog box. Picking the **OK** button to exit the **View Manager** dialog box does not necessarily activate the view that you just created or modified. The view must be set current to see the effects of the changes to the background. A view can be set current using the drop-down list in the **Views** panel on the **Visualize** tab of the ribbon, the **View Controls** flyout in the viewport controls, or the **View Manager** dialog box.

Exercise 17-5

www.g-wlearning.com/CAD/

Complete the exercise on the companion website.

Figure 17-26.
Changing the background of an existing, named view.

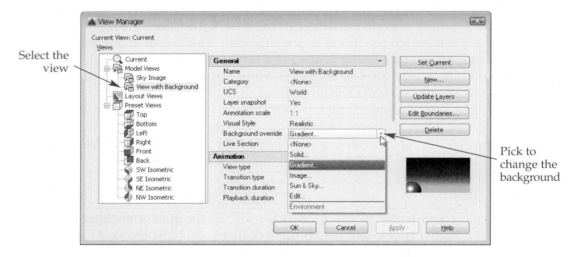

Chapter Review

Answer the following questions using the information in this chapter.

1. Compare and contrast *ambient light*, *distant lights*, *point lights*, *spotlights*, and *weblights*.
2. Define *angle of incidence*.
3. Define *angle of reflection*.
4. A smooth surface has a(n) _____ specular factor.
5. Describe *hotspot* and *falloff*. Which lights have these properties?
6. What is *attenuation*?
7. What are the three types of lighting in AutoCAD?
8. Briefly describe *photometric lighting*.
9. If you want to use the **GEOMAPIMAGE** command to embed a portion of a geographically located map in the drawing, what view must you set?
10. What are the four types of light objects in AutoCAD?
11. What are light glyphs and which lights have them?
12. Which type of light object allows you to incorporate light distribution data from the Illuminating Engineering Society?
13. What must be created before a background can be added to a scene?
14. What are the four types of backgrounds in AutoCAD, other than the default background?
15. Why would you draw a line between the "from" point and "to" point of a distant light?
16. Describe how a gradient background can be used to represent a clear blue sky.

Drawing Problems

1. **Architectural** In this problem, you will draw some basic 3D shapes to create a building similar to an ancient structure, place lights in the drawing, and render the scene.
   A. Begin a new drawing and set the units to architectural. Save the drawing as P17-1.
   B. Draw a 32′ × 22′ planar surface to represent the floor. Using the materials browser, attach a material of your choice to the floor.
   C. Draw cylinders to represent pillars. Make each ⌀2′ × 15′ tall. There are 10 pillars per side. Attach a suitable material to the pillars.
   D. The roof is 32′ × 22′ and 5′ tall at the ridge. Attach an appropriate material.
   E. Create a perspective viewpoint looking into the building.
   F. Turn on sunlight and turn off the default lighting. Set the geographic location to Athens, Greece. Change the date and time to whatever you wish to get the proper lighting and shadows.
   G. Render the scene.
   H. Save the drawing.

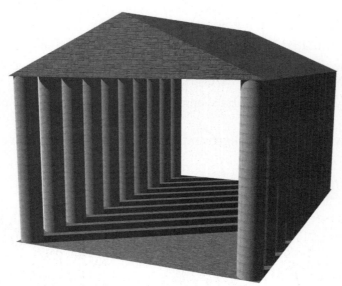

A full-color image is provided on the companion website.   www.g-wlearning.com/CAD/

2. **Architectural** Using the drawing from Problem 1, you will experiment with different lighting types.
   A. Open drawing P17-1 and save it as P17-2.
   B. Turn off the sun.
   C. Place three point lights inside of the building. Evenly space the lights along the centerline of the ceiling. Adjust the light intensity so that the interior is not washed out. Set the color of the middle light to white. Set the color of the outside lights to red or blue. Render the scene.
   D. Turn off the point lights.
   E. Place two spotlights, one pointing from the front corner to the rear corner and the other pointing between the pillars on the left side of the building. Target them at the floor. Render the scene. Adjust the intensity, hotspot, and falloff as needed.
   F. Turn the point lights back on and render the scene with all six lights active. Adjust the light intensities again if the rendering is too washed out with light.
   G. Save the drawing.

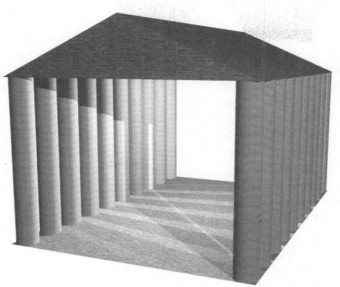

A full-color image is provided on the companion website. www.g-wlearning.com/CAD/

3. **Mechanical** Using a previously created mechanical model, you will apply materials and lights to make it ready for presentation.
   A. Open drawing P6-5 created in Chapter 6. Save it as P17-3.
   B. Draw a planar surface below the flange to represent a tabletop.
   C. Attach an appropriate material, such as a wood or tile material, to the surface.
   D. Open the materials editor and create a new material based on the Metal material type. Experiment with different finish settings until you find one you like. Apply the material to the flange.
   E. Place two spotlights in the drawing and target them at the flange from different angles. Adjust their hotspot, falloff, and intensity to get the proper lighting.
   F. Create a perspective view of the scene. Then, render the scene.
   G. Save the drawing.

A full-color image is provided on the companion website. www.g-wlearning.com/CAD/

Drawing Problems - Chapter 17

4. **Architectural** The building shown will be used to study passive solar heating at different times of the year. Model the building using the overall dimensions given. Use your own dimensions for everything else. The side with the windows should be facing South (–Y in AutoCAD, by default).

A. Set the geographic location to a city in the northern hemisphere.
B. Set the date to midsummer and the time to noon.
C. Turn on sunlight and turn the default lighting off.
D. Render the scene and note the location of the shadows inside the building.
E. Change the date to late winter, render the scene again, and note the new location of the shadows. You can easily switch between the rendered images in the **Render** window by selecting each rendering in the history and statistics pane (this is discussed more in the next chapter).
F. Observing the changes in the shadow locations, what design changes can be made to maximize sun exposure in the cold winter months? What design changes can be made to minimize sun exposure in the heat of summer?
G. Change the geographic location to somewhere closer to the equator. Then, render the scene in summer and winter. How do the shadows compare to those in the previous location?
H. Save the drawing as P17-4.

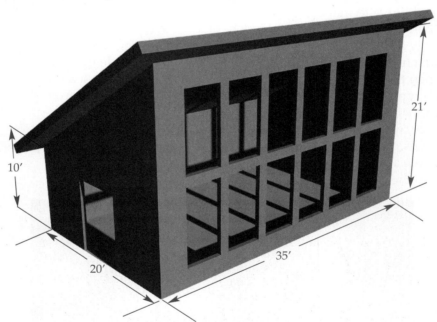

A full-color image is provided on the companion website.   www.g-wlearning.com/CAD/

5. **Architectural** In this problem, you will be adding lights to a model and controlling the light properties to create a pleasing scene. Model the courtyard shown. The overall dimensions are 15′ × 16′ × 4′ (wall height). Use your own dimensions for everything else. Add lights as follows.

A. Turn on the sun and turn off the default lighting. Set the time to late in the day so that the sun is close to the horizon.

B. In the **Sky Properties** category of the **Sun Properties** palette, select Sky Background and Illumination for the Status property.

C. Add a point light at the center of each sphere.

D. Attach a glass material to the spheres that will make them look like they are glowing. There are materials in the Glass category of the Autodesk Library that can be used as a base for creating a new material.

E. Create a fill light above to illuminate the scene. This can be a point light or spotlight.

F. Render the scene using the Low preset to see the lighting effects.

G. Adjust the sun properties to create the look that you want.

H. Adjust the properties of the point lights and any other lights in the scene. You may have to increase the intensity of the lights to properly illuminate the scene.

I. When the scene is illuminated the way you want it, render the scene using the Medium or High preset. If time allows, render the scene using the Coffee-Break Quality or Lunch Quality preset.

J. Save the drawing as P17-5.

A full-color image is provided on the companion website.   www.g-wlearning.com/CAD/

6. **Architectural** In Chapter 8, you completed the kitchen chair model that you started in Chapter 2. In this problem, you will be adding lights to the model and attaching materials to the various components in the chair.
   A. Open P8-8 from Chapter 8.
   B. Create a layer for each of the chair components: seat, legs/crossbars, seatback bow, and seatback spindles.
   C. Draw a planar surface to represent the floor. Place this on its own layer.
   D. Assign materials to each of the layers. You can use materials from the Autodesk Library or create your own materials.
   E. Set the Realistic visual style current.
   F. Adjust material mapping as needed.
   G. Add lighting to the scene.
   H. Render the scene. If you experience problems with the materials, try attaching by object instead of by layer.
   I. Save the drawing as P17-6.

A full-color image is provided on the companion website.   www.g-wlearning.com/CAD/

CHAPTER 18

Advanced Rendering

Learning Objectives

After completing this chapter, you will be able to:

✓ Make advanced rendering settings.
✓ Manage and create render presets.
✓ Save a rendering to an image file.
✓ Add a render environment to a scene.

In Chapter 15, you learned how to create a view of your scene that is more realistic than a visual style. In that chapter, you used AutoCAD's sunlight feature to create a simple rendering with mostly default settings. In this chapter, you will discover how to make a rendering look truly realistic, or *photorealistic*. Some of the advanced rendering features add significantly to the rendering time and you must learn how to balance the quality of the rendering with an acceptable time frame to get the job done.

Render Window

By default, a drawing is rendered in the **Render** window. The **Render** window allows you to inspect the rendering, save it to a file, compare it with previous renderings, and take note of the statistics. See **Figure 18-1**. There are two main areas of the **Render** window—the image pane and the history and statistics pane.

Image Pane

As AutoCAD processes the scene, the image begins to appear in the image pane in its final form. As discussed in Chapter 15, the rendering is processed over a number of levels. The quality of the image improves as the levels are processed. You can cancel the rendering at any time by pressing [Esc] or picking the **Cancel Rendering** button at the top of the image pane.

Immediately below the image pane is the progress meter. The progress meter displays the progress of the rendering. The current level being processed is shown on the left end of the bar, the percentage of completion is shown in the middle, and the elapsed time is shown on the right end. The rendering is completed when the progress meter reaches 100%.

Figure 18-1.
The **Render** window.

Pick to
zoom

Pick to print
a hard copy

Pick to
save the
rendering

Progress
meter

History and
statistics
pane

Right-click
to access
shortcut
menu

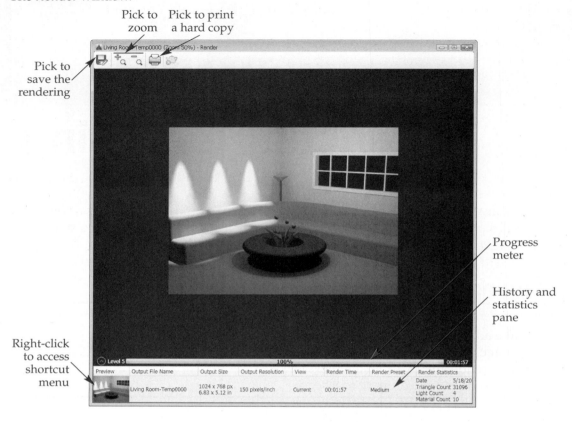

While the rendering is being processed, you can zoom in and out to inspect the image. Use the mouse scroll wheel or the zoom buttons at the top of the window. When the rendering is completed, you can save it to an image file. Pick the save button at the top of the window to display a standard save dialog box. The rendering can be saved to the BMP, TGA, TIF, JPEG, or PNG format. After selecting the file format and picking **Save**, you are prompted to select the quality of the image. The settings are different for each image type.

To print a hard copy of the rendered image, pick the print button at the top of the window. Specify the print device in the **Print** dialog box. This dialog box is similar to the **Plot** dialog box used for printing drawings from model space or paper space.

History and Statistics Pane

The history and statistics pane does not initially appear when the **Render** window is first displayed. To expand or collapse the pane, pick the vertical arrow button to the left of the progress meter. Refer to **Figure 18-1**. The history and statistics pane contains a list of all of the renderings completed in the drawing in the current drawing session. The items in this list are called *history entries*. Information and details about each history entry appear under the column headings at the top of the list.

The Preview column displays a thumbnail view of the rendered image. The Output File Name column lists the name of the rendered image. If the rendering has not been saved, the name is displayed in the format FileName-Temp*n* with *n* representing an incremental number. The Output Size column lists the width and height of the image in pixels followed by inches. The Output Resolution column lists the resolution of the image in pixels per inch. The View column lists the name of the view that was rendered. If a named view was not rendered, the listing is Current. The Render Time column lists the amount of time taken to complete the rendering in hours, minutes, and seconds. The

Render Preset column lists the name of the render preset used to process the rendering. The Render Statistics column lists the date and time of the rendering and other data, including the triangle count, the number of lights active, and the number of materials used on the objects in the scene.

Selecting one of the history entries redisplays it in the image pane. Right-clicking on a history entry displays a shortcut menu. The options in this menu can be used to save the image, render the image again, and manage the entry. The options in the shortcut menu are:

- **Render Again.** Renders the scene again using the same settings. A new entry is not added to the history pane.
- **Save.** Saves the rendered image to a file using a standard save dialog box. The output file name is changed to the new path and file name.
- **Save Copy.** Saves the rendered image to a new file without changing the current entry.
- **Make Render Settings Current.** Makes all of the rendering settings of the entry the current rendering settings in the drawing. This allows you to render the current scene using the settings of the entry.
- **Remove From the List.** Deletes the entry from the history and statistics pane, but any image files saved from the entry remain.
- **Delete Output File.** Deletes the image file created by saving the entry. The entry remains in the history and statistics pane and any image files that were created as copies are retained.

Exercise 18-1

www.g-wlearning.com/CAD/

Complete the exercise on the companion website.

Render Presets

The quickest and easiest way to control the quality of a rendering is with render presets. AutoCAD provides six standard render presets: Low, Medium, High, Coffee-Break Quality, Lunch Quality, and Overnight Quality. See **Figure 18-2**.

The Low, Medium, and High presets are based on the number of rendering levels that are processed. The higher the number of levels, the better the rendering quality. Higher scene complexity will increase rendering times when using these presets.

- Low. This preset produces the lowest quality, but the quickest rendering time. It applies only 1 rendering level with a raytracing depth of 3 (raytracing is discussed later in this chapter). This is good for a quick view of the rendered scene. Do not expect materials, lights, and shadows to be correctly displayed.
- Medium. This preset will take a little longer than the Low preset but produces better quality. It applies 5 rendering levels with a raytracing depth of 5. Lights, shadows, and materials display nicely, providing a quick, nice-quality rendering.
- High. This preset improves upon the Medium preset. It applies 10 rendering levels with a raytracing depth of 7. You may notice a grainy quality to the scene that can be corrected by using one of the timed rendering presets.

The Coffee-Break Quality, Lunch Quality, and Overnight Quality presets are based on the length of time over which rendering occurs. Image quality improves the longer the scene is rendered. Scene complexity does not increase rendering time, but may impact the quality of the rendering.

Figure 18-2.
A comparison of renderings made with the six standard render presets in AutoCAD. A—Low. B—Medium. C—High. D—Coffee-Break Quality. E—Lunch Quality. F—Overnight Quality.

- Coffee-Break Quality. This preset takes 10 minutes to complete and has a low rendering accuracy with a raytracing depth of 3. The quality of the rendering may be similar to that produced by the Medium preset.
- Lunch Quality. This preset improves upon the Coffee-Break Quality preset. It takes 1 hour to complete and has a low rendering accuracy with a raytracing depth of 5.
- Overnight Quality. This is the highest-quality rendering preset. It takes 12 hours to complete. It applies the highest rendering accuracy with a raytracing depth of 7.

The render presets can be selected in the **Render** panel on the **Visualize** tab of the ribbon or from the drop-down list in the **Render Presets Manager** palette. The **Render Presets Manager** palette also allows you to create and use your own render presets, as discussed in the next section.

Render Settings

The **Render Presets Manager** palette is used to control basic render settings. See **Figure 18-3**. The **RPREF** command opens the palette. The palette can also be displayed by picking the dialog box launcher button at the lower-right corner of the **Render** panel on the **Visualize** tab of the ribbon. The settings in this palette are discussed in the following sections.

Output Settings

The settings at the top of the **Render Presets Manager** palette specify the location where AutoCAD processes the rendering and the output size of the rendered image. The same settings are available in the **Render** panel on the **Visualize** tab of the ribbon. Picking the **Render** button at the top-right corner of the palette renders the scene with the current settings.

There are three settings in the **Render in:** drop-down list. The **Window** setting renders the scene in the **Render** window. This is the default setting. The **Viewport** setting renders the scene directly to the viewport. Because the area of the rendering will most likely be larger, expect the rendering to take longer to process. However, it will be easier to see the details of the rendered scene. Zooming, panning, or regenerating will cause the rendering to disappear. You can use the **SAVEIMG** command to save the rendered image to an image file. However, the quality of the result is dependent on the resolution of the monitor, which may produce a low-resolution image. The **Region** setting also renders the scene directly to the viewport. When using this setting, you are prompted to pick two points defining a rectangular region to render. AutoCAD only processes the objects in this region. This is an excellent way to quickly perform a test rendering and check on lighting, materials, and shadows. As when rendering with the **Viewport** setting, the rendering is temporary and will disappear once you zoom or pan in the viewport.

Figure 18-3.
The **Render Presets Manager** palette.

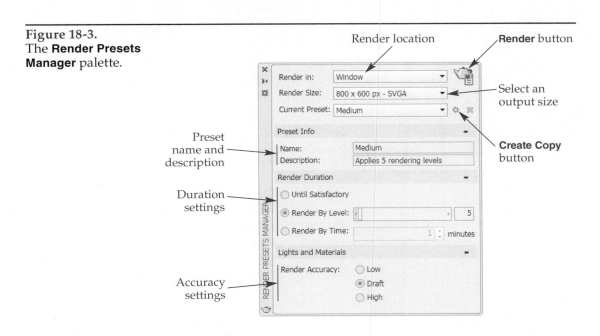

The **Render Size:** drop-down list is available if you are rendering to the **Render** window. The selections in this drop-down list specify the output size of the rendered image in pixels. The seven settings in the top section range from 800×600 (SVGA) to 1920×1080 (full HDTV). The four settings below the top section are designed for printing the image at 300 dpi (dots per inch) on 8 1/2″ × 11″, 11″ × 17″, ISO A4, and ISO A3 size sheets.

Selecting **More Output Settings…** opens the **Render to Size Output Settings** dialog box, which provides additional settings for file output. See **Figure 18-4**. The **Image Size** drop-down list contains the same output sizes available in the **Render Presets Manager** palette, but you can enter your own width and height using the **Width** and **Height** text boxes. Sizes can be entered in pixels, inches, or centimeters by selecting the appropriate format from the **Units** drop-down list. To maintain the same ratio of width to height when changing the size, check the **Lock Image Aspect** check box. The **Resolution** text box setting determines the image resolution. Higher values produce a larger file size, but improve the quality of the image when printed. The resolution can be set in pixels per inch or centimeter, which can be thought of as dots per inch or centimeter. A good rule of thumb is to set the resolution close to the maximum resolution of the printer that you will be using. The approximate file size resulting from the size and resolution settings is displayed below the **Resolution** text box.

The **Save Image** area is used to save an image file with a specified name and file path. If the **Automatically save rendered image** check box is checked, a file is automatically saved when the rendering is generated. Check the check box and then pick the **Browse…** button to specify a file name and type.

Picking the **Render To Size** button at the bottom of the dialog box renders the scene using the specified output settings. If a custom output size has been defined, it appears in the list of sizes in the **Render Presets Manager** palette and the **Render to Size** drop-down list in the ribbon after the rendering is completed. To save the custom output size and close the **Render to Size Output Settings** dialog box without rendering, pick the **Save** button.

Figure 18-4.
The **Render to Size Output Settings** dialog box.

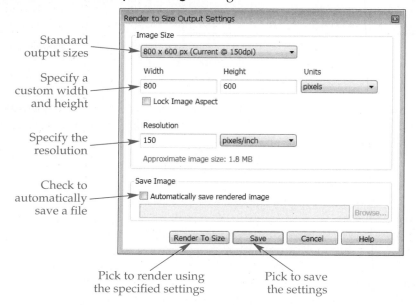

Current Preset and Preset Information

The **Current Preset:** drop-down list in the **Render Presets Manager** palette is used to select one of the six standard render presets or a custom render preset. Picking the **Create Copy** button next to the drop-down list creates a custom preset, as discussed later in this chapter.

The entries in the **Name:** and **Description:** text boxes in the **Preset Info** area correspond to the currently selected preset in the **Current Preset:** drop-down list. These entries are read-only when a standard preset is selected. The **Name:** and **Description:** text box entries can be edited when creating or editing a custom preset.

Render Duration Settings

The settings in the **Render Duration** area allow you to control the duration of the rendering in three different ways. The current setting in this area corresponds to the currently selected preset in the **Current Preset:** drop-down list. If you select a different radio button when a standard preset is current, a custom preset is created. Creating a custom preset is discussed later in this chapter.

The **Until Satisfactory** setting is used to render without a predefined number of processing levels or set time limit. With this setting, you can simply observe the rendering and then cancel when it reaches a point where the quality is good enough. Picking the **Cancel Rendering** button at the top of the **Render** window ends processing and allows you to save the image to file. The **Render By Level:** setting is used to set the number of levels that you want AutoCAD to process. Enter a value in the text box or use the slider to set the value. The **Render By Time:** setting is used to specify the amount of time the rendering takes to complete. Enter the number of minutes in the text box or use the up and down arrows to set the value.

Render Accuracy Settings

The settings in the **Lights and Materials** area are used to specify the quality of the rendering based on a specific lighting model. The current **Render Accuracy:** setting corresponds to the currently selected preset in the **Current Preset:** drop-down list. If you select a different radio button when a standard preset is current, a custom preset is created. Creating a custom preset is discussed in the next section.

The **Low** setting uses the simplified lighting model. This produces quick results but is not very realistic. Reflections and refractions are limited and global illumination is inactive. Global illumination is discussed later in this chapter.

The **Draft** setting uses the basic lighting model. Global illumination is active and reflections and refractions are more realistic. Using this setting may be good enough for some situations.

The **High** setting uses the advanced lighting model. This takes longer to complete, but produces the most realistic results. Global illumination is active and reflections and refractions are calculated at a high level.

Creating a Custom Render Preset

Custom render presets are created in the **Render Presets Manager** palette. The six standard presets can be used as a starting point when creating a custom render preset. Set the appropriate preset current and then pick the **Create Copy** button next to the **Current Preset:** drop-down list. The name of the new preset appears in the **Current Preset:** drop-down list in the format Name-Copy *n* with *n* representing an incremental number. Use the **Name:** and **Description:** text boxes to change the default name and description. Then, adjust the rendering settings as desired. The new custom render preset is available in the **Current Preset:** drop-down list and in the **Render Presets** drop-down list in the **Render** panel on the **Visualize** tab of the ribbon.

Another way to create a custom render preset is to select a standard preset in the **Current Preset:** drop-down list and then change one of the settings in the **Render Duration** or **Lights and Materials** area, as previously discussed. To delete a custom render preset, select the preset in the **Current Preset:** drop-down list and pick the **Delete** button next to the **Create Copy** button. The standard render presets cannot be deleted.

Rendering Fundamentals

Rendering requires significant computer processing resources and much is happening when your computer is processing the scene. As previously discussed, rendering times increase at higher resolution and accuracy settings. The following sections will help you understand the process and troubleshoot problems when they arise.

Sampling

Sampling is a process that tests the scene color at each pixel and then determines what the final color should be. This is most important in transition areas, such as edges of objects or shadows. Increasing the sampling will smooth out the jagged edges and incorrect coloring, but increase rendering time. Increasing the number of samples computed at each pixel means that the color displayed by each pixel will be more accurate. You may also notice thicker lines in the final rendering. Higher render accuracy settings will have higher sampling rates.

Raytracing

Raytracing is a method of calculating reflections, refractions, and shadows by tracing the path of the light rays from the light sources. The render accuracy settings previously described specify the maximum number of times that a ray can be reflected or refracted. See **Figure 18-5.** Try to imagine a light ray bouncing off reflective objects or traveling through transparent objects in your scene. Count how many surfaces the object must contact before it comes to rest. This can help you determine if you will need to increase the render accuracy settings.

Indirect Illumination

Indirect illumination is an AutoCAD mechanism that simulates natural, bounced light. If indirect illumination is turned off and light does not directly strike an object, the object is black. Without indirect illumination enabled, other lights must be added to the scene to simulate indirect illumination. If you have placed lights in your scene and you render the scene using the Low render preset, you will notice that only the objects directly in the light's path are illuminated. See **Figure 18-6.** This is not how lights work in the real world.

Global Illumination

Global illumination (GI) is an indirect illumination process. Bounced light is simulated by generating photon maps on surfaces in the scene. These maps are created by tracing photons from the light source. Photons bounce around the scene from one object to the next until they finally strike a diffuse surface. When a photon strikes a surface, it is stored in the photon map. This map is calculated for each scene you render. The lights you create and the materials added to the objects in your scene determine the complexity of this map.

Figure 18-5.
A—If the raytracing depth is limited to 3, a light ray will bounce three times and then stop at the next surface. B—If transparent objects exist in the scene, refractions need to be calculated as well as reflections. The light ray will still make it to the back wall as long as the render accuracy settings are high enough. C—Lower render accuracy settings may mean that the light ray will not make it to the back wall.

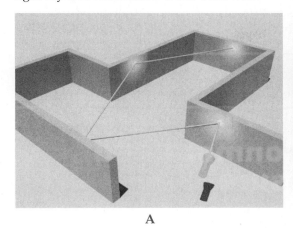

A

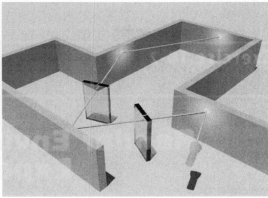

B

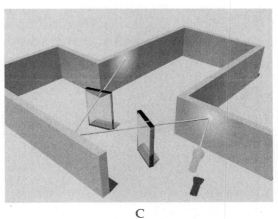

C

Figure 18-6.
A—This scene has a single point light and has been rendered with the Low render preset. The portion of the floor in the shadow behind the box primitive is not in the direct path of light and is not illuminated. B—Using a higher render preset activates indirect illumination. Notice that the portion of the floor in the shadow is illuminated and visible. The Coffee-Break Quality render preset is used in this example.

A

B

Final Gathering

The global illumination process produces a photon map that may have dark and light areas. In other words, it may be splotchy. *Final gathering* increases the number of rays in the rendering and cleans up these artifacts. It will also greatly increase rendering time. If you render a scene with one of the higher render presets and watch it all the way through, you may actually see this phase. Final gathering works best with scenes that contain overall diffuse lighting. See **Figure 18-7**.

Exercise 18-2

www.g-wlearning.com/CAD/

Complete the exercise on the companion website.

Render Environment and Exposure

The render environment settings allow for the addition of lighting and a background image based on a predefined environment. The **RENDEREXPOSURE** or **RENDERENVIRONMENT** command displays the **Render Environment & Exposure** palette. See **Figure 18-8**. In this palette, you can quickly set up a scene for rendering

Figure 18-7.
A—This scene has a single point light. B—Global illumination is active. Notice the unevenness of the lighting. This can be seen especially on the sofa and in the corner of the walls.
C—Final gathering cleans up artifacts and provides a more even illumination.

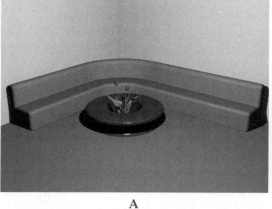

A

B

C

RENDEREXPOSURE

Ribbon
Visualize
> Render

Render Environment and Exposure

Type
RENDER-EXPOSURE
RENDER-ENVIRONMENT

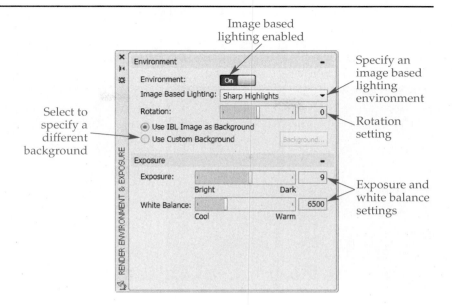

Figure 18-8.
The **Render Environment & Exposure** palette.

Image based lighting enabled

Specify an image based lighting environment

Rotation setting

Select to specify a different background

Exposure and white balance settings

without having to create and position lights or adjust sunlight settings. You can also adjust the exposure settings for lighting in the scene. The **Render Environment & Exposure** palette is divided into two areas: **Environment** and **Exposure**. The settings in these areas are discussed in the following sections.

Environment Settings

To use the render environment settings, the switch at the top of the **Render Environment & Exposure** palette must be set to **On**. When the switch is set to **Off**, AutoCAD uses one of the other types of lighting for the scene (default lighting, sunlight, or user-created lighting, as discussed in Chapter 17).

The settings in the **Environment** area are used to apply image based lighting and a background image to the scene. *Image based lighting* simulates a 360° environment around the scene. The lighting used to illuminate the scene is an image lighting map associated with a specific environment, such as an outdoor scene or an interior setting. The environment can be rotated to reposition the light source and change how the scene is illuminated. This provides a way to quickly adjust the lighting to enhance the scene and achieve different results.

The **Image Based Lighting:** drop-down list contains 11 different environments from which to choose. The Gypsum Crater, Dry Lake Bed, Plaza, Snow Field, and Village environments provide a panoramic background image that is applied to the scene when rendered. These environments are designed for exterior scenes and are described below. See **Figure 18-9.**

- Gypsum Crater. In this environment, the ground is covered in grayish rock. There are hills covered with green foliage in the background and a blue sky appears beyond. The lighting is similar to bright sunshine.
- Dry Lake Bed. This is a flat sandy desert environment with bluffs and mountains in the background. The sky is pale blue and the lighting is bright and harsh.
- Plaza. This environment sets the scene in the center of a cobblestone plaza in the late afternoon. Buildings with European architecture surround the plaza. The lighting has a hazy quality to it.
- Snow Field. This environment sets the scene in the middle of a flat snow-covered plane. A road below has been plowed and the sky is mostly overcast with gray wispy clouds in the background. Lighting is even and diffuse.

Figure 18-9.
A comparison of the image based lighting environments for exterior scenes used with an outdoor cooker model. A—Gypsum Crater. B—Dry Lake Bed. C—Plaza. D—Snow Field. E—Village.

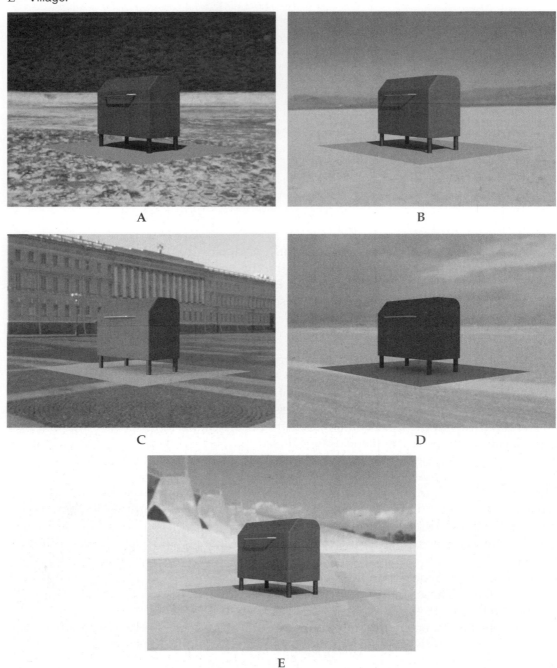

- Village. The name of this environment may be misleading. The scene appears to be at the end of a concrete spillway in bright sunshine. The structures in the background may create an interesting scene.

The Sharp Highlights, Rim Highlights, Grid Lights, Cool Light, Warm Light, and Soft Light environments simulate an indoor setting and have a background that resembles the inside of a room. These environments provide different lighting choices and are very useful for creating presentations of consumer products, mechanical parts, or vehicles. They are described as follows.

- Sharp Highlights. This environment provides a dark and light background with a strong low light from one direction. See **Figure 18-10**.
- Rim Highlights. This environment provides a gray indistinct background with diffuse overhead lighting.
- Grid Lights. In this environment, the scene is positioned inside a room with uniform grid lighting from above.
- Cool Light. This environment is similar to the Grid Lights environment with light panels providing highlights on objects in the scene.
- Warm Light. This environment is the same as the Cool Light environment, but the light is more like incandescent lighting.
- Soft Light. This environment is similar to the Grid Lights environment, but the lighting is dim and diffuse. Very few highlights are produced on objects in the scene.

NOTE

The projection must be set to perspective when using one of the image based lighting environments. An image based lighting background cannot be displayed in the viewport, but displays when the scene is rendered.

The **Rotation:** setting below the **Image Based Lighting:** drop-down list is used to rotate the environment around your scene. The light source rotates with the background, allowing you to position it to your liking. Use the slider to change the rotation or enter a value in the text box.

If the **Use IBL Image as Background** radio button is selected, the scene is rendered with the image based lighting background. Selecting the **Use Custom Background** radio button allows you to use a background of your choice. You can set the background to a solid, gradient, or image. Pick the **Background...** button to display the **Image Based Lighting Background** dialog box. Select a background type from the **Type:** drop-down list and then make the appropriate settings. The background is displayed in the viewport and is used with the specified image based lighting when the scene is rendered.

A full-color image is provided on the companion website.   www.g-wlearning.com/CAD/

Figure 18-10.
A preliminary battery pack design rendered using the Sharp Highlights image based lighting environment.

Exposure Settings

The **Exposure:** and **White Balance:** settings in the **Exposure** area of the **Render Environment & Exposure** palette are used to adjust the brightness and color temperature of lighting in the scene. These are the same settings as the **Light Exposure** and **Light White Balance** settings in the expanded **Lights** panel on the **Visualize** ribbon tab, discussed in Chapter 17. The **Exposure:** setting is used to brighten or darken the scene and the **White Balance:** setting is used to make the lighting warmer or cooler. After you specify environment settings and render the scene, you can use the exposure and white balance settings to make fine-tuning adjustments to the lighting. These are global controls that affect the entire scene.

PROFESSIONAL TIP

Renderings can take a long time to process and may require great amounts of computer resources such as memory and powerful graphics hardware. The Autodesk A360 rendering service is available for generating renderings using online resources. Moving the processing online allows you to save time and free up your computer for other tasks. A user account is required to use the service.

To access the Autodesk A360 rendering service, pick the **Render in Cloud** button located on the **A360** panel on the **Visualize** tab of the ribbon. Selecting the **Render Gallery** button on the **A360** panel allows you to display your online rendering status and completed renderings.

Exercise 18-3

www.g-wlearning.com/CAD/

Complete the exercise on the companion website.

CHAPTER 19

Using Show Motion to View a Model

Learning Objectives

After completing this chapter, you will be able to:

✓ Explain the show motion tool.
✓ Create still shots of 3D models.
✓ Create walk shots of 3D models.
✓ Create cinematic shots of 3D models.
✓ Replay single shots and a sequence of shots.
✓ Change the properties of a shot.

AutoCAD's *show motion* tool is a powerful function that allows you to create named views, animated shots, and basic walkthroughs. It can quickly display a variety of named shots. It is also used to create basic animated presentations and displays. This capability is especially useful for animating 3D models that do not require the complexity and detail of fully textured, rendered animations and walkthroughs. Walkthroughs are presented in Chapter 20.

A *view* is a single-frame display of a model or drawing from any viewpoint. A *shot* is the manner in which the model is put in motion and the way the camera *moves* to that view. Therefore, a single, named view can be modified to create several different shots using camera motion and movement techniques. Using show motion, you can create shots in one of three different formats:

- Still.
- Walk.
- Cinematic.

A *still shot* is exactly the same as a named view, but show motion allows you to add transition effects to display it. A *walk shot* requires that you use the **Walk** tool to define a camera motion path to create an animated shot. A *cinematic shot* is a single view to which you can add camera motion and movement effects to display the view. These shots are all created using the **New View/Shot Properties** dialog box. This is the same dialog box used to create named views.

Understanding Show Motion

Show motion is simply a means for creating, manipulating, and displaying named views. The **NAVSMOTION** command is used for show motion. The process involves the creation of shots using the **ShowMotion** toolbar and the **New View/Shot Properties** dialog box. Once a shot is created, you can give it properties that allow it to be displayed in many different ways. If a saved shot does not display in the manner you desire, it is easily modified.

A *view category* is a heading under which different views are filed. It is not necessary to create view categories, especially if you will be making just a few views. On the other hand, if you are working on a complex model and need to create a number of views with a variety of cinematic and motion characteristics, it may be wise to create view categories.

The process of using show motion to create, modify, and display shots begins by first using the **ShowMotion** toolbar. All of your work with shots will be performed using the **New View/Shot Properties** dialog box. Options in this dialog box change based on the type of shot that is selected. These options are discussed in the sections that follow relating to each kind of shot.

PROFESSIONAL TIP

As you first start working with show motion, you may want to create view categories named Still, Cinematic, and Walk. As you create shots, file each under its appropriate category name. This will assist you in seeing how the different shot-creation techniques work. Later, you can apply the view categories to specific components of a complex model. For example, a subassembly of a 3D model may have a view category that contains all three types of shots for use in different types of modeling work or presentations. Creating and using view categories is discussed in detail at the end of this chapter.

Show Motion Toolbar

NAVSMOTION

Type
NAVSMOTION MOTION

Toolbar
Navigation Bar

ShowMotion

The **ShowMotion** toolbar is displayed at the bottom-center of the screen when the **NAVSMOTION** command is entered. See **Figure 19-1**. The quickest way to display the **ShowMotion** toolbar is to pick the **ShowMotion** button on the navigation bar. The **ShowMotion** toolbar provides controls for creating and manipulating views:

- **Unpin ShowMotion/Pin ShowMotion**
- **Play All**
- **Stop**
- **Turn on Looping**
- **New Shot...**
- **Close ShowMotion**

When the toolbar is pinned, it remains displayed if you execute other commands, minimize the drawing, change ribbon panels, or switch to another software application. The **Unpin ShowMotion** button is used to unpin the toolbar. The **Pin ShowMotion** button is then displayed in its place. If you unpin the toolbar, you must execute the **NAVSMOTION** command each time you wish to use show motion.

Pick the **Play All** button to play all of the shots displayed as thumbnails above the toolbar. The playback of these shots will loop (repeat) if the **Turn on Looping** button is selected. If looping is turned on, a shot, category, or all categories are displayed in a

---

Figure 19-1.
The **ShowMotion**
toolbar is displayed
at the bottom of the
screen and provides
controls for creating
and manipulating
shots.

ShowMotion toolbar

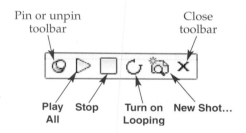

Pin or unpin
toolbar

Close
toolbar

Play   Stop
All

Turn on   New Shot...
Looping

loop whenever the **Play All** button is selected. The **Turn on Looping** button is a toggle. The button image changes to indicate whether looping is turned on.

Picking the **New Shot...** button opens the **New View/Shot Properties** dialog box. This dialog box is discussed in the next section. Picking the **Close ShowMotion** button closes the **ShowMotion** toolbar.

New View/Shot Properties Dialog Box

All shot creation takes place inside of the **New View/Shot Properties** dialog box. It is accessed by using the **ShowMotion** toolbar as previously described or by using the **NEWSHOT** command. The dialog box can also be displayed from within the **View Manager** dialog box by picking the **New...** button. The **New View/Shot Properties** dialog box is shown in **Figure 19-2**.

You must supply a shot name and a shot type. A view category is not required, but can help organize shots. This feature is discussed later in the chapter. The dialog box provides two tabs containing options that allow you to define the overall view and then specify the types of movement and motion desired in the shot. These tabs are discussed in the next sections.

Toolbar

ShowMotion

New Shot...

NEWSHOT

Figure 19-2.
The **New View/Shot Properties** dialog box is used to create the three types of shots: still, walk, and cinematic.

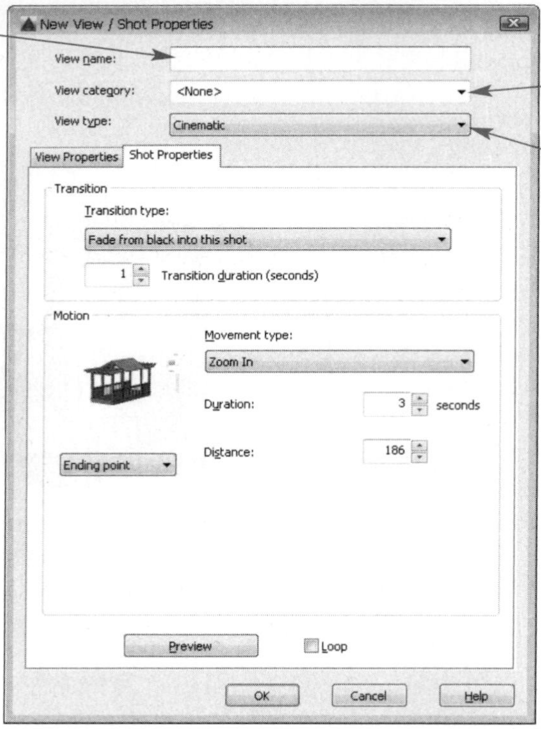

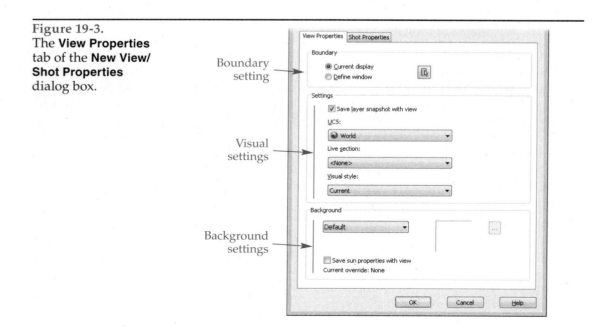

Enter a name

Select a category

Select a view type

View Properties Tab

The **View Properties** tab of the **New View/Shot Properties** dialog box is composed of three areas in which you can specify the overall presentation of the model view. See **Figure 19-3**. The overall presentation includes the view boundary, visual settings, and background.

Boundary Settings

The setting in the **Boundary** area of the **View Properties** tab determines what is displayed in the shot. The **Current display** radio button is on by default. This means the shot will be composed of what is currently shown in the drawing area.

Figure 19-3.
The **View Properties** tab of the **New View/Shot Properties** dialog box.

Boundary setting

Visual settings

Background settings

You can adjust the view by picking the **Define window** radio button. This temporarily closes the dialog box so you can draw a rectangular window to define the view. After picking the second corner, you can adjust the view by picking the first and second corners of the window again. Press [Enter] to accept the window and return to the dialog box.

Visual Settings

The **Settings** area of the **View Properties** tab contains options that apply to the overall display of the model in the shot. When the **Save layer snapshot with view** check box is checked, all of the current layer visibility settings are saved with the new shot. This is checked by default.

Any UCS currently defined in the drawing can be selected for use with the new shot. Use the **UCS:** drop-down list to select the UCS. When the shot is restored, that UCS is restored, too. If you select <None> in the drop-down list, there is no UCS associated with the shot.

Live sectioning is a tool that allows you to view the internal features of 3D solids that are cut by a section plane object. This feature is covered in detail in Chapter 14. When a section plane object is created, it is given a name. Therefore, if a model contains one or more section plane objects, their names appear in the **Live section:** drop-down list. If a section plane object is selected in this list, the new shot shows the live sectioning for that plane.

The **Visual style:** drop-down list contains all visual styles in the drawing plus the options of Current and <None>. Selecting a visual style from the list results in that style being set current when the shot is played. Selecting Current or <None> will cause the shot to be displayed in the visual style currently displayed on the screen. A detailed discussion of visual styles is provided in Chapter 15.

NOTE

A shot created with live sectioning on will be displayed in that manner even though live sectioning may be currently turned off in the drawing. However, subsequent displays of the model that were created with live sectioning off are shown with it on. Keep this in mind as you develop shots for show motion.

Background Settings

The **Background** area of the **View Properties** tab provides options for changing the background of the new shot. The drop-down list in this area allows you to select a solid, gradient, image, or sun and sky background. You can also choose to retain the default background. If you pick Solid, Gradient, or Image, the **Background** dialog box appears. If you select Sun & Sky, the **Adjust Sun & Sky Background** dialog box appears. These dialog boxes are used to set the background and are discussed in detail in Chapter 17.

The **Save sun properties with view** setting is used to apply the sunlight data to the view. If you are displaying an architectural model using sunlight, you will likely want this check box checked for show motion. Sunlight and geographic location are discussed in detail in Chapter 17.

Shot Properties Tab

Settings in the **Shot Properties** tab of the **New View/Shot Properties** dialog box provide options for controlling the transition and motion of shots. There are numerous movement and motion options in this tab. The specific options available are based on the type of shot selected: still, walk, or cinematic. In addition, the cinematic shot contains a variety of motions that can be applied to the shot and each type of motion contains a number of variables. The options in the **Shot Properties** tab are discussed later in this chapter as they apply to different shots, movements, and motions.

Creating a Still Shot

A still shot is the same as a named view, but with a transition. Open the **New View/ Shot Properties** dialog box and enter a name in the **View name:** text box. Next, pick in the **View category:** text box and type a category name or select an existing category using the drop-down list. Remember, it is not necessary to create or select a category at this time. Now, select Still from the **View type:** drop-down list.

NOTE

The Still view type is also used to create named views. A *named view* is a single-frame display of a model from a given viewpoint. Named views can be used for any purpose in AutoCAD, not just for show motion. For example, it is useful to create named views for use in rendering. Named views are saved with the drawing and can be displayed by accessing the drop-down list in the **Views** panel on the **Visualize** tab of the ribbon. A named view can be displayed at any time, and can be used for any show motion view.

In the **Shot Properties** tab, select a transition from the **Transition type:** drop-down list. You can select from one of three transitions:
- **Fade from black into this shot.** The screen begins totally black and fades into the current background color.
- **Fade from white into this shot.** The screen begins totally white and fades into the current background color.
- **Cut to shot.** The view is immediately displayed without a transition and the shot movements are applied.

The fade transitions will not function unless hardware acceleration is enabled. This feature is normally activated by default upon software installation. However, you can confirm this by checking the appearance of the **Graphics Performance** tool in the status bar. See **Figure 19-4**. The tool is shaded blue if hardware acceleration is turned on. Hover over the tool to display the tooltip, which should read **Hardware acceleration - On**. If hardware acceleration is not on, right-click on the tool to display the shortcut menu and select **Graphics Performance** to open the **Graphics Performance** dialog box. Refer to **Figure 19-4B**. In the **Graphics Performance** dialog box, pick the **Hardware Acceleration** button to enable or disable hardware acceleration. Other graphics effects can also be enabled or disabled in this dialog box. See **Figure 19-5**.

After hardware acceleration has been enabled, display the **New View/Shot Properties** dialog box, enter a name, and select a transition. Next, in the **Transition duration (seconds)** text box, enter a length of time over which the transition will occur.

Figure 19-4.
Hardware acceleration must be activated for the show motion fade transitions to function. A—Hover over the **Graphics Performance** tool to confirm that hardware acceleration is turned on. B—The shortcut menu displayed by right-clicking on the **Graphics Performance** tool.

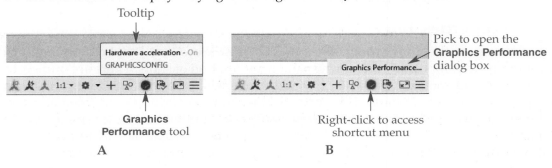

Figure 19-5.
The **Graphics Performance** dialog box.

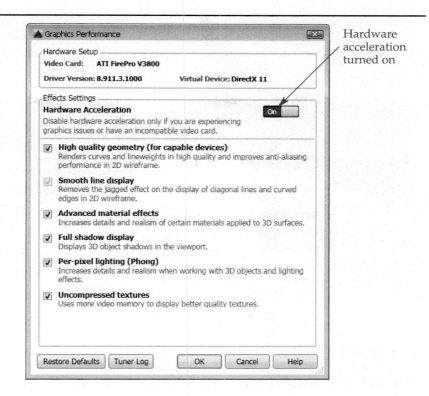

Hardware acceleration turned on

If you want a fade transition to be complete, be sure to enter a value in the **Duration:** text box that is equal to or greater than the transition duration. Pick the **Preview** button to view the shot, and then edit the transition and motion values as needed.

When finished, pick the **OK** button to close the **New View/Shot Properties** dialog box. The thumbnail image for the new shot is displayed above the **ShowMotion** toolbar. See **Figure 19-6A**. The large thumbnail image represents the view category. Since there was no view category selected for the view, the name <None> is displayed. The small thumbnail image represents the shot just created. Its name may be truncated. Move the cursor into the shot thumbnail image and a large image is displayed. The name of the shot should appear in its entirety. The category thumbnail image is reduced to a small image. See **Figure 19-6B**.

NOTE

If the **Loop** check box at the bottom of the **Shot Properties** tab is checked, the shot will continuously loop through the transition during playback. This is similar to picking the **Turn on Looping** button in the **ShowMotion** toolbar. All three shot types have this option.

Exercise 19-1   www.g-wlearning.com/CAD/

Complete the exercise on the companion website.

Creating a Walk Shot

To create a walk shot, open the **New View/Shot Properties** dialog box, enter a name, and select a category, if needed. Next, select Recorded Walk from the **View type:** drop-down list. In the **View Properties** tab, set up the view as described for a still shot.

Figure 19-6.
A—The large thumbnail image on the bottom is for the category. The small thumbnail image is for the shot just created. Its name may be truncated. B—Move your cursor into the shot thumbnail image and it converts to a large image. The category thumbnail image is reduced to a small image.

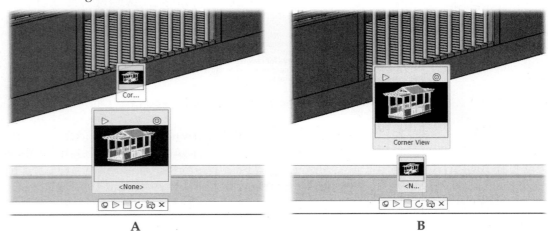

A   B

In the **Shot Properties** tab, set up the transition as described for a still shot. The only option in the **Motion** area of the tab is the **Start recording** button. See **Figure 19-7.** The **Duration:** text box is grayed out because the value is based on how long you record the walk. The camera drop-down list below the preview image is also grayed out because the walk begins at the current display and ends at the point where you terminate it.

This type of shot uses the same **Walk** tool found in the steering wheels, discussed in Chapter 3. After picking the **Start recording** button, the **New View/Shot Properties** dialog box is hidden and the **Walk** tool message is displayed. Pick and drag to activate the **Walk** tool. As soon as you pick, the center circle icon is displayed. If needed, you

Figure 19-7.
Creating a walk shot.

Transition settings

Pick to record the walk

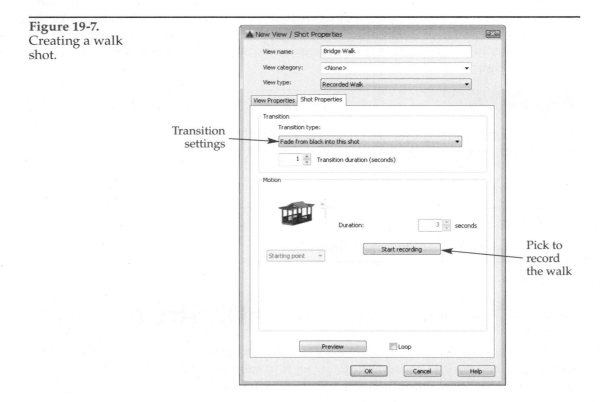

can hold down the [Shift] key to move the view up or down. When the [Shift] key is released (with the mouse button still held down), you can resume walking. The mouse button must be depressed the entire time to record all movements. As soon as you release the button, the recording ends and the dialog box is redisplayed.

Finally, preview the shot. If you need to re-record it, pick the **Start recording** button and begin again. You cannot add to the shot; you must start over. Pick the **OK** button to save the shot and a new thumbnail is displayed in the **ShowMotion** toolbar.

NOTE

The walk shot capability of show motion is limited in its ability to produce a true "walk-through." The path of your "walk" is inexact and it may take several times to create the effect you need. Should you wish to create a genuine walkthrough of an architectural or structural model, it is better to use tools such as **3DWALK** (walkthrough), **3DFLY** (flyby), or **ANIPATH** (motion path animation). These powerful commands are used to create professional walkthroughs and animations and are covered in Chapter 20.

Exercise 19-2

www.g-wlearning.com/CAD/

Complete the exercise on the companion website.

Creating a Cinematic Shot

To create a cinematic shot, open the **New View/Shot Properties** dialog box, enter a name, and select a category, if needed. Next, select Cinematic from the **View type:** drop-down list. In the **View Properties** tab, set up the view as described for a still shot. In the **Shot Properties** tab, pick a transition type and duration as described for a still shot.

The specific options in the **Motion** area of the **Shot Properties** tab are discussed in the next sections, **Figure 19-8**. Pick a type of movement in the **Movement type:** drop-down list. The movement options available will change depending on the type of movement you select. Refer to the chart in **Figure 19-9** as a quick reference for movement options when creating a cinematic shot. Finally, pick the current position of the camera from the camera drop-down list below the preview image.

Pick the **Preview** button to view the shot. Make any changes required before picking the **OK** button to save the shot. Once the **OK** button is picked, the new view is displayed as a small thumbnail image above the large category thumbnail image. See **Figure 19-10**.

Cinematic Basics

The motion and movement options available for a cinematic shot allow you to create a final display that appears to move into position as if the camera is traveling in a path toward the object. The **Motion** area of the **Shot Properties** tab contains a variety of options for creating an array of cinematic shots. Refer to **Figure 19-8**. The options can be confusing unless you understand some basics about essential components of a cinematic shot. The most important aspect is the preview of the current view. All of the motion actions revolve around this view. This image tile is the current position of the camera and is also referred to as the *key position* of a shot. This is the position that is displayed when you pick the **Go** button on a thumbnail image above the **ShowMotion** toolbar.

Figure 19-8.
Creating a cinematic shot.

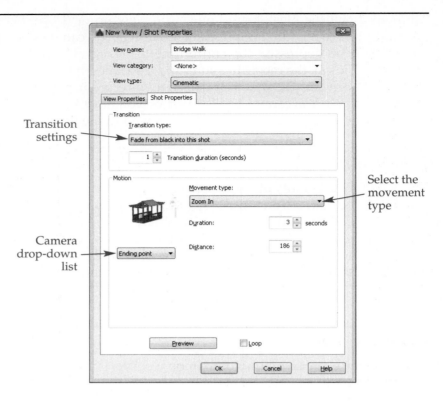

Transition settings

Camera drop-down list

Select the movement type

Figure 19-9.
This chart shows the different cinematic movement options based on the movement type selected in the **Movement type:** drop-down list in the **New View/Shot Properties** dialog box.

	Duration	Distance	Look at Camera Point	Distance Up	Distance Back	Distance Down	Distance Forward	Shift Left/Right	Degrees Left/Right	Degrees Up/Down
Zoom In	X	X								
Zoom Out	X	X								
Track Left	X	X	X							
Track Right	X	X	X							
Crane Up	X		X	X	X			X		
Crane Down	X		X			X	X	X		
Look	X								X	X
Orbit	X								X	X

Figure 19-10.
The thumbnail image for the new cinematic shot is displayed above the category thumbnail image.

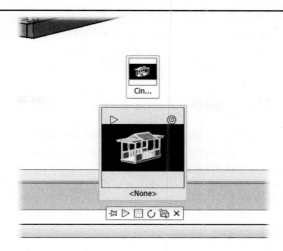

The two elements of a cinematic shot are motion and movement. *Motion* relates to the behavior of the object and how it appears to be in motion during the cinematic shot. In addition, it refers to the position of the model at a specified point in the animation. *Movement* in a cinematic shot is the manner in which the camera moves in relation to the object.

PROFESSIONAL TIP

If your goal is to create a series of shots that blend together, it is a good idea to first develop a storyboard of the entire sequence. This could be as simple as a few notes indicating how you want the shots to move or even a few sketches noting the required movements and motion values. Planning your shots will save time when you begin creating them in AutoCAD.

Camera Drop-Down List

The camera drop-down list is located below the preview image in the **Motion** area of the **Shot Properties** tab. The preview image represents the position of the camera based on the option selected in the camera drop-down list. The following three options are available in the drop-down list, **Figure 19-11.**

- **Starting point.** The view in the preview image is the display that will be shown at the start of the cinematic shot. All movement options begin at this point.
- **Half-way point.** The view in the preview image is the display that will be shown at the half-way point of the cinematic shot.
- **Ending point.** The view in the preview image is the display that will be shown at the end of the cinematic shot. All movement options take the shot to this point.

Figure 19-11.
Selecting what the key position represents.

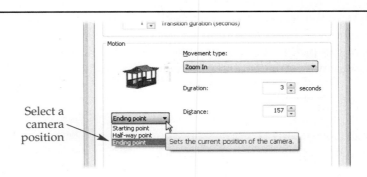

These options, in part, determine how the cinematic shot is created. If Starting point is selected, the cinematic shot begins with the current view. During the animation, the model may move off the screen. If this happens, you may want to select Ending point so the model appears to move into position and stay there.

PROFESSIONAL TIP

The **Preview** button in the **New View/Shot Properties** dialog box is the best way to test any changes you make to the motion options. Each time you make a change, pick the button to see if the effect is what you want. Previewing each change is far more efficient than making several changes before you examine the results.

Movement Type

The **Movement type:** drop-down list in the **Shot Properties** tab of the **New View/Shot Properties** dialog box is used to select the motion for the cinematic shot. There are eight types of camera movement that can be used with a cinematic shot:

- Zoom in.
- Zoom out.
- Track left.
- Track right.
- Crane up.
- Crane down.
- Look.
- Orbit.

The options available in the **Motion** area of the tab are based on which movement type is selected. The movement types and their options are discussed in the next sections.

Zoom In

When Zoom In is selected in the **Movement type:** drop-down list, the camera appears to zoom into the model in the shot. This movement type has two options, **Figure 19-12**. The value in the **Duration:** text box is the length of time over which the animation is recorded. The value in the **Distance:** text box is the distance the camera travels during the animation. The camera zooms in to cover the distance in the specified duration of time. When Zoom In is selected, the camera drop-down list is automatically set to Ending point. Keep in mind that with this option the current position of the camera represents the final display after the cinematic shot is complete.

Zoom Out

When Zoom out is selected in the **Movement type:** drop-down list, the camera appears to zoom away from the model in the shot. This movement type has **Duration:** and **Distance:** text box options, as described for the Zoom In movement type.

When Zoom out is selected, the camera drop-down list is automatically set to Starting point. With this setting, the camera will zoom out the specified distance and the final image in the shot will be smaller than the preview image. If this is not the effect you want, it may be better to pick Ending point for the current position of the camera. Then, the model will appear to move from behind your view to stop at the current view.

Figure 19-12.
The settings for the Zoom In movement type.

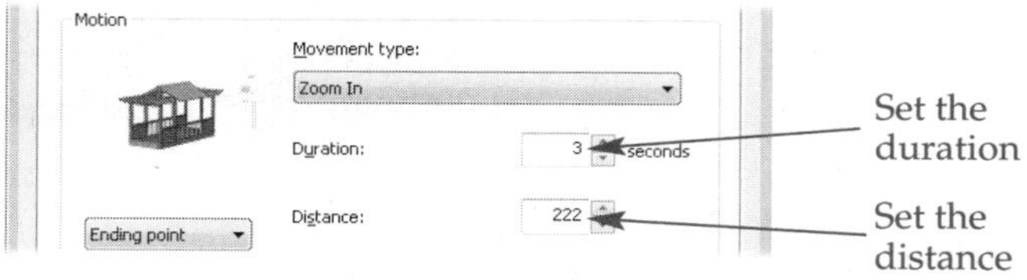

Set the duration

Set the distance

Track Left

When Track left is selected in the **Movement type:** drop-down list, the camera will move from right to left. This results in the view moving from left to right. This movement occurs over the specified distance and duration. The Track left movement type has **Duration:** and **Distance:** text boxes as described previously and the **Always look at camera pivot point** check box, **Figure 19-13**.

If the **Always look at camera pivot point** check box is checked, the center of the view remains stationary. As a result, the view in the shot appears to rotate about this point instead of sliding across the screen. The best way to visualize this motion is to use the **Preview** button with the option checked and then unchecked. This option is available for all "track" and "crane" movement types.

When Track left is selected, the camera drop-down list is automatically set to Half-way point. This means the preview image is the middle point of the animation. It will be displayed at the midpoint of the **Duration:** value.

PROFESSIONAL TIP

If the **Distance:** value is large, the screen may be blank for a few moments until the camera moves enough to bring the model into view. If this is not what you want, decrease the **Distance:** value or check the **Always look at camera pivot point** check box.

Track Right

When Track right is selected in the **Movement type:** drop-down list, the camera will move from left to right. This results in the view moving from right to left. This movement type has **Distance:** and **Duration:** text boxes and the **Always look at camera pivot point** check box. These options function in the same manner as described for the Track left movement type. When Track right is selected, the camera drop-down list is automatically set to Half-way point.

Crane Up

Where the "track" movement types move the view left and right, the "crane" movement types move the view up and down. When Crane up is selected in the **Movement type:** drop-down list, the camera will move from bottom to top and then backward. This results in the view moving from top to bottom and zooming out. This movement type has the **Duration:** text box and the **Always look at camera pivot point** check box described previously. The **Always look at camera pivot point** check box is checked by default.

This movement type has three additional options, **Figure 19-14**. The value in the **Distance Up:** text box is the distance the camera is moved upward. The value in the **Distance Back:** text box is the distance the camera is moved backward. The backward movement is typically short compared to the upward movement.

Figure 19-13.
The settings for the Track left movement type.

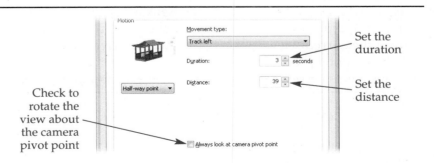

Check to rotate the view about the camera pivot point

Set the duration

Set the distance

Figure 19-14.
The settings for the Crane up movement type.

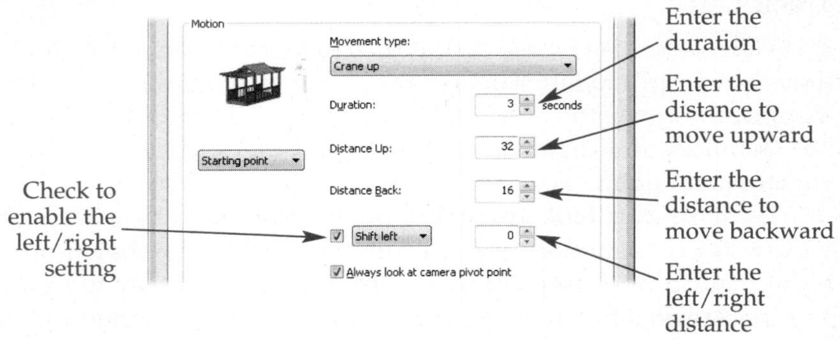

Enter the duration

Enter the distance to move upward

Enter the distance to move backward

Check to enable the left/right setting

Enter the left/right distance

You can add more interest to the motion in the shot by shifting the view left or right. First, enable the shift option by checking the check box below the **Distance Back:** setting. Refer to **Figure 19-14.** Then, select either Shift left or Shift right from the drop-down list. Finally, enter a distance in the text box next to the drop-down list. The camera will be shifted left or right for the distance specified in this text box, resulting in the view shifting in the opposite direction. This shifting option is available for both "crane" motion types.

When Crane up is selected, the camera drop-down list is automatically set to Starting point. This means the preview image shows the beginning of the shot before the movement is applied. If the **Always look at camera pivot point** check box is not checked, and depending on the movement settings, the model may move off the screen in the shot.

Crane Down

When Crane down is selected in the **Movement type:** drop-down list, the camera will move from top to bottom and then forward. This results in the view moving from bottom to top and zooming in. This movement type has the **Duration:** text box and the **Always look at camera pivot point** check box described previously. The **Always look at camera pivot point** check box is checked by default. This movement type also has the left/right shifting option described in the previous section.

Instead of **Distance Up:** and **Distance Back:** settings, this movement type has **Distance Down:** and **Distance Forward:** settings. The value in the **Distance Down:** text box is the distance the camera cranes down in the shot. The value in the **Distance Forward:** text box is the distance the camera is moved forward in the shot.

When Crane down is selected, the camera drop-down list is automatically set to Ending point. This means the preview image shows the end of the shot before the movement is applied. If the **Always look at camera pivot point** check box is not checked, and depending on the movement settings, the model may start off the screen in the shot.

NOTE

All distance values related to the Crane up and Crane down movement types represent how far away from the view the camera must begin before the cinematic shot is started. For example, a large **Distance Back:** value means that the camera may have to begin at a point beyond or even around the view in order to travel the distance back to display the current view (when Ending point is selected in the camera drop-down list). It is always good practice to make a single change to the movements, and then preview the shot. This is especially true for the Crane up and Crane down movement types.

Look

When Look is selected in the **Movement type:** drop-down list, the camera pans based on the values for left/right and up/down to display the view, **Figure 19-15.** For example, if the movement is set to look up 45° and Ending point is selected in the camera drop-down list, then the camera begins at a 45° angle below the view and looks up to display it.

The value in the **Duration:** text box is the length of time over which the animation is recorded. This option is the same as described for the other movement types.

The first drop-down list below the **Duration:** text box is used to specify either left or right movement. Select Degrees left in the drop-down list to have the camera move from right to left, resulting in the view moving from left to right. Select Degrees right to have the camera move from left to right and the view right to left. Next, specify the angular value for this movement in the degrees text box to the right of the drop-down list. For example, suppose you select Degrees left and enter an angle of 15°. In this case, the camera will start 15° to the *right* of the model and rotate to the left in the shot.

The second drop-down list below the **Duration:** text box is used to specify either up or down movement. Select Degrees up in the drop-down list to have the camera move from bottom to top, resulting in the view moving from top to bottom. Select Degrees down to have the camera and view move in the opposite direction. Specify the angular value for this movement in the degrees text box to the right of the drop-down list.

When Look is selected in the **Movement type:** drop-down list, the camera drop-down list is automatically set to Starting point. This means the shot will start with the current view and then apply the movement settings.

Orbit

When Orbit is selected in the **Movement type:** drop-down list, the camera rotates in place based on the values set for left/right and up/down movements. The Orbit movement type has the same options as the Look movement type, as described in the previous section.

When Orbit is selected in the **Movement type:** drop-down list, the camera drop-down list is automatically set to Starting point. This means the shot will start with the current view and then apply the movement settings. Unlike the Look movement type, the view center remains stationary in the shot. As a result, you do not need to worry about the model moving off the screen.

Exercise 19-3

www.g-wlearning.com/CAD/

Complete the exercise on the companion website.

Figure 19-15.
The settings for the Look movement type.

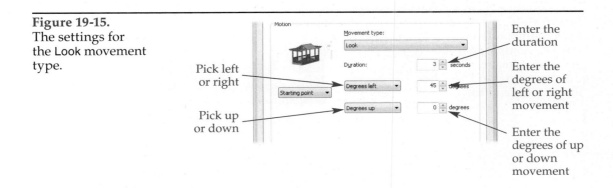

Pick left or right

Pick up or down

Enter the duration

Enter the degrees of left or right movement

Enter the degrees of up or down movement

Displaying or Replaying a Shot

As each shot is created, its thumbnail image is placed above the **ShowMotion** toolbar. The shot name is displayed below the thumbnail image. By default, the category thumbnail image is large and each shot thumbnail image is small. If the cursor is moved over one of the shot thumbnail images, all of the shot thumbnail images are enlarged and the category thumbnail image is reduced in size.

Each thumbnail image is composed of an image of the view in the shot, the shot name, and viewing controls. See **Figure 19-16**. The viewing controls are only displayed when the cursor is moved over the thumbnail image. The three viewing controls are:

- **Play.** Plays the shot. If looping is enabled, the shot repeats. Otherwise, the shot is played once. When this button is picked, it changes to **Pause**. You can also play the shot by picking anywhere on its thumbnail image. The **VIEWPLAY** command can be used to replay a shot.
- **Pause.** Pauses the shot. When this button is picked, it changes to **Play**. Picking anywhere inside the image or on the **Play** button restarts playing of the shot.
- **Go.** Displays the key position of the view without playing the shot. The **VIEWGO** command also restores a named view.

You can play all of the shots in sequence by picking the **Play** button in the view category thumbnail image or by picking anywhere inside of the view category thumbnail image. Additionally, you can move to the key position of the first shot in a view category by picking the **Go** button on the view category thumbnail image.

Exercise 19-4

www.g-wlearning.com/CAD/

Complete the exercise on the companion website.

Thumbnail Shortcut Menu

Right-clicking on a shot or view category thumbnail image displays the shortcut menu shown in **Figure 19-17**. This menu provides quick access for modifying and manipulating shots and their thumbnail images.

Picking the **New View/Shot...** entry displays the **New View/Shot Properties** dialog box. Picking the **Properties...** entry displays the **View/Shot Properties** dialog box. This dialog box is the same as the **New View/Shot Properties** dialog box. However, all of the settings of the shot are displayed and can be changed. This option is only available to change the properties of a shot, not a category.

Figure 19-16.
Each thumbnail image is composed of the shot image, the shot name, and the viewing controls.

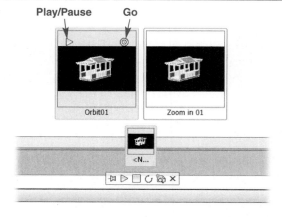

Figure 19-17.
Right-click on a shot
or view category
thumbnail image
in the **ShowMotion**
toolbar to display
this shortcut menu.

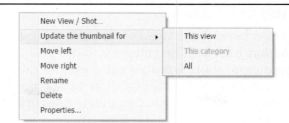

To rename a shot, pick the **Rename** entry in the shortcut menu. The name is high-lighted below the thumbnail image. Type the new name and press [Enter]. To delete a shot, select **Delete** from the shortcut menu. There is no warning; the shot is simply deleted. The **UNDO** command can reverse this action.

If there is more than one shot in a category, you can rearrange the order. Right-click on a thumbnail image and pick either **Move left** or **Move right** in the shortcut menu. This moves the shot one step in the selected direction.

When a change is made to the model, it is not automatically reflected in the thumbnail images. If you do not update thumbnail images, they will remain in their original format, regardless of how many changes are made to the model. Selecting **Update the thumbnail for** in the shortcut menu displays a cascading menu with options for updating the thumbnail image:

- **This view.** Updates only the thumbnail image for the shot on which you right-clicked.
- **This category.** Updates all thumbnail images in the category. This option is only enabled when you right-click on a category thumbnail image.
- **All.** This option updates all thumbnail images in all categories and shots.

Creating and Using View Categories

A *view category* is a grouping that can be created in order to separate different types of shots. A view category can also contain shots arranged in a sequence so that they appear connected as they are played together. This is an optional feature, but an efficient method of separating different types of shots or grouping shots to use for a specific purpose. When a new category is created, it is represented by a category thumbnail image displayed above the **ShowMotion** toolbar.

To create a new view category, pick in the **View category:** drop-down list text box in the **New View/Shot Properties** dialog box. Then, enter a name. See **Figure 19-18**. The name is added to the drop-down list. If you wish to organize shots by categories, be sure to select the view category from the drop-down list before picking the **OK** button to exit the dialog box and create the shot.

You must create a shot to create a view category. If you delete the only shot in a view category, you also delete the category. The view category will no longer be available in the **New View/Shot Properties** dialog box.

Figure 19-18.
To create a new view category, type its name in the **View category:** drop-down list text box.

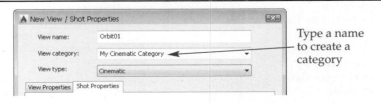

View Category Basics

After a new view category is created, it is represented by a thumbnail image above the **ShowMotion** toolbar. The category name is shown below the thumbnail image. As you create more shots within a category, the thumbnail images for the new shots are placed to the right of existing shots above the category thumbnail image. The thumbnail image for the first shot in a view category is displayed as the view category thumbnail image.

An entire view category and all of the views included in it can be quickly deleted. Simply right-click on the view category thumbnail image and pick **Delete** in the shortcut menu. An alert box appears asking you to confirm the deletion, **Figure 19-19**. The shots in a deleted view category cannot be recovered, so be sure there are no shots in the category you wish to save before deleting. However, you can use the **UNDO** command to reverse the deletion.

PROFESSIONAL TIP

Remember, you can change the order of the shots in a view category using the shortcut menu. Picking **Move left** or **Move right** moves the shot one step. Continue moving shots until the order is appropriate.

Playing and Looping Shots in a View Category

A view category is a useful tool for grouping shots to be played together in a sequence. To play the shots in a view category, move the cursor over the view category thumbnail image above the **ShowMotion** toolbar. Then, pick the **Play** button in the thumbnail image. All shots in the category will be played. You can also use the **SEQUENCEPLAY** command to play the shots in a view category.

If you want the shots in a category to run on a continuous loop, pick the **Turn on Looping** button in the **ShowMotion** toolbar. Then, when the **Play** button is picked, the shots in the view category will display until the **Pause** button is picked either on the **ShowMotion** toolbar or in the view category thumbnail image. Pressing the [Esc] key also stops playback. To return to single-play mode, pick the **Turn off Looping** button on the **ShowMotion** toolbar. This button replaces the **Turn on Looping** button.

Changing a Shot's View Category and Properties

If you put a shot in the wrong view category, it is simple to move the shot to a different category. Right-click on the shot thumbnail image and select **Properties...** in the shortcut menu. This displays the **View/Shot Properties** dialog box. Select the proper category name in the **View category:** drop-down list and pick the **OK** button. The shot thumbnail image will move into position above its new view category thumbnail image.

Figure 19-19.
To remove a view category, right-click on its thumbnail image in the **ShowMotion** toolbar and pick **Delete** in the shortcut menu. This alert box appears to confirm the deletion.

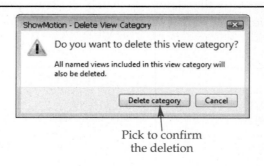

Pick to confirm the deletion

To change shot properties, right-click on the shot thumbnail image and pick **Properties...** in the shortcut menu. The **View/Shot Properties** dialog box is displayed. All properties of the shot can be changed. Change the settings as needed. Always preview the shot before you pick the **OK** button to save it.

PROFESSIONAL TIP

Try this procedure for creating a series of shots in a view category that are to be played in sequence.
1. Play each shot to determine where it should be located in the sequence.
2. Move shots left or right in the category as needed to create the proper sequence.
3. Play the category to determine if the shots properly transition from one to the next.
4. If a subsequent shot does not begin where the previous shot ended, right-click on that shot, pick **Properties...**, and make the necessary adjustments in the **View/Shot Properties** dialog box.
5. Play the category again.
6. Repeat the editing process with each shot until the category plays smoothly.

Exercise 19-5

www.g-wlearning.com/CAD/

Complete the exercise on the companion website.

Chapter Review

Answer the following questions using the information in this chapter.

1. For what is the *show motion tool* used?
2. Define *view* as it relates to the show motion tool.
3. Define *shot* as it relates to the show motion tool.
4. List the formats in which a shot can be created.
5. List the six buttons on the **ShowMotion** toolbar.
6. What is *live sectioning* and how can it be included in a shot?
7. Which type of shot is the same as a named view, but with a transition?
8. Which type of shot requires you to navigate through the view as you record the motion?
9. What is the *key position* of a cinematic shot?
10. Define *motion* and *movement* as they relate to a cinematic shot.
11. What is the purpose of the camera drop-down list in the **Shot Properties** tab of the **New View/Shot Properties** or **View/Shot Properties** dialog box?
12. List the eight types of camera movement for a cinematic shot.
13. Briefly describe two ways to play a single shot.
14. What is a *view category*?
15. How is an entire category of shots replayed?

Drawing Problems

1. **General** Open one of your 3D drawings from Chapter 17 or 18. Do the following.
   A. Display a pictorial view of the drawing.
   B. Create a still shot. Use settings of your choice. Create a view category and place the shot in it.
   C. Display a different view of the drawing and create another still shot. Place the shot in the view category you created.
   D. Display a third view of the drawing and create a third still shot. Place the shot in the view category you created.
   E. Play each shot. If necessary, rearrange the shots. Then, play all shots in the category.
   F. Save the drawing as P19-1.

2. **General** Open one of your 3D drawings from Chapter 7 or 8. Do the following.
   A. Display the objects in the Conceptual or Realistic visual style.
   B. Toggle the projection from parallel to perspective.
   C. Create two still shots and four cinematic shots of the model. Use a different motion type with each of the cinematic shots.
   D. Create two new view categories and place one still and two cinematic shots in each view category.
   E. Edit the shots in each view category to create a smooth motion sequence.
   F. Save the drawing as P19-2.

3. **General** Open one of your 3D drawings from Chapter 6. Do the following.
   A. Display a pictorial view.
   B. Create a walk shot. Place it in a view category named Walk Shots.
   C. Play the shot. How does this compare to a cinematic shot as far as ease of creation?
   D. Save the drawing as P19-3.

4. **General** Open drawing P19-3 and do the following.
   A. Create a new cinematic shot using the Orbit motion type. Place it in a view category named Cinematic Shots.
   B. Create another cinematic shot using the Track left or Track right movement type. Check the **Always look at camera pivot point** check box. Place the view in the Cinematic Shots category.
   C. Create a third cinematic shot using the Crane up or Crane down movement type. Make sure the **Always look at camera pivot point** check box is checked. Place the view in the Cinematic Shots category.
   D. Play the Cinematic Shots category. Edit the shots as needed to create a smooth display.
   E. Create three still shots. Place them in a view category named Still Shots. Try creating three different gradient backgrounds to simulate dawn, noon, and dusk.
   F. Play the Still Shots category. Edit the shots as needed to create a smooth display.
   G. Save the drawing as P19-4.

CHAPTER 20

Cameras, Walkthroughs, and Flybys

Learning Objectives

After completing this chapter, you will be able to:

✓ Create a camera to define a static 3D view.
✓ Activate and adjust front and back clipping planes.
✓ Record a walkthrough of a 3D model to a movie file.
✓ Record a flyby of a 3D model to a movie file.
✓ Create walkthroughs and flybys by following a path.
✓ Control the viewpoint, speed, and quality of the animation.

Once you have a 3D design complete, or even while still in the conceptual phase of design, you may want to take a stroll through the model and have a look around. You may also want to strap on some wings and fly over and around the model to see it from above. A *walkthrough animation* shows a scene as a person would view it walking through the scene. Walkthroughs are typically used to show the interior of a building, but can be created for exterior scenes as well. A *flyby animation* is similar to a walkthrough, except that the person is not bound by gravity. In other words, the scene is viewed as a bird flying above would see it. Flybys often show the exterior of a building.

The **3DWALK** command is used to create a walkthrough by recording views as a camera "walks" through the scene. The **3DFLY** command is very similar, but the movement of the camera is not limited to a single Z value. The **ANIPATH** command allows you to draw a path and link the camera to the path. This chapter discusses these commands and other methods used to create the animation you need. In addition, this chapter discusses creating and using cameras.

Creating Cameras

Cameras are used in AutoCAD to store a viewpoint and easily recall it later when needed for viewing or rendering the scene. After the camera is established, you can zoom, pan, and orbit as needed and then come back to the camera view. It is not necessary to create a camera before using the **3DWALK**, **3DFLY**, and **ANIPATH** commands (discussed later in this chapter) because these commands create their own cameras.

The **CAMERA** command allows you to add a camera to the scene. Cameras are normally placed in the plan view of the scene to make it easy for you to pick where you want to "stand" and where you want to "look." Once the command is selected, you are first prompted to specify the camera location. A camera glyph is placed in the

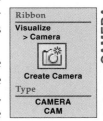

Ribbon
Visualize
> Camera

Create Camera
Type
CAMERA
CAM

CAMERA

scene at the camera location, **Figure 20-1.** Next, you must specify the target location. As you move the cursor before picking the target location, a pyramid-shaped field of view indicates what will be seen in the view. Once you select the target location, the command remains active for you to select an option:

Enter an option [?/Name/LOcation/Height/Target/LEns/Clipping/View/eXit]<eXit>:

The list, or ?, option allows you to list the cameras in the drawing. Select this option and type an asterisk (*) to show all of the cameras in the drawing. You can also enter a name or part of a name and an asterisk. For example, entering HOUSE* will list all of the cameras whose name begins with HOUSE, such as HOUSE_SW, HOUSE_SE, and HOUSE_PLAN.

The **Name** option allows you to change the name of the camera as you create it. If you do not rename the camera, it is given a default, sequential name, such as Camera1, Camera2, Camera3, and so on. It is always a good idea to provide meaningful names for cameras. Names such as Living Room_SW, Corner, or Hallway_Looking East leave no doubt as to what the camera shows. If you choose not to rename the camera at this point, it can be renamed later using the **Properties** palette.

The **Location** option allows you to change the placement of the camera. Enter the option and then specify the new location. You can enter coordinates or pick a location in the drawing.

The **Height** option allows you to change the vertical location of the camera. Enter the option and then enter the height of the camera. The value you enter is the number of units from the current XY plane. If you are placing the camera in a plan view, this option is used to tilt the view up or down from the current XY plane.

The **Target** option allows you to change the placement of the camera target. Enter the option and then specify the new location. You can enter coordinates or pick a location in the drawing.

The **Lens** option allows you to change the focal length of the camera lens. If you change the lens focal length, you are really changing the field of view, or the area of the drawing that the camera covers. The lower the lens focal length, the wider the field of view angle. The focal length is measured in millimeters.

The **Clipping** option is used to turn the front and back clipping planes on or off. These planes are used to limit what is shown in the camera view. Clipping planes are discussed later in this chapter.

The **View** option is used to change the current view to that shown by the camera. This option has two choices—**Yes** or **No**. If you select **Yes**, the active viewport switches to the camera view and the **CAMERA** command ends. If you select **No**, the previous prompt returns.

Once you have made all settings, press [Enter] or select the **Exit** option to end the command. The view (camera) is listed with the other saved views in the drop-down list in the **Views** panel on the **Visualize** tab of the ribbon and in the **View Controls** flyout of the viewport controls. Selecting the view makes it the current view in the active

Figure 20-1.
A camera is represented by a glyph. When the camera is selected, the field of view (shown in color) and grips are displayed.

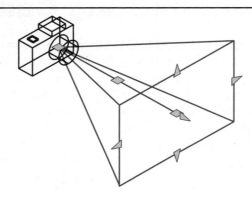

viewport. The view is also listed under the Model Views branch in the **View Manager** dialog box. It can be made current by selecting the view, picking the **Set Current** button, and then picking the **OK** button.

PROFESSIONAL TIP

In addition to the camera name, many other camera properties can be changed in the **Properties** palette. The camera and target locations can be changed, the lens focal length and field of view can be adjusted, and the clipping planes can be modified. Also, you can change the roll angle, which is the rotation about a line from the camera to the target, and set the camera glyph to plot.

Camera System Variables

The **CAMERADISPLAY** system variable controls the visibility of camera glyphs. When set to 1, camera glyphs are displayed. When set to 0, camera glyphs are not displayed. Creating a camera automatically sets the variable to 1. The **Camera Display** button in the **Camera** panel on the **Visualize** tab of the ribbon toggles the display of camera glyphs off and on.

When creating a camera, if you pick the camera and target locations without using object snaps, you may assume that the camera and target are located on the XY plane (Z coordinate of 0) of the current UCS. This may or may not be true. The **CAMERAHEIGHT** system variable determines the default height of the camera if a Z coordinate is not provided. It is a good idea to set this variable to a typical eye height before placing cameras. There is no corresponding system variable for the target because the target is usually placed by snapping to an object of interest. If X and Y coordinates are entered for the target location, but a Z coordinate is not provided, the Z value is automatically 0. If a camera was previously created in the drawing session and the **Height** option was used, that height value becomes the default camera height.

Cameras Tool Palette

The **Cameras** tool palette provides a quick way to add a camera, but the default tools do not allow for the options described earlier. The **Normal Camera** tool creates a camera with a 50 mm focal length. This camera simulates normal human vision. The **Wide-angle Camera** tool creates a camera with a 35 mm focal length. This type of view is commonly used for scenery or interior views where it is important to show as much as possible with minimal distortion. The **Extreme Wide-angle Camera** tool creates a camera with a 6 mm focal length. This camera produces a fish-eye view, which is very distorted and mainly useful for special effects.

Changing the Camera View

Once the camera is placed, it is easy to manipulate. If you select a camera, the **Camera Preview** window is displayed by default. This window shows the view through the camera, **Figure 20-2**. The view in the window can be displayed in any of the 3D visual styles or any named visual style. Select the visual style in the drop-down list in the window. If the **Display this window when editing a camera** check box at the bottom of the window is unchecked, the window is not displayed the next time a camera is selected. The next time the drawing is opened, this setting is restored (checked).

When a camera is selected, grips are displayed. Refer to **Figure 20-1**. If you hover the cursor over a grip, a tooltip appears indicating what the grip will alter. Picking the base grip on the camera allows you to reposition the camera in the scene. If the **Camera Preview** window is open, watch the preview as you move the camera to help

Figure 20-2.
The **Camera Preview** window is displayed, by default, when a camera is selected.

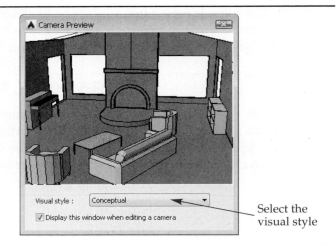

Select the visual style

guide you. Selecting the grip on the target allows the target to be repositioned. Again, use the preview in the **Camera Preview** window as a guide. The grip at the midpoint between the camera and target can be used to reposition the camera and target at the same time. If you pick and move one of the arrow grips on the end of the field of view, the lens focal length and field of view are changed.

Camera Clipping Planes

Clipping planes allow you to suppress objects in the foreground or background of your scene. Picture these clipping planes as flat, 2D objects perpendicular to the line of sight that can be moved closer to or farther from the viewer. Only the objects between the front and back clipping planes and within the field of view are seen in the camera view. This is helpful for eliminating walls, roofs, or any other clutter that may take away from the focus of the scene. Clipping planes can be set while creating the camera or later using the **Properties** palette. Clipping planes can be adjusted using grips.

To set the clipping planes while creating the camera, enter the **Clipping** option. You are prompted:

Enable front clipping plane? [Yes/No] <No>:

To enable the front clipping plane, select **Yes**. You are then asked to specify the offset from the target plane. The target plane is described next. Once you enter the offset, or if you answer **No**, you are prompted:

Enable back clipping plane? [Yes/No] <No>:

To enable the back clipping plane, select **Yes** and then specify the offset from the target plane.

The *target plane* is the 2D plane that is perpendicular to the line of sight and passing through the camera's target point. Offsets for both front and back clipping planes are from this plane. Positive values place the clipping planes between the camera and the target plane. Negative values place the planes on the opposite side of the target plane from the camera. You can place the clipping planes anywhere in the scene from the camera location to infinity. You cannot, however, place the back clipping plane between the front clipping plane and the camera.

The best way to adjust clipping planes is using the **Properties** palette or grips. Create the camera and then display a plan view of the camera and target (an approximate plan view is okay). Select the camera and open the **Properties** palette. In the **Clipping** category, select the Clipping property. In the property drop-down list, select Front on, Back on, or Front and back on to turn on the appropriate clipping plane(s). Notice that the clipping planes are visible in the viewport, **Figure 20-3**. Next, enter offset values for the Front plane and Back plane properties as appropriate or use grips to set the locations of the clipping planes. By displaying a plan view of the camera and target, you can see where the clipping planes are located and visualize their effect on the scene. If the **Camera Preview** window is open, the clipping is displayed in the preview.

Exercise 20-1

www.g-wlearning.com/CAD/

Complete the exercise on the companion website.

Animation Preparation

The tools presented in this chapter make it easy to lay out a path, plan camera angles, and record the movement of the camera. The resulting animation can be directly output to a number of movie file types that can be shared with others. However, there are some decisions to make first.

It is important to plan exactly what you want to see in the animation. Think like a movie director and plan the "shots." Ask these questions:
*   What will be visible from each camera angle?
*   Is there a background in place?
*   Is the lighting appropriate?
*   Will a simple walkthrough suffice or will a flyby be necessary?
*   How close is the viewer (camera) going to be to the objects in the scene?

The answers to these questions will help determine the modeling detail required. Do not model anything that will not be seen. Also, do not place detailed materials on objects that are not the focus of the animation. Processing the animation may take a

Figure 20-3.
Adjusting the clipping planes for a camera.

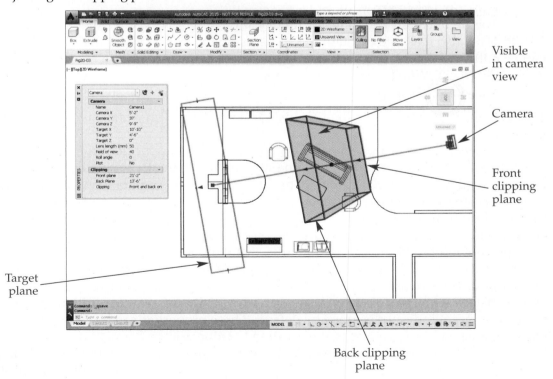

long time. Unnecessary detail may bog down the computer. In addition, walkthroughs and flybys must be created in views with perspective, not parallel, projection.

The "visual quality" of the scene has the biggest impact on the time involved in rendering the animation. An animation can be rendered in any visual style or using any render preset that is available in the drawing. It is a natural tendency to render at the highest level to make the animation look the best. However, a computer animation has a playback rate of 30 frames per second (fps). If a single frame (view) takes three minutes to render, how long will it take to render a 30 second animation? An animation 30 seconds in length has 900 frames (30 fps × 30 seconds). If each frame takes three minutes to render, the entire animation will take 2700 minutes, or 45 hours, to render.

Are you willing to wait two or three days for a 30 second movie? How about your boss or your client? There are trade-offs and concessions to be made. Perform test renderings on static views and note the rendering time. Then, decide on the acceptable level of quality versus rendering time and move ahead with it.

Walking and Flying

The process for creating a walkthrough or a flyby is the same. First, the command is initiated. Then, the movement is defined and recorded. Finally, the recorded movement is saved to an animation file. Note: the **Animations** panel in the **Visualize** tab on the ribbon may not be displayed by default.

When using the **3DWALK** and **3DFLY** commands, you can move through the scene using the arrow keys or the [W], [A], [S], and [D] keys on the keyboard to control your movements. Once either command is initiated, a message appears from the **Help** flyout in the **InfoCenter**, if balloon notifications are turned on. If you expand this message, the key movements are explained. See **Figure 20-4**. To redisplay this message while the command is active, press the [Tab] key.

- **Move forward.** Up arrow or [W].
- **Move left.** Left arrow or [A].
- **Move right.** Right arrow or [D].
- **Back up.** Down arrow or [S].
- **Toggle between walk and fly modes.** [F].

You can also navigate through the scene using the mouse. Press and hold the left mouse button and then drag the mouse in the active viewport to "steer" through the scene. With the **3DWALK** command, the camera remains at the same Z value. With the **3DFLY** command, the Z position of the camera can change. The steps for creating a walkthrough or flyby are provided at the end of this section.

Position Locator

When the **3DWALK** or **3DFLY** command is initiated, the **Position Locator** palette appears. See **Figure 20-5**. The preview in this palette shows a plan view of the scene. The purpose of this window is to provide an overview of the scene, in plan, while you develop the animation. It does not need to be displayed to create an animation and can be closed if it takes up too much space or slows down the rendering.

Position and target indicators appear in the plan view to show the location of the camera and its target. The green triangular shape displays the field of view. The *field of view (FOV)* is the area within the camera's "vision." The field of view indicator is only displayed when the target indicator is displayed. By default, the position indicator is red. The target indicator is green by default. These properties can be changed in the **General** category at the bottom of the **Position Locator** palette.

You can reposition the camera and the target in the plan view simply by picking and dragging either indicator. The effect of the change is visible in the active viewport.

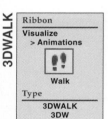

3DWALK

Ribbon
Visualize
> Animations

Walk

Type
3DWALK
3DW

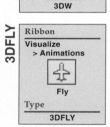

3DFLY

Ribbon
Visualize
> Animations

Fly

Type
3DFLY

Figure 20-4.
This message from the **InfoCenter** shows the keys that can be used to navigate through an animation.

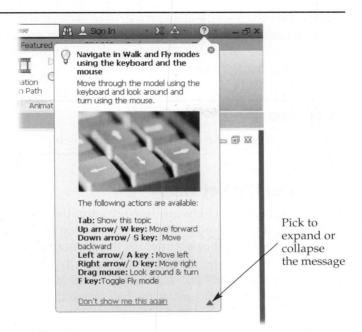

Pick to expand or collapse the message

Moving the position and target indicators closer together reduces the field of view. Picking the field of view lines and dragging moves the position and target indicators at the same time. The Position Z property in the **General** category sets the Z coordinate value for the position indicator. The Target Z property in the **General** category sets the Z coordinate value for the target indicator. The Z coordinate value determines eye level.

In addition to changing the color of the position and target indicators, you can use the properties in the **General** category to modify the display in the **Position Locator** palette. The Position indicator size property determines if both indicators are displayed small, medium, or large. If the Position indicator blink property is set to On, both indicators flash on and off in the preview. The Preview visual style property sets the visual style for the preview. This setting does not affect the current viewport or the animation. The Preview transparency property is set to 50% by default, but can be changed to whatever you want. If the view in the **Position Locator** palette is obscured by something (a roof, perhaps), you may want to set the Preview visual style property to Hidden

Figure 20-5.
The **Position Locator** palette.

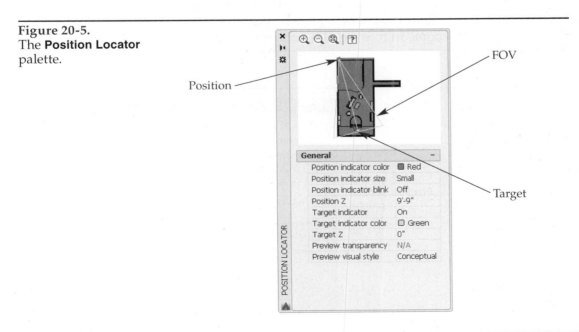

Position

FOV

Target

and the Preview transparency property to 80% or 90%. This will make the objects under the roof visible. If hardware acceleration is on, then the Preview transparency property is disabled. The **Graphics Performance** tool in the status bar indicates the on/off status of hardware acceleration.

Exercise 20-2

www.g-wlearning.com/CAD/

Complete the exercise on the companion website.

Walk and Fly Settings

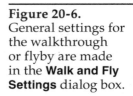

Ribbon

Visualize
> Animations

Walk and Fly Settings

Type

WALKFLY-SETTINGS

WALKFLYSETTINGS

General settings for walkthroughs and flybys are made in the **Walk and Fly Settings** dialog box. See **Figure 20-6**. Open this dialog box by picking the **Walk and Fly Settings** button in the **Animations** panel on the **Visualize** tab of the ribbon (in the **Walk** flyout). This panel may not be displayed by default. The dialog box can also be displayed by picking the **Walk and Fly...** button in the **3D Modeling** tab of the **Options** dialog box.

The three radio buttons at the top of the dialog box are used to determine when the message shown in **Figure 20-4** is displayed. The check box determines if the **Position Locator** palette is automatically displayed when the **3DWALK** or **3DFLY** command is entered.

The text boxes in the **Current drawing settings** area determine the size of each step and the number of steps per second. The **Walk/fly step size:** setting controls the **STEPSIZE** system variable. This is the number of units that the camera moves in one step. The **Steps per second:** setting controls the **STEPSPERSEC** system variable. This is the number of steps the camera takes each second. Together, these two settings determine how fast the camera moves in the animation.

PROFESSIONAL TIP

You will have to experiment with step size and steps per second values to make an animation that is easy to watch. Start with low numbers (for slow movement) and work your way up. Fast movement is disorienting and makes the viewer feel as if something was not seen, or missed. The viewer should be able to watch at a comfortable pace and get a good look at your design.

To get a feel for the proper speed for a walkthrough, pay attention to the next movie or TV show that you watch. When the director wants you to get a good look at the setting for the scene, the camera very slowly pans around the room. To emphasize distance, the camera slowly zooms in to a target object or person.

Figure 20-6.
General settings for the walkthrough or flyby are made in the **Walk and Fly Settings** dialog box.

Check to automatically display the **Position Locator** palette

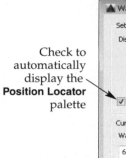

Camera Tools

The expanded **View** panel on the **Home** tab of the ribbon contains some tools for quickly adjusting the camera before starting the animation. See **Figure 20-7**. The **Lens length** slider controls how much of the scene is seen by the camera. The *lens length* refers to the focal length of the camera lens. The higher the number, the closer you are to the subject. The range is from about 1 to 100,000; 50 is a good starting point. There are stops on the slider for standard lens lengths. You can enter a specific value for the lens length or field of view by selecting the text in the slider, typing the value, and pressing [Enter].

Below the **Lens length** slider are text boxes for the camera and target positions. These can be used to change the X, Y, and Z coordinates of the camera or target before starting the animation.

NOTE

The view of the current viewport is changed when using the camera tools in the expanded **View** panel on the **Home** tab of the ribbon. If you change the current view and want to save it for future use, save it as a new view.

Animation Tools

The **Animations** panel on the **Visualize** tab of the ribbon contains the tools for controlling the recording and playback of the animation. See **Figure 20-8**. Remember, this panel may not be displayed by default. The **Record Animation** button is used to initiate recording of camera movement. After the **3DWALK** or **3DFLY** command is activated, pick the button to start recording. Make sure that you are ready to start moving when you pick the button because recording starts as soon as it is picked.

Figure 20-7.
Camera tools are located in the expanded **View** panel on the **Home** tab of the ribbon.

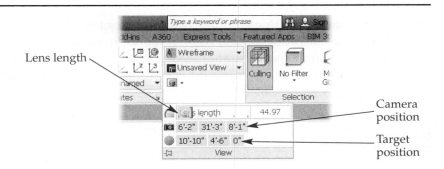

Lens length

Camera position

Target position

Figure 20-8.
The **Animations** panel on the **Visualize** tab of the ribbon is where you can record and play back the walkthrough or flyby animation. This panel may not be displayed by default.

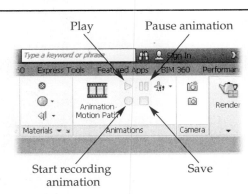

Play

Pause animation

Start recording animation

Save

Picking the **Pause Animation** button temporarily stops recording. This allows you to adjust the view without recording the actions. When you are ready to begin recording again, pick the record button to resume.

Picking the **Play Animation** button stops the recording and opens the **Animation Preview** dialog box, where the animation is played. See **Figure 20-9**. The controls in this dialog box can be used to rewind, pause, and play the animation. The slider can be dragged to preview part of the animation or move to a specific frame. The visual style can also be set using the drop-down list. If the animation is created using a render preset, the file must be played in Windows Media Player or another media player to view the rendered detail. Render presets are available in the **Animation Settings** dialog box, as discussed in the next section.

Picking the **Save Animation** button in the **Animation** panel stops recording and opens the **Save As** dialog box. Name the animation file, navigate to a location, and pick the **Save** button.

CAUTION

While the **3DWALK** or **3DFLY** command is active and the record button is on, you are creating an animation. If you move the camera in the **Position Locator** palette and start re-recording the animation to correct a problem, but do not first exit the current **3DWALK** or **3DFLY** command session, you are adding another segment to the animation you just previewed. To start over, first exit the current command session.

Exercise 20-3

www.g-wlearning.com/CAD/

Complete the exercise on the companion website.

Animation Settings

The **Animation Settings** dialog box may contain the most important settings pertaining to walkthroughs and flybys. See **Figure 20-10**. These animation settings determine how good the animation looks, how long it is going to take to complete, and how big the file will be. The dialog box is displayed by picking the **Animation settings...** button in the **Save As** dialog box displayed when saving an animation.

Figure 20-9.
The animation is played in the **Animation Preview** dialog box.

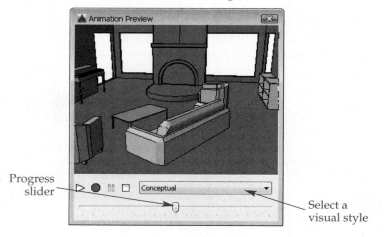

Progress slider

Select a visual style

Figure 20-10.
The **Animation Settings** dialog box contains important settings pertaining to walkthroughs and flybys.

Select a visual style option or the **Rendered** option

Set the resolution

Set the frame rate

Select the output file type

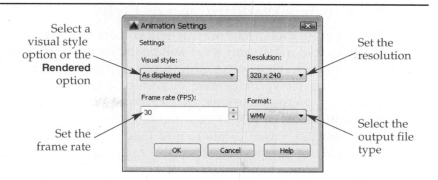

The **Visual style:** drop-down list is used to set the shading level in the animation. The name of this drop-down list is a little misleading because a **Rendered** option is included along with the available visual styles. Using a higher shading level or the **Rendered** option will result in a longer rendering time and larger file size. If you have numerous lights casting shadows and detailed materials, settle in for a long wait. A simple, straight-ahead walkthrough of 10 or 15 feet can easily result in 300 frames of animation. If each frame takes about five seconds to render, that equals 1500 seconds, or 25 minutes, to create an animation file that is only 10 seconds long.

The value in the **Frame rate (FPS):** text box sets the number of frames per second for the playback. In other words, this sets the speed of the animation playback. The default is 30 fps, which is a common playback rate for computers.

The **Resolution:** drop-down list offers standard options for resolution, from 160 × 120 to 1024 × 768. These are measured in pixels × pixels. Remember, higher resolutions mean longer processing times and larger file sizes.

The **Format:** drop-down list is used to select the output file type. The file type must be set in this dialog box. It cannot be changed in the **Save As** dialog box. The output file type options are:

- **WMV.** The standard movie file format for Windows Media Player.
- **AVI.** Audio-Video Interleaved is the Windows standard for movie files.
- **MPG.** Moving Picture Experts Group (MPEG) is another very common movie file format.

Depending on the configuration of your computer, you may not have all of these file type options or you may have additional options not listed here.

PROFESSIONAL TIP

Other AutoCAD navigation modes may be used to create a walkthrough or flyby. Any time that you enter constrained, free, or continuous orbit, you can pick the **Record Animation** button to record the movements. After you activate the **3DWALK** or **3DFLY** command, right-click and experiment with some of the other options as you record your animation. You can even combine some of these navigation modes with walking or flying. For example, use **3DFLY** to zoom into a scene, pause the animation, switch to constrained orbit, restart the recording, and slowly circle around your model. Animation settings are also available in this shortcut menu.

Exercise 20-4

www.g-wlearning.com/CAD/

Complete the exercise on the companion website.

Steps to Create a Walkthrough or Flyby

1. Plan your animation. Determine where you are moving from and to, what you are going to be looking at, and what will be the focal point of the scene.
2. Set up a multiple-viewport configuration of three or four viewports.
3. In one of the viewports, create or restore a named view with the appropriate starting viewpoint. Make sure a background is set up, if desired.
4. Start the **3DWALK** or **3DFLY** command. Note in the **Position Locator** palette the camera location, target location, and field of view. Adjust these in the **Position Locator** palette preview area or the expanded **View** panel on the ribbon, if necessary.
5. Position your fingers over the navigation keys on the keyboard.
6. Pick the **Record Animation** button.
7. Practice navigating through the view and then pick the **Play Animation** button to preview your animation. When you are done practicing, make sure to cancel the command and reposition the camera at the starting point.
8. Pick the **Record Animation** button to start over.
9. Start navigating through the view. Try to keep the movements as smooth as possible. Any jerks and shakes will be visible in the animation.
10. When you are done, stop moving forward and then pick the stop (**Save Animation**) button.
11. In the **Save As** dialog box, pick the **Animation Settings...** button. In the **Animation Settings** dialog box, select the desired visual style, frame rate, resolution, and output file format. Pick the **OK** button to close the **Animation Settings** dialog box. In the **Save As** dialog box, name and save the file.
12. The **Creating Video** dialog box is displayed as AutoCAD processes the frames, **Figure 20-11**.
13. When the **Creating Video** dialog box is automatically closed, the animation file is saved and you can view it. Pick the **Play Animation** button and watch the animation in the **Animation Preview** window. You can also locate the file using Windows Explorer. Then, double-click on the file to play the animation in Windows Media Player (or whichever program is associated with the file type).
14. Exit the command. If you are not satisfied with the results and want to try it again, make sure to exit the command before you attempt another walkthrough or flyby.

Figure 20-11.
The **Creating Video** dialog box is displayed as AutoCAD generates the animation.

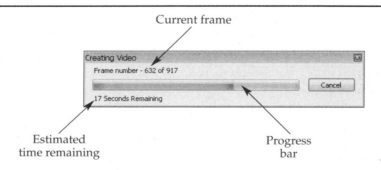

Current frame

Estimated time remaining

Progress bar

Motion Path Animation

You may have found it difficult to create smooth motion using the keyboard and mouse. Fortunately, AutoCAD provides an easy way to create a nice, smooth animated walkthrough or flyby. This is done using a motion path. A *motion path* is simply a straight or curved path along which the camera, target, or both travel during the animation.

One method of using a motion path is to link the camera and target to a single path. The camera and its line of sight then follow the path as a train follows tracks. See **Figure 20-12**.

Another option when using a motion path is to link the camera to a single point in the scene and the target to a path. For example, the target can be set to follow a circle or arc. The camera swivels on the point and "looks at" the path as if it is being rotated on a tripod. See **Figure 20-13**.

A third way to use a motion path is to have the camera follow a path, but have the target locked onto a stationary point. This is similar to riding in a vehicle and watching an object of interest on the side of the road. As the vehicle moves, your gaze remains fixed on the object. See **Figure 20-14**.

Figure 20-12.
A—The camera and target are linked to the same path (shown in color). B—The camera looks straight ahead as it moves along the path.

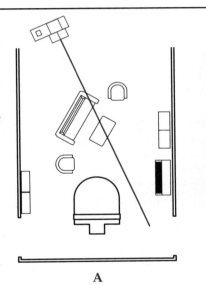

A

B

Figure 20-13.
A—The camera is linked to a point so it remains stationary. The target is linked to the circle. B—The camera view rotates around the room as if the camera is on a swivel tripod.

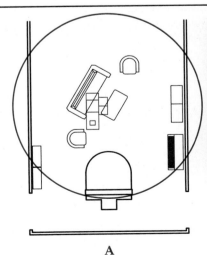

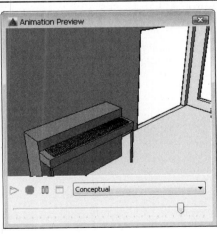

A

B

Figure 20-14.
A—The camera is linked to the spline path and the target is linked to the point (shown in color). B—As the camera moves along the path, it always looks at the point.

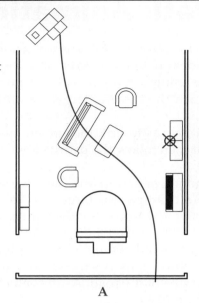

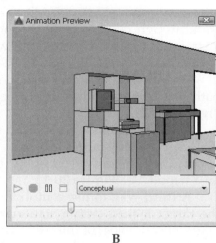

A   B

The fourth method of using a motion path is to have both the camera and target follow separate paths. Picture yourself walking into an unfamiliar room. As you walk into the center of the room, your gaze sweeps left and right across the room. In this case, the camera (you) follows a straight-line path and the target (your gaze) follows an arc from one side of the room to the other.

The **ANIPATH** command is used to assign motion paths. The command opens the **Motion Path Animation** dialog box. See **Figure 20-15**. This dialog box has three main areas: **Camera**, **Target**, and **Animation settings**. These areas are described in detail in the next sections. The steps for creating a motion path animation are provided at the end of this section.

ANIPATH

Ribbon
**Visualize
> Animations**

**Animation Motion
Path**

Type
ANIPATH

NOTE

Selecting a motion path automatically creates a camera. You cannot add a motion path to an existing camera.

Figure 20-15.
The **Motion Path Animation** dialog box is used to create an animation that follows a path.

Camera settings

Target settings

Animation settings

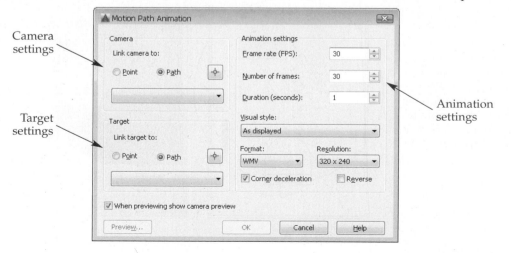

Camera Area

The camera can be linked to a path or a point. To select a path, pick the **Path** radio button and then pick the "select" button next to the radio buttons. The dialog box is temporarily closed for you to select the path in the drawing. The path may be a line, arc, circle, ellipse, elliptical arc, polyline, 3D polyline, spline, or helix, but it must be drawn before the **ANIPATH** command is used. Splines are nice for motion paths because they are smooth and have gradual curves. The camera moves from the first point on the path to the last point on the path, so create paths with this in mind. Once you select the path, you will be prompted to specify a name for the path. If you do not name it, AutoCAD will give it a default, sequential name, such as Path1, Path2, etc.

To select a stationary point, pick the **Point** radio button. Then, pick the "select" button next to the radio buttons. When the dialog box is hidden, specify the location in the drawing. You can use object snaps or enter coordinates. It may be a good idea to have a point drawn and use object snaps to select the point. As when selecting a path, once you select the point, you will be prompted to specify a name for the point.

The camera must be linked to either a path or a point. If neither is selected, the command cannot be completed. If you want the camera to remain stationary as the target moves, select the **Point** radio button and then pick the stationary point in the drawing.

Once a point or path has been selected, it is added to the drop-down list. All named motion paths and selected motion points in the drawing appear in this list. Instead of using the "select" button, you can select the path or point in this drop-down list.

Target Area

The target is the location where the camera points. Like the camera, the target can be linked to a point or a path. To link the target to a path, select the **Path** radio button. Then, pick the "select" button and select the path in the drawing. If the camera is linked to a point, the target must be linked to a path. If the camera is set to follow a path, then you actually have three choices for the target. It can be linked to a path, point, or nothing. To link the target to a point, pick the **Point** radio button. Then, pick the "select" button to select the point in the drawing. The None option, which is selected in the drop-down list, means that the camera will look straight ahead down the path as it moves.

Animation Settings Area

Most of the settings in the **Animation settings** area have the same effect as the corresponding settings in the **Animation Settings** dialog box discussed earlier. However, there are four settings unique to the **Motion Path Animation** dialog box.

The **Number of frames:** text box is used to set the total number of frames in the animation. Remember, a computer has a playback rate of 30 fps. Therefore, if the frame rate is set to 30, set the number of frames to 450 to create an animation that is 15 seconds long ($30 \times 15 = 450$).

The value in the **Duration (seconds):** text box is the total time of the animation. This value is automatically calculated based on the frame rate and number of frames. However, you can enter a duration value. Doing so will automatically change the number of frames based on the frame rate.

By default, the **Corner deceleration** check box is checked. This slows down the movement of the camera and target as they reach corners and curves on the path. If this is unchecked, the camera and target move at the same speed along the entire path, creating very jerky motion on curves and at corners. It is natural to decelerate on curves.

Checking the **Reverse** check box simply switches the starting and ending points of the animation. If the camera (or target) travels from the first endpoint to the second endpoint, checking this check box makes the camera (or target) travel from the second endpoint to the first.

Previewing and Completing the Animation

To preview the animation, pick the **Preview...** button at the bottom of the **Motion Path Animation** dialog box. The camera glyph moves along the path in all viewports. If the **When previewing show camera preview** check box is checked, the **Animation Preview** window is also displayed and shows the animation.

To finish the animation, pick the **OK** button in the **Motion Path Animation** dialog box. The **Save As** dialog box is displayed. Name the file and specify the location. If you need to change the file type, pick the **Animation settings...** button to open the **Animation Settings** dialog box. Change the file type, close the dialog box, and continue with the save.

Steps to Create a Motion Path Animation

1. Plan your animation. Determine where you are moving from and to, what you are going to be looking at, and what will be the focal point of the scene.
2. Draw the paths and points to which the camera and target will be linked. Draw the path in the direction the camera should travel (first point to last point). Do not draw any sharp corners on the paths and make sure that the Z value (height) is correct.
3. Start the **ANIPATH** command.
4. Pick the camera path or point.
5. Pick the target path or point (or None).
6. Adjust the frames per second, number of frames, and duration to set the length and speed of the animation.
7. Select a visual style, the file format, and the resolution.
8. Preview the animation. Adjust settings, if needed.
9. Save the animation to a file.

Exercise 20-5

www.g-wlearning.com/CAD/

Complete the exercise on the companion website.

Chapter Review

Answer the following questions using the information in this chapter.

1. Which system variable controls the display of camera glyphs?
2. Name the three camera tools available on the **Cameras** tool palette and explain the differences between them.
3. When is the **Camera Preview** window displayed, by default?
4. From where is the offset distance for the camera clipping planes measured?
5. What is the difference between the **3DWALK** and **3DFLY** commands?
6. How do you "steer" your movement when creating a walkthrough or flyby animation?
7. What is the *field of view*?
8. What is the purpose of the **Position Indicator** palette?
9. In the **Walk and Fly Settings** dialog box, which settings combine to control the speed of the animation?
10. How do you start recording a walkthrough or flyby?
11. What must be done before correcting a motion error in a walkthrough or flyby?
12. Motion path animation involves linking a camera or target to _____ or _____.
13. Which types of objects may be used as a motion path?
14. If None is selected as the target "path," what does the camera do in the animation?
15. Explain *corner deceleration*.

Drawing Problems

1. **General** In this problem, you will create and manipulate a camera in a drawing from a previous chapter.
   A. Open drawing P18-2 from Chapter 18 and save it as P20-1.
   B. Create at least two viewports and display a plan view in one of them.
   C. Use the **CAMERA** command to create a camera looking at the objects from the southwest quadrant. Change the camera settings as needed to display a pleasing view of the scene.
   D. Name the camera SW View.
   E. Turn on both the front and back clipping planes. Adjust them to eliminate one object in the front and one object in the back.
   F. Open the **Cameras** tool palette and, using the tools in the palette, create three more cameras looking at the scene from various locations. Change their names to Normal, Wide-angle, and Fish-eye to match the type of camera.
   G. Save the drawing.

2. **Architectural** In this problem, you will draw some basic 3D shapes to represent equipment in a small workshop. Then, you will create an animated walkthrough.
   A. Start a new drawing and set the units to architectural. Save it as P20-2.
   B. Draw a planar surface that is 15′ × 30′.
   C. Draw three 9′ tall walls enclosing the two long sides and one short side.
   D. Use boxes and a cylinder to represent equipment and shelves. Refer to the illustration shown. Use your own dimensions.
   E. Use the **3DWALK** command to create an animation of walking into the workshop. Turn and look at the shelves at the end of the animation.
   F. Set the visual style to Conceptual and the resolution to 640 × 480.
   G. Save the animation as P20-2.avi (or the file format of your choice).
   H. Save the drawing.

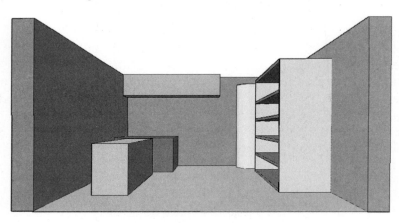

3. **Architectural** In this problem, you will create a motion path animation for the workshop drawn in Problem 20-2.
   A. Open drawing P20-2 and save it as P20-3.
   B. Draw a line and an arc similar to those shown (in color). The dimensions are not important.
   C. Move both objects so they are 4′ off the floor.
   D. Using the **ANIPATH** command, link the camera to the line and the target to the arc. Set the resolution to 640 × 480.
   E. Preview the animation. Adjust the animation settings as necessary. You may need to slow down the animation quite a bit. How do you do this?
   F. Save the animation as P20-3.wmv (or the file format of your choice).
   G. Save the drawing.

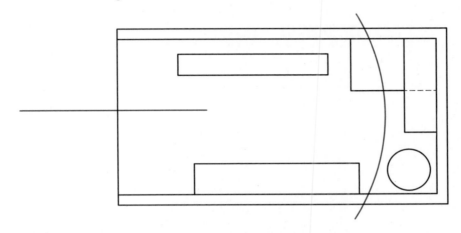

4. **Mechanical** In this problem, you will create a motion path animation for the presentation of a mechanical drawing from a previous chapter.
   A. Open drawing P17-3 and save it as P20-4.
   B. Draw a circle centered on the flange with a radius of 300.
   C. Move the circle 200 units in the Z direction.
   D. Using the **ANIPATH** command, link the camera to the circle and the target to the center of the flange.
   E. Preview the animation and adjust the animation settings as necessary. Due to the materials in the scene, the preview may play slowly, depending on the capabilities of your computer.
   F. Select a visual style that your computer can handle. Set the resolution to 320 × 240.
   G. Save the animation as P20-4.avi (or the file format of your choice).
   H. Save the drawing.

5. **Architectural** In this problem, you will create a flyby of the building that you created in Chapter 17.
   A. Open drawing P17-1 and save it as P20-5.
   B. Create a perspective view of the scene that shows the building from slightly above it. Save the view.
   C. Start the **3DFLY** command. Practice with the movement keys to make sure you know how to fly around the building. Then, cancel the command.
   D. Restore the starting view and select the **3DFLY** command.
   E. Record the flyby and save the animation as P20-5.wmv (or the file format of your choice).
   F. Save the drawing.

INDEX

D

Index - Advanced

Index - Advanced

N

Index - Advanced

Index - Advanced

Index - Advanced

Index - Advanced

W

Index - Advanced

X

Z